Rules of Thumb for Chemical Engineers

To my wife, Barbara Ann,
who's perspectives shape
all of my thoughts.

RULES OF THUMB FOR CHEMICAL ENGINEERS

Fifth Edition

Stephen Hall

ELSEVIER

AMSTERDAM • BOSTON • HEIDELBERG • LONDON • NEW YORK • OXFORD
PARIS • SAN DIEGO • SAN FRANCISCO • SINGAPORE • SYDNEY • TOKYO

Butterworth-Heinemann is an imprint of Elsevier

Butterworth-Heinemann is an imprint of Elsevier
The Boulevard, Langford Lane, Kidlington, Oxford OX5 1GB, UK
225 Wyman Street, Waltham, MA 02451, USA

Fourth edition published by Gulf Professional Publishing (an imprint of Elsevier), 2005

Fifth edition 2012

Notice
No responsibility is assumed by the publisher for any injury and/or damage to persons or property as a matter of products liability, negligence or otherwise, or from any use or operation of any methods, products, instructions or ideas contained in the material herein. Because of rapid advances in the medical sciences, in particular, independent verification of diagnoses and drug dosages should be made

British Library Cataloguing in Publication Data
A catalogue record for this book is available from the British Library

Library of Congress Cataloging-in-Publication Data
A catalog record for this book is availabe from the Library of Congress

ISBN: 978-0-12-387785-7

For information on all Butterworth-Heinemann publications
visit our web site at books.elsevier.com

Printed and bound in the United States of America

12 13 14 15 16   10 9 8 7 6 5 4 3 2 1

Working together to grow
libraries in developing countries

www.elsevier.com | www.bookaid.org | www.sabre.org

ELSEVIER   BOOK AID International   Sabre Foundation

Contents

10: Cooling Towers, 182

11: Refrigeration, 190

12: Closed Loop Heat Transfer Systems, 203

13: Biopharmaceutical Systems, 220

14: Vacuum Systems, 233

15: Pneumatic Conveying, 244

16: Blending and Agitation, 257

Publisher's Note

Excel spreadsheet workbooks accompany many of the chapters in this book. You can download them from the companion web site: http://www.elsevierdirect.com/companion.jsp?ISBN=9780123877857

You can contact Stephen Hall through his website: www.pipesizingsoftware.com where the workbooks may also be found.

1

Fluid Flow

1

Introduction

Chemical engineers who are designing plants and specifying equipment probably face more fluid flow problems than any other. Pressure drop calculations help the engineer size pipes and ducts, determine performance requirements for pumps and fans, and specify control valves and meters. And although the underlying theory is rather simple, its practical application can be confusing due to the empirical nature of important correlations, multiple methods for expressing parameters, many variable inputs, and alternative units of measurement.

This chapter presents formulae and data for sizing piping systems for incompressible and compressible flow:

- Friction factor, which is an empirical measure of the resistance to flow by a pipe or duct.
- Equivalent Length of a pipe segment, which characterizes bends and fittings as an equivalent length of straight pipe.
- Pressure drop due to friction for liquids and gases, both isothermal and adiabatic.
- Restriction orifices.
- Control valves.
- Two-phase flow.

A successful design requires much more analysis than a set of mathematical results. Consider these questions:

- What is the required flow range, now and in the future? Plant systems often operate at a range of flows, due to production rate variances, start-up differences,

multiple uses of the same system (such as production followed by flushing and cleaning), and future de-bottlenecking. Ensure that all of the components, such as control valves, account for these variations.
- How might the chemical composition and temperature change? Again, different production scenarios may result in wide variances in chemicals in use, and in temperatures. It may be prudent to create material and energy balances for different scenarios, and perform fluid flow calculations on several of these. Gaseous systems are especially sensitive to composition and temperature changes.
- Are the piping specifications fixed, or is there flexibility to optimize material selections? Construction materials may affect the pressure drop calculations due to differences in their surface roughness and dimensions. Specified wall thicknesses may be adjusted to save money (thinner walls reduce initial capital cost and potentially operating cost since the pressure drop will be lower at otherwise identical flow conditions).
- What are the status of flow diagrams, piping and instrumentation diagrams (P&IDs), general arrangement drawings, and elevation drawings? Each of these are needed for a thorough analysis. But preliminary results from fluid flow calculations may be needed to complete the drawing.

An Excel workbook with VBA function routines accompanies this chapter.

Data Required

Design Documents

Block flow diagram (minimum requirement)
Process Flow Diagram (PFD) (highly recommended)
P&ID (desirable)
General Arrangement drawing with major equipment from the P&IDs shown, in scale, in plan and elevation
Piping specifications. Minimum requirements are materials of construction, pressure and temperature limits, and dimensional specifications (e.g., acceptable diameters, wall schedule vs. diameter, long or short elbow radius).

Material and Energy Balance

For each pipe in the analysis, the material composition, flow, and temperature are needed. Various scenarios might be needed.

Physical Properties of the Materials

For liquids, the density, coefficient of thermal expansion, and viscosity at flowing temperature(s) are required.

For gases, provide molecular weight, ratio of specific heats (C_p/C_v), and compressibility factor, Z. Z is needed at high pressures and/or low temperatures. Chapter 27 discusses the determination of Z.

General Procedure

1. List the pipes that are included in the analysis. They are usually the lines shown on the P&IDs.

2. Using the General Arrangement drawings, estimate the length of each line and the number of each type of fitting (elbow, tee, etc.). If the drawings are already completed in CAD, and piping is already drawn, then an actual take-off of the piping should be obtainable from the CAD system. For conceptual or preliminary work, a very quick take-off of each pipe is made by roughly measuring the distance from the origin to the destination of the pipe using x-y-z coordinates, then adding a contingency factor of about 25%. See sidebar "Quickly estimating pipe lengths."

3. If the piping is not yet sized, use velocity to tentatively select pipe diameters.

4. Calculate pressure drop due to friction for selected scenarios. This step is used to determine the optimum size for the pipe during early design work, and to support sizing pumps and control valves during detailed design.

5. Size pumps, fans, and control valves. Iterate through all of the steps as information is developed.

6. Create system curves (see Chapter 5).

7. Update the P&IDs and General Arrangement drawings.

8. Conduct Process Hazards Analyses (PHAs), especially when flammable or toxic chemicals are being processed (see Chapter 20).

Quickly Estimating Pipe Lengths

Here's a simple way to get started on a project. The objective is to create a list of pipe segments, with size and length for each, which can be used for preliminary calculations and cost estimation.

1. From a Process Flow Diagram and conceptual or preliminary General Arrangements, make a sketch that mimics the PFD in both Elevation and Plan.
2. Write down measurements in x-y-z coordinates.
3. Use those measurements to estimate the physical length of each pipe segment.
4. Add a factor of 25% to account for errors in this highly conceptual procedure.
5. Add an additional factor of 50% to 100% to account for fittings. Alternatively, if fittings are considered to be fairly well known (elbows, tees, valves, instruments, and orifices) then refer to the Equivalent Length section in this chapter to obtain data.

Example: Refer to the sketch in Figure 1-0.

Line 215 length = 2 m + 2.5 m = 4.5 m (ignore liquid level in R-200)

Line 216 length = 6 m + 2.5 m + 6 m + 10 m + 1.5 m for = 26 m

Line	Physical Length	Contingency	Subtotal	Fittings	Total Equiv L
215	4.5 m	1 m	5.5 m	3 m	8.5 m
	(14 ft)	(3 ft)	(17 ft)	(10 ft)	(27 ft)
216	26 m	5 m	30 m	15 m	45 m
	(85 ft)	(15 ft)	(100 ft)	(50 ft)	(150 ft)

Recommended Velocity

Engineers determine pipe sizes by analyzing performance and economic parameters. Over the years, an enormous number of systems have been designed, installed, and operated. Those systems often share similar characteristics since they were built from the same catalog of available equipment, such as pumps and control valves. It's reasonable, then, to begin a new design project using existing system designs as a starting point.

Therefore, a Rule of Thumb is to use tables of suggested velocity for an initial determination of pipe size.

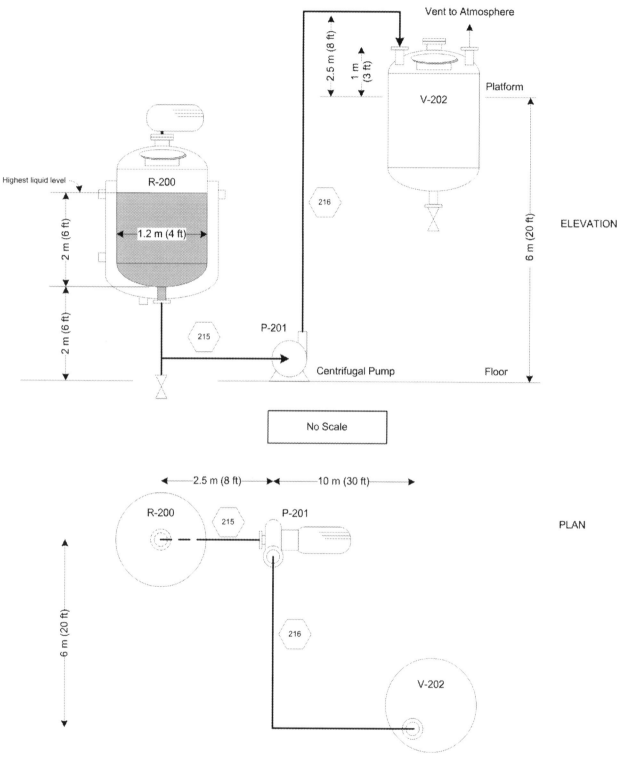

Figure 1-0.

The values in the tables have been widely disseminated, and have long since lost their original source. Also, the suggested velocities are often given as ranges. You are cautioned to use the information judiciously and perform your own analysis as your piping system design develops.

Factors to consider regarding velocity:

- Low velocity may indicate a larger pipe diameter than is necessary which raises cost.
- Low calculated velocity may result in the pipe running partially full in horizontal runs.
- Low velocity can lead to laminar flow conditions which may promote fouling and will definitely hinder heat transfer (if applicable).
- High velocity may be noisy.
- High velocity can cause damage to the pipe due to erosion.
- Certain components such as check valves and control valves are designed to operate with a specific flow range; manufacturers of the components recommend minimum lengths of pipe of specified diameter upstream and downstream of their device.

Some more comments by type of service:

Clean single-phase fluids (gas or liquid) tolerate the widest range of velocity. Generally use 1.5 to 4.0 m/s (5 to 12 ft/s) for liquids, or 15 to 40 m/s (50 to 120 ft/s) for gases as a starting point. Use a Net Present Value (NPV) economic analysis to balance the initial capital cost for the piping with energy cost for moving the fluid.

Steam systems are generally designed with higher velocity in the headers than the branch lines. Recommendations are based on comparison with sonic velocity and vary with the system pressure.

Hot fluids, whose vapor pressure is near the system pressure, require careful analysis to ensure the fluids won't flash inside the pipe. This is particularly important on the suction side of pumps, where a Net Positive Suction Head (NPSH) calculation must be performed. See the Pumps chapter for more information on NPSH.

Two-phase systems include solid-liquid, liquid-gaseous, and liquid-liquid. Generally, higher velocity threatens pipe damage due to erosion, especially at bends and components such as valves. Lower velocities may allow the heavy phase to drop out of the flow and collect at low points or coat the pipe surface. The velocities recommended for clean fluids apply, but try hard to avoid excursions. Erosion is proportional to the impact velocity

to the power n, where n is usually 2 or 3 depending on the properties of the sediment.

Corrosive fluids, defined as those that chemically attack the pipe *if the naturally protective layer is absent* (such as with passivated stainless steel), should be treated as if they are erosive. It's important that the flowing fluid does not scour the passivation layer.

Pipe material must be factored into the analysis. Softer materials, such as thermoplastics and copper, are more susceptible to erosion than hard materials, such as austenitic stainless steels.

Tip: For gaseous flow problems, be absolutely sure that velocity is calculated using the pressure and temperature at flowing conditions. Reduce ambiguity by expressing the flow rate in mass, rather than volumetric, units. See Table 1-1.

Table 1-1
Suggested starting point for pipe sizing using fluid velocity or pressure drop criteria

Service	Velocity, m/s or Other Criteria	Velocity, ft/s or Other Criteria
Air, compressed	20 to 30	65 to 100
Gas, dry	15 to 40	50 to 120
Gas, wet	10 to 18	30 to 60
Petrochemicals	1.5 to 4	5 to 12
Sodium hydroxide, 0–30%	1.8	6
Sodium hydroxide, 30–50%	1.5	5
Sodium hydroxide, 50–73%	1.2	4
Steam, dry, high pressure (> 2 bar, superheated)	50	150
Steam, saturated, low pressure (<= 2 bar)	30	100
Steam, small branch lines	15	50
Steam, wet	10 to 15	30 to 50
Vacuum, below 50 mm Hg absolute pressure	Max 5% pressure loss	
Vapor lines, general	Up to 0.3 Mach	
Water, average service	1.5 to 3	5 to 10
Water, boiler feed	1.5 to 4.6	5 to 15
Water, pump suction	0.3 to 1.5	1 to 5
Water, sea and brackish	1.5 to 4	5 to 12
Water, wastewater, pump suction	1 to 1.8	3 to 6
Water, wastewater, pump discharge	1 to 2.5	3 to 8
Water, wastewater, gravity	0.6 to 2.5	2 to 8

Compressible Flow of a Gas

The maximum velocity of a gas is equal to the speed of sound at flowing conditions. This is generally at the discharge end of a pipe, but may be at an intermediate location where there is a flow restriction such as a valve or orifice. It's good practice to design for a velocity that is less than 30% of the speed of sound (0.3 Mach). For adiabatic flow, the equation is:

$$U_{max} = \sqrt{\frac{Z \gamma g_c R T}{M}} \qquad (1\text{-}1)$$

Z = Compressibility Factor = 1 for an ideal gas
γ = Ratio of C_p/C_v

Oil/Gas Mixtures

A widely used formula is published in API 14E [1], but this has been challenged by Salama as being overly conservative [21]. It is intended to give the maximum velocity for crude oil/gas mixtures, flowing in two phases (that may contain sand sediment), to avoid excessive erosion.

$$U = \frac{C_1}{\sqrt{\rho}} \qquad (1\text{-}2)$$

U = velocity, m/s or ft/s
ρ = gas/liquid mixture density, kg/m^3 or lb/ft^3
C_1 = coefficient specific to piping material and quality of the fluid. Coefficient for SI units (multiply by 0.82 for US Customary units):
  Steel pipe, clean fluid (no sand), continuous service: 122
  Steel, clean fluid (no sand), intermittent service: 152
  Steel, clean fluid, non-corrosive, continuous service: 180 to 240
  Use lower (unspecified) values for sand-containing fluids

API 14E states that the minimum velocity for two-phase flow should be about 3 m/s (10 ft/s).

Check Valves

Check valves are a special case, as reported in detail by Crane [10]. There is a certain minimum velocity needed to keep check valves fully open. If the velocity drops below this, then the valve will partially close. Crane states that:

most of the difficulties encountered with check valves, both lift and swing types, have been found to be due to oversizing which results in noisy operation and premature wear of the moving parts.

The formula for minimum pipe velocity (determined by the nominal pipe size of the check valve, not the flow opening in the valve) is:

$$U_{min} = C \beta^2 \sqrt{V} \qquad (1\text{-}3)$$

C = coefficient, see Table 1-2.
β = ratio of flow opening in the valve to pipe size
$\overline{V}$ = specific volume of fluid (inverse of density)

Table 1-2
Coefficients for equation 1-3 [10]

Check Valve Type	C, $kg^{0.5}/m^{0.5}$ s	C, $lb^{0.5}/ft^{0.5}$ s
Swing Check, angled disk	43	35
Swing Check, vertical disk	74	60
Lift Check, horizontal plug	49	40
Lift Check, angled plug	170	140
Tilting Disk, 5° angle	98	80
Tilting Disk, 15°angle	37	30
Globe Stop Check, straight	67 to 73	55 to 60
Globe Stop Check, angle	73 to 91	60 to 75

Equivalent Length

There are two prevalent methods to characterize the flow resistance of a pipeline with its fittings.

The much easier and more convenient Equivalent Length method computes a pressure drop that is 10% to 20% higher for a typical pipeline compared to the K coefficient method.

In the Equivalent Length method, fittings are assigned a resistance value expressed in terms of length of straight pipe. Alternatively, fittings are assigned a resistance expressed as equivalent length divided by diameter. A long radius elbow in a DN 50 (2-inch) pipe offers the same resistance to flow as about 1 meter (3.4 feet) of straight pipe. Sum the equivalent length of all fittings in the pipe, add to the actual length of the pipe (including distance through fittings) and use the resultant as the Length factor in the equations in the following sections.

For new designs, where fittings have not yet been determined, a rule of thumb is to estimate the total length of the pipe (see sidebar) then add 50% to 100% as a factor for fittings [2].

In the K coefficient method, each fitting is evaluated according to the velocity head equation:

$$\Delta P = K \left(\frac{\rho}{g_c} \right) \left(\frac{U^2}{2} \right) \tag{1-4}$$

K = geometry and size dependent loss coefficient

Coefficients are found in the literature, such as Crane [10], or are provided by the fitting manufacturer.

The pressure drops for all of the fittings are summed and added to the pressure drop calculated for the total length of pipe (including distance through fittings) as described in the following sections. For systems where the density and velocity remain constant, simply sum the K factors for all of the fittings, then multiply by the density and velocity terms.

K coefficients are more accurate than equivalent length, because the flow resistance of a fitting is nearly insensitive to the friction factor and the K coefficients are computed using a standardized fully turbulent flow assumption. Results from the equivalent length method are proportional to the friction factor. Note that for compressible flow, velocity increases as pressure decreases. Therefore, a rigorous analysis requires fitting-to-fitting pressure drop calculations to enable evaluation of the velocity at each fitting.

Convert K to equivalent length with this relationship:

$$L = \frac{K D}{f_T} \tag{1-5}$$

Where L and D are in the same units (m or ft) and f_T is the Darcy friction factor with fully turbulent flow, given by:

$$f_T = \frac{0.25}{\left[\log_{10} \dfrac{\varepsilon/D}{3.7} \right]^2}$$

ε = surface roughness, m or ft
D = pipe diameter, m or ft

From Crane [10], the resistance coefficients for various fittings are listed in Table 1-3. Note that for most types of fittings the coefficient is related to f_T. Therefore, the equivalent length of those fittings is the factor multiplied by the pipe diameter (e.g., for a standard elbow, $L = 30 D$).

Example: Equivalent Length

What is the equivalent length of a 4-inch, Schedule 40 pipe segment that contains two elbows ($r/d = 1.5$), one full-port gate valve, and a length of 25 feet as measured through the centerline of the pipe? Use the data in Table 1-3.

Table 1-3
Resistance coefficients [10]

Fitting	Description	Resistance Coefficient, K
90° Elbow	Standard, threaded or socket welded	$K = 30 f_T$
90° Bend	Flanged or butt-welded, with radius/diameter ratio of	
	$r/d = 1$	$K = 20 f_T$
	$r/d = 1.5$	$K = 14 f_T$
	$r/d = 2$	$K = 12 f_T$
45° Elbow	Standard, threaded or socket welded	$K = 16 f_T$
TEE	Flow through the straight run	$K = 20 f_T$
	Flow through the branch (same size as straight run)	$K = 60 f_T$
	Crane reports new research that accounts for differences between converging or diverging flow, diameter ratio of the straight and branch legs of the tee, and angle of the branch leg [10]	
180° Return Bend	Close pattern, threaded or socket welded	$K = 50 f_T$
Pipe Entrance	Inward projecting	$K = 0.78$
	Flush with sharp edge	$K = 0.5$
	Flush with entrance radius/diameter ratio of $r/d = 0.1$	$K = 0.09$
Pipe Exit	Projecting, sharp edge, or rounded	$K = 1.0$

(Continued)

Table 1-3
Resistance coefficients [10]—cont'd

Fitting	Description	Resistance Coefficient, K
Pipe Coil	Multiple 90° bends connected together to form a continuous coil. This K includes the physical length of the coil. The term n is the number of 90° bends comprising the coil. K_1 is the coefficient for a single bend which is dependent on the radius/diameter ratio. The formula reported by Crane is: $K = (n-1)(0.25 f_T \pi r/d + 0.5 K_1) + K_1$	
Valve	Any valve for which the manufacturer has specified a flow coefficient	$K = \left[\dfrac{0.04\ d^2}{K_v}\right]^2$
	SI Units: valve coefficient K_v = flow of water at 20 °C in m³/h with a pressure drop of 1 bar (d = pipe ID, mm) US Units: valve coefficient C_v = flow of water at 60°F in gal/min with a pressure drop of 1 psi (d = inches)	$K = \left[\dfrac{29.84\ d^2}{C_v}\right]^2$
Gate Valve	Full port valve	$K = 8 f_T$
Globe Valve	Full port valve with port 90° to pipe direction	$K = 340 f_T$
Angle Valve	Full port valve with 90° connections (i.e., flanges)	$K = 150 f_T$
Ball Valve	Full port valve	$K = 3 f_T$
Swing Check	Disc is vertical in the pipe	$K = 50 f_T$
Lift Check	Disc is horizontal in the pipe, full port	$K = 600 f_T$

Solution: The equivalent length method is based on the nominal diameter of the pipe, so the wall thickness is irrelevant. The frictional term cancels (f_T). Therefore, the equivalent length of this segment is: L = (2) (14) (4/12) + (8) (4/12) + 25 = 37 feet.

Darby [12] recommends his more rigorous 3-K method that makes an adjustment to the K coefficients depending on the fitting size and the flowing Reynolds number. The practical difference between the K methods is insignificant for flows at high Reynolds numbers, but is important for laminar flow situations with many fittings.

Calculate K from the three parameters in Table 1-4.

$$K = \frac{K_m}{N_{Re}} + K_i\left(1 + \frac{K_d}{d_{nom}^{0.3}}\right) \tag{1-6}$$

Where d_{nom} is the nominal pipe diameter in inches (e.g., 1 inch, not 1.049 the actual inside diameter).

Table 1-4
3-K Constants for loss coefficients [20]

Fitting	Description	r/d	K_m	K_i	K_d
Elbows − 90°	Threaded, standard	1	800	0.14	4.0
	Threaded, long radius	1.5	800	0.075	4.2
	Flanged, welded, or bends	1	800	0.091	4.0
		2	800	0.056	3.9
		4	800	0.066	3.9
		6	800	0.075	4.2
	Mitered, 1 weld (90°)		1000	0.27	4.0
	Mitered, 2 welds (45°)		800	0.136	4.1
	Mitered, 3 welds (30°)		800	0.105	4.2
Elbows − 45°	Threaded, standard	1	500	0.071	4.2
	Threaded, long radius	1.5	500	0.052	4.0
	Mitered, 1 weld (45°)		500	0.086	4.0
	Mitered, 2 weld (22.5°)		500	0.052	4.0
Elbows − 180°	Threaded, close return bend	1	1000	0.23	4.0
	Flanged	1	1000	0.12	4.0
	Threaded, flanged, or welded	1.5	1000	0.10	4.0
Tees	Through branch (as elbow)				
	Threaded	1	500	0.274	4.0
	Threaded	1.5	800	0.14	4.0

Table 1-4
3-K Constants for loss coefficients [20]—cont'd

Fitting	Description	r/d	K_m	K_i	K_d
	Flanged	1	1000	0.28	4.0
	Stub-in branch		1000	0.34	4.0
	Run-through				
	Threaded		200	0.091	4.0
	Flanged		150	0.05	4.0
	Stub-in branch		100	0	0
Angle Valve	45° angle, full line size	β = 1	950	0.25	4.0
	90° angle, full line size	β = 1	1000	0.69	4.0
Globe Valve	Standard	β = 1	1500	1.70	3.6
Plug Valve	Branch flow		500	0.41	4.0
	Straight through		300	0.084	3.9
	Three-way (flow through)		300	0.14	4.0
Gate Valve	Standard	β = 1	300	0.037	3.9
Ball Valve	Standard	β = 1	300	0.017	3.5
Diaphragm	Dam-type		1000	0.69	4.9
Swing Check	See velocity requirement, p. 6		1500	0.46	4.0
Lift Check	See velocity requirement, p. 6		2000	2.85	3.8

Shortcut Equation for Pressure Drop Due to Friction

A handy relationship for pressure drop in commercial steel pipe is:

For SI Units, use equation 1-7 with units bar/100 m, kg/h, mPa-s, mm, and kg/m^3:

$$\Delta P_F = \frac{4150\, W^{1.8} \mu^{0.2}}{d^{4.8} \rho} \qquad (1\text{-}7)$$

For US Units, use equation 1-8 with units psi/100 ft, lb/h, cP, in., and lb/ft^3:

$$\Delta P_F = \frac{W^{1.8} \mu^{0.2}}{20000\, d^{4.8} \rho} \qquad (1\text{-}8)$$

This relationship holds for Reynolds numbers above 2,100 (i.e., turbulent flow) and is applicable for liquids and compressible fluids if the pressure drop is less than 10% of the inlet pressure, and choking does not occur. For smooth tubes, the SI factor should be 3,610 instead of 4,150 and the US factor 23,000 instead of 20,000.

The equation was derived from the Fanning equation for US units:

$$\Delta P_F = \frac{2 f\, U^2 L \rho}{32.2\, D}$$

and the approximate relationship [24]:

$$f = \frac{0.0054}{N_{\text{Re}}^{0.2}}$$

Reynolds Number

The Reynolds number, N_{Re}, is a dimensionless number that relates inertial and viscous forces. It is used in the friction factor correlation, to determine the resistance to flow by a pipe.

$$N_{\text{Re}} = \frac{\rho\, D\, U}{\mu} \qquad (1\text{-}9)$$

D = pipe diameter, m or ft

U = average fluid velocity, m/s or ft/s = $\dfrac{G}{\rho}$

μ = fluid dynamic viscosity, kg/m-s or lb/ft-h

ρ = density of liquid, kg/m^3 or lb/ft^3, or gas = $\dfrac{P M}{R T}$

Friction Factor

The friction factor is a measure of the resistance to flow by a pipe. It was developed in the 1930s and remains the basis for computing pressure drop due to friction [18].

There are two "flavors" of friction factor; the Darcy (or Moody) friction factor is equal to four times the Fanning friction factor. It's imperative to specify which factor you are using to avoid confusion. Darcy friction factors are often in the range of 0.01 to 0.03 while Fanning friction factors are 0.002 to 0.008.

The Darcy friction factor is used throughout this book.

Churchill [9] developed an expression for the friction factor that spans all flow regimes (laminar, turbulent, and transitional). It agrees with the original Colebrook-White equation, which requires an iterative or graphical solution, while also obtaining the correct result for Reynolds numbers below 2,000 (laminar flow regime). The Darcy friction factor according to Churchill's equation is:

$$f = 8\left[\left(\frac{8}{N_{Re}}\right)^{12} + \frac{1}{(a+b)^{3/2}}\right]^{1/12}$$ (1-10)

Where:

$$a = \left[2.457 \ln \frac{1}{(7/N_{Re})^{0.9} + (0.27\ \varepsilon/d)}\right]^{16}$$

$$b = \left(\frac{37530}{N_{Re}}\right)^{16}$$

ε = surface roughness, m or ft

Surface Roughness

The surface roughness value used in Equation 1-10 is empirically derived and should not be confused with

Table 1-5
Recommended surface roughness values for various piping materials

Pipe Material	Surface Roughness, m	Surface Roughness, ft
Copper, drawn, tubing	0.000002	0.0000067
Glass, drawn tubing	0.000002	0.0000067
Plastic, drawn tubing	0.000002	0.0000067
Brass, drawn	0.000002	0.0000067
Iron, cast – new	0.0003	0.0021
Iron, wrought – new	0.000045	0.00017
Iron, galvanized	0.00015	0.0005
Iron, asphalt coated	0.00015	0.0005
Steel, new	0.000045	0.00015
Steel, lightly corroded	0.0003	0.00125
Steel, heavily corroded	0.002	0.0067
Steel, galvanized	0.00015	0.0005
Steel, polished (hygienic)	0.000002	0.0000067
Steel, stainless, drawn tubing	0.000002	0.0000067
Sheet metal ductwork, smooth joints	0.00003	0.0001
Concrete, very smooth	0.00004	0.00013
Concrete, wood floated, brushed	0.0003	0.001
Concrete, rough, visible form marks	0.002	0.0067
Rubber, smooth tubing	0.00001	0.000033
Rubber, wire reinforced	0.001	0.0033

Never use a value less than 0.0000015 m or 0.000005 ft, which are the limiting values that define "smooth" pipe [11].

actual measurement of the roughness of a piping material. Use Table 1-5 for the values. Contingency factors may be added, especially if the pipe is old and has signs of wear or corrosion. For example, in a new piping system with steel pipe, compare the results of using the tabulated value with those when the tabulated value is inflated by, say, 30%. Consider specifying a pump that could handle the higher calculated demand, possibly with an impeller change.

Incompressible Flow

Use this incompressible flow equation for liquids; it can also be used for gases when the pressure drop is less than 10 percent of the upstream pressure.

$$\Delta P = \frac{f\ L\ U^2 \rho}{2g_c D}$$ (1-11)

For gases, better accuracy is obtained by using the velocity and density computed at the average pressure (upstream and downstream). First compute the pressure drop using upstream conditions. Then, average the upstream and computed downstream pressure. Re-evaluate velocity and density at the average. Finally, calculate the pressure drop again using the new values.

For liquid applications with significant temperature change, in heat exchangers for example, use the density of the fluid averaged from inlet and outlet temperatures.

Additional pressure changes due to elevation differences of the pipe ends and pressure at the upstream reservoir or downstream receiver may also need to be accounted for. This is discussed in Chapter 5.

Compressible Flow – Isothermal

Most gaseous pipe flow problems can be treated as isothermal. Temperature change due to steady flow is minimal unless the critical velocity is approached or there is a sudden acceleration through a nozzle, valve or, orifice. The isothermal equation [17] is easier to solve than the adiabatic equations presented in the next section. When solving for G (given P_1 and P_2), an initial guess determines the Reynolds number and friction factor. A few iterations on the guess will refine the result.

$$\Delta P = \frac{R T Z G^2}{\overline{P} M g_c}\left[\frac{f L}{2 D} + \ln\left(\frac{P_1}{P_2}\right)\right] \qquad (1\text{-}12)$$

G = mass flux, kg/s-m^2 or lb/s-ft^2 = $\dfrac{W}{3600\,A}$

$\overline{P}$ = average pressure = $\dfrac{P_1 + P_2}{2}$

Z = Compressibility Factor = 1 for an ideal gas (see page 415)

Compressible Flow – Transmission Equations

Three equations commonly encountered in the natural gas industry are the Panhandle A, Panhandle B, and Weymouth equations. Often cited for use in calculating pressure drop in long transmission pipelines, they each incorporate a friction factor term so the friction factor is not calculated independently. The advantage of doing this is that the pressure drop is correlated directly with flow rate. The drawback is that the friction factor term is relatively inaccurate when compared with the Darcy factor, especially at low flow rates (i.e., in the laminar region). But, in each case, at low flow they are conservative and under-predict flow while at high flow they are overly optimistic and over-predict flow. The primary difference between the equations is where the low and high flow are defined [23].

The equations assume isothermal flow. If temperature change is defined (due to heat transfer through the pipe wall, not from adiabatic expansion), the average temperature is used in the equation. To account for surface roughness in the pipe, as well as other factors such as bends and fittings, an "efficiency" factor is introduced, typically ranging from 0.85 to 0.92. As a pipe corrodes, the efficiency decreases. This is because the equations are arranged to return the flow rate for a given set of conditions, including pressure drop. Thus, rougher pipe means lower flow.

Weymouth Equation

US Units:

$$Q = 433.49\,\frac{T_b}{P_b}\,D^{8/3}\,e\left(\frac{P_1{}^2 - P_2{}^2 - H_c}{L\,G\,T_a\,Z_a}\right)^{0.5} \qquad (1\text{-}13)$$

Panhandle A Equation

$$Q = C\left(\frac{T_b}{P_b}\right)^{1.0788} D^{2.6182}\,e\left(\frac{P_1{}^2 - P_2{}^2 - H_c}{L\,G^{0.8538}\,T_a\,Z_a}\right)^{0.5394} \qquad (1\text{-}14)$$

Where $C = 435.87$ for US units, and 0.0045965 for SI units

Panhandle B Equation

$$Q = C\left(\frac{T_b}{P_b}\right)^{1.02} D^{2.53}\,e\left(\frac{P_1{}^2 - P_2{}^2 - H_c}{L\,G^{0.961}\,T_a\,Z_a}\right)^{0.51} \qquad (1\text{-}15)$$

Where $C = 737$ for US units, and 0.010019 for SI units

Nomenclature

D = pipe inside diameter, mm or in.

e = efficiency, dimensionless

G = gas specific gravity (compared to air), dimensionless

$$H_c = \frac{C2\, G\, (H_2 - H_1)\, P_a}{Z_a\, T_a}$$

= head correction, kPa or psia

Where $C2 = 0.0375$ for US units, and 0.06835 for SI units

H_1 = elevation of pipeline at origin, m or ft

H_2 = elevation of pipeline at terminus, m or ft

L = pipe length, km or miles

$$P_a = \frac{2}{3}\left[P_1 + P_2 - \left(\frac{P_1\, P_2}{P_1 + P_2}\right)\right]$$

= average pressure for calculating H_c and Z_a

P_b = pressure base, standard conditions, kPa or psia

P_1 = inlet pressure, kPa or psia

P_2 = outlet pressure, kPa or psia

Q = flow rate, standard m^3/day or standard ft^3/day

T_a = average temperature, K or R

T_b = temperature base, standard conditions, K or R

Z_a = average compressibility factor, dimensionless

Compressible Flow – Adiabatic

This section gives equations for computing pressure drop due to friction for adiabatic flow [17]. An iterative solution is required, but this is easy using the Solver tool in Excel, or in a VBA subroutine. Use the following procedure to solve for the flow rate (given upstream and downstream pressures) or one of the pressures (given the flow rate and the other pressure).

1. Calculate the *Mach number*, defined as the ratio of the velocity of the gas in a pipe to the speed of sound in the gas at flowing conditions (temperature and pressure). Use the upstream (given) temperature for the initial guess of the downstream temperature.

$$N_{Ma} = \sqrt{\frac{U^2}{g_c\, \gamma\, R\, T\, Z/M}} \qquad (1\text{-}16)$$

2. Calculate an intermediate value for each of the two Mach numbers:

$$X = 1 + N_{Ma}^2\left[\frac{(\gamma - 1)}{2}\right] \qquad (1\text{-}17)$$

3. Calculate the downstream temperature:

$$T_2 = T_1 \frac{X_1}{X_2} \qquad (1\text{-}18)$$

4. Solve the following equation by iteration of the unknown flow or pressure, simultaneously with equations 1-16, 1-17, and 1-18.

$$\frac{f\, L}{D} = \frac{1}{\gamma}\left(\frac{1}{N_{Ma1}^2} - \frac{1}{N_{Ma2}^2} - \frac{(\gamma + 1)}{2}\ln\left(\frac{N_{Ma2}^2 X_1}{N_{Ma1}^2 X_2}\right)\right) \qquad (1\text{-}19)$$

The next section includes a VBA function that solves this problem.

Visual Basic Subroutines for Pressure Drop Due to Friction

The VBA functions listed here are used in conjunction with an Excel spreadsheet to solve for pressure drop due to friction in circular pipes running full. There are three functions, designed to work in SI units. Incompressible and adiabatic compressible flow problems are modeled in a single function, with two of three values as input and the third as output (P_1, P_2, or W). The functions and examples are provided in the accompanying Excel workbook.

After entering the code into a VBA Module, call the functions from an Excel worksheet using the syntax given. Each function returns a single value as described.

Reynolds Number

Syntax:	$=NReSI(W, \mu, d, \rho, t, M, P)$
Returns:	Reynolds number
Required:	mass flow rate, viscosity, pipe diameter

Either density *or* temperature, molecular weight and pressure

Example:	Figure 1-1

Listing

```
Function NReSI(W, mu, d, Optional ro, Optional Tin,
Optional Mw, Optional p)
  ' W = Flowrate in kg/h
  ' mu = Viscosity in mPa-s
  ' d = PipeID in mm
  ' ro = density in kg/m3 (required for liquid)
  ' Tin = temperature, deg C (required for gas) - default
20 deg C
  ' Mw = molecular weight (required for gas) - default 29
  ' p = pressure, kPa (required for gas) - default 1000 kPa
  ' NRe = ro U L/mu
  ' ro = density
  ' U = velocity
  ' L = length (or pipe diameter)
  ' mu = viscosity
  ' Convert viscosity to units kg/m-h
  mu = mu * 0.001 * 3600
  ' Convert pipe diameter to meters
  d = d/1000
  Pi = 3.141592654
  R1 = 8314.47 ' gas constant
  If IsMissing(ro) Then ' assume this is a gas calculation
  If IsMissing(Tin) Then Tin = 20
  If IsMissing(Mw) Then Mw = 29
```

```
  If IsMissing(p) Then p = 1000 'kPa
  ' Convert temperature to deg K
  Tin = Tin + 273.15
  ' Convert pressure to Pa
  p = p * 1000
  ro = p * Mw/(R1 * Tin)
  End If
  If Val(ro) = 0 Then
  If IsMissing(Tin) Then Tin = 20
  If IsMissing(Mw) Then Mw = 29
  If IsMissing(p) Then p = 1000 'kPa
  ' Convert temperature to deg K
  Tin = Tin + 273.15
  ' Convert pressure to Pa
  p = p * 1000
  ro = p * Mw/(R1 * Tin)
  End If
  ' Calculate velocity
  U = W/ro/(Pi * (d/2) ^ 2)
  NReSI = ro * U * d/mu
End Function
```

Darcy Friction Factor

Syntax:	$= FrictionSI(\varepsilon, N_{RE}, d)$
Returns:	Darcy friction factor
Required:	pipe roughness, Reynolds number, pipe diameter
Example:	Figure 1-2

Listing

```
Function FrictionSI(epsilon, NRe, d)
  ' epsilon = Surface roughness is in units m
  ' d = PipeID is in units mm
```

	A	B	C	D	E	F	G
5							
6	**Inputs**						
7		Parameter	Units	Example 1	Example 2		
8		Mass Flow Rate	kg/h	10,000.0	1400		
9		Viscosity	mPa-s	1.2	0.011		
10		Pipe Diameter	mm	38.1	26.6		
11		Density	kg/m3	961.0			
12		Temperature	C		40		
13		Molecular Weight	kg/kgmol		16.04		
14		Pressure	kPa, absolute		2200		
15							
16							
17	**Output**						
18		Reynolds Number	dimensionless	77,357.3	1,692,238		
19							
20							
21							
22				=NReSI(D8,D9,D10,D11)	=NReSI(E8,E9,E10,,E12,E13,E14)		
23							
24							

Figure 1-1.

	A	B	C	D
6	**Inputs**			
7		Parameter	Units	Example 3
8		Mass Flow Rate	kg/h	10,000.0
9		Viscosity	mPa-s	1.2
10		Pipe Diameter	mm	38.1
11		Density	kg/m3	961.0
12		Temperature	C	
13		Molecular Weight	kg/kgmol	
14		Pressure	kPa, absolute	
15				
16		Pipe Roughness	m	0.0000457
17				
18	**Output**			
19		Reynolds Number	dimensionless	77,357
20				
21		Darcy Friction Factor	dimensionless	0.0234
22				
23				
24				
25			=FrictionSI(D16,D19,D10)	
26				

Figure 1-2.

```
' convert diameter from mm to m
d = d/1000
' Churchill, S.W., "Friction Factor Equation Spans all
Flow Regimes,"
' Chemical Engineering, 84:24, p 91, 1977.
a = (2.457 * Log(1/((7/NRe) ^ 0.9 + (0.27 * epsilon/
d)))) ^ 16
b = (37530/NRe) ^ 16
FrictionSI = 8 * ((8/NRe) ^ 12 + 1/(a + b) ^ 1.5) ^ (1/12)
End Function
```

Pressure Drop Due to Friction

Syntax: = PD(W, P_1, P_2, d, L, f, ρ, T_1, M, γ)
Function PD(W, Pin, Pout, d, L, f, Optional Density, Optional Tin, Optional Mw, Optional Gamma)
Returns: Downstream pressure, upstream pressure, or flow
Required: Two of the three parameters: downstream pressure, upstream pressure, or flow
            Pipe parameters diameter, equivalent length, and friction factor. When solving for flow, an additional iteration is needed on the Reynolds number. See the example for an explanation.
            Either density *or* temperature, molecular weight and gamma

Note: As written, limited to perfect gas behavior. The compressibility factor, Z, could be added wherever the term R * T appears (change to R * T * Z); in this case, add Z to the list of arguments.

Example, see Figure 1-3:

Create a cell formula that computes the difference between the guess and the calculated (E26) flow rate.

Then, use Goal Seek to find a value for Guess that equals the calculated rates.

Actually, this is overkill given the inherent uncertainty in the friction factor to begin with. There is nothing wrong with calculating a friction factor based on an initial guess then just going with it. An example flow rate calculation is shown in Figure 1-4.

Listing

```
Function PDSI(W, Pin, Pout, d, L, f, Optional Density,
Optional Tin, Optional Mw, Optional Gamma, Optional
Isothermal)
    ' Pressure Drop due to friction in a round pipe
    ' with the following arguments
    ' Specify two of the following three; function will
compute the third
    ' W = mass flow rate, kg/h
    ' Pin = inlet, or upstream, pressure, kPa
    ' Pout = outlet, or downstream pressure, kPa
    ' Pipe properties
    ' d = pipe diameter, mm
    ' L = pipe length, m
    ' f = Darcy friction factor
    ' Fluid properties
    ' Density (optional) -- specify for liquids, kg/m3
    ' Tin (optional) -- specify for gas, inlet temperature,
deg C (default to 20)
    ' Mw (optional) -- specify for gas, molecular weight
(default to 29 for air)
    ' Gamma (optional) -- specify for gas, ratio of Cp/Cv
(default to 1.4)
    ' Isothermal (optional) -- any value forces isothermal
compressible calc, if missing then adiabatic
    ' Establish constants
    gc = 1 ' conversion constant, m/sec2
    R1 = 8314.47 ' gas constant, m3-Pa/kgmol-K
    Pi = 3.1415926
    ' Convert d to meters
    d = d/1000
    ' convert temperature to deg K
    If IsMissing(Tin) Then Tin = 20
    Tin = Tin + 273.15
    ' Determine which unknown to solve for - Flow, Inlet
Pressure, or Outlet Pressure
    On Error Resume Next
    If IsMissing(W) Then W = 0
    If IsMissing(Pin) Then Pin = 0
    If IsMissing(Pout) Then Pout = 0
    On Error GoTo 0
    If W = 0 Then opt = opt + 1
    If Pin = 0 Then opt = opt + 3
    If Pout = 0 Then opt = opt + 5
    If opt = 0 Then
    PDSI = "Input Err"
    GoTo PDEnd
    End If
    ' Limited input checking
```

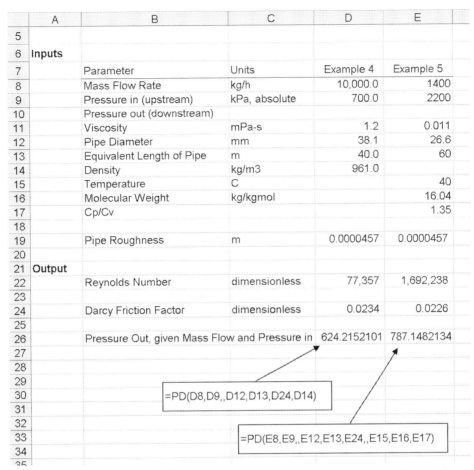

	A	B	C	D	E
5					
6	Inputs				
7		Parameter	Units	Example 4	Example 5
8		Mass Flow Rate	kg/h	10,000.0	1400
9		Pressure in (upstream)	kPa, absolute	700.0	2200
10		Pressure out (downstream)			
11		Viscosity	mPa-s	1.2	0.011
12		Pipe Diameter	mm	38.1	26.6
13		Equivalent Length of Pipe	m	40.0	60
14		Density	kg/m3	961.0	
15		Temperature	C		40
16		Molecular Weight	kg/kgmol		16.04
17		Cp/Cv			1.35
18					
19		Pipe Roughness	m	0.0000457	0.0000457
20					
21	Output				
22		Reynolds Number	dimensionless	77,357	1,692,238
23					
24		Darcy Friction Factor	dimensionless	0.0234	0.0226
25					
26		Pressure Out, given Mass Flow and Pressure in		624.2152101	787.1482134
27					
28					
29					
30		=PD(D8,D9,,D12,D13,D24,D14)			
31					
32					
33		=PD(E8,E9,,E12,E13,E24,,E15,E16,E17)			
34					

Figure 1-3. Calculation of downstream pressure given upstream pressure and mass flow rate.

```
' If Density input is greater than 30 then assume this is
a liquid pressure drop calculation
  If Not IsMissing(Density) Then
  If Density > 50 Then ' do the liquid pressure drop
calculation then exit the function
  ' hf = f (L/D) x (v2/2gc)
  ' Constant for the liquid pressure drop equation
  ' to gather conversion factors and gc together
  ' converting hf in feet of liquid to psi
  PDConst = (1/3600 ^ 2) * (1/(Pi * (d/2) ^ 2)) ^ 2 * (1/(2 *
gc)) * 0.001
  Select Case opt
  Case 1 'solve for W (liquid)
  W = ((Pin – Pout)/(PDConst * (f * L/d/Density))) ^ 0.5
  PDSI = W
  Case 3 'solve for Pin (liquid)
  Pin = Pout + PDConst * f * L/d * W ^ 2/Density
  PDSI = Pin
  Case 5 'solve for Pout (liquid)
  Pout = Pin - PDConst * f * L/d * W ^ 2/Density
  PDSI = Pout
  Case Else

  PDSI = "Input Err - Liq"
  End Select
  GoTo PDEnd
  End If
  End If
  '========================================
  ' The rest of the function performs compressible flow
calculations
  If IsMissing(Mw) Or Mw > 300 Then Mw = 29
  If Mw > 300 Or Mw < 2 Then Mw = 29
  If IsMissing(Gamma) Then Gamma = 1.4
  If Gamma > 1.8 Or Gamma < 1.1 Then Gamma = 1.4
  If IsMissing(Isothermal) Then Isothermal = "No"
  If Isothermal = -1 Then Isothermal = "No"
  If Isothermal <> "No" Then 'ANY value results in
isothermal calculation
    opt = opt + 1
  End If
  Select Case opt
  ' Solve for W =========================
  Case 1 'solve for W, adiabatic
  ' convert pressure to Pa
```

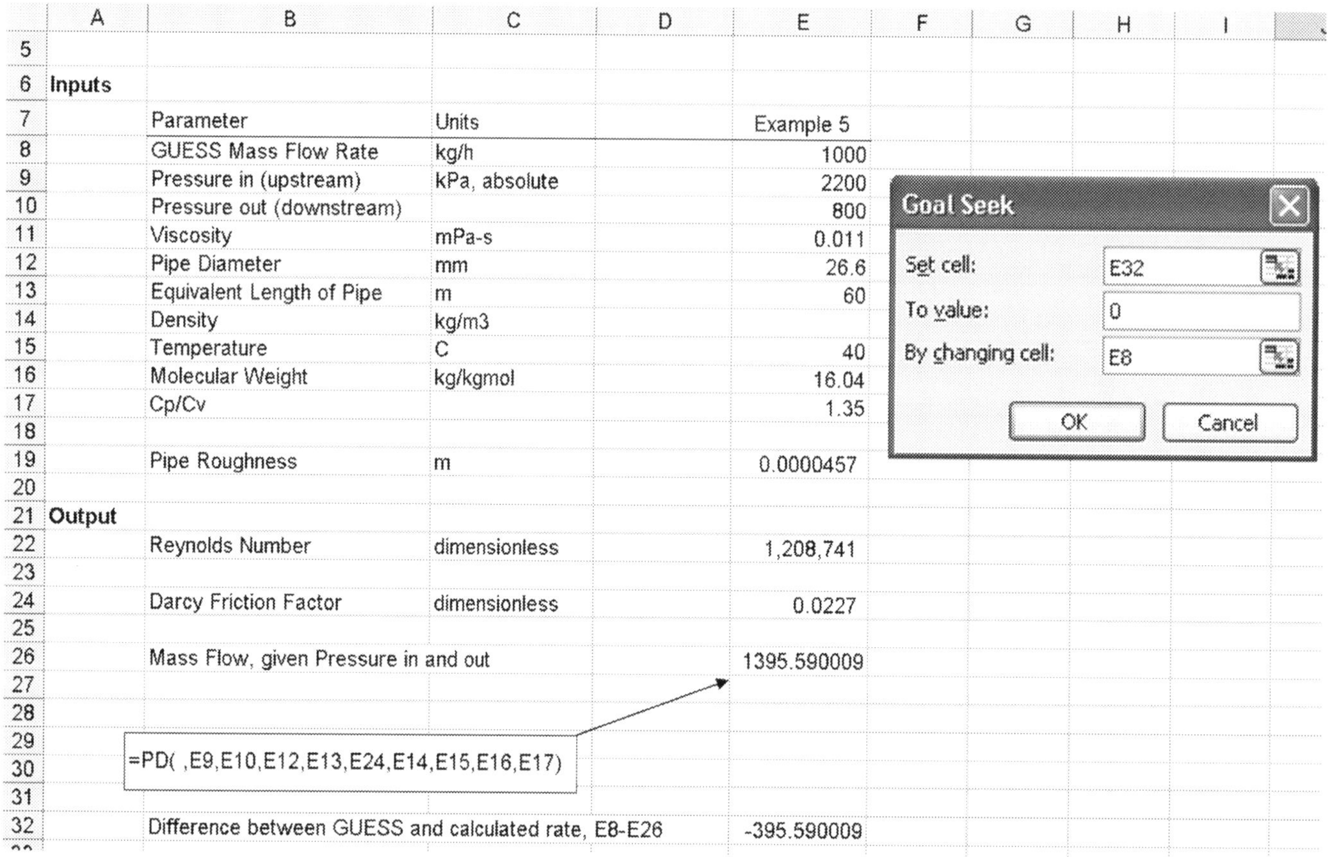

	A	B	C	D	E	F	G	H	I
5									
6	Inputs								
7		Parameter	Units		Example 5				
8		GUESS Mass Flow Rate	kg/h		1000				
9		Pressure in (upstream)	kPa, absolute		2200				
10		Pressure out (downstream)			800				
11		Viscosity	mPa-s		0.011				
12		Pipe Diameter	mm		26.6				
13		Equivalent Length of Pipe	m		60				
14		Density	kg/m3						
15		Temperature	C		40				
16		Molecular Weight	kg/kgmol		16.04				
17		Cp/Cv			1.35				
18									
19		Pipe Roughness	m		0.0000457				
20									
21	Output								
22		Reynolds Number	dimensionless		1,208,741				
23									
24		Darcy Friction Factor	dimensionless		0.0227				
25									
26		Mass Flow, given Pressure in and out			1395.590009				
27									
28									
29									
30		=PD(,E9,E10,E12,E13,E24,E14,E15,E16,E17)							
31									
32		Difference between GUESS and calculated rate, E8-E26			-395.590009				

Goal Seek

Set cell: E32

To value: 0

By changing cell: E8

OK   Cancel

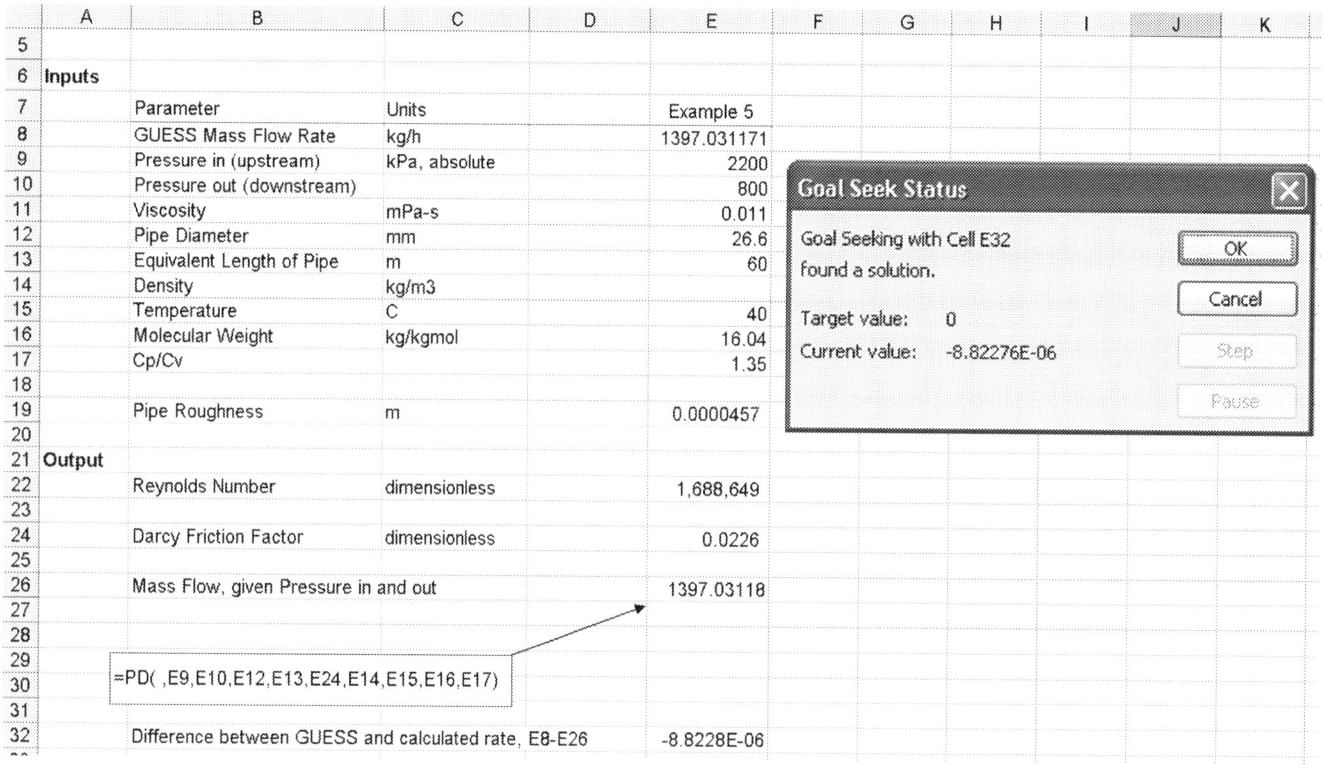

	A	B	C	D	E	F	G	H	I	J	K
5											
6	Inputs										
7		Parameter	Units		Example 5						
8		GUESS Mass Flow Rate	kg/h		1397.031171						
9		Pressure in (upstream)	kPa, absolute		2200						
10		Pressure out (downstream)			800						
11		Viscosity	mPa-s		0.011						
12		Pipe Diameter	mm		26.6						
13		Equivalent Length of Pipe	m		60						
14		Density	kg/m3								
15		Temperature	C		40						
16		Molecular Weight	kg/kgmol		16.04						
17		Cp/Cv			1.35						
18											
19		Pipe Roughness	m		0.0000457						
20											
21	Output										
22		Reynolds Number	dimensionless		1,688,649						
23											
24		Darcy Friction Factor	dimensionless		0.0226						
25											
26		Mass Flow, given Pressure in and out			1397.03118						
27											
28											
29											
30		=PD(,E9,E10,E12,E13,E24,E14,E15,E16,E17)									
31											
32		Difference between GUESS and calculated rate, E8-E26			-8.8228E-06						

Goal Seek Status

Goal Seeking with Cell E32 found a solution.

Target value: 0
Current value: -8.82276E-06

OK   Cancel   Step   Pause

Figure 1-4. Calculation of flow rate given upstream and downstream pressures. Two-step method is illustrated. First, guess the flow rate (Cell E8); the accuracy of the guess is unimportant. The Reynolds number and friction factor are calculated using the guessed flow rate.

```
Pin = Pin * 1000
Pout = Pout * 1000
' Tout can be calculated directly for adiabatic flow
Tout = Tin * (Pout/Pin) ^ ((Gamma - 1)/Gamma)
' start by estimating W using the isothermal
calculation (Case 2)
A = f * L/(2 * d)
Pavg = (Pin + Pout)/2
' calculate mass flux, kg/s-m2
G = ((Pin - Pout) * Pavg * Mw/(A + Log(Pin/Pout))/(R1 *
Tin)) ^ 0.5
' solve for W
WGuess = G * 3600 * (Pi * (d/2) ^ 2)
' Assume answer is within 50% of the initial guess
Wup = WGuess * 1.5
Wlow = WGuess/1.5
' Calculate left hand side of equation, fL/d
LHS = f * L/d
j = 0
A = 1/Gamma
Do
j = j + 1 'counts iterations
' calculate mass flux, kg/s-m2
G = WGuess/3600/(Pi * (d/2) ^ 2)
' Mach number at inlet conditions
DensityIn = Pin * Mw/(R1 * Tin)
Vin = G/DensityIn
MachIn = (Vin ^ 2/(gc * Gamma * Tin * R1/Mw)) ^ 0.5
' Mach number at outlet conditions
DensityOut = Pout * Mw/(R1 * Tout)
Vout = G/DensityOut
MachOut = (Vout ^ 2/(gc * Gamma * Tout * R1/Mw)) ^ 0.5
' With downstream pressure, calculate Mach and then RHS
' where the RHS is comprised of 4 parts
' RHS = A * (B - C - DD)
' A = 1/Gamma
' B = 1/MachIn^2
' C = 1/MachOut^2
' DD = (Gamma+1)/2 * ln(num/den)
' num = MachOut^2 * (1 + ((Gamma-1)/2) * MachIn^2)
' den = MachIn^2 * (1 + ((Gamma-1)/2) * MachOut^2)
' Iteratively solve for Tguess and FlowGuess
' starting with Tguess = Tin
B = 1/MachIn ^ 2
c = 1/MachOut ^ 2
num = MachOut ^ 2 * (1 + ((Gamma - 1)/2) * MachIn ^ 2)
den = MachIn ^ 2 * (1 + ((Gamma - 1)/2) * MachOut ^ 2)
DD = (Gamma + 1)/2 * Log(num/den)
RHS = A * (B - c - DD)
If RHS > LHS Then
Wlow = WGuess
WGuess = (Wup + WGuess)/2
Else
Wup = WGuess
WGuess = (Wlow + WGuess)/2
End If
Loop While (Abs(RHS - LHS)/LHS) > 0.00000001 And j < 100

' Be sure convergence was reached, otherwise return an
error message
If j >= 100 Then WGuess = "Not converged!"
PDSI = WGuess
'=========================================
Case 2 'Solve for W, isothermal
' convert pressure to Pa
Pin = Pin * 1000
Pout = Pout * 1000
A = f * L/(2 * d)
Pavg = (Pin + Pout)/2
' calculate mass flux, kg/s-m2
G = ((Pin - Pout) * Pavg * Mw/(A + Log(Pin/Pout))/(R1 *
Tin)) ^ 0.5
' solve for W
W = G * 3600 * (Pi * (d/2) ^ 2)
' Mach number at outlet and choking W conditions
' for this purpose, assume isothermal
DensityOut = Pout * Mw/(R1 * Tin)
Vout = G/DensityOut
MachOut = (Vout ^ 2/(gc * Gamma * Tin * R1/Mw)) ^ 0.5
If MachOut > 1 Then
PDSI = "Choked W"
GoTo PDEnd
End If
PDSI = W
' Solve for Pin ============================
Case 3 'Solve for Pin, adiabatic
' convert pressure to Pa
Pout = Pout * 1000
' calculate mass flux, kg/s-m2
G = W/3600/(Pi * (d/2) ^ 2)
' Mach number at outlet and choking W conditions
' for this purpose, assume isothermal
DensityOut = Pout * Mw/(R1 * Tin)
Vout = G/DensityOut
MachOut = (Vout ^ 2/(gc * Gamma * Tin * R1/Mw)) ^ 0.5
If MachOut > 1 Then
PDSI = "Choked W"
GoTo PDEnd
End If
' Calculate left hand side of equation, fL/d
LHS = f * L/d
' With a guessed upstream pressure, calculate Mach and
then RHS
' where the RHS is comprised of 4 parts
' RHS = A * (B - C - DD)
' A = 1/Gamma
' B = 1/MachIn^2
' C = 1/MachOut^2
' DD = (Gamma+1)/2 * ln(num/den)
' num = MachOut^2 * (1 + ((Gamma-1)/2) * MachIn^2)
' den = MachIn^2 * (1 + ((Gamma-1)/2) * MachOut^2)
' Iteratively solve for TOut and MachOut
' starting with TOut = Tin
Pin = Pout * 200 ' Arbitrarily high initial guess
Pup = Pout * 400 ' Arbitrarily higher limit
```

```
Plow = Pout ' PIn cannot be less than POut
j = 0
A = 1/Gamma
' Iterative solution for Pin
j = 0
Do
j = j + 1
' Calculate Tout from Adiabatic Expansion equation
Tout = Tin * (Pout/Pin) ^ ((Gamma - 1)/Gamma)
' Using Pout and Tout, calculate the velocity
Vout = G/(Pout * Mw/(R1 * Tout))
' Now calculate MachOut
MachOut = (Vout ^ 2/(gc * Gamma * R1 * Tout/Mw)) ^ 0.5
' Using Pin and Tin, calculate the inlet velocity
Vin = G/(Pin * Mw/(R1 * Tin))
' Now calculate MachIn
MachIn = (Vin ^ 2/(gc * Gamma * R1 * Tin/Mw)) ^ 0.5
B = 1/MachIn ^ 2
c = 1/MachOut ^ 2
num = MachOut ^ 2 * (1 + ((Gamma - 1)/2) * MachIn ^ 2)
den = MachIn ^ 2 * (1 + ((Gamma - 1)/2) * MachOut ^ 2)
DD = (Gamma + 1)/2 * Log(num/den)
RHS = A * (B - c - DD)
If RHS < LHS Then
Plow = Pin
Pin = (Pup + Plow)/2
Else
Pup = Pin
Pin = (Pup + Plow)/2
End If
Loop While (Abs(RHS - LHS)/LHS) > 0.00000001 And j < 100
Pin = Pin/1000
' Be sure convergence was reached, otherwise return an
error message
If j >= 100 Then Pin = "Not converged!"
PDSI = Pin
'=========================================
Case 4 ' Isothermal Equation for Pin
' convert pressure to Pa
Pout = Pout * 1000
' calculate mass flux, kg/s-m2
G = W/3600/(Pi * (d/2) ^ 2)
' solve iteratively, starting with initial guess that
Pin = Pout * 20
Pup = Pout * 20
Plow = Pout
Pin = (Pup + Plow)/2
A = f * L/(2 * d)
RHS = R1 * Tin * G ^ 2/Mw
j = 0
Do
j = j + 1
Pavg = (Pin + Pout)/2
LHS = (Pin - Pout) * Pavg/(A + Log(Pin/Pout))
If RHS > LHS Then
Plow = Pin
Pin = (Pup + Plow)/2
```

```
Else
Pup = Pin
Pin = (Plow + Pup)/2
End If
Loop While Abs((RHS - LHS)/LHS) > 0.0000001 And j < 1000
Pin = Pin/1000
' Be sure convergence was reached, otherwise return an
error message
If j >= 100 Then Pin = "Not converged!"
PDSI = Pin
' Solve for Pout =======================
Case 5 'solve for Pout, adiabatic
' convert pressure to Pa
Pin = Pin * 1000
' calculate mass flux, kg/s-m2
G = W/3600/(Pi * (d/2) ^ 2)
' Mach number at inlet W conditions
DensityIn = Pin * Mw/(R1 * Tin)
Vin = G/DensityIn
MachIn = (Vin ^ 2/(gc * Gamma * Tin * R1/Mw)) ^ 0.5
' Stagnation temperature
Tst = Tin * (1 + MachIn ^ 2 * (Gamma - 1)/2)
' Calculate left hand side of equation, fL/d
LHS = f * L/d
A = 1/Gamma
B = 1/MachIn ^ 2
' RHS = A * (B - C - DD)
' A = 1/Gamma
' B = 1/MachIn^2
' C = 1/MachOut^2
' DD = (Gamma+1)/2 * ln(num/den)
' num = MachOut^2 * (1 + ((Gamma-1)/2) * MachIn^2)
' den = MachIn^2 * (1 + ((Gamma-1)/2) * MachOut^2)
LHS = LHS/A - B
' Now, RHS = -C-DD
' multiply both sides by -1
LHS = LHS * (-1)
' Now, RHS = C+DD
' Assume an initial value for Pout and its limits
Pup = Pin
Plow = Pin/((Gamma + 1)/2) ^ (Gamma/(Gamma - 1))
Pout = (Pup + Plow)/2
Tout = Tin * (Plow/Pin) ^ ((Gamma - 1)/Gamma)
' Iterative solution for Pout
j = 0
Do
j = j + 1
' Calculate Tout from Adiabatic Expansion equation
Tout = Tin * (Pout/Pin) ^ ((Gamma - 1)/Gamma)
' Using Pout and Tout, calculate the velocity
Vout = G/(Pout * Mw/(R1 * Tout))
' Now calculate MachOut
MachOut = (Vout ^ 2/(gc * Gamma * R1 * Tout/Mw)) ^ 0.5
c = 1/MachOut ^ 2
num = MachOut ^ 2 * (1 + ((Gamma - 1)/2) * MachIn ^ 2)
den = MachIn ^ 2 * (1 + ((Gamma - 1)/2) * MachOut ^ 2)
DD = (Gamma + 1)/2 * Log(num/den)
```

```
RHS = c + DD
If RHS < LHS Then
Plow = Pout
Pout = (Pup + Plow)/2
Else
Pup = Pout
Pout = (Pup + Plow)/2
End If
Loop While (Abs(RHS - LHS)/LHS) > 0.00000001 And j < 100
PDSI = Pout/1000
'========================================
Case 6 ' Isothermal Equation for Pout
' convert pressure to Pa
Pin = Pin * 1000
' calculate mass flux, kg/s-m2
G = W/3600/(Pi * (d/2) ^ 2)
' Pavg = (Pin+Pout)/2
' (Pin - Pout) = RTG^2/PavgM (fL/2D + ln(Pin/Pout))
' solve iteratively, starting with initial guess that
Pout = Pin/2
Pout = Pin/2
Pup = Pin
Plow = 0
A = f * L/(2 * d)
RHS = R1 * Tin * G ^ 2
```

```
j = 0
Do
j = j + 1
Pavg = (Pin + Pout)/2
LHS = (Pin - Pout) * (Pavg * Mw)/(A + Log(Pin/Pout))
If RHS < LHS Then
Plow = Pout
Pout = (Pup + Pout)/2
Else
Pup = Pout
Pout = (Plow + Pout)/2
End If
Loop While (Abs(RHS - LHS)/LHS) > 0.0000001 And j < 100
Pout = Pout/1000
' Be sure convergence was reached, otherwise return an
error message
If j >= 100 Then Pout = "Not converged!"
PDSI = Pout
'========================================
Case Else 'error
PDSI = "Input Error"
End Select
' Output the answer
PDEnd:
End Function
```

Orifices

The Bernoulli principle shows that pressure and velocity are related. When fluid is accelerated through an orifice, the pressure is reduced. Downstream of the orifice the flow's velocity slows again and the pressure increases, to near where it was upstream of the orifice. This principle is used for inferring flow rate from pressure readings upstream and at the orifice, and is also applied to other flow instruments such as venturi meters and flow nozzles.

The *permanent* pressure drop caused by an orifice, largely due to friction as the fluid passes through, is important to know. Whether the orifice is primarily intended as a flow meter, or if it is designed to purposely restrict flow, the permanent pressure drop should be considered when designing the piping system.

Incompressible Flow (Liquids)

For preliminary design work, assume the orifice has a coefficient of discharge, $C = 0.62$. Then, calculate the approximate permanent pressure drop through the orifice with Equation 1-20 (derived from equations 5-12 and 5-24 in Ref [19]). Rearrange this equation to determine the

orifice diameter required to achieve a specific pressure drop. The relationship holds for orifices from 20% to 80% of the pipe diameter.

The first term in the equation is the pressure difference between taps located one pipe diameter upstream and half a pipe diameter downstream of the front face of the orifice (i.e., radius taps). The second term is the portion of pressure loss that is permanent. When the Reynolds number in the pipe upstream of the orifice is greater than about 20,000, the pressure difference between the taps is insensitive to the orifice hole diameter. It doesn't matter if the orifice actually has radius taps; Equation 1-20 is simply using the radius tap locations in the computation for permanent pressure drop.

$$\Delta P = \frac{\left(\frac{w}{C\,A}\right)^2}{2\,g_c\,\rho}\left(1 - \beta^2\right) \tag{1-20}$$

ΔP = permanent pressure loss, Pa or lb/ft^2
w = mass flow rate, kg/s or lb/s

C = coefficient of discharge (use 0.62 or calculate with Equation 1-21 until manufacturer's data is known)
A = orifice hole cross sectional area, m^2 or ft^2
g_c = conversion factor = 1 for SI units or 32.17 ft/s^2
ρ = density, kg/m^3 or lb/ft^3
β = orifice diameter divided by pipe diameter

The coefficient of discharge can be calculated using the Stolz equation [25]. Although replaced in ISO [16] by a more complex relationship, the Stolz formula is more than adequate for normal design work.

$$C = 0.5959 + 0.0312\,\beta^{2.1} - 0.184\,\beta^8$$

$$+ 0.0029\,\beta^{2.5}\left(\frac{10^6}{N_{\mathrm{Re}}}\right) + 0.09\,L_1\frac{\beta^4}{1-\beta^4} - L_2\,\beta^3 \tag{1-21}$$

For radius taps, $L_1 = 1$ and $L_2 = 0.47$.

Crane [10] gives a relationship for the flow coefficient, K, that can be used along with K coefficients for fittings:

$$K = \left[\frac{\sqrt{1 - \beta^4(1 - C^2)}}{C\,\beta^2} - 1\right]^2 \tag{1-22}$$

Compressible Flow (gases)

Calculate the critical pressure ratio with:

$$\left(\frac{P_2}{P_1}\right)_{crit} = \left[\frac{2}{(\gamma + 1)}\right]^{\gamma/(\gamma-1)} \tag{1-23}$$

P_1 = Pressure upstream of orifice, Pa or $lb/in.^2$
P_2 = Pressure at minimum pressure point at orifice discharge
γ = ratio of specific heats, C_p/C_v

When $\gamma = 1.4$, the critical ratio is 0.53. Thus, if the orifice is discharging to atmosphere, the upstream pressure must be less than about 1 barg to avoid choking flow. If the pressure is higher, then flow is choked by the sonic velocity and further increases in upstream pressure will not result in increased flow through the orifice.

For critical flow, estimate the mass quantity using physical property values (temperature and pressure) at the point just upstream of the orifice.

$$w = C\,A\,P_1\sqrt{\left(\frac{\lambda\,M}{Z\,R\,T}\right)\left(\frac{2}{\gamma+1}\right)^{(\gamma+1)/(\gamma-1)}} \tag{1-24}$$

For subsonic compressible flow, equation 1-20 applies with the addition of the expansion factor, Y.

$$\Delta P = \frac{\left(\frac{w}{Y\,C\,A}\right)^2}{2\,g_c\,\rho}(1 - \beta^2) \tag{1-25}$$

$$Y = 1 - (0.41 + 0.35\,\beta^4)\,\frac{\Delta P/(1 - \beta^2)}{\gamma\,P_1} \tag{1-26}$$

Control Valves

Valves are usually installed for one of the following purposes:

1. Open/closed. The valve is either fully open or fully closed and is intended to allow flow through a pipe. Gate, plug, butterfly, and ball valves are most common in this service.
2. Diverting. The valve is used to split a flow between two branch lines, or to fully divert the flow to one branch or the other. Three-way plug or ball valves are used in small diameter piping, about DN 100

(4-inch) and smaller. For larger pipe sizes, it's usually more economical to use two open/closed valves, one on each branch.
3. Regulating. The valve is partially closed and either actively regulated or manually set so the flow through the pipe is maintained at a certain rate. This installation may be designed to balance the flow through a complex piping system, or to allow the use of a slightly oversized pump, taking up the "slack" with the valve and anticipating increasing

pressure drop through the piping over time as it corrodes or fouls.

4. Controlling. A control loop actively adjusts the valve opening so the flow or pressure downstream of the valve meets a specific set point. Every type of valve may be used, but flow characteristics may favor one type over another. This is discussed in this section.

Valve manufacturers specify the pressure drop with a *flow coefficient* that is specific to a particular valve. In SI Units the flow coefficient, K_v = flow of water at 20 °C in m^3/h with a pressure drop of 1 bar. In US Units the flow coefficient, C_v = flow of water at 60°F in gpm with a pressure drop of 1 psi. See Table 1-3.

To determine the desired flow coefficient from a known pressure drop:

$$K_v(or\ C_v) = Q\sqrt{\frac{SG}{\Delta P}} \qquad (1\text{-}27)$$

Where SG is the specific gravity of the flowing fluid relative to water.

Rules of thumb for control valve selection include:

- Assume the valve will dissipate, at the maximum controlled flow rate, 10% to 15% of the total pressure drop through the pipeline, or 70 kPa (10 psi), whichever is greater.
- Select a valve that operates between 10% and 80% open at anticipated flow rates.
- Choose a valve that is no smaller than half the pipe size.
- A typical globe-type control valve has a rangeability – the ratio of maximum to minimum controllable flow rate – of 50:1 [22].
- Circumstances that can lead to oversizing a control valve include [22]:
  - An additional "safety factor" has been added to the system calculations (see recommendations on page 3).
  - Sizing routines include operational factors such as an overzealous allowance for fouling or corrosion.
  - The calculated valve coefficient is only slightly higher than the coefficient for a standard valve and the next larger size must be selected.

Partially Full Horizontal Pipes

The equations in the previous sections are, of course, intended for use with full pipes. Durand [13] provides a rapid way to estimate whether a horizontal pipe carrying liquid is full. This method is intended for gravity drains, not pumped systems that have no entry point for vapors. With Q = flow rate in gallons per minute, and d = pipe diameter in inches, the criteria are:

If $\dfrac{Q}{d^{2.5}} \geq 10.2$ the pipe is full.

If $\dfrac{Q}{d^{2.5}} < 10.2$ do a partially full flow analysis as follows:

1. Let $x = \ln\left(\dfrac{Q}{d^{2.5}}\right)$ and find the height of the liquid in the pipe by:

$$\frac{H}{D} = 0.446 + 0.272\,x + 0.0397\,x^2 - 0.0153\,x^3 - 0.003575\,x^4$$

2. Find the equivalent diameter by:

$$\frac{D_e}{D} = -0.01130 + 3.040\left(\frac{H}{D}\right)$$
$$- 3.461\left(\frac{H}{D}\right)^2 + 4.108\left(\frac{H}{D}\right)^3 - 2.638\left(\frac{H}{D}\right)^4$$

This is an empirical way to avoid getting D_e from

$$D_e = \frac{cross\ sectional\ area}{wetted\ perimeter}$$

Note that for $1.0 > H/D > 0.5$ that $D_e/D > 1$. My calculations and all references confirm this. D_e is substituted for D in subsequent flow analysis.

D = diameter of pipe
H = liquid level in the pipe

Example:

Given:	horizontal pipe, d = 4 inches ID, Q = 100 gpm
Find:	Is the pipe full? If not, what is the liquid height? Also, what is the pipe's equivalent diameter?

Calculations:

$Q/d^{2.5} = 100/32 = 3.125$

Not full since $Q/d^{2.5} < 10.2$

$X = \ln(3.125) = 1.1394$

$H/D = 0.779$

$H = 0.779\ (4) = 3.12$ in

$D_e/D = 1.227$

$D_e = 1.227\ (4) = 4.91$ in

Two-Phase Flow

Two-phase flow is difficult to model and extreme caution is recommended. Published pressure drop correlations are applicable for specific situations. Blindly applying a correlation may result in orders of magnitude error.

For gas-liquid flows, consider how the stream might behave as it travels through the piping system:

• Elevation changes are important due to the great difference in density between the vapor and liquid phases. With a single-phase liquid flow, only the terminal point elevations are considered since

potential energy is reversible. This is usually not true in two-phase gas-liquid flows.

• Due to the large density difference between vapor and liquid, buoyancy greatly affects the flow regime, distribution of void fraction, and pressure drop. Flow regimes found in horizontal flow are illustrated in Figure 1-5.

• Low pressure zones at pump suctions and behind orifice plates and valves are prone to localized vaporization and condensation, called cavitation, which can cause excessive wear, vibration, and noise. See Chapter 5 for the discussion of NPSH.

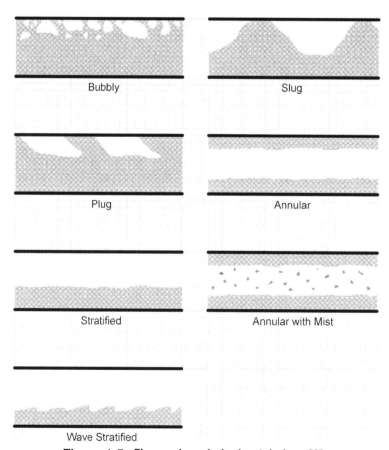

Figure 1-5. Flow regimes in horizontal pipes [6].

- Flashing flow may cause choking. For example, if saturated water is flowing at about 6 m/s velocity, choking conditions occur if about 35% of this flashes to steam. This is the reason that relief systems, especially those designed for runaway reactions, require scrutiny. See Chapter 20, "Safety," for information on this topic.
- Condensing flow improves flowability as vapor converts to much denser liquid.
- Gas-liquid systems that are essentially immiscible, such as air-water, flow in different patterns depending on the relative quantities of the gas and liquid phases, the flow rate, and the direction of flow relative to earth's gravity.

It's also important to understand how the relative quantity of the two phases is determined. For example, the two-phase flow of a pure compound, such as water or refrigerant, is often characterized by its "quality," defined as the mass fraction of flow that is vapor. Pressure drop due to friction varies linearly with the fluid quality; a small error in estimating the vapor content greatly affects the predicted result.

This section presents four different models for calculating two-phase (liquid-vapor) pressure drop due to friction. Each has its strengths and weaknesses, but for well-behaved flows where the velocity is less than 30% of critical velocity they return very similar results. After the presentation of the four models, an apples-to-apples comparison is given.

Nomenclature in the models includes subscripts that are defined as follows:

g	vapor portion of the flow
l	liquid portion of the flow
m	homogenous mixture

Lockhart-Martinelli

The Lockhart-Martinelli model [15] is probably the most well-known method, commonly used in refrigeration and wet steam calculations and recommended by ASHRAE [2].

First, calculate the pressure drop in the pipe considering the liquid and vapor components separately. In each case, the calculation is performed with the full pipe diameter as if the other phase disappears. Thus, if the total flow is 100 and the quality is 0.2, calculate the pressure drop for a liquid flow of 80 and a vapor flow of 20.

Then,

$$X = \left(\frac{\Delta P_l}{\Delta P_g}\right)^{1/2} \tag{1-28}$$

$$Y = 4.6\,X^{-1.78} + 12.5\,X^{-0.68} + 0.65$$

[8]

$$\Delta P = Y\,\Delta P_l \tag{1-30}$$

Homogeneous Model

Awad and Muzychka [4] showed that a liquid-vapor stream can be treated as a homogenous fluid. This is conceptually true if the vapor is considered to be uniformly dispersed in the liquid as tiny bubbles, and the two phases are flowing together at the same velocity. The model treats the two phases as a liquid with average fluid properties that depend on the relative quantity of vapor and liquid (i.e., quality).

Using the inlet pressure, or an average pressure in the pipe segment (obtained iteratively), calculate the homogenous mixture density and viscosity:

$$\rho_m = \left(\frac{x}{\rho_g} + \frac{1-x}{\rho_l}\right)^{-1} \tag{1-31}$$

$$\mu_m = \left(\frac{x}{\mu_g} + \frac{1-x}{\mu_l}\right)^{-1} \tag{1-32}$$

Next, calculate the Reynolds number (page 9) and Darcy Friction Factor, f_m, for the homogenous mixture. Use the total flow rate (vapor + liquid), the full pipe size, and the density and viscosity for the mixture.

Finally, using the following form of the incompressible flow formula (page 10), calculate the pressure drop:

$$\Delta P = \frac{f_m\,L\,G^2}{2\,D\,\rho_m} \tag{1-33}$$

Split Bounds Model

This is a separated flow model, where the vapor and liquid are treated separately. Awad and Muzychka [5] presented equations to represent the highest and lowest pressure drops, or bounds, that a system is likely to experience. They suggest using the arithmetic average of the bounds as an acceptable prediction of pressure drop.

The lower bound is:

$$\Delta P_{lower} = \frac{0.158 \, L \, G^{1.75}(1-x)^{1.75} \mu_l^{0.25}}{D^{1.25} \rho_l}$$

$$\times \left[1 + \left(\frac{x}{1-x} \right)^{0.7368} \left(\frac{\rho_l}{\rho_g} \right)^{0.4211} \left(\frac{\mu_g}{\mu_l} \right)^{0.1053} \right]^{2.375}$$

(1-34)

The upper bound is:

$$\Delta P_{upper} = \frac{0.158 \, L \, G^{1.75}(1-x)^{1.75} \, \mu_l^{0.25}}{D^{1.25} \rho_l}$$

$$\times \left[1 + \left(\frac{x}{1-x} \right)^{0.4375} \left(\frac{\rho_l}{\rho_g} \right)^{0.25} \left(\frac{\mu_g}{\mu_l} \right)^{0.0625} \right]^{4}$$

(1-35)

Asymptotic Model

Awad and Muzychka [3] presented a semi-theoretical method for modeling two-phase flow that is especially suitable for systems for which measured data are available.

In this method a fitting parameter, p, is used to calibrate the prediction to actual data. The fitting parameter must be determined using a least squares fit. Awad and Muzychka give examples where p ranges from 0.25 to 0.8.

First, calculate the pressure drop in the pipe considering the liquid and vapor components separately. In each case, the calculation is performed with the full pipe diameter as if the other phase disappears. Thus, if the total flow is 100 and the quality is 0.2, calculate the pressure drop for a liquid flow of 80 and a vapor flow of 20.

The asymptotic pressure drop is:

$$\Delta P = \left[\Delta P_l^p + \Delta P_g^p \right]^{1/p}$$

(1-36)

Comparison of the Two-Phase Models

Using the exact same input data, Figure 1-6 and Figure 1-7 chart results from the four models. For the split bound model the graph only includes the average value. The models allow results even when velocity in the pipe would exceed the critical velocity, so it is clear that extreme care must be taken when using them. Figure 1-7 is on a log-log scale; the Lockhart-Martinelli prediction is at least 50% greater than the Awad and Muzychka values.

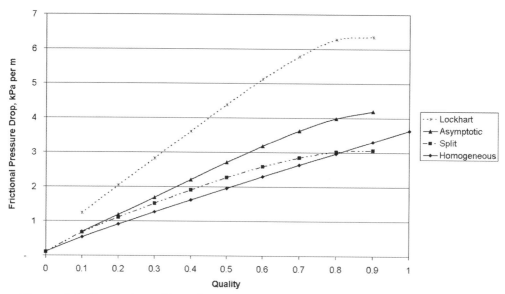

Comparison of Two-Phase Models
R12, 6 Bar pressure, 1000 kg/m²-s in 50 mm smooth pipe

Figure 1-6. Comparison of two-phase models over a range of vapor quality from 0 to 1.

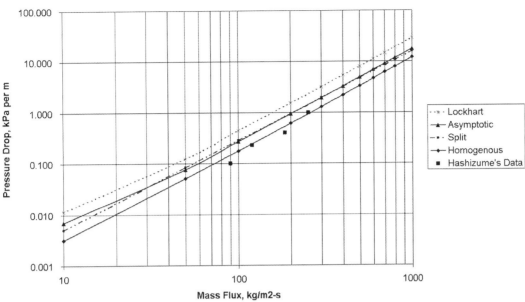

Figure 1-7. Comparison of two-phase models over a range of flow rates at constant vapor quality, including actual measured data from Hashizume [5].

Nomenclature

A = cross-sectional area of pipe, m^2 or ft^2

C = coefficient of discharge, dimensionless

C_p = heat capacity, constant pressure

C_v = heat capacity, constant volume

d = pipe diameter, mm or in.

D = pipe diameter, m or ft

f = Darcy friction factor, dimensionless

G = mass flux, kg/s-m^2 or lb/s-ft^2 $= \dfrac{W}{3600\,A}$

g_c = conversion factor, 1 m/s^2 or 32.17 ft/s^2

H = liquid level, m or ft

K = flow coefficient, dimensionless

L = pipe equivalent length, m or ft

M = molecular weight

N_{Ma} = Mach number, dimensionless

N_{Re} = Reynolds number, dimensionless

P = absolute pressure, Pa or psia

Q = volumetric flow rate, m^3/h or gal/min

R = gas constant, 8314.5 m^3-Pa/kgmol-K or 10.73 ft^3-psi/lbmol-R

t = temperature, °C or °F

T = absolute temperature, °K or °R

U = average fluid velocity at local conditions, m/s or ft/s $= \dfrac{G}{\rho}$

w = mass flow rate, kg/s or lb/s

W = mass flow rate, kg/h or lb/h

X = Lockhart-Martinelli two-phase parameter

Y = Lockhart-Martinelli two-phase multiplier

Z = compressibility factor =1 for a perfect gas

β = diameter ratio, dimensionless

ε = surface roughness, m or ft

γ = ratio of specific heats, C_p/C_v

μ = fluid dynamic viscosity, kg/m-s or lb/ft-h

ρ = density of gas at local conditions, kg/m^3 or lb/ft^3 $= \dfrac{P\,M}{R\,T}$

References

[1] American Petroleum Institute. Recommended Practice for Design and Installation of Offshore Production Platform Piping Systems, API Recommended Practice 14E, March 2007.

[2] American Society of Heating, Refrigerating and Air-Conditioning Engineers. *1997 ASHRAE® Handbook, Fundamentals.*

[3] Awad M, Muzychka Y. A Simple Asymptotic Compact Model for Two-Phase Frictional Pressure Gradient in Horizontal Pipes. *Proceedings of IMECE04 2004 ASME International Mechanical Engineering Congress and Exposition.* Anaheim: California, USA; November 13–20, 2004.

[4] Awad M, Muzychka Y. A Simple Two-Phase Frictional Multiplier Calculation Method. *Proceedings of IPC2004 International Pipeline Conference.* Calgary: Alberta, Canada; October 4–8, 2004.

[5] Awad M, Muzychka Y. Bounds on Two-Phase Flow, Part 1 – Frictional Pressure Gradient in Circular Pipes. *Proceedings of IMECE2005 ASME International Mechanical Engineering Congress and Exposition.* Orlando: Florida, USA; November 5–11, 2005.

[6] Balasubramaniam R, Ramé E, Kizito J, Kassemi M. Two Phase Flow Modeling: Summary of Flow Regimes and Pressure Drop Correlations in Reduced and Partial Gravity. CR-2006–214085, NASA, January 2006.

[7] Branan CR. *The Process Engineer's Pocket Handbook,* vol. 1. Gulf Publishing Co; 1976.

[8] Branan CR. *The Process Engineer's Pocket Handbook,* vol. 2. Gulf Publishing Co; 1983.

[9] Churchill SW. Friction Factor Equation Spans all Flow Regimes. *Chemical Engineering,* 1977;84:24–91.

[10] Crane Company. Flow of Fluids Through Valves, Fittings, and Pipe. Technical Paper No. 410 2009.

[11] Darby R. *Chemical Engineering Fluid Dynamics.* 2nd ed. New York: Marcel Dekker; 2001.

[12] Darby R. Correlate Pressure Drops Through Fittings. *Chemical Engineering,* April 2001:127.

[13] Durand AA, Marquez-Lucer M. Determining Sealing Flow Rates in Horizontal Run Pipes. *Chemical Engineering,* March 1998;129:127.

[14] Geankoplis CJ. *Transport Processes and Separation Process Principles (Includes Unit Operations),* 4th ed. Prentice Hall; 2003.

[15] Lockhart R, Martinelli R. Proposed Correlation of Data for Isothermal Two-Phase, Two-Component Flow in Pipes. *Chemical Engineering Progress Symposium Series,* 1949;45(1):39–48.

[16] International Organization of Standards (ISO 5167-1) Amendment 1. Measurement of fluid flow by means of pressure differential devices, Part 1: Orifice plates, nozzles, and Venturi tubes inserted in circular cross-section conduits running full. Reference number: ISO 5167-1:1991/Amd.1:1998(E); 1998.

[17] McCabe WL, Smith J, Harriott P. *Unit Operations of Chemical Engineering.* 7th ed. McGraw-Hill; 2005.

[18] Moody LE. Friction Factors for Pipe Flow. Transactions ASME 1944;V66:671–84.

[19] Perry RH, Green DW, editors. *Perry's Chemical Engineers' Handbook.* 6th ed. McGraw-Hill; 1984.

[20] Personal communication with Dr. Ron Darby. Data is an updated version of Table 1 in Reference 12.

[21] Salama M. Influence of sand production on design and operation of piping systems *NACE International Corrosion 2000 (conference).* Orlando: Florida, USA; March 26–31, 2000.

[22] Sarco-Spirax, Ltd., Steam Engineering Tutorials, at http://www.spiraxsarco.com/resources/steam-engineering-tutorials.asp, downloaded August, 2010.

[23] Schroeder D. A Tutorial on Pipe Flow Equations, at http://www.psig.org/papers/2000/0112.pdf; August 16, 2001.

[24] Simpson LL. Sizing Piping for Process Plants. *Chemical Engineering,* June 17, 1968:197.

[25] Stolz J. An Approach toward a General Correlation of Discharge Coefficients of Orifice Plate Flowmeters. ISO/TC30/SC2 (France 6), 1975:645.

2

Heat Exchangers

Introduction

Heat exchangers are critical elements in every process plant. While the majority of exchangers are the shell-and-tube type, there are several additional important types. The major types of heat transfer equipment are:

- Shell-and-tube
- Finned tube
- Bare tube
- Plate-and-frame
- Spiral
- Plate coil

This chapter focuses on shell-and-tube exchangers, covering topics of interest to typical process engineers. Plate-and-frame and spiral exchangers are also discussed.

Four factors impact the performance, longevity, and maintenance requirements for heat-transfer equipment and related components [22]:

- Initial knowledge and documentation of all the operating parameters. Without correct operating parameters and application information, proper sizing and selection of heat exchangers is impossible, and all aspects of performance will be compromised.
- Codes and design specifications. Specifying a TEMA designation and an ASME pressure and temperature requirement will enhance all heat transfer selections.
- Installation. Following appropriate installation recommendations can eliminate most premature failures and greatly enhance the performance and efficiency of the heat transfer unit.
- Evaluation. Always evaluate the selections in terms of a ten-year operational period, considering all factors.

An Excel workbook accompanies this chapter. The workbook performs calculations for a liquid-liquid shell-and-tube heat exchanger and completes the associated TEMA datasheet.

TEMA

Describe shell-and-tube heat exchangers using nomenclature from the Standards of the Tubular Exchanger Manufacturers Association (TEMA). Figure 2-1 illustrates the front head, shell, and rear head types and lists letter designations corresponding to each. Figure 2-2 shows six typical heat exchanger configurations, with their corresponding TEMA designation (e.g., BEM). The various parts of the exchangers are called out with the key to the parts listed in Table 2-1.

In addition to the exchanger configurations, TEMA provides design and construction standards for three major classes of exchanger, called R, C, and B. Table 2-2 compares attributes of the three exchanger classes. The three classes are listed in order of decreasing cost (and mechanical performance).

Use datasheets to tabulate the primary process and mechanical requirements for a heat exchanger. TEMA datasheets are recommended because they are well known by engineers and fabricators. Versions with SI and US units are given in Figure 2-3 and Figure 2-4. Similar datasheets from other sources, such as heat exchanger manufacturers and engineering companies, may also be used. Enter the TEMA designation (e.g., BEM) into the cell labeled "Type" on line 6. Enter the TEMA Class (e.g., R) on line 54.

The process engineer usually works closely with the exchanger manufacturer to complete the datasheet. Heat exchanger design is often a trial-and-error process, with different combinations of shell diameter, tube size, length, tube passes, and other attributes being tested. All heat exchanger manufacturers use sophisticated software for thermal and mechanical design, and they are usually more than happy to assist customers by running multiple design cases.

Although computers solve the design equations for most new exchangers, engineers may want to do some preliminary work using the manual methods as described later in this chapter. Sophisticated software such as the HTRI Xchanger Suite [11] performs rigorous incremental calculations that account for the highly dynamic nature of heat exchangers. The manual calculation methods use physical properties averaged across the exchanger, and provide heat transfer and pressure drop approximations for various zones within the exchanger.

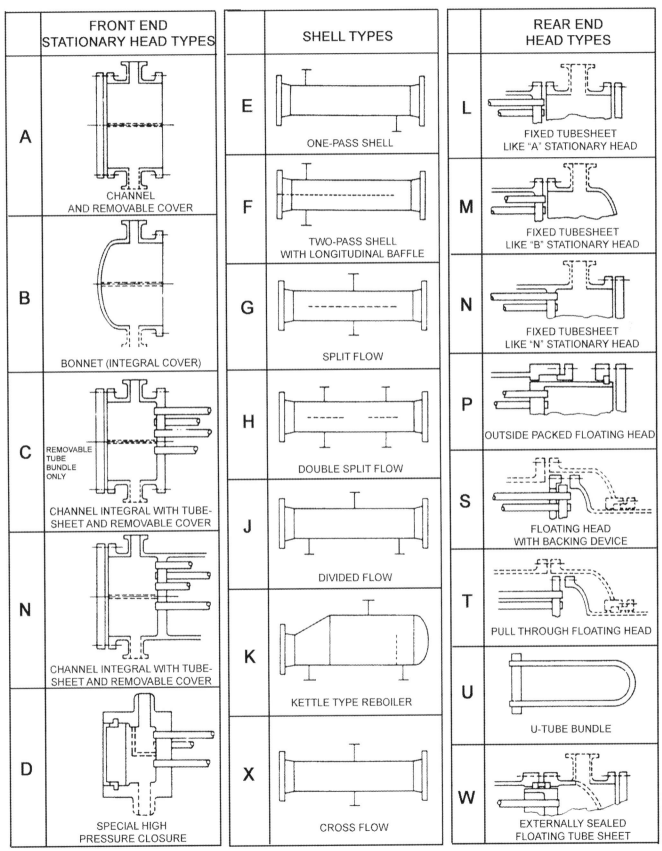

FRONT END STATIONARY HEAD TYPES	SHELL TYPES	REAR END HEAD TYPES
A CHANNEL AND REMOVABLE COVER	**E** ONE-PASS SHELL	**L** FIXED TUBESHEET LIKE "A" STATIONARY HEAD
B BONNET (INTEGRAL COVER)	**F** TWO-PASS SHELL WITH LONGITUDINAL BAFFLE	**M** FIXED TUBESHEET LIKE "B" STATIONARY HEAD
C REMOVABLE TUBE BUNDLE ONLY — CHANNEL INTEGRAL WITH TUBESHEET AND REMOVABLE COVER	**G** SPLIT FLOW	**N** FIXED TUBESHEET LIKE "N" STATIONARY HEAD
N CHANNEL INTEGRAL WITH TUBESHEET AND REMOVABLE COVER	**H** DOUBLE SPLIT FLOW	**P** OUTSIDE PACKED FLOATING HEAD
	J DIVIDED FLOW	**S** FLOATING HEAD WITH BACKING DEVICE
D SPECIAL HIGH PRESSURE CLOSURE	**K** KETTLE TYPE REBOILER	**T** PULL THROUGH FLOATING HEAD
	X CROSS FLOW	**U** U-TUBE BUNDLE
		W EXTERNALLY SEALED FLOATING TUBE SHEET

Figure 2-1. Nomenclature for shell-and-tube heat exchangers [24].

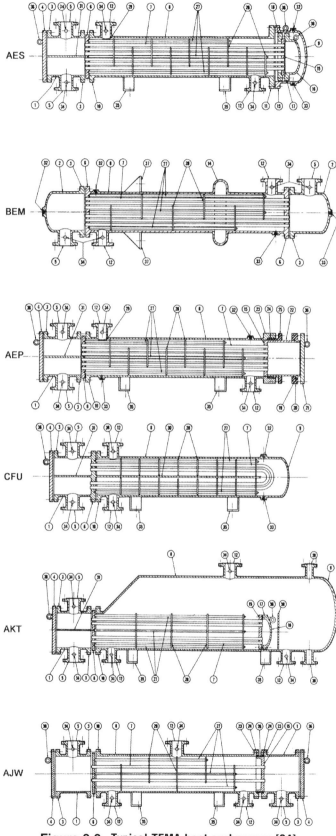

Figure 2-2. Typical TEMA heat exchangers [24].

Table 2-1
Heat exchanger parts and connections (for Figure 2-2)

1. Stationary Head — Channel	20. Slop-on Backing Flange
2. Stationary Head — Bonnet	21. Floating Head Cover — External
3. Stationary Head Flange — Channel or Bonnet	22. Floating Tubesheet Skirt
4. Channel Cover	23. Packing Box Flange
5. Stationary Head Nozzle	24. Packing
6. Stationary Tubesheet	25. Packing Follower Ring
7. Tubes	26. Lantern Ring
8. Shell	27. Tie Rods and Spacers
9. Shell Cover	28. Transverse Baffles or Support Plates
10. Shell Flange — Stationary Head End	29. Impingement Baffle
11. Shell Flange — Rear Head End	30. Longitudinal Baffle
12. Shell Nozzle	31. Pass Partition
13. Shell Cover Flange	32. Vent Connection
14. Expansion Joint	33. Drain Connection
15. Floating Tubesheet	34. Instrument Connection
16. Floating Head Cover	35. Support Saddle
17. Floating Head Flange	36. Lifting Lug
18. Floating Head Backing Device	37. Support Bracket
19. Split Shear Ring	38. Weir
	39. Liquid Level Connection

Table 2-2
Comparison of TEMA class R, C, and B heat exchangers. Cost decreases from left to right [23]

Attribute	Class R	Class C	Class B
Application	Generally severe requirements such as petroleum and related processing applications	Generally moderate requirements such as commercial and general process applications	General process service
Corrosion allowance on carbon steel	0.125 in. (3.2 mm)	0.0625 in (1.6 mm)	0.0625 in (1.6 mm)
Tube diameters, OD	¾, 1, 1¼, 1½, and 2 in.	R + ¼, ⅜, ½, and ⅝ in.	R + ⅝ in.
Tube pitch and minimum cleaning lane	1.25 x tube OD ¼ inch lane	R + ⅜ tubes may be located 1.2 x tube OD	R + lane may be ³⁄₁₆ inch in 12 inch and smaller shells for ⅝ and ¾ in tubes
Minimum shell diameter	8 inch, tabulated	6 inch, tabulated	6 inch tabulated
Longitudinal baffle thickness	¼ inch minimum	⅛ inch alloy, ¼ inch carbon steel	⅛ inch alloy, ¼ inch carbon steel
Floating head cover cross-over area	1.3 x tube flow area	Same as tube flow area	Same as tube flow area
Lantern ring construction	375 °F maximum 300 psi up to 24 inch diameter shell 150 psi for 25 to 42 in. 75 psi for 43 to 60 in.	600 psi maximum	375 °F maximum 300 psi up to 24 inch diameter shell 150 psi for 25 to 42 in. 75 psi for 43 to 60 in.
Gasket materials	Metal jacketed or solid metal for a) internal floating head cover, b) 300 psi and up, c) all hydrocarbons	Metal jacketed or solid metal for a) internal floating head, b) 300 psi and up	Metal jacketed or solid metal for a) internal floating head, b) 300 psi and up
Peripheral gasket contact surface	Flatness tolerance specified	No tolerance specified	No tolerance specified
Minimum tubesheet thickness with expanded tube joints	Outside diameter of the tube	0.75 x tube OK for 1 inch and smaller ⅞ inch for 1¼ OD 1 inch for 1½ OD 1.25 inch for 2 OD	0.75 x tube OK for 1 inch and smaller ⅞ inch for 1¼ OD 1 inch for 1½ OD 1.25 inch for 2 OD
Tube hole grooving	Two grooves	Above 300 psi design pressure or 350 °F design temperature: 2 grooves	Two grooves
Length of expansion	Smaller of 2 inch or tubesheet thickness	Small of 2 x tube OD or 2 inch	Smaller of 2 inch or tubesheet thickness
Tubesheet pass partition grooves	³⁄₁₆ inch deep grooves required	Over 300 psi: ³⁄₁₆ inch deep grooves required or other	Over 300 psi: ³⁄₁₆ inch deep grooves required or other

(Continued)

Table 2-2
Comparison of TEMA class R, C, and B heat exchangers. Cost decreases from left to right [23]—cont'd

Attribute	Class R	Class C	Class B
		suitable means for retaining gaskets in place	suitable means for retaining gaskets in place
Pipe tap connections	6000 psi coupling with bar stock plug	3000 psi coupling	3000 psi coupling with bar stock plug
Pressure gage connections	Required in nozzles 2 inch and up	Specified by purchaser	Required in nozzles 2 inch and up
Thermometer connections	Required in nozzles 4 inch and up	Specified by purchaser	Required in nozzles 4 inch and up
Nozzle construction	No reference to flanges	No reference to flanges	All nozzles larger than one inch must be flanged
Minimum bolt size	¾ inch	½ inch recommended; smaller bolting may be used	⅝ inch

#							
1				Job No.			
2	Customer			Reference No			
3	Address			Proposal No			
4	Plant Location			Date		Rev.	
5	Service of Unit			Item No.			
6	Size	Type	(Hor/Vert)	Connected in		Parallel	Series
7	Surf/Unit (Gross/Eff.)		Sq m; Shells/Unit	Surf/Shell (Gross/Eff.)			sq m
8	**PERFORMANCE OF ONE UNIT**						
9	Fluid Allocation			Shell Side		Tube Side	
10	Fluid Name						
11	Fluid Quantity Total		kg/Hr				
12	Vapor (In/Out)						
13	Liquid						
14	Steam						
15	Water						
16	Noncondensable						
17	Temperature (In/Out)		°C				
18	Specific Gravity						
19	Viscosity, Liquid		Cp				
20	Molecular Weight, Vapor						
21	Molecular Weight, Noncondensable						
22	Specific Heat		J/kg °C				
23	Thermal Conductivity		W/m °C				
24	Latent Heat		J/kg @ °C				
25	Inlet Pressure		kPa(abs.)				
26	Velocity		m/sec				
27	Pressure Drop, Allow. /Calc		kPa	/		/	
28	Fouling Resistance (Min.)		Sq m °C / W				
29	Heat Exchanged			W MTD (Corrected)			°C
30	Transfer Rate, Service			Clean			W/Sq m °C
31	**CONSTRUCTION OF ONE SHELL**				Sketch (Bundle/Nozzle Orientation)		
32			Shell Side	Tube Side			
33	Design / Test Pressure		kPag	/	/		
34	Design Temp. Max/Min		°C	/	/		
35	No. Passes per Shell						
36	Corrosion Allowance		mm				
37	Connections In						
38	Size & Out						
39	Rating Intermediate						
40	Tube No. OD mm;Thk (Min/Avg) mm;Length mm;Pitch mm						◆ 30 △ 60 ⊞ 90 ◇ 45
41	Tube Type			Material			
42	Shell ID OD mm			Shell Cover		(Integ.)	(Remov.)
43	Channel or Bonnet			Channel Cover			
44	Tubesheet-Stationary			Tubesheet-Floating			
45	Floating Head Cover			Impingement Protection			
46	Baffles-Cross Type			%Cut (Diam/Area)	Spacing: c/c	Inlet	mm
47	Baffles-Long			Seal Type			
48	Supports-Tube U-Bend				Type		
49	Bypass Seal Arrangement			Tube-to-Tubesheet Join			
50	Expansion Joint			Type			
51	pv²-Inlet Nozzle		Bundle Entrance		Bundle Exit		
52	Gaskets-Shell Side			Tube Side			
53	Floating Head						
54	Code Requirements			TEMA Class			
55	Weight / Shell		Filled with Water	Bundle			kg
56	Remarks						
57							
58							
59							
60							
61							

Figure 2-3. Data Sheet for shell-and-tube heat exchanger, SI units [24].

1				Job No.		
2	Customer			Reference No		
3	Address			Proposal No		
4	Plant Location			Date		Rev.
5	Service of Unit			Item No.		
6	Size	Type	(Hor/Vert)	Connected in		Parallel Series
7	Surf/Unit (Gross/Eff.)		sq ft; Shells/Unit	Surf/Shell (Gross/Eff.)		sq ft
8	**PERFORMANCE OF ONE UNIT**					
9	Fluid Allocation			Shell Side		Tube Side
10	Fluid Name					
11	Fluid Quantity Total		lb/hr			
12	Vapor (In|Out)					
13	Liquid					
14	Steam					
15	Water					
16	Noncondensable					
17	Temperature		°F			
18	Specific Gravity					
19	Viscosity, Liquid		cP			
20	Molecular Weight, Vapor					
21	Molecular Weight, Noncondensable					
22	Specific Heat		BTU / lb °F			
23	Thermal Conductivity		BTU ft / hr sq ft °F			
24	Latent Heat		BTU / lb @ °F			
25	Inlet Pressure		psia			
26	Velocity		ft / sec			
27	Pressure Drop, Allow. /Calc		psi	/		/
28	Fouling Resistance (Min.)		hr sq ft °F / BTU			
29	Heat Exchanged		BTU / hr MTD (Corrected)			°F
30	Transfer Rate, Service		Clean			BTU / hr sq ft °F
31	**CONSTRUCTION OF ONE SHELL**				Sketch (Bundle/Nozzle Orientation)	
32			Shell Side	Tube Side		
33	Design / Test Pressure	psig	/	/		
34	Design Temp. Max/Min	°F	/	/		
35	No. Passes per Shell					
36	Corrosion Allowance	in				
37	Connections In					
38	Size & Out					
39	Rating Intermediate					
40	Tube No. OD in;Thk (Min/Avg)		in;Length	ft;Pitch	in	◁ 30 △ 60 ⊟ 90 ◇ 45
41	Tube Type			Material		
42	Shell ID OD in		Shell Cover		(Integ.) (Remov.)	
43	Channel or Bonnet		Channel Cover			
44	Tubesheet-Stationary		Tubesheet-Floating			
45	Floating Head Cover		Impingement Protection			
46	Baffles-Cross Type		%Cut (Diam/Area)	Spacing: c/c Inlet		in
47	Baffles-Long		Seal Type			
48	Supports-Tube U-Bend			Type		
49	Bypass Seal Arrangement		Tube-to-Tubesheet Joint			
50	Expansion Joint		Type			
51	ρv²-Inlet Nozzle Bundle Entrance			Bundle Exit		
52	Gaskets-Shell Side		Tube Side			
53	Floating Head					
54	Code Requirements		TEMA Class			
55	Weight / Shell Filled with Water		Bundle			lb
56	Remarks					
57						
58						
59						
60						
61						

Figure 2-4. Data Sheet for shell-and-tube heat exchanger, US units [24].

Selection Guides

The following factors should be considered when choosing the type of heat exchanger to use for a particular application:

- Operating conditions: service requirements (e.g., phase change), thermal duty, temperature approach
- Cleanliness of the streams

- Maximum design pressure and temperature
- Heating or cooling application
- Maintenance requirements
- Material compatibility with process fluids: wetted surfaces and gaskets

Shell-and-Tube Heat Exchangers

This is the most common type of heat exchanger used in the chemical process industries. It is often the lowest cost option, especially when made of carbon steel. Off-the-shelf models are available in fixed tubesheet and U-tube design configurations in smaller sizes, and are usually used for liquid-liquid, reboiling, and gas cooling applications.

TEMA Class exchangers are used for most custom designs, with TEMA B (chemical industry service) being the most common. TEMA guidelines are limited to a shell diameter of 1524 mm (60 in.), working pressure of 207 bar (3,000 psig), and product of shell diameter times pressure not exceeding 315,000 mm-bar (60,000 in.-psig).

Plate-and-Frame Heat Exchangers

In appropriate circumstances, plate-and-frame heat exchangers offer many advantages compared with shell-and-tube designs. The plate-and-frame units have higher heat transfer coefficients – often three to four times that of a shell-and-tube exchanger. They are compact, cost effective, and can handle certain fouling fluids. The most efficient design is achieved when the hot and cold fluid flow rates are approximately the same, resulting in similar velocities on both sides of the plates. This may require different process parameters (i.e., outlet temperature) to a shell-and-tube exchanger that is specified for the same service where the engineer specifies a high shellside flow rate to maximize the shellside film coefficient.

The design of plate-and-frame exchangers is highly specialized and often proprietary. Manufacturers provide some curves and software for use by end users (for example, see Ref [10]), but detailed design is normally left to the manufacturers.

Spiral Heat Exchangers

Increased turbulent heat transfer, reduced fouling, easier maintenance, and smaller size characterize the performance of spiral heat exchangers when compared with shell-and-tube exchangers. These are true countercurrent units. Moretta has summarized the design calculations for heat transfer and pressure drop [17].

Table 2-3
Shell-and-tube exchanger selection guide (cost increases from left to right) [1]

Type of Design	"U" Tube	Fixed Tubesheet	Floating Head Outside Packed	Floating Head Split Backing Ring	Floating Head Pull-Through Bundle
Provision for differential expansion	Individual tubes free to expand	Expansion joint in shell	Floating head	Floating head	Floating head
Removable bundle	Yes	No	Yes	Yes	Yes
Replacement bundle possible	Yes	Not practical	Yes	Yes	Yes
Individual tubes replaceable	Only those in outside row	Yes	Yes	Yes	Yes
Tube interiors cleanable	Difficult to do mechanically, can do chemically	Yes, mechanically or chemically	Yes, mechanically or chemically	Yes, mechanically or chemically	Yes, mechanically or chemically
Tube exteriors with triangular pitch cleanable	Chemically only	Chemically only	Chemically only	Chemically only	Chemically only
Tube exteriors with square pitch cleanable	Yes, mechanically or chemically	Chemically only	Yes, mechanically or chemically	Yes, mechanically or chemically	Yes, mechanically or chemically
Number of tube passes	Any practical even number possible	Normally no limitations	Normally no limitations	Normally no limitations	Normally no limitations
Internal gaskets eliminated	Yes	Yes	Yes	No	No

Table 2-4
Compact heat exchanger attributes

Exchanger Type	Attributes
Shell-and-tube	Up to 650 °C (1200 °F); 310 bar (4,500 psig) in the shell, 1380 bar (20,000 psig) in the tubes
	Up to 4650 m^2 (50,000 ft^2) heat transfer area
	Typical maximum sizes
	Floating Head Fixed Head or U-Tube
	Diameter 1524 mm (60 in.) 2000 mm (80 in.)
	Length 9m (30 ft) horizontal 12 m (40 ft) 25 m (75 ft) vertical
	Area 1270 m^2 (13,650 ft^2) 4310 m^2 (46,400 ft^2)
Gasketed plate-and-frame	Up to 180 °C (350 °F) and 20 bar (300 psig); fatigue characteristics of the metal plate may be limiting if temperature or pressure cycling is a process characteristic
	Up to 2800 m^2 (30,000 ft^2) heat transfer area in a single unit
	Typically designed with 70 kPa to 100 kPa (10 to 15 psi) pressure drop
	Maximum flow 2500 m^3/h (11,000 gpm)
	Minimum velocity 0.1 m/s (0.3 ft/s)
	Plates 0.5 to 1.2 mm (0.02 to 0.05 in.) thick
	0.03 to 2.2 m^2 (0.32 to 23.7 ft^2) area per plate
	1.5 to 5.0 mm (0.06 to 0.2 in.) spacing between plates
	Typically used in clean service (no particles larger than 2.5 mm), although "deep groove" or "wide gap" plate designs can tolerate up to 18 mm particles [14]. Usually only used for liquid-liquid service.
	Operates efficiently with crossing temperatures and close approach temperatures
	Only the plate edges are exposed to atmosphere, so little or no insulation is required
	Consider when a high-grade, expensive construction material (e.g., tantalum) is required, when space is tight, or when enhanced energy recovery is important
	High turbulence
	High heat transfer coefficients
	High fouling resistance
	Not available in carbon steel
	Hot and cold side channels have nearly identical geometry, so hot and cold fluids should have roughly equivalent flow rates
	Significant size reduction and weight savings compared with shell-and-tube
	Gasketed exchangers may be unsuitable for use in highly aggressive media or when leakage is not tolerable
Welded, brazed, or fusion-sealed plate-and-frame	Up to 450 °C (850 °F) and 40 bar (600 psig); fatigue characteristics of the metal plate may be limiting if temperature or pressure cycling is a process characteristic
	Other characteristics are similar to the gasketed plate-and-frame exchangers
Spiral	Up to 500 °C (930 °F) and 25 bar (360 psig); limits vary depending on size and material of construction
	Up to 350 m^3/h (1500 gpm); limited due to single channel
	0.5 to 500 m^2 (5 to 5400 ft^2) heat transfer area in one spiral body
	Countercurrent design allows for very deep temperature cross and close approach
	High turbulence reduces fouling and, especially, sedimentation (compared with shell-and-tube)
	Particularly effective in handling sludges, viscous liquids, and liquids with solids in suspension

Design Recommendations

For conceptual and preliminary design work, engineers can easily model liquid-liquid shell-and-tube heat exchangers. Where process fluids undergo a change in state (condensers and boilers), the design calculations are much more complex, and specialized software and training are recommended.

Process engineers should start with a full understanding of the duty requirements. After collecting and tabulating thermodynamic properties for the major fluid components, create heat and material balances for normal operating conditions (including start-up and turndown scenarios). There may be design trade-off decisions and it is usually the process engineer's responsibility to address potential performance differences among alternative design solutions.

Here are guidance questions for the process engineer:

- Which of the following parameters can float? To close the heat balance, at least one parameter is determined from the other five: hot and cold stream inlet temperature, outlet temperature, and flow rate. The answer is often flexible, meaning that two or three of the parameters may be safely varied within ranges. For example, if a liquid-liquid compact heat exchanger is anticipated, the flow rate of the two streams should be within about 20% of each other.

- What variation in temperature of the fluids is expected? This is particularly pertinent for cooling tower water that has a temperature that varies with the outside dew point temperature.
- What are the maximum allowable pressure drops through the equipment for the two streams? Be sure that unintended vaporization would not occur as pressure is reduced.
- Are there conditions that could result in freezing, precipitation, or fouling? If the hot stream flow is stopped while continuing the cold stream flow, what would happen as the temperature of the stagnant fluid in the heat exchanger cools?
- Similarly, what outcome is expected if the cold stream flow stops without interrupting the hot stream?
- Are thermodynamic properties for the hot and cold streams available, or can they be predicted from the properties of the pure components? There are many miscible liquids that behave rationally when mixed; for instance, the mixed liquid viscosity is a logarithmic average of the components' mass-weighted viscosities (see Equation 27-3 in Chapter 27). However, other mixtures deviate widely such as polar liquids (e.g., water, alcohols) and non-Newtonian emulsions and slurries.
- Is a temperature cross expected and if so can it be avoided? A temperature cross occurs when the outlet temperature of the hot fluid is lower than the outlet temperature of the cold fluid. It is physically possible in true counter-current equipment such as a spiral heat exchanger, a double-pipe exchanger, and a single-pass type BEM shell-and-tube unit. In many instances, to use shell-and-tube equipment, multiple shells are required.
- Are there physical limitations? Consider the available space for installation (including logistics of rigging the exchanger into place), maintenance (with an allocation for removing tubes), and elevation requirements (the relationship with associated equipment such as columns and pumps).
- Is this a batch or continuous process? Operating efficiency, in terms such as pumping cost and maintenance, is usually more important for exchangers that are in continuous operation for months, or years, between shutdowns.
- How will the fluid flow rates be controlled? If it's planned to control the flow rate of cooling water, for example, a reduction in duty due to process variations, or a lower than planned cooling water flow rate due to oversizing the exchanger may result in excessive fouling.

Evaluate the design problem using physical properties appropriate to the temperature of the fluids. This is especially important for viscosity which is highly temperature dependent, is a major contributor to the heat transfer coefficient, and plays a central role in pressure drop calculations. For preliminary design work, properties evaluated at the average temperature for each stream are fine.

Calculate the total duty for the exchanger in Watts, or Btu/h. Add a safety factor of 10% which includes fouling and uncertainty (or another factor depending on the specific design problem). Then use the tabulated "typical" heat transfer coefficients to compute the required heat transfer area. This is conceptual. The actual required heat transfer area depends on the mechanical design of the exchanger and will be determined later.

At this point the top part of the datasheet can be completed and sent to a vendor or heat exchanger engineer to design an exchanger using one of the sophisticated computer programs they have at their disposal. However, the process engineer may also (or instead) use the approximate methods and procedure given below to come up with a reasonable design solution. The calculations can be solved with spreadsheets to provide a platform for evaluating alternatives or rating existing exchangers without involving vendors or consultants.

Pick either the hot or cold fluid to flow inside the tubes (for a shell-and-tube exchanger). Assume a tube diameter (usually start with ¾ inch) and calculate the total length of tubes to achieve the surface area based on the assumed overall heat transfer coefficient. Then manipulate the exchanger length and number of tube passes, calculating the pressure drop through the tubes until a combination results in an acceptable pressure drop.

Pick a shell type based on the process requirements. Determine its diameter by the tube layout and passes. Estimate the pressure drop through the shell using the method given in this chapter.

Iterate the preceding two steps using different assumptions (e.g., tube diameter, pressure drop, swapping the fluids between tube and shell side, etc.) to find a reasonable design. What is "reasonable?" There's no one "correct" answer which is why experience and expertise are important characteristics for the designer.

Calculate heat transfer film coefficients for the tube and shell side and combine with the tube resistance and assumed fouling factors to compute an overall heat transfer coefficient. Compare with the original assumption and iterate, using the newly computed coefficient in place of the assumption, through the design steps if necessary.

The proper selection of a heat exchanger depends on interrelated factors; typically, many design solutions are compared before a final design is accepted. Factors include:

- Heat transfer rate ("U")
- Cost (operating and maintenance over the expected life of the exchanger or 10 years)
- Pumping power
- Size and weight
- Materials of construction
- Miscellaneous factors such as leak-tightness, safety, reliability, and noise

Process Data

The Excel spreadsheet accompanying this chapter steps through the design steps for a shell-and-tube exchanger in liquid-liquid service. The worksheet called "Fluid Data" tabulates temperature-correlated coefficients for vapor pressure, viscosity, density, specific heat, and thermal conductivity. It also has point values for molecular weight, heat of vaporization, and flash point.

The fundamental process parameters – flow and temperature – are entered on the "Process Data" worksheet. There are input cells for all six flow and temperature values even though at least one of these must be adjusted to satisfy the heat balance. There are also inputs for

pressure, allowable pressure drop, and fouling resistance. See Figure 2-5.

The change in enthalpy for each stream is evaluated using the equation:

$$\Delta H = W C_p (t_{out} - t_{in}) \qquad (2\text{-}1)$$

Where:

ΔH = enthalpy change, kJ/h or Btu/h
W = mass flow rate, kg/h or lb/h
C_p = specific heat, kJ/kg-°C or Btu/lb-°F
t_{out} = temperature at exchanger outlet, °C or °F
t_{in} = temperature at exchanger inlet, °C or °F

	A	B	C	D	E	F	G
6	Inputs						
7	Hot Side						
8		Fluid name		Process Stream			
9		Flow Total	lb/h	15,000		○ Hot Side Flow	
10		Temperature, in	°F	104		○ Hot Side Temperature In	
11		Temperature, out	°F	74			
12		Pressure, in	psig	75		○ Hot Side Temperature Out	
13		Press Drop Allowed	psig	5			
14		Fouling Resistance	ft²-h-°F/Btu	0.0030		Solve Heat Balance	
15							
16	Cold Side						
17		Fluid name		Chilled Water			
18		Flow Total	lb/h	25,000		○ Cold Side Flow	
19		Temperature, in	°F	39		○ Cold Side Temperature In	
20		Temperature, out	°F				
21		Pressure, in	psig	60		● Cold Side Temperature Out	
22		Press Drop Allowed	psig	10			
23		Fouling Resistance	ft²-h-°F/Btu	0.0010			

Figure 2-5. Fundamental process data includes flow and temperature information for the hot and cold streams. One of the values is calculated based on the other five to close the heat balance; the radio buttons identify the unknown.

Note that the specific heat is equal to the average of the values at inlet and outlet temperatures.

The two results (for hot and cold streams) are added in a cell named "HeatBalance." When the heat balance is satisfied, ΔH for the hot side is a negative value and it is positive for the cold side. Therefore, HeatBalance has a zero value, and Excel's GoalSeek function is used to find the unknown variable. In this example the cold stream temperature is found to be 10 °C (50 °F), and 79,000 W (270,000 Btu/h) are transferred.

The heat balance can also be solved algebraically if the heat capacity is assumed to be constant (which is a good assumption). GoalSeek is used by the spreadsheet because it is easy to implement and allows for changing of the heat capacity variable with temperature.

The stream properties are evaluated as follows.

Density, specific heat, and thermal conductivity are evaluated for each component of the hot and cold streams at the inlet and outlet temperatures for each stream. They are multiplied by the mass fraction of the component in the stream then summed. This gives an estimate for the properties at the inlet and outlet of the exchanger; as the temperatures are changed during the design procedure, the properties are immediately updated.

Viscosity is also tabulated and the logarithmic average is taken, weighted by the mass fraction of the components (see Equation 27-3 in Chapter 27).

Heat Exchanger Configuration and Area

Pick either the hot or cold stream to flow through the tubes. Rules of thumb to help decide include:

- If one fluid is highly corrosive, put it inside the tubes to reduce cost. Then only the tubes, tubesheets (sometimes just faced), tube channels, and piping need to be made of the corrosion-resistant alloy.
- If one fluid is at a much higher pressure than the other, put it inside the tubes.
- If one fluid is much more severely fouling than the other place it in the tubes. Tubes are easier to clean than shells, especially when mechanical means such as brushes are used.
- If one fluid has a very limited allowable pressure drop, put it in the shell.

Characterize the tube side by assuming an overall heat transfer coefficient (see Table 2-8 on page 47) and a safety factor primarily to account for fouling. Select a tube size (Table 2-2), wall thickness (start with 14 BWG), length (typically 4 ft, 8 ft, 12 ft, 15 ft, or 20 ft), and number of passes (either 1-pass or an even number up to about 14).

After completing all of the calculations in the following sections, return to this step and update the assumed overall heat transfer coefficient to equal that which was determined by the procedure. Iterate until the calculated overall coefficient equals the assumed one.

The heat transfer area is related to the heat duty, overall heat transfer coefficient, and mean temperature difference:

$$A = \frac{Q}{U \, \Delta T_{mean}}$$

(2-2)

A = heat transfer area, usually calculated at the outside tube diameter, m^2 or ft^2

Q = heat transferred, W or Btu/h

U = overall heat transfer coefficient, W/m^2-°C or Btu/h-ft^2-°F

ΔT_{mean} = mean temperature difference (MTD) between hot and cold streams, °C or °F

Determine the mean temperature difference (MTD) by calculating the log-mean temperature difference (LMTD) then applying a correction factor that is based on the number of tube and shell passes. For a strict cocurrent flow design (single pass shell and tube), there is no correction factor and this equation applies:

$$\Delta T_{mean} \, (cocurrent) = \frac{(T_{in} - t_{in}) - (T_{out} - t_{out})}{\ln \dfrac{(T_{in} - t_{in})}{(T_{out} - t_{out})}}$$

(2-3)

Other designs use the following formula for LMTD and a correction factor read from graphs corresponding to different shell and tube configurations. $F = 1.0$ for a true countercurrent exchanger (shell passes = tube passes). If the correction factor is less than about 0.80 then consider adding shells to achieve a result that is closer to countercurrent design.

$\Delta T_{mean}\ (countercurrent)$

$$= F\ \frac{(T_{in} - t_{out}) - (T_{out} - t_{in})}{\ln \dfrac{(T_{in} - t_{out})}{(T_{out} - t_{in})}} \qquad (2\text{-}4)$$

$F =$ LMTD configuration correction factor, dimensionless (see next section)
$T =$ inlet and outlet temperatures of the hot stream, °C or °F

$t =$ inlet and outlet temperatures of the cold stream, °C or °F

From the tube outside diameter, heat transfer area, and safety factor, calculate the total tube length:

$$L_{All\ tubes} = \frac{A\ F_{safety}}{\pi\ d_o} \qquad (2\text{-}5)$$

Determine the minimum number of tubes by dividing the total length, $L_{Alltubes}$, by tube length and rounding up to the next integer that is evenly divisible by the number of tube passes. This Excel formula gives the answer:

$n_{tubes} = \text{ROUND(LengthOfAllTubes/(TubeLength}^*\text{TubePasses)}) + 0.5, 0)^*\text{TubePasses}$

Determining the LMTD Configuration Correction Factor

Many references present F factors in graphical form (for example: Perry's). Bowman compiled formulae that accurately represent the graphs for every configuration of shell-and-tube exchanger system [4]. Fakheri then collapsed the correlations into a single algebraic equation that is applicable to shell-and-tube heat exchangers with N shell passes and 2NM tube passes per shell (for example, with 2 shell passes there may be any multiple of 2N tube passes or 4, 8, 12, etc. tube passes) [6].

$$F = \frac{S \ln W}{\ln \dfrac{1 + W - S + SW}{1 + W + S - SW}} \qquad (2\text{-}6)$$

Where:

$$S = \frac{\sqrt{R^2 + 1}}{R - 1}$$

$$W = \left(\frac{1 - PR}{1 - P}\right)^{1/N}$$

$$R = \frac{T_{in} - T_{out}}{t_{out} - t_{in}}$$

$$P = \frac{t_{out} - t_{in}}{T_{in} - t_{in}}$$

For the special case when $R = 1$ (and the logarithms cannot be evaluated):

$$F = \frac{\sqrt{2}\,\dfrac{1 - W'}{W'}}{\ln \dfrac{\dfrac{W'}{1 - W'} + \dfrac{1}{\sqrt{2}}}{\dfrac{W'}{1 - W'} - \dfrac{1}{\sqrt{2}}}} \qquad (2\text{-}7)$$

Where:

$$W' = \frac{N - NP}{N - NP + P}$$

And:

$$\Delta T_{mean} = F\,(T_{out} - t_{in})$$

Assumptions for the F factor equations and charts are:

- The overall heat transfer coefficient, U, is constant throughout the heat exchanger
- The rate of flow of each fluid is constant
- The specific heat of each fluid is constant
- There is no condensation of vapor or boiling of liquid in a part of the exchanger
- Heat losses are negligible
- There is equal heat transfer surface area in each pass
- The temperature of the shell-side fluid in any shell-side pass is uniform over any cross section
- There is no leakage of fluid or heat across the transverse baffle separating two shell passes

Tubeside Pressure Drop

Calculate the pressure drop in two parts then add together:

1. Using the mass flow rate per tube, use equations 2-9, 2-10, and 2-11 to compute pressure drop through the tubes.
2. From the velocity in the tubes and number of tube passes, estimate the pressure drop for turning the flow through the heads or channels with [15]:

$$\Delta P_t = \frac{2\,(n_p - 1)\,\rho\,u^2}{g_c} \tag{2-8}$$

Where:

ΔP_t = pressure drop through turns, Pa or psf (divide by 144 for psi)
n_p = number of passes
ρ = density, kg/m^3 or lb/ft^3
u = velocity in tubes, m/s or ft/s
g_c = conversion factor, 1 m/s^2 or 32.17 ft/s^2

Compare the calculated and allowable pressure drops. Adjust physical parameters (tube size, exchanger length, and number of tube passes) and repeat the calculations for heat exchanger area, total tube length, and pressure drop; iterate until a "reasonable" configuration is attained.

The "Tube Pressure Drop" and "F Factor" worksheets do the calculations just described.

Tube Side Film Coefficient

Compute the tube side film coefficient from physical properties evaluated at the average fluid temperature. Use the correlation that corresponds to the flow regime (laminar, transitional, or turbulent) for the tube side film coefficient.

1. Calculate the mean wall temperature, then evaluate the viscosity at that temperature. The formula uses the overall heat transfer coefficient, expressed in terms of the surface area inside the tubes, and the inside film coefficient. Neither of these values is known until the calculations for both the tube side and shell side are complete, so use an assumed value for both then iterate through all of the calculations until the assumed values match the calculated ones. The overall coefficient was already assumed to estimate the heat transfer area; it was based on the outside area of the tubes (see page 38). A good initial guess for the film coefficient is about 2,000 W/m^2-°C or 400 Btu/ft^2-°F.

$$\overline{T}_w = \overline{t} + \frac{U_i}{h_i}\,(\overline{T} - \overline{t}) \tag{2-9}$$

Where:
$\overline{T}_w$ = average inside wall temperature, °C or °F
$\overline{t}$ = average temperature, tube-side fluid, °C or °F

$\overline{T}$ = average temperature, shell side fluid, °C or °F
U_i = overall heat transfer coefficient based on inside area, W/m^2-°C or Btu/ft^2-°F

$$= U_o\,\frac{d_o}{d_i}$$

h_i = inside film coefficient, W/m^2-°C or Btu/ft^2-°F

2. Use the Hausen correlation for laminar flow (Reynolds number $<= 2000$) [2]:

$$\overline{h}_i =$$

$$\frac{k}{d_i}\left[3.66 + \frac{0.0668\,N_{Re}\,N_{Pr}\,(d_i/L)}{1 + 0.40\,[N_{Re}\,N_{Pr}\,(d_i/L)]^{2/3}}\right]\left(\frac{\mu}{\mu_w}\right)^{0.14} \tag{2-10}$$

Where the properties are evaluated at the average fluid temperature and L is the length for the tube pathway (e.g., if there are 10 tubes per pass then L is the total length of tubing divided by 10).

$$N_{Pr} = \text{Prandtl Number} = \frac{c_p\,\mu}{k}$$

μ = viscosity, mPa-s or lb$_m$/ft-h

3. Use the Sieder Tate equation for turbulent flow (Reynolds number $>= 10,000$) [2]

$$h_i = 0.023 \frac{k}{d_i} N_{\text{Re}}^{0.8} N_{\text{Pr}}^{1/3} \left(\frac{\mu}{\mu_w}\right)^{0.14} \quad (2\text{-}11)$$

$$(h_i)_T = \overline{h_i} + (h_i - \overline{h_i})\left(\frac{N_{\text{Re}} - 2000}{8000}\right) \quad (2\text{-}12)$$

4. Avoid the transition region if possible because the heat transfer coefficient is very unpredictable and there is a possibility of flow oscillations. However, the transition coefficient is bounded by the laminar and turbulent coefficients, and a plausible equation, based on the laminar and turbulent equations, is [2]:

The "Tubes htc" worksheet calculates the film coefficient using the formulae in this section. Input an assumed value for the film coefficient in Cell D7; the spreadsheet uses this to calculate the wall temperature and evaluate the viscosity at that temperature. Note the calculated coefficient in Cell D44 and make one or two iterations by changing the assumed value to equal the calculated result.

Shell Diameter

The shell diameter is related to the number of tubes, tube passes, tube diameter, tube pitch, tube pitch layout, and tube omissions to allow space for impingement baffles or to decrease the number of tubes in the baffle windows. TEMA and many others publish tables that list the number of tubes that will fit into shells of standard diameters.

For a quick estimation which should suffice for preliminary design work, use this procedure (easily implemented in Excel):

1. Calculate the cross-sectional area occupied by each tube. For triangular pitch, draw the equilateral triangle with vertices at the center of three tubes. The area of the triangle is one-half of the area required to accommodate one tube. Similarly, for square pitch draw the square with corners at the center of four tubes. The area of the square is equal to the area required to accommodate one tube.

$$Area_{1\ tube,\ triangular} = 2\,(PR\,d_o)^2\,\frac{\sqrt{3}}{4} \quad (2\text{-}13)$$

$$Area_{1\ tube,\ square} = (PR\,d_o)^2 \quad (2\text{-}14)$$

Where:

PR = tube pitch ratio (usually 1.25, 1,285, 1.33, or 1.5)
d_o = outside diameter of tubes, mm or in.

2. Calculate the diameter of a circle that equates to the area for all tubes in the shell.

$$D_{tight} = 2\left(\frac{N_t\,Area_{tube}}{\pi}\right)^{0.5} \quad (2\text{-}15)$$

n_t = number of tubes in the shell

3. For each tube pass greater than one, add cross sectional area to account for the pass partition by multiplying the tube diameter by D_{tight}.

$$A_{corrected} = D_{tight}\,d_o\,(n_p - 1) + (N_t\,Area_{tube}) \quad (2\text{-}16)$$

n_p = number of tube passes in the shell

4. Calculate the minimum shell diameter by adding two tube diameters to the circle equating to $A_{corrected}$.

$$D_{s,\min} = 2\left(\frac{A_{corrected}}{\pi}\right)^{0.5} + 2\,d_o \quad (2\text{-}17)$$

5. Finally, round up to the next standard shell size. For example, if $D_{s,\ minimum} = 20.5$ inches, use the next standard size which is 21.25 inches (inside diameter)

Ideal Shell Side Film Coefficient

Use the Bell-Delaware method to compute the shell side film coefficient, as described by Bejan and Kraus [1] and many others. The Bell-Delaware method computes the heat transfer film coefficient for an ideal bank of tubes, then applies correction factors to account for baffle cut and spacing, baffle leakage effects, bundle bypass flow, variable baffle spacing in the inlet and outlet sections, and adverse temperature gradient build-up if laminar flow.

$$h_o = h_{ideal} \, J_c \, J_l \, J_b \, J_s \, J_r \qquad (2\text{-}18)$$

Implied by the nature of the correction factors, many geometrical properties of the shell such as baffle cut, baffle spacing, shell diameter, and outside diameter of the tube bundle must be known or estimated. The procedure uses the geometrical properties to calculate each factor.

If the geometrical properties are unknown, then a total correction of 0.60 may be used ($h_o = 0.6 \, h_{ideal}$) since this has "long been used as a rule of thumb" [16].

Calculate the ideal heat transfer coefficient for pure crossflow in an ideal tube bank from [13]:

$$h_{ideal} = J_{ideal} \, c_{ps} \left(\frac{w_s}{A_s} \right) \left(\frac{k_s}{c_{ps} \, \mu_s} \right)^{2/3} \left(\frac{\mu_s}{\mu_{s,w}} \right)^{0.14}$$

$$\qquad (2\text{-}19)$$

Where:

J_{ideal} = the Colburn factor for an ideal tube bank

The subscript s stands for physical properties at the average temperature of the shell side fluid; subscript w is at the wall temperature.

W_s = mass flow rate of shell side fluid across the tube bank

A_s = bundle crossflow area at the centerline of the shell between two baffles

For $30°$ and $90°$ tube layout bundles, $45°$ layout with $p_t/d_o \geq 1.707$, and $60°$ layout with $p_t/d_o \geq 3.732$:

$$A_s = L_{bc} \left[D_s - D_{otl} + (D_{otl} - d_o) \left(\frac{p_n - d_o}{p_n} \right) \right]$$

For a $45°$ and $60°$ layouts with ratios less than 1.707 and 3.732 respectively, the equation is:

$$A_s = L_{bc} \left[D_s - D_{otl} + (D_{otl} - d_o) \left(\frac{p_t - d_o}{p_n} \right) \right]$$

$p_t = PR \, d_o$, Pitch, which is the Pitch Ratio x tube OD
p_n = pitch normal to the flow direction (see Table 2-6)
L_{bc} = baffle spacing

Table 2-5
Correlation coefficients for J_{ideal} and f_{ideal} [13]

Pitch Layout	Reynolds Number	a_1	a_2	a_3	a_4	b_1	b_2	b_3	b_4
30	0–10	1.4	−0.667	1.45	0.519	48	−1	7	0.5
30	10–100	1.36	−0.657	1.45	0.519	45.1	−0.973	7	0.5
30	100–1000	0.593	−0.477	1.45	0.519	4.57	−0.476	7	0.5
30	1000–10000	0.321	−0.388	1.45	0.519	0.486	−0.152	7	0.5
30	10000+	0.321	−0.388	1.45	0.519	0.372	−0.123	7	0.5
45	0–10	1.55	−0.667	1.93	0.5	32	−1	6.59	0.52
45	10–100	0.498	−0.656	1.93	0.5	26.2	−0.913	6.59	0.52
45	100–1000	0.73	−0.5	1.93	0.5	3.5	−0.476	6.59	0.52
45	1000–10000	0.37	−0.396	1.93	0.5	0.333	−0.136	6.59	0.52
45	10000+	0.37	−0.396	1.93	0.5	0.303	−0.126	6.59	0.52
60	0–10	1.4	−0.667	1.45	0.519	48	−1	7	0.5
60	10–100	1.36	−0.657	1.45	0.519	45.1	−0.973	7	0.5
60	100–1000	0.593	−0.477	1.45	0.519	4.57	−0.476	7	0.5
60	1000–10000	0.321	−0.388	1.45	0.519	0.486	−0.152	7	0.5
60	10000+	0.321	−0.388	1.45	0.519	0.372	−0.123	7	0.5
90	0–10	0.97	0.667	1.187	0.37	35	−1	6.3	0.378
90	10–100	0.9	−0.631	1.187	0.37	32.1	−0.0963	6.3	0.378
90	100–1000	0.408	−0.46	1.187	0.37	6.09	−0.602	6.3	0.378
90	1000–10000	0.107	−0.266	1.187	0.37	0.0815	−0.022	6.3	0.378
90	10000+	0.37	−0.395	1.187	0.37	0.391	−0.148	6.3	0.378

Table 2-6
Tube geometry as a function of tube pitch, p_t

Tube Layout	Pitch Normal to Flow, p_n	Pitch Parallel to Flow, p_p
30° Triangular Staggered Array	p_t	$\left(\dfrac{\sqrt{3}}{2}\right) p_t$
60° Rotated Triangular Staggered Array	$\sqrt{3}\, p_t$	$\dfrac{p_t}{2}$
90° Square Inline Array	p_t	p_t
45° Rotated Square Staggered Array	$\sqrt{2}\, p_t$	$\dfrac{p_t}{\sqrt{2}}$

The Colburn factor is a function of the shell side Reynolds number:

$$N_{Re,s} = \frac{d_o\, W_s}{\mu_s\, A_s} \tag{2-20}$$

Calculate J_{ideal} from the following relationship:

$$J_{ideal} = a_1 \left(\frac{1.33}{PR/d_o}\right)^a N_{Re,s}{}^{a_2} \tag{2-21}$$

Where:

$$a = \frac{a_3}{1 + 0.14\, N_{Re,s}{}^{a_4}}$$

The coefficients, listed in Table 2-5, depend on the tube pitch layout and Reynolds number.

Shell Side Film Coefficient Correction Factors

This section describes each of the five Bell-Delaware correction factors. Some of the equations require additional information about the construction of the heat exchanger, as noted.

Baffle Cut and Spacing, *Jc*

This factor takes into account the heat transfer rate that occurs in the baffle window where the shell side fluid flows more longitudinally, deviating from the ideal cross-flow arrangement. It is related to the shell diameter, tube diameter, and baffle cut. The value ranges from about 0.53 for a large baffle cut up to 1.15 for small windows with a high window velocity. If there are no tubes in the window $J_c = 1.0$ [13]. It is expressed as a fraction of the number of tubes in cross flow, F_c [1]; the equation assumes single segmental baffles:

$$J_c = 0.55 + 0.72\, F_c \tag{2-22}$$

Where:

$$F_c = \frac{1}{\pi} \left[\pi + 2\,\phi\,\sin(\arccos\phi) - 2\arccos\phi\right]$$

$$\phi = \frac{D_s - 2\, l_c}{D_{otl}}$$

l_c = baffle cut = distance from the baffle to the inside of the shell, mm or in.
D_{otl} = outside diameter of the tube bundle, mm or in.

Baffle Leakage Effects, *J_L*

This factor includes tube-to-shell and tube-to-baffle leakage, where the shell fluid bypasses the normal flow path. If baffles are too closely spaced, the fraction of flow in the leakage stream increases compared with cross flow. It is typically between 0.7 and 0.8 [13]. Use this formula [1]:

$$J_l = 0.44\,(1 - r_a) \\ + [1 - 0.044\,(1 - r_a)]\exp(-2.2\, r_b) \tag{2-23}$$

Where:

$$r_a = \frac{A_{sb}}{A_{sb} + A_{tb}}$$

$$r_b = \frac{A_{sb} + A_{tb}}{A_w}$$

Calculate A_{sb}, A_{tb}, and A_w as follows:

$$A_{sb} = \frac{1}{2} (\pi - \theta_1) D_s \delta_{sb}, \tag{2-24}$$

shell-to-baffle leakage area

Where:

$$\theta_1 = \arccos \left(1 - \frac{2 l_c}{D_s}\right)$$

$\delta_{sb} = D_s - D_b$, shell-to-baffle spacing. See Table 2-7.
D_b = baffle diameter

$$A_{tb} = \frac{\pi d_o (1 - F_w) N_t \delta_{tb}}{4}, \tag{2-25}$$

tube-to-baffle leakage area

Where:

$$F_w = \frac{\theta_3 - \sin \theta_3}{2 \pi}, \text{ fraction of the total number of}$$
tubes in one window

$$\theta_3 = 2 \arccos \frac{D_s - 2 l_c}{D_s - C_1}$$
$C_1 = D_s - D_{otl}$, shell-to-outer tube limit distance
δ_{tb} = baffle-hole diameter − tube OD (usually 0.8 mm or 0.03125 in., but may be reduced to 0.4 mm or

0.0156 in to reduce the leak stream between tube and baffle hole [19]),

$$A_w = A_{wg} - A_{wt}, \tag{2-26}$$

free area for fluid flow in one window section

Where:

$$A_{wg} = \frac{D_s}{8} (\theta_2 - \sin \theta_2), \text{ gross window area}$$

$$\theta_2 = \arccos \left(\frac{1 - 2 l_c}{D_s}\right)$$

$$A_{wt} = \frac{\pi}{4} n_{tw} d_o, \text{ area occupied by tubes in one window}$$

$n_{tw} = F_w n_t$, number of tubes in the window

Bundle and Partition Bypass Effects, *Jb*

This factor corrects for flow that bypasses the tube bundle due to clearance between the outermost tubes and the shell and pass dividers. For exchangers with very small clearances the factor is about 0.9, but larger clearances are required for a pull-through floating head where the factor is about 0.7. Sealing strips can increase the value [13]. A rule of thumb is to use one pair of sealing strips for approximately every six tube rows [2]. Use these formulae to calculate J_b [1]:

$$J_b = \exp \left[-C r_c (1 - 2 \zeta^{1/3})\right] \quad \text{for } \zeta < \frac{1}{2} \tag{2-27}$$

$$\text{Or} \quad J_b = 1 \quad \text{for } \zeta \geq \frac{1}{2}$$

Table 2-7
Diametric shell-to-baffle clearance, based on TEMA class R [24]

Nominal Shell Diameter		Shell Type	Difference in Shell-to-Baffle Diameter	
DN	Inches		Millimeters	Inches
200 to 325	8 to 13	Pipe	2.540	0.100
350 to 425	14 to 17	Pipe	3.175	0.125
450 to 575	18 to 23	Pipe	3.810	0.150
600 to 975	24 to 39	Rolled	4.445	0.175
1000 to 1350	40 to 54	Rolled	5.715	0.225
1375 to 1500	55 to 60	Rolled	7.620	0.300

This parameter strongly influences the calculation of J_l. The clearance may be reduced to 0.0035 to 0.004 times the shell diameter limit the baffle-to-shell leak stream, but only for rolled shells and only if necessary since it is hard to guarantee compliance [19].

Where:

$C = 1.35$ for $N_{RE,s} <= 100$ or 1.25 for $N_{RE,s} > 100$

$r_c = \dfrac{A_{bp}}{A_s}$

$\zeta = \dfrac{n_{ss}}{n_{r,cc}}$ (API Standard 660 requires a seal device from 25 mm to 75 mm, 1 in to 3 in., from the baffle tips and for every 5 to 7 tube pitches thereafter [19], leading to the rule of thumb of 0.17 for this parameter)

n_{ss} = number of sealing strip pairs

$n_{r,cc} = \dfrac{D_s - 2\,l_c}{p_p}$

p_p = longitudinal tube pitch

$A_{bp} = L_{bc}\,(D_s - D_{otl} + 0.5\,n_{dp}\,w_p)$

L_{bc} = central baffle spacing, mm or in.

n_{dp} = number of bypass divider lanes that are parallel to the crossflow stream

w_p = width of the bypass divider lane (if unknown, assume 2 x Tube OD)

Variations in Baffle Spacing, *Js*

When baffle spacing is increased at the ends of the exchanger to accommodate the nozzles, local decreases in flow velocity occur. This factor accounts for the consequent decrease in heat transfer, and typically ranges from 0.85 to 1.0 [13]. Calculate J_s with [25]:

$$J_s = \frac{n_b - 1 + (L_i^*)^{(1-n)} + (L_o^*)^{(1-n)}}{n_b - 1 + (L_i^*) + (L_o^*)} \qquad (2\text{-}28)$$

Where:

n_b = number of baffles in the exchanger

$L_i^* = \dfrac{L_{bi}}{L_{bc}}$

$L_o^* = \dfrac{L_{bo}}{L_{bc}}$

$n = {}^{3/5}$ for turbulent flow or ${}^{1/3}$ for laminar flow
And L_{bi}, L_{bo}, and L_{bc} are baffle spacing at inlet, outlet, and central respectively

Temperature Gradient for Laminar Flow Regime, *Jr*

The final correction factor is used when the Reynolds number on the shell side is less than 100. It is equal to 1.0 for $N_{RE,s} >= 100$. If $N_{RE,s} <= 20$:

$$J_r = \left(\frac{10}{n_{r,cc}}\right)^{0.18} \qquad (2\text{-}29)$$

Where $n_{r,cc}$ is the number of effective tube rows crossed through one crossflow section. For $20 < N_{RE,s} < 100$, perform a linear interpolation between the two extreme values [1].

Overall Heat Transfer Coefficient

Given the tube (inside) and shell (outside) film coefficients, fouling factors, and tube wall thermal conductivity, calculate the overall heat transfer coefficient for both the clean and fouled conditions. The clean coefficient is:

$$U_{o,clean} = \frac{1}{\dfrac{d_o}{d_i\,h_i} + \dfrac{d_o \ln(d_o/d_i)}{2\,k} + \dfrac{1}{h_o}} \qquad (2\text{-}30)$$

And the coefficient in the fouled condition is:

$$U_{o,fouled} = \frac{1}{\dfrac{d_o}{d_i\,h_i} + \dfrac{d_o\,R_{f,i}}{d_i} + \dfrac{d_o \ln(d_o/d_i)}{2\,k} + R_{f,o} + \dfrac{1}{h_o}}$$

$$(2\text{-}31)$$

Where:

U_o = overall heat transfer coefficient based on the outside area of the tubes

d_o and d_i = outside and inside tube diameter, respectively

h_o and h_i = outside and inside film coefficients, respectively

$R_{f,o}$ and $R_{f,i}$ = fouling factors on the shell and tube side, respectively

k = thermal conductivity of the tube material (see Table 2-9)

It is good practice to limit the reduction in heat transfer due to fouling to about 80% of the clean heat transfer coefficient. This is done by instituting a cleaning schedule that removes accumulations before they become too severe.

Use this calculated overall heat transfer coefficient to update the assumed coefficient (page 18) and iterate the calculations until the values are in reasonable agreement.

Shell Side Pressure Drop

The Bell-Delaware method accounts for tube bundle bypass and baffle leakage effects. It computes a pressure drop that is 20% to 30% of that calculated without the bypass and leakage effects.

1. The crossflow section between the interior baffles.

Use the b coefficients in Table 2-5 to compute the friction factor for an ideal tube bank, which depends on the tube layout and Reynolds number:

$$f_{ideal} = b_1 \left(\frac{1.33}{PR/d_o} \right)^b N_{Re,s}^{b_2} \tag{2-32}$$

Where:

$$b = \frac{b_3}{1 + 0.14 N_{Re,s}^{b_4}}$$

The pressure drop for one ideal crossflow section is:

$$\Delta P_{b,ideal} = \frac{4 f_{ideal} W_s^2 n_{r,cc}}{2 \rho_s g_c A_s} \left(\frac{\mu_w}{\mu} \right)_s^{0.14} \tag{2-33}$$

The bundle bypass correction factor uses parameters determined for J_b, the film coefficient correction factor for bundle and partition bypass effects; it typically ranges from 0.5 to 0.8 [13]. For a Reynolds number $<= 100$, $C_{bp} = 4.5$; Reynolds number > 100, $C_{bp} = 3.7$. The limit for R_b is 1.0 for $\zeta >= 0.5$.

$$R_b = \exp[-C_{bp} r_c (1 - \sqrt[3]{2\zeta})] \tag{2-34}$$

The baffle leakage correction factor is a function of r_a and r_b (see page 28); it typically ranges from 0.4 to 0.5.

$$R_l = \exp[-1.33 (1 + r_a) r_b^c] \tag{2-35}$$

$$c = -0.15 (1 + r_a) + 0.8$$

2. The baffle windows.

For an ideal window, calculate the pressure drop using the equation corresponding to the flow regime.

For $N_{Re} >= 100$:

$$\Delta P_{w,ideal} = \frac{W_s (2 + 0.6 n_{tw})}{2 g_c A_s A_w \rho_s} \tag{2-36}$$

If $N_{Re} < 100$:

$$\Delta P_{w,ideal} = 26 \frac{\mu_s W_s}{\sqrt{A_s A_w} \rho} \left(\frac{n_{r,tw}}{p_t - d_o} + \frac{L_{bc}}{D_{w^2}} \right) + \frac{W_s}{A_s A_w \rho} \tag{2-37}$$

$$D_w = \frac{4 A_w}{\pi d_o n_{tw} + D_s \theta_2 / 2}$$

$$n_{r,tw} = \frac{0.8 [l_c - 0.5 (D_s - D_{otl} + d_o)]}{p_p}$$

3. The entrance and exit sections, from the nozzle to the first baffle window.

Combined with the crossflow and baffle window findings, the total pressure drop through the exchanger (excluding the nozzles) is:

$$\Delta P_s = [(n_b - 1) (\Delta P_{b,ideal}) R_b + n_b \Delta P_{w,ideal}] R_l + 2 \Delta P_{b,ideal} R_b \left(1 + \frac{n_{r,tw}}{n_{r,cc}} \right) \tag{2-38}$$

Heat Transfer Coefficients

Table 2-8
Approximate overall heat transfer coefficients [21]

Hot Fluid	Cold Fluid	U W/m²-°C	U Btu/h-ft²-°F
Sensible Heat Transfer (No Change of Phase)			
Water	Water	850–1700	150–300
Organic solvents	Water	280–850	50–150
Gases	Water	20–280	3–50
Light oils	Water	340–900	60–160
Heavy oils	Water	60–280	10–50
Organic solvents	Light oil	110–400	20–70
Water	Brine	570–1140	100–200
Organic solvents	Brine	170–510	30–90
Gases	Brine	20–280	3–50
Organic solvents	Organic solvents	110–340	20–60
Heavy oils	Heavy oils	50–280	8–50
Heaters			
Steam	Water	1400–4300	250–750
Steam	Light oils	280–850	50–150
Steam	Heavy oils	60–450	10–80
Steam	Organic solvents	570–1140	100–200
Steam	Gases	30–280	5–50
Dowtherm	Gases	20–230	4–40
Dowtherm	Heavy oils	50–340	8–60
Flue gas	Aromatic HC and Steam	30–85	5–15
Evaporators			
Steam	Water	2000–4300	350–750
Steam	Light oils	450–1000	80–180
Steam	Heavy oils (vacuum)	140–430	25–75
Steam	Organic solvents	570–1140	100–200
Water	Refrigerants	430–850	75–150
Organic solvents	Refrigerants	170–570	30–100
Condensers			
Steam (pressure)	Water	2000–4300	350–750
Steam (vacuum)	Water	1700–3400	300–600
Saturated organic solvents near atmos.	Water	570–1140	100–200
Saturated organic solvents with some non-cond	Water, brine	280–680	50–120
Organic solvents, atmospheric and high non-condensable	Water, brine	280–680	50–120
Aromatic vapors, atmospheric with non-condensables	Water	30–170	5–30
Organic solvents, vacuum and high non-condensables	Water, brine	60–280	10–50
Low boiling hydrocarbon, atmospheric	Water	450–1140	80–200
High boiling hydrocarbon, vacuum	Water	60–170	10–30

Fouling Resistances

The following are the more common fouling mechanisms [5]:

- Crystallization. Certain salts commonly present in natural waters have a lower solubility in warm water than in cold. Therefore, when cooling water is heated, particularly at the tube wall, these dissolved salts will crystallize on the surface in the form of scale. Common solution: reducing the temperature of the heat transfer surface often softens the deposits.
- Sedimentation. Depositing of dirt, sand, rust, and other small particles is also common when fresh water is used. Common solution: velocity control.

- Biological growth. Common solution: material selection. Smooth surfaces (e.g., chrome plated) and copper or copper alloys reduce biological growth.
- Chemical reaction coking. This appears where hydrocarbons deposit in a high temperature application. Common solution: reducing the temperature between the fluid and the heat transfer surface.
- Corrosion. Common solution: material selection.
- Freezing fouling. Overcooling at the heat transfer surface can cause solidification of some of the fluid stream components. Common solution: reducing the temperature gradient between the fluid and the heat transfer surface.

Plate-and-frame heat exchangers are usually less prone to fouling than shell-and-tube units. Also, because they have much higher overall heat transfer coefficients, using the same fouling resistance values as for a shell-and-tube exchanger has a proportionally greater effect on the calculated overall U. This is a common engineering error that leads to oversizing the plate-and-frame exchanger. The general practice is to specify plate-and-frame exchangers with no fouling factor, but to specify a percent of excess surface area instead. Also, select a frame size that will accommodate additional plates in the event that more surface is needed because of a loss of performance due to fouling.

Recent research by HTRI [11] shows that fouling in crude oil preheat service depends primarily on velocity, surface temperature, and the composition of the stream. Nesta outlined a "no foul design method" that is applicable to medium through high boiling point liquid hydrocarbon mixtures with API gravity less than 45 [19]. By increasing the velocity of the hydrocarbon above threshold values and providing little or no excess surface area (that normally is allocated for fouling), the method provides much longer run time than traditional designs. Here is a summary of the no-foul design method from Nesta:

1. Tube side: minimum velocity 2 m/s (6.6 ft/s) for 19 mm (0.75 in.) and 25.4 mm (1 in.) tubes; minimum velocity 2.2 m/s (7.2 ft/s) for 31.75 mm (1.25 in.) and 38.1 mm (1.5 in.) tubes.
2. Shell side: minimum cross-flow stream velocity 0.6 m/s (2 ft/s).
3. Maximum temperature at the tube wall: 300 °C (570 °F).
4. Shell design should use single segmental baffles with 20% cut, oriented horizontally for TEMA Type E and J shells. Where impingement protection is required, use impingement rods, not plates.
5. Provide up to 20% excess surface area when both streams are within the scope of this design practice, but do not apply a fouling factor.
6. Provide pressure drop as required to achieve the minimum velocities.

Building on the no-foul design method, Bennett, et.al. provided this "most basic" design algorithm [3]:

1. Check company experience with the heat exchanger to be designed
2. Decide on fouling factors. If a stream is determined to be non-fouling, do not use a fouling factor for that stream. If a stream is known to foul, use a fouling factor in accordance with the company's best practices.
3. Place the most heavily fouling stream on the tube-side to facilitate cleaning, if necessary, and to avoid the areas of low velocity that occur on the shellside
4. Design for high velocities within erosion and vibration limits (per the no-foul design method). Exceptions to this general high-velocity rule for fouling mitigation include corrosion, geothermal brines, and slurries that present an erosion limit.
5. Keep overdesign between 0% and 20%.

Installation Recommendations

Here are some installation tips for typical shell-and-tube heat exchangers [12] and [22]:

- Provide sufficient clearance for removing the tube bundle at the head end of the exchanger. For exchangers with fixed tube sheets, allow enough room to remove the heads and clean the tubes

(consider the possibility of using brushes that would be at least as long as the tubes).
- Provide valves and bypasses in the piping system for both the shell and tube sides. Ball valves with locking handles are recommended if available for the pipe sizes.

- Provide thermowells and pressure gage connections in the piping at each inlet and outlet, located as close to the unit as practicable. Some exchangers are designed with these features, in which case they can be omitted from the piping.
- Provide valves to allow venting of gas vapor from the exchanger, and vacuum breakers for exchangers in steam service. The normal locations are close to the steam inlet or on the top portion of the shell.
- Ensure that foundations are adequately sized. In concrete footings, foundation bolts set in pipe sleeves of larger size than the bolt size will allow for adjustment after the foundation has set.
- Loosen foundation bolts at one end of unit to allow free expansion and contraction of the heat exchanger shell.
- Exchangers in condensing steam duty should be installed at a 3° to 4° slope, toward the shell outlet, to facilitate drainage of condensate. Heat exchangers should be installed to promote gravity drainage with no vertical lift before or after steam traps. Condensate accumulating in the exchanger results in water hammer and poor temperature control; corrosion problems may also occur.
- Condensate drainage pipes should have a vertical drop-leg of at least 18 inches from the exchanger to the trap.

- For condensate capacities of 3,500 kg/h (8,000 lb/h) or less, use a steam trap; for capacities higher than that use a control valve with level controller.
- If the steam supply is modulated with a control valve, all condensate drains must flow by gravity to a collection tank or pumping system to return the condensate to the boiler. Install a condensate drip pocket with a steam trap in front of the steam control valve. Install a strainer in front of the control valve. Locate the valve at least 10 pipe diameters away from the exchanger, and use a pipe size equal to or larger than the inlet connection to the unit.
- Do not pipe drain connections to a common closed manifold.
- Install a gage glass in a vapor or gas space to indicate possible flooding due to faulty trap operation.
- Quick-opening and closing valves controlling fluids to or from an exchanger may cause water-hammer, and care should be taken for proper selection of such equipment.
- Re-torque all external bolted joints after installation and again after the exchanger has been heated to prevent leaks and blowing out of gaskets.
- Insulate all heat-transfer-exposed surface areas.

Thermal Conductivity of Metals

Use the values in Table 2-9 when computing overall heat transfer coefficients (page 45). Thermal conductivity is the quantity of heat transferred through a unit thickness.

Table 2-9
Thermal conductivity of metals used in heat exchangers

Heat Exchanger Tube Material	k, W/m-K	k, Btu/h-ft-°F
Aluminum	147	85
Brass, Admiralty	111	64
Brass, Red	159	92
Carbon steel (0.5% C)	54 @ 20°C	31 @ 68°F
Carbon steel (1.5% C)	36 @ 20°C	21 @ 68°F
	33 @ 400°C	19 @ 750°F
Copper	386	223
Hastelloy C	8.7	5
Inconel	14.5	8.4
Monel	26	15
Nickel	90	52
Tantalum	54	31
Titanium	21	12
Type 316 stainless steel	16.3	9.4
Type 410 stainless steel	24.9	14.4

Vacuum Condensers

This section provides tips for designing overhead condensers for vacuum distillation [20].

Outlet Temperature and Pressure. It is important to have proper subcooling in the vent end of the unit to prevent large amounts of process vapors from going to the vacuum system along with the inerts.

Control. It is necessary to have some over-surface and to have a proper baffling to allow for pressure control during process swings, variable leakage of inerts, etc. One designer adds 50% to the calculated length for the over-surface. The condenser must be considered part of the control system (similar to extra trays in a fractionator) to allow for process swings not controlled by conventional instrumentation.

The inerts will "blanket" a portion of the tubes. The blanketed portion has very poor heat transfer. The column pressure is controlled by varying the percentage of the tube surface blanketed. When the desired pressure is exceeded, the vacuum system will suck out more inerts, and lower the percentage of surface blanketed. This will increase cooling and bring the pressure back down to the desired level. The reverse happens if the pressure falls below that desired. This is simply a matter of adjusting the heat transfer coefficient to heat balance the system.

Figure 2-6 shows typical baffling. The inerts move through the first part of the condenser as directed by the baffles. The inerts then pile up at the outlet end lowering heat transfer as required by the controller. A relatively

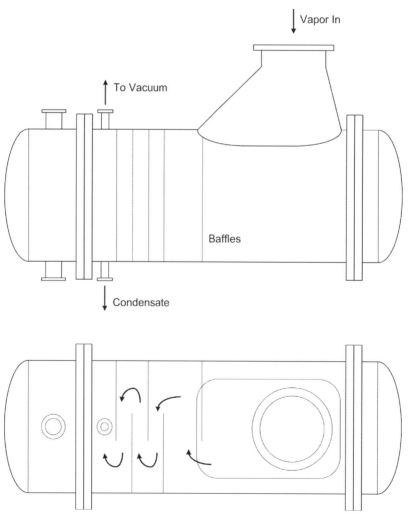

Figure 2-6. Baffling and inlet "bathtub" are shown in this typical vacuum condenser design. The vapor inlet nozzle is expanded to five times its area.

large section must be covered by more or less stagnant inerts which are subcooled before being pulled out as needed. Without proper baffles, the inerts build up in the condensing section and decrease heat transfer until the pressure gets too high. Then the vacuum valve opens wider, pulling process vapor and inerts into the vacuum system. Under these conditions pressure control will be very poor.

Pressure Drop. Baffling must be designed to keep the pressure drop as low as possible. The higher the pressure drop the higher the energy consumption and the harder the job of attaining proper vent end subcooling. Pressure drop is lower at the outlet end because of smaller mass flow.

Bypassing. Baffles should prevent bypass of inlet vapor into the vent. This is very important.

Typical Condenser. Figure 2-6 illustrates an inlet "bathtub" used for low vacuums to limit pressure drop at entrance to exchanger and across first rows of tubes. Note the staggered baffle spacing with large spacing at inlet, and the side-to-side (40% cut) baffles. Enough baffles must be used in the inlet end for minimum tube support. In the last 25% of the outlet end a spacing of 1/10 of a diameter is recommended.

Air-cooled Heat Exchangers: Forced vs. Induced Draft

Air-cooled heat exchangers are classified as forced draft when the tube section is located on the discharge side of the fan, or induced draft when the tube section is located on the suction side of the fan. Forced draft units are more common.

Typically, 25.4-mm (1-in.) OD carbon steel tubes are fitted with aluminum fins, 12.7 to 15.9 mm high (½ to ⅝ inch), providing outside surface area about 14 to 21 times greater than the area of the bare tubes. The process stream, flowing inside the tubes, can be cooled to about 10 °C to 15 °C (20 °F to 30 °F) above the dry-bulb temperature of the air. Air flows at a velocity of 3 to 6 m/s (10 to 20 ft/s).

Table 2-10
Comparison of forced draft and induced draft air-cooled heat exchangers [8]

Attribute	Forced Draft	Induced Draft
Distribution of air across section	Poor distribution of air over the section	Better
Effluent air recirculation to intake	Greatly increased possibility of hot air recirculation due to low discharge velocity and absence of stack	Lower possibility because fan discharges air upward, away from the tubes, at about 2½ times the intake velocity, or about 450 m/min (25 ft/s)
Influence of weather conditions	Total exposure of tubes to sun, rain, and hail	Less effect from sun, rain, and hail because 60% of face is covered
Freezing conditions	Easily adaptable for warm air recirculation during freezing conditions	Warm discharge air not recirculated
Result of fan failure	Low natural draft capability on fan failure due to small stack effect	Natural draft stack effect is greater than forced draft type
Power requirement	Slightly lower fan power because the fan is located in the cold air stream (air has higher density)	Slightly higher fan power because the fan is located in the hot air stream (air has lower density)
Temperature limit — discharge air stream	No limit	Limited to about 95 °C (200 °F) to prevent potential damage to fan blades, bearings, belts, and other components in the air stream
Temperature limit — tubeside process fluid	Limited by tube components	Limited to 175 °C (350 °F) because fan failure could subject fan blades and bearings to excessive temperatures
Maintenance	Better access to mechanical components	Mechanical components are more difficult to access because they are above the tubes

Air-cooled Heat Exchangers: Air Data

The overall heat transfer coefficient is governed by the air film heat transfer, which is generally in the order of 60 W/m^2-$°C$ (10 Btu/h-ft^2-$°F$). Air-cooled exchangers transfer less than 10% of that of water-cooled shell-and-tube units. Also, the specific heat of air is only 25% that of water (on a mass basis). As a result, air coolers are very large relative to water coolers. On the other hand, the finned tubes partially offset the poor thermal performance because they provide an external surface area about 20 times that of plain tubes.

The performance of air-coolers is tied to the dry-bulb air temperature, which varies considerably throughout the year. Assume a design temperature that is exceeded during 2% to 5% of the annual time period, but calculate the performance of the cooler at the higher end of the temperatures that are known to occur at the plant site, in order to obtain a feel for the performance range to expect.

Obtain the following data to get a realistic estimate of the design air temperature [7]:

- Annual temperature-probability curve
- Typical daily temperature curves
- Duration-frequency curves for the occurrence of the maximum dry-bulb temperature

The air density affects fan design (flow, head, and power). Table 2-11 gives values for correction factors for altitude and temperature.

Air data should include environmental characteristics. Marine air or sulfur dioxide content can be corrosive to fans, fins, tubes, and structures. Dusty atmospheres may lead to increased fouling, indicating incorporation of fouling factors in the design and possibly suggesting design accommodations such as increased tube pitch. Wind and rain patterns should also be considered [7].

Table 2-11
Approximate correction factor for air density as a function of altitude and temperature

Altitude, m (ft)	Air Temperature			
	− 20 °C (0 °F)	20 °C (70 °F)	40 °C (100 °F)	90 °C (200 °F)
0	1.15	1.00	0.92	0.80
300 (1,000)	1.11	0.96	0.91	0.77
600 (2,000)	1.07	0.93	0.88	0.75
900 (3,000)	1.03	0.90	0.85	0.72
1,200 (4,000)	0.99	0.86	0.82	0.69
1,500 (5,000)	0.96	0.83	0.79	0.67
1,800 (6,000)	0.92	0.80	0.76	0.64
2,100 (7,000)	0.89	0.77	0.73	0.62
2,400 (8,000)	0.86	0.74	0.70	0.60

Air-cooled Heat Exchangers: Thermal Design

Thermal performance calculations are analogous with those for shell-and-tube exchangers. The process fluid flows inside the tubes, and the inside heat transfer film coefficient is calculated exactly the same way as with shell-and-tube units. The air flows on the outside of the tubes; calculation of the air side film coefficient is complicated; some guidance is given later in this section. For the heat balance, $Q = U A MTD$, the corrected log-mean temperature difference is determined from charts

(Figure 2-7 and Figure 2-8). For four or more tube passes the correction factor is 1; it is slightly less than 1 for three-pass units. Use the charts for one- and two-pass coolers. If the factor is less than 0.8 then strongly consider changing the design temperatures or number of passes to obtain a good design.

Engineers can juggle at least nine variables when optimizing the design of an air-cooled heat exchanger. Mukherjee discussed each of these variables in terms of

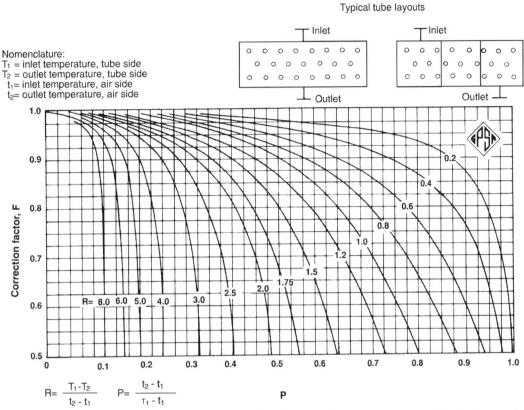

Figure 2-7. MTD correction factors for air-cooled heat exchangers (1-pass, cross-flow, both fluids unmixed) [8].

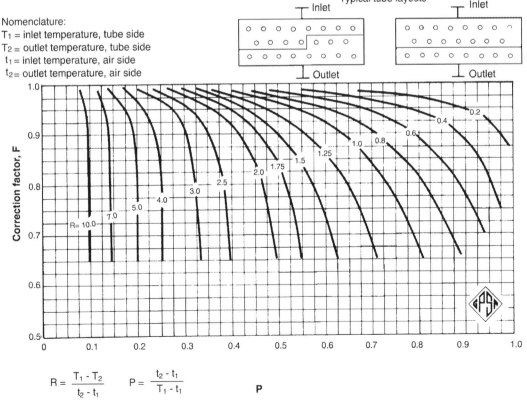

Figure 2-8. MTD correction factors for air-cooled heat exchangers (2-pass, cross-flow, both fluids unmixed) [8].

economic impact; highlights are given in Table 2-12 [18].

Ganapathy has described a procedure for designing an air-cooler [7]:

1. Identify all process and site data.
2. Assume the layout of the tube bundle, air temperature rise or mass flowrate, and fin geometry.
3. For the assumed values, calculate film coefficients and overall heat transfer coefficient, effective

temperature difference, and surface area; check this surface against the assumed layout.
4. When the required surface fits the assumed layout, calculate the tube-side pressure drop and check this against the allowable pressure drop.
5. When surface and tube-side pressure drop are verified, calculate the air-side pressure drop and fan horsepower.

Table 2-12
Variables that must be optimized for air-cooled heat exchanger design [18]

Variable	Considerations
Air flow rate	Rule of thumb for face velocity approaching the tube bundle (total flow divided by total area of bundle): — 3 row coil: 240 to 275 m/min (800 to 900 ft/min) — 4 row coil: 150 to 210 m/min (500 to 700 ft/min) — 5 row coil: 140 to 180 m/min (450 to 600 ft/min) — 6 row coil: 100 to 150 m/min (350 to 500 ft/min) Air-side film coefficient varies to the 0.5 power of air mass velocity Air-side pressure drop varies to the 1.75 power of air mass velocity
Tube length	Length is established in conjunction with the bundle width. There are usually two bundles in a section, and two fans per section. Bundle width normally limited to 3.2 m to 3.5 m (10 ft to 11.5 ft); fans are commonly 3.6 m to 4.3 m (12 ft to 14 ft) in diameter. API 661 specifies minimum fan coverage of 40%. Therefore, tubes are typically in the range of 8 m to 10 m long (26 ft to 33 ft).
Tube outside diameter	Cost of exchanger is lower with smaller diameter tubes Cleaning is more difficult with smaller diameter Minimum recommended (and most common) tube size is 25 mm (1 in) OD Optimize with pressure drop by adjusting the number of passes and tube size
Fin height	Usual fin heights are 9.5 mm, 12.7 mm, and 15.9 mm (3/8 in., 1/2 in., and 5/8 in.) Selection depends on relative values of air-side and tube-side film coefficients With higher fins, fewer tubes can be accommodated per row Typically, use higher fins for steam condensers and water coolers Typically, use lower fins for gas coolers and viscous liquid hydrocarbon coolers
Fin spacing	Spacing usually varies between 276 to 433 fins/m (7 to 11 fins/in.) Typically, use higher density for steam condensers and water coolers Typically, use lower density for gas coolers and viscous liquid hydrocarbon coolers
Tube pitch	Staggered pattern almost invariably employed Designers tend to use the following combinations of bare-tube OD, finned-tube OD, and tube pitch: 25 mm / 50 mm /60 mm (1 in / 2 in / 2.375 in.) 25 mm / 57 mm / 67 mm (1 in / 2.25 in / 2.625 in.) As tube pitch is decreased, air-side pressure drop and power consumption increase more rapidly than the air-side heat transfer coefficient
Number of tube rows	Most exchangers have four to six tube rows, but can range from three to ten Air-side film coefficient varies inversely with number of tube rows More rows advantage: more heat transfer area in the same bundle width, reducing number of bundles and sections More rows disadvantage: increases fan horsepower for the same air velocity and lowers the Mean Temperature Difference Typically, four or five tube rows for steam condensers and water coolers Typically, six or seven tube rows for gas coolers and viscous liquid hydrocarbon coolers
Number of tube passes	Distribution of tubes in the various passes need not be uniform; especially useful in condensers where the flow area in each pass can be gradually reduced as the liquid fraction increases progressively Optimize to obtain uniform pressure drop in each pass
Fan power consumption	Power varies directly with volumetric air flow rate and pressure drop Fan horsepower varies to the 2.75 power of the air mass velocity Optimum air mass velocity is higher when air-side heat transfer coefficient is highly controlling (e.g., steam condensers and water coolers) Exchangers are usually designed with a pressure drop between 0.3 in. H_2O and 0.7 in. H_2O

Air-Side Heat Transfer Coefficient

The Briggs and Young correlation (as reported in [2]) solves for the air-side film coefficient, h_o. It was developed empirically using data from tube diameters from 11 mm to 41 mm (0.44 in. to 1.61 in.) and fin heights from 1.4 mm to 16.6 mm (0.056 in. to 0.652 in.). Fin spacings ranged from 0.9 mm to 3 mm (0.035 in. to 0.117 in.); the tubes were in equilateral triangular pitch tube banks with pitches up to 4.5 in.

$$h_o = C \frac{k_{air}}{d_o} \left(\frac{d_o \, \rho_{air} \, u_{max}}{\mu_{air}}\right)^{0.68} (N_{Pr})^{1/3} \left(\frac{H}{s}\right)^{-0.2} \left(\frac{Y}{s}\right)^{-0.12}$$

(2-39)

Where:

h_o = air-side heat transfer film coefficient, W/m^2-C or Btu/h-ft^2-F

C = coefficient (includes units conversion), 0.000231 (SI) or 0.134 (US)

k_{air} = thermal conductivity of air, 0.026 W/m-C or 0.015 Btu/h-ft-F

d_o = outside diameter of tube (without fins), m or ft

ρ_{air} = density of air, 1.23 kg/m^3 or 0.0765 lb/ft^3 (see Table 2-11)

u_{max} = maximum velocity of air, m/h or ft/h

u_{max} is related to the face velocity of the air approaching the tube bundle by the ratio of total face area to open area between tubes.

μ_{air} = viscosity of air, 0.0000181 Pa-s or 0.0438 lb$_m$/ft-h

N_{Pr} = Prandtl number, dimensionless = $\dfrac{c_p \, \mu_{air}}{k_{air}}$

c_p = heat capacity of air, 1005 J/kg-C or 0.24 Btu/lb-F

H = height of fin, mm or in.

s = spacing between fin centers, mm or in.

Y = thickness of fin, mm or in.

Air-cooled Heat Exchangers: Pressure Drop, Air Side

Calculate the air side pressure drop with the Robinson and Briggs correlation (as reported in [2]). Exchangers are usually designed with a pressure drop between 75 Pa and 175 Pa (0.3 in H$_2$O and 0.7 in H$_2$O). First, calculate the friction factor in consistent units:

$$f = 9.47 \left(\frac{d_o \, \rho_{air} \, u_{max}}{\mu_{air}}\right)^{-0.32} \left(\frac{p_t}{d_o}\right)^{-0.93}$$

(2-40)

Then:

$$\Delta P_{air} = \frac{2 f \, n \, \rho_{air} \, (u_{max})^2}{g_c}$$

(2-41)

Where:

p_t = tube pitch, m or ft

n = number of tube rows in the bundle

g_c = conversion factor, 1 m/s^2 or 32.17 ft/s^2

The other variables are the same as for Equation 2-39, but be sure the units are consistent, especially for u_{max}. Results will be kg/m^2 (x 9.81 = Pa) or lb$_f$/ft^2 (x 0.192 = in. H$_2$O).

Air-cooled Heat Exchangers: Temperature Control

Various methods are used to control the process fluid outlet temperature: switching fans on and off, use of two-speed or variable-speed motors, use of variable pitch fan blades, and adjustable shutters mounted above the tube sections. The manufacturer of the heat exchanger will normally recommend the best solution after consulting with the buyer and designing the unit.

Nomenclature

A	=	heat transfer area, usually calculated at the outside tube diameter, m^2 or ft^2
A_{bp}	=	tube bundle bypass area
A_s	=	free flow area through one crossflow section evaluated at centerline
A_{sb}	=	shell to baffle leakage area for a single baffle
A_{tb}	=	tube to baffle leakage area for a single baffle
A_w	=	area available for flow through a single baffle window
A_{wg}	=	flow area through a single baffle window with no tubes
A_{wt}	=	window area that is occupied by tubes
c_p	=	heat capacity, kJ/kg- °C or Btu/lb- °F
D_b	=	baffle diameter
D_{otl}	=	outside diameter of the tube bundle, mm or in.
D_s	=	inside diameter of the shell
D_w	=	effective diameter of a baffle window
d_i	=	inside tube diameter, consistent units
d_o	=	outside tube diameter, consistent units
F	=	LMTD configuration correction factor, dimensionless
F_c	=	fraction of cross sectional area in the crossflow section
F_w	=	fraction of cross sectional area in the baffle window
f	=	friction factor
g_c	=	conversion factor, 1 m/s^2 or 32.17 ft/s^2
H	=	height of fin, mm or in.
h	=	film coefficient, W/m^2- °C or Btu/h-ft^2- °F
J	=	Bell Delaware correction factor (various subscripts)
k	=	thermal conductivity, W/m- °C or Btu/ft- °F
L	=	tube length
L_{bc}	=	central baffle spacing
L_{bi}	=	baffle spacing at inlet
L_{bo}	=	baffle spacing at outlet
l_c	=	baffle cut = distance from the baffle to the inside of the shell, mm or in.
N_{Pr}	=	Prandtl number $= \dfrac{c_p \, \mu}{k}$
N_{Re}	=	Reynolds number $= \dfrac{d \, \rho \, u}{\mu}$

n_b	=	number of baffles in the exchanger
$n_{r,cc}$	=	effective tube rows crossed through one crossflow section
$n_{r,tw}$	=	effective tube rows crossed in the window section
n_t	=	number of tubes
n_{ss}	=	number of sealing strip pairs
n_{tw}	=	number of tubes in a baffle window
n_p	=	number of passes
n_{dp}	=	number of bypass dividers parallel to crossflow stream
ΔP_t	=	pressure drop through turns, Pa or psf (divide by 144 for psi)
PR	=	pitch ratio
p_t	=	tube pitch
p_n	=	tube pitch normal to the flow direction
p_p	=	tube pitch parallel to the flow direction
Q	=	heat transferred, W or Btu/h
R_f	=	fouling factor
r_a	=	$A_{sb}/(A_{sb} + A_{tb})$
r_b	=	$(A_{sb} + A_{tb})/A_w$
r_c	=	A_{bp}/A_s
s	=	spacing between fin centers, mm or in.
T	=	inlet and outlet temperatures of the hot stream, °C or °F
t	=	inlet and outlet temperatures of the cold stream, °C or °F
ΔT_{mean}	=	mean temperature difference between hot and cold streams, °C or °F
U	=	overall heat transfer coefficient, W/m^2- °C or Btu/h-ft^2- °F
u	=	velocity in tubes, m/s or ft/s
W	=	mass flow rate
w_p	=	width of bypass divider lanes that are parallel to the crossflow stream
Y	=	thickness of fin, mm or in.
ρ	=	density, kg/m^3 or lb/ft^3
μ	=	viscosity, cP
ζ	=	ratio of sealing strip pairs to tube rows in crossflow section

References

[1] Bejan A, Kraus A. *Heat Transfer Handbook*. John Wiley & Sons; 2003.

[2] Bell K, Mueller A. *Wolverine Engineering Data Book*. Huntsville, Alabama: Wolverine Tube, Inc. published online at www.wlv.com/products; 2001.

[3] Bennett C, Kistler RS, Lestina T, King D. Improving Heat Exchanger Designs. *Chemical Engineering Progress*, April, 2007:40.

[4] Bowman R, Mueller A, Nagle W. Mean Temperature Difference in Design. *Transactions of the American Society of Mechanical Engineers*, May 1940;62:283.

[5] Delta T Heat Exchangers. Fouling in Heat Exchangers, downloaded from www.deltathx.com in January, 2011.

[6] Fakheri A. A General Expression for the Determination of the Log Mean Temperature Correction Factor for Shell and Tube Heat Exchangers. *Journal of Heat Transfer, American Society for Mechanical Engineers (ASME)*, June, 2003;125:527.

[7] Ganapathy V. Process-design Criteria. *Chemical Engineering*, March 27, 1978:112–9.

[8] Gas Processors Suppliers Association (GPSA). *Engineering Data Book, SI Version*. 12th ed., vol. 1;2004.

[9] Gunnarsson J, Sinclair I, Alanis F. Compact Heat Exchangers: Improving Recovery. *Chemical Engineering*, February 2009:44–7.

[10] Haslego C, Polley G. Designing Plate-and-Frame Heat Exchangers. *Chemical Engineering Progress*, September, 2002:32.

[11] Heat Transfer Research, Inc. (HTRI), www.htri.net

[12] "Installation and Operating Instructions for Armstrong Shell-and-Tube Heat Exchangers", File 138.65, Armstrong Pumps, Inc., www.armstrongpumps.com, May, 2007.

[13] Kakaç S, Liu H. Heat Exchangers: *Selection, Rating, and Thermal Design*. 2nd ed. Boca Raton: CRC Press; 2002.

[14] Kerner J. Plate Heat Exchangers: Avoiding Common Misconceptions. *Chemical Engineering*, February 2009:40–3.

[15] Kern D, Kraus A. *Extended Surface Heat Transfer*. New York: McGraw-Hill; 1972.

[16] Leong K, Leong Y, Toh K. Shell and Tube Heat Exchanger Design Software for Educational Applications. *International Journal of Engineering Education*, 1998;14(3):217–24.

[17] Moretta A. Spiral Heat Exchangers: Sizing Units for Cooling Non-Newtonian Slurries. *Chemical Engineering*, May, 2010:44.

[18] Mukherjee R. Effectively Design Air-Cooled Heat Exchangers. *Chemical Engineering Progress*, February, 1997:26–47.

[19] Nesta J. Reduce fouling in shell-and-tube heat exchangers. *Hydrocarbon Processing*, July, 2004: 77–82.

[20] Personal communications between Carl Branan, Jack Hailer, and Guy Z. Moore while all were employed at El Paso Products Company.

[21] Pfaudler Corporation, Rochester, NY.

[22] Plant Support & Evaluations, Inc., "Industrial Steam System Heat-Transfer Solutions", A Best Practices Steam Technical Brief published by the Industrial Technologies Program, Energy Efficiency and Renewable Energy, U.S. Department of Energy (DOE), DOE/GO-102003–1738, June 2003.

[23] Rubin F. What's the Difference Between TEMA Exchanger Classes. *Hydrocarbon Processing*, June 1980;9:92.

[24] TEMA. *Standards of the Tubular Exchanger Manufacturers Association*. 9th ed. www.tema.org; 2007.

[25] Than STM, Lin KA, Mon MS. Heat Exchanger Design. World Academy of Science, Engineering and Technology, www.waset.org, October, 2008; (46):604.

3

Fractionators

Introduction

More than any other unit operation, fractionation defines chemical engineering. Fractionation is the theory and practice of separating mixtures into their pure components, usually by a distillation process. It's a key process for manufacturing chemicals, taught only to chemical engineers.

The many shortcut design methods and rules of thumb for fractionation are intended to simplify what is in reality an exceptionally complex process. In the early to mid 20th century, chemical engineers developed expressions that estimated minimum reflux ratio or column diameter before the advent of HySim, Aspen, and other computer simulators. Graphical design methods were used when the engineer's toolbox consisted of a slide rule, pencil, and graph paper.

Today's chemical engineer is tempted to assign the design of a fractionation system to a technician who is skilled in using a computerized simulation package. It would be a mistake, however, to bend to this temptation. Although the use of simulators is mandatory for most system designs, simulators are like instruments that are played by well-trained and informed chemical engineers.

Shortcut methods continue to play an important role. Engineers can rapidly compare alternative concepts with them. They check the results from their simulators. And they design columns for simple separations without resorting to simulators at all. But perhaps the best reason for teaching and using shortcut methods is that they provide chemical engineers with a fundamental understanding of the fractionation process – which leads to more intelligent use of the tools that are enabled by computers.

This chapter provides practical equations for solving fractionation problems, and guidance for constructing solutions using McCabe-Thiele diagrams (in Excel rather than on graph paper). Simulators require that the engineer selects or provides an appropriate set of data; guidance is provided for this critical step. Then, the actual equipment used for fractionation columns – trays and packing – are explored.

An Excel workbook, containing worked examples, accompanies this chapter.

Separation Factor

The separation factor is defined with the desired concentration of light key and heavy key components in the overheads and bottoms. This factor is typically between about 500 and 2,000, but ranges to about 10,000 for sharp separations. The number of trays will be roughly proportional to the log of the separation factor for a given system [13].

$$S_F = \left(\frac{x_D}{x_B}\right)_{LK} \left(\frac{x_B}{x_D}\right)_{HK} \qquad (3\text{-}1)$$

Relative Volatility

The relative volatility of mixtures of two or more components is equal to their relative vapor pressures *for ideal mixtures*. In a binary system:

$$\alpha_{ij} = \frac{P_i}{P_j} \qquad (3\text{-}2)$$

α_{ij} = relative volatility of component i and j
P_i = vapor pressure of pure component i
P_j = vapor pressure of pure component j

The relative volatility is temperature dependent, although for similar components (two alkanes, for example) it is fairly constant over the temperature range of a distillation column. Strictly, the relative volatility is the ratio of K factors for the components, where for each component, i, $K_i = y_i/x_i$. For ideal components, $K_1/K_2 = P_1/P_2$.

Vapor-Liquid Equilibrium

The vapor-liquid equilibrium (VLE) determines the performance of a fractionation column. For ideal mixtures, VLE is simply correlated to the partial pressures of the components. Non-ideal mixtures – and it should be assumed that real-world separations are non-ideal unless known otherwise – require the use of correction factors, called *activity coefficients*, which are temperature and pressure dependent. Activity coefficients are *specific* to a particular mixture (e.g., the coefficients for ethanol-water are different than those for methanol-water or ethanol-methanol-water).

Given a mixture of liquids, with molar concentrations x_1, x_2, ... x_i, the vapor concentrations y_1, y_2, ... y_i are determined from the relationship:

$$y_i = \gamma_i \, x_i \, P_i \qquad (3\text{-}3)$$

Where:

$\gamma_i =$ the activity coefficient specific to the particular system

$P_i =$ partial pressure

Assuming ideal behavior, which is fine for certain mixtures as explained later, then $\gamma_i = 1$.

A VLE curve is normally constructed over the entire range of liquid concentrations *at constant pressure*. Since the sum of the partial pressures ($x_i \, P_i$) equals the total pressure, it's important to determine the bubble point temperature at each concentration used in constructing the curve.

Example

This example, for the binary system ethanol-methanol, explains the method for constructing a VLE curve and assumes ideal behavior. We are creating the curve using Microsoft Excel at a total system pressure of 1 atmosphere ($= 760$ mm Hg).

1. Antoine coefficients for correlating vapor pressure with temperature. See "Vapor Pressure" in the Physical Properties chapter. For this example, the coefficients reported in Perry's Handbook are used.

$$P = 10^{\left(A - \frac{B}{t+C}\right)} \qquad (3\text{-}4)$$

	A	B	C	D
1				
2		A	B	C
3	Methanol	8.08097	1,582.271	239.726
4	Ethanol	8.11220	1,592.864	226.184
5				

Figure 3-1. Antoine coefficients.

Where:

$P =$ vapor pressure

A, B, C are Antoine coefficients, shown in Figure 3-1.

$t =$ temperature, °C

2. Create a table that relates the liquid mole fraction of methanol (X) with vapor pressure and partial pressure of both methanol and ethanol. Calculate the values for vapor pressure from the temperature in Column C and the Antoine coefficients. Since there are no entries for temperature yet, the vapor pressures compute to the value at 0 °C.

   The formulae, copied down the columns, are:
   D9 = 10^($B$3−$C$3/($C9+$D$3))
   E9 = 10^($B$4−$C$4/($C9+$D$4))
   F9 = A9*D9
   G9 = (1−A9)*E9

   This is shown in Figure 3-2.

3. The next step is to determine the bubble point temperature for each liquid concentration. This is an iterative calculation. By changing the temperature in Column C, the vapor pressures and therefore the partial pressures change. The goal is to have the partial pressures sum to 760 mm Hg.

   To achieve this, first enter this formula in H9 and copy it down the column:
   H9 = (F9+G9)−760

   Next, use the Goal Seek function in Excel to find the temperature that results in column H being zero. In this screen shot, Goal Seek has already been applied to Row 9, and it is set up to do Row 10. Repeat for each row in the table (see Figure 3-3).

   Automate this with a macro. If Cell C8 is named "Temperature" and Cell H8 is named "Diff," the macro is:

```
Sub FindTemperature()
' Macro finds the Bubble Point Temperature
' for each entry in the table
For i = 1 To 11
Range("Diff").Offset(i, 0).GoalSeek goal:=0, _
changingcell:=Range("Temperature").Offset(i, 0)
Next i
End Sub
```

	A	B	C	D	E	F	G
7				VP, mm Hg		Partial Pressure	
8	X		Temp, deg C	Methanol	Ethanol	Methanol	Ethanol
9	0.00			30.24	11.75	0.00	11.75
10	0.10			30.24	11.75	3.02	10.57
11	0.20			30.24	11.75	6.05	9.40
12	0.30			30.24	11.75	9.07	8.22
13	0.40			30.24	11.75	12.10	7.05
14	0.50			30.24	11.75	15.12	5.87
15	0.60			30.24	11.75	18.15	4.70
16	0.70			30.24	11.75	21.17	3.52
17	0.80			30.24	11.75	24.20	2.35
18	0.90			30.24	11.75	27.22	1.17
19	1.00			30.24	11.75	30.24	0.00
20							

Figure 3-2. Initial setup.

	A	B	C	D	E	F	G	H
7				VP, mm Hg		Partial Pressure		
8	X		Temp, deg C	Methanol	Ethanol	Methanol	Ethanol	Difference
9	0.00		78.2981922	1,275.42	760.00	0.00	760.00	0.00
10	0.10			30.24	11.75	3.02	10.57	(746.40)
11	0.20			30.24	11.75	6.05	9.40	(744.55)
12	0.30					9.07	8.22	(742.71)
13	0.40					12.10	7.05	(740.86)
14	0.50					15.12	5.87	(739.01)
15	0.60					18.15	4.70	(737.16)
16	0.70					21.17	3.52	(735.31)
17	0.80					24.20	2.35	(733.46)
18	0.90					27.22	1.17	(731.61)
19	1.00			30.24	11.75	30.24	0.00	(729.76)

Goal Seek

Set cell: H10
To value: 0
By changing cell: C10

OK   Cancel

Figure 3-3. Goal seek to find bubble point temperature.

4. Calculate the molar vapor concentration by multiplying the liquid mole fraction by the partial pressure at the bubble point temperature. Partial pressure = vapor pressure divided by total pressure. Therefore:

B9 = A9*D9/760 (Then copy down the column – see Figure 3-4.)

5. Chart the results using a Scatter Diagram. This is shown later in the chapter in the section about graphical methods.

	A	B	C	D	E	F	G	H
7	Mole Fraction MeOH			VP, mm Hg		Partial Pressure		
8	X	Y	Temp, deg C	Methanol	Ethanol	Methanol	Ethanol	Difference
9	0.00	0.00	78.2981922	1,275.42	760.00	0.00	760.00	0.00
10	0.10	0.16	76.62503425	1,200.44	711.06	120.04	639.96	0.00
11	0.20	0.30	75.03043977	1,132.40	666.90	226.48	533.52	0.00
12	0.30	0.42	73.50920122	1,070.50	626.93	321.15	438.85	0.00
13	0.40	0.53	72.05646331	1,014.03	590.65	405.61	354.39	0.00
14	0.50	0.63	70.66771301	962.37	557.63	481.19	278.81	0.00
15	0.60	0.72	69.33873788	915.01	527.49	549.00	210.99	(0.00)
16	0.70	0.80	68.06573628	871.47	499.91	610.03	149.97	(0.00)
17	0.80	0.88	66.84506279	831.35	474.62	665.08	94.92	(0.00)
18	0.90	0.94	65.67341435	794.29	451.36	714.86	45.14	(0.00)
19	1.00	1.00	64.54773241	760.00	429.92	760.00	0.00	(0.00)

Figure 3-4. Completed table.

Activity Coefficients

Activity coefficients are used in so-called "solution models" to adjust to non-ideal interactions between components in the liquid phase. There are several different solution models in use that correlate activity coefficients to experimental data, predict the coefficients from molecular structure, or a combination of both. The most common models are: Wilson, Margules, van Laar, NRTL, and UNIQUAC.

Each model uses an expression with two or three parameters that computes the activity coefficient. The parameters are specific to the components in the liquid phase and the system pressure. Parameters for very common systems, such as ethanol-water, can be readily found in open literature. The vast majority of data are stored in databases that are sold with the simulation programs.

Because the activity coefficients are used with vapor pressure, it's important to use Antoine vapor pressure parameters that are consistent with the solution model's parameters. In other words, the two van Laar parameters work together with the three Antoine parameters. Expect reduced accuracy if vapor pressure data is independent from the solution model parameters.

It is useful and instructive to construct a VLE curve using activity coefficients, shown below in an example. For real-world problem solving where a simulator is utilized, the engineer is responsible for deciding which solution model to use. Since the complexity of the model is invisible to the simulation user, the choice is normally to use the most accurate model available for the particular separation being simulated.

Find information in the simulation software that, for any model chosen, reports how the parameters were obtained and their applicable range of pressure and concentration.

For fractionations operating at higher than about 10 atmospheres, where the reduced temperature in the vapor (temperature divided by critical temperature) is greater than about 0.75, the vapor phase is less likely to behave like a perfect gas and additional corrections are required. These corrections are modeled with Equations of State (EOS) which provide another adjustment coefficient. There are multiple EOS models which are beyond the scope of this book.

Data are available from several sources, including DECHEMA, and within simulation software packages such as Aspen.

Example

This example, for the binary system ethanol-water, explains the method for constructing a VLE curve using activity coefficients. We are creating the curve using Microsoft Excel at a total system pressure of 1 atmosphere (= 760 mm Hg).

The method is exactly the same as in the ideal VLE example, except that partial pressures are calculated by multiplying the liquid mole fraction and vapor pressure by the activity coefficient.

1. Begin by entering the Antoine coefficients for ethanol and water into the spreadsheet (see Figure 3-5).
2. Create a table that relates the liquid mole fraction of ethanol (X) with the vapor pressure, activity coefficient, and partial pressure of both ethanol and water. Calculate the values for vapor pressure from the temperature in Column C and the Antoine coefficients. Since there are no entries for temperature yet, the vapor pressures compute to the value at $0\,°C$. Enter a value of 1 for the activity coefficients.

   The formulae, copied down the columns are:
   D9 = 10^($B$3−$C$3/($C9+$D$3))
   E9 = 10^($B$4−$C$4/($C9+$D$4))
   H9 = A9*F9*D9
   I9 = (1−A9)*G9*E9

   This is shown in Figure 3-6.
3. Enter the parameters for the Margules model into the spreadsheet, and calculate the activity coefficients.

   The parameters in Cells F3 and G3 are specific to the binary system (in this case, ethanol-water). Calculate the activity coefficients from the liquid mole fraction and the parameters.

   The formulae, copied down the columns, are (see Figure 3-7):
   F9 = EXP(($F$3+2*($G$3−$F$3)*A9)
   *(1−A9)^2)
   G9 = EXP(($G$3+2*($F$3−$G$3)*(1−A9))
   *A9^2)

	A	B	C	D
2		A	B	C
3	Ethanol	8.11220	1,592.864	226.184
4	Water	8.07131	1,730.630	233.426
5				

Figure 3-5. Antoine coefficients.

	A	B	C	D	E	F	G	H	I
7	Mole Fraction EtOH			VP, mm Hg		Margules		Partial Pressure	
8	X		Temp, deg C	Ethanol	Water	γ_1	γ_2	Ethanol	Water
9	0.00			11.75	4.54	1.000	1.000	0.00	4.54
10	0.10			11.75	4.54	1.000	1.000	1.17	4.09
11	0.20			11.75	4.54	1.000	1.000	2.35	3.63
12	0.30			11.75	4.54	1.000	1.000	3.52	3.18
13	0.40			11.75	4.54	1.000	1.000	4.70	2.73
14	0.50			11.75	4.54	1.000	1.000	5.87	2.27
15	0.60			11.75	4.54	1.000	1.000	7.05	1.82
16	0.70			11.75	4.54	1.000	1.000	8.22	1.36
17	0.80			11.75	4.54	1.000	1.000	9.40	0.91
18	0.90			11.75	4.54	1.000	1.000	10.57	0.45
19	1.00			11.75	4.54	1.000	1.000	11.75	0.00

Figure 3-6. Table relates mole fraction with vapor pressure, activity coefficient, and partial pressure; data not yet computed.

	F	G
1	Margules Parameters	
2	A_{12}	A_{21}
3	1.6022	0.7947
4		
5		
6		
7	Margules	
8	γ_1	γ_2
9	4.964	1.000
10	3.212	1.023
11	2.268	1.087
12	1.729	1.189
13	1.411	1.326
14	1.220	1.493
15	1.107	1.680
16	1.043	1.872
17	1.012	2.045
18	1.001	2.170
19	1.000	2.214

Figure 3-7. Margules parameters and activity coefficients.

4. Determine the bubble point temperature for each liquid concentration using Excel's Goal Seek function. The partial pressures sum to 760 mm Hg.
The formula in the Difference column is:
H9 = (H9+I9)−760

This is shown in Figure 3-8.

5. After finding all of the temperatures (use the same macro shown in the first VLE example), calculate the molar vapor concentration by multiplying the liquid mole fraction by the partial pressure at the bubble point temperature. This is equivalent to multiplying the mole fraction by the activity coefficient and vapor pressure divided by total pressure.

Therefore, the formula for y is:

B9 = A9*D9*F9/760 (then copy down the column – see Figure 3-9).

Sources for Activity Coefficients

1. The DETHERM database provides thermophysical property data for more than 30,000 pure compounds and 100,000 mixtures. Search the database for free; pay only for downloaded data. http://www.dechema.de/en/detherm.html

2. *Fluid Phase Equilibria*, Elsevier, http://www.sciencedirect.com/science/journal/03783812

	A	B	C	D	E	F	G	H	I	J
1			Antoine Constants			Margules Parameters				
2		A	B	C		A_{12}	A_{21}			
3	Ethanol	8.11220	1,592.864	226.184		1.6022	0.7947			
4	Water	8.07131	1,730.630	233.426						
5										
6										
7	Mole Fraction EtOH			VP, mm Hg		Margules		Partial Pressure		
8	X		Temp, deg C	Ethanol	Water	γ_1	γ_2	Ethanol	Water	Difference
9	0.00		100.00	1,693.65	760.00	4.964	1.000	0.00	760.00	(0.00)
10	0.10			11.75	4.54	3.212	1.023	3.77	4.18	(752.05)
11	0.20			11.75	4.54	2.268	1.087	5.33	3.95	(750.72)
12	0.30					1.729	1.189	6.09	3.78	(750.13)
13	0.40					1.411	1.326	6.63	3.61	(749.76)
14	0.50					1.220	1.493	7.16	3.39	(749.45)
15	0.60					1.107	1.680	7.80	3.05	(749.15)
16	0.70					1.043	1.872	8.58	2.55	(748.87)
17	0.80					1.012	2.045	9.51	1.86	(748.63)
18	0.90					1.001	2.170	10.59	0.99	(748.43)
19	1.00			11.75	4.54	1.000	2.214	11.75	0.00	(748.25)

Goal Seek
Set cell: j10
To value: 0
By changing cell: c10
OK   Cancel

Figure 3-8. Bubble point temperatures determined using Goal Seek function.

	A	B	C	D	E	F	G	H	I	J
7	Mole Fraction EtOH			VP, mm Hg		Margules		Partial Pressure		
8	X	Y	Temp, deg C	Ethanol	Water	γ_1	γ_2	Ethanol	Water	Difference
9	0.00	0.00	100.00	1,693.65	760.00	4.964	1.000	0.00	760.00	(0.00)
10	0.10	0.44	86.59	1,045.95	460.65	3.212	1.023	335.99	424.01	0.00
11	0.20	0.54	82.94	910.71	399.01	2.268	1.087	413.01	346.99	0.00
12	0.30	0.59	81.48	860.98	376.43	1.729	1.189	446.65	313.35	0.00
13	0.40	0.62	80.65	833.52	363.99	1.411	1.326	470.40	289.60	0.00
14	0.50	0.65	79.99	812.40	354.43	1.220	1.493	495.48	264.52	(0.00)
15	0.60	0.69	79.42	794.26	346.23	1.107	1.680	527.37	232.63	(0.00)
16	0.70	0.75	78.94	779.53	339.57	1.043	1.872	569.33	190.67	(0.00)
17	0.80	0.82	78.60	769.19	334.90	1.012	2.045	623.03	136.97	(0.00)
18	0.90	0.91	78.41	763.22	332.21	1.001	2.170	687.92	72.08	(0.00)
19	1.00	1.00	78.30	760.00	330.76	1.000	2.214	760.00	0.00	0.00

Figure 3-9. Completed table.

Publishes high quality papers dealing with experimental, theoretical and applied research related to equilibrium and transport properties of fluid and solid phases.

3. *Journal of Chemical and Engineering Data*, ACS Publications, http://pubs.acs.org/journal/jceaax. Reports on experimental, evaluated, and predicted data on the physical, thermodynamic, and transport properties of well-defined materials including complex mixtures of known compositions, and systems of environmental and biochemical interest.

4. *The Journal of Chemical Thermodynamics*, Elsevier, http://www.sciencedirect.com/science/journal/00219614. For dissemination of significant new measurements in experimental thermodynamics and thermophysics including bio-thermodynamics, calorimetry, phase equilibria, equilibrium thermodynamic properties, and transport properties.

Establishing the Column Pressure

In most systems, the engineer may dictate the column pressure, within limits, and achieve the desired separation. Do not use a lower pressure than necessary since lowering the pressure results in a larger column diameter and possibly more equilibrium stages (decreased efficiency). Here is a procedure for selecting the pressure.

1. From available VLE data, determine if a certain pressure or pressure range is required to achieve the desired separation. Ideal systems perform identically at any pressure, but non-ideal systems may have pinch points at certain pressures and different (or no) pinch points at other pressures. If pressure limits exist, then confine evaluations in the following steps to that range. The temperature of the bottom product may not exceed the critical temperature; the pressure may not exceed the critical pressure of the overhead product. For temperature-sensitive materials, the reboiler temperature may be limited.

2. Based on process conditions upstream and downstream of the column, determine if there is a preferred operating pressure for the column. If so, the upstream and/or downstream equipment pressures should be evaluated in parallel with the present column. For example, if a gaseous overhead product is compressed downstream of the column, a higher operating pressure may be desired to reduce the cost of compression.

3. Tabulate the available heating and cooling media for the column. For example, there may be steam available at three pressures and cooling water at two temperatures. Apply judgment or the plant's standard practices to assign the maximum permissible boiling temperature (for the reboiler) and minimum permissible condensing temperature (for the overhead condenser) for each of the heating and cooling streams. The boiling temperature must be less than the temperature of the heating medium; condensing must be at a higher temperature than the coolant temperature. See Table 3-1.

4. For each heating source, calculate the system pressure that corresponds to the "maximum permissible boiling temperature." These are the *maximum* column pressures that are compatible with each of the heating media. Multiply the bottoms mole fraction and vapor pressure for each of the components then sum to get the total pressure.

$$P = x_1 P_1 + x_2 P_2 + \ldots + x_i P_i$$

Table 3-1
Example data for heating and cooling media

Heating or Cooling Media	Media Temperature at the Column	Maximum Permissible Boiling Temperature	Minimum Permissible Condensing Temperature
Low Pressure Steam (1 bar)	121 °C (250 °F)	106 °C (222 °F)	NA
Medium Pressure Steam (5 bar)	159 °C (319 °F)	144 °C (292 °F)	NA
High Pressure Steam (15 bar)	202 °C (395 °F)	187 °C (368 °F)	NA
Air-Cooled Heat Exchanger	40 °C (104 °F)	NA	50 °C (122 °F)
Tower Water	30 °C (86 °F)	NA	40 °C (104 °F)
Chilled Water	5 °C (41 °F)	NA	15 °C (59 °F)

5. For each cooling source, calculate the system pressure that corresponds to the "minimum permissible condensing temperature." These are the *minimum* column pressures that are compatible with each of the cooling media. For a liquid overhead product, the condenser operates at the bubble point. For a gaseous product the condenser operates at the dew point.
6. Finally, apply engineering judgment to assign the column pressure within the minimum and maximum

values, including those identified in steps 1 and 2. Beyond the limits already described, capital cost and energy cost generally lead to the result. Capital costs include utility piping, upgrading utility capacity (if necessary), a vacuum system (if applicable), and process controls. Energy cost is differentiated into varying costs assigned to each of the heating and cooling sources (e.g., tower water is cheaper than chilled water) and the use of vacuum. Other factors include safety, operability, and reliability.

Minimum Reflux – Binary

The Reflux Ratio is the amount of liquid returned to the column from the overhead condenser (L) divided by the distillate removed from the column (D), expressed in molar units. Therefore, R = L / D. The minimum reflux ratio, R_m, is the value for R that is needed to achieve a desired distillate composition in a column with an infinite number of separation stages, or trays.

For a binary or near binary minimum reflux ratio, use the following Underwood equations [31]:

Bubble Point Liquid Feed

$$R_m = \frac{1}{(\alpha - 1)}\left[\frac{x_{LD}}{x_{LF}} - \frac{\alpha(1 - x_{LD})}{(1 - x_{LF})}\right] \quad (3-5)$$

Dew Point Vapor Feed

$$R_m = \frac{1}{(\alpha - 1)}\left[\frac{\alpha x_{LD}}{x_{LF}} - \frac{(1 - x_{LD})}{(1 - x_{LF})}\right] \quad (3-6)$$

Where:

R_m = minimum reflux ratio, L/D
α = relative volatility of the light component to the heavy component
x_{LD} = mole fraction of the light component in the distillate
x_{LF} = mole fraction of the light component in the feed

Example

Estimate the minimum reflux ratio for a binary feed consisting of 40 mole% methanol and 60 mole% ethanol at its bubble point. The desired distillate composition is 98 mole% methanol.

1. Using the method described in the VLE example, determine the bubble point temperature (Figure 3-10).
2. Calculate the relative volatility, α

$$\alpha = \frac{P_1}{P_2} = 1014/591 = 1.717$$

	A	B	C	D	E	F	G	H
1			Antoine Constants					
2		A	B	C				
3	Methanol	8.08097	1,582.271	239.726				
4	Ethanol	8.11220	1,592.864	226.184				
5								
6	**Molar Composition**		**Bottoms**	**Feed**	**Distillate**			
7	Methanol			0.40	0.98			
8	Ethanol			0.60	0.02			
9								
10								
11	Mole Fraction MeOH			VP, mm Hg		Partial Pressure		
12	X	Y	Temp, deg C	Methanol	Ethanol	Methanol	Ethanol	Difference
13	0.40	0.53	72.06	1,014.03	590.65	405.61	354.39	0.00

Figure 3-10. Bubble point temperature.

3. Apply the Underwood equation to find R_m

$$R_m = \frac{1}{(\alpha - 1)}\left[\frac{x_{LD}}{x_{LF}} - \frac{\alpha(1 - x_{LD})}{(1 - x_{LF})}\right]$$

$$= 1/(1.717 - 1)(0.98/0.40 - 1.717(1 - 0.98)/(1 - 0.40))$$

$$= 3.34$$

Minimum Reflux – Multicomponent

The Underwood Method will provide a quick estimate of minimum reflux requirements. It is a good method to use when distillate and bottoms compositions are specified. Although the Underwood Method is outlined here, other good methods exist – such as the Brown-Martin [21] and Colburn [8] methods. These and other methods are discussed and compared in Van Winkle's book [32]. A method to use for column analysis when distillate and bottoms compositions are not specified is discussed by Smith [27]. The Underwood Method involves finding a value for a constant, q, that satisfies the equation:

$$\sum_{1}^{n}\frac{x_{iF}\,\alpha_i}{\alpha_i - \theta} = 1 - q$$

$$= \frac{x_{1F}\,\alpha_1}{\alpha_1 - \theta} + \frac{x_{2F}\,\alpha_2}{\alpha_2 - \theta} + \ldots \quad (3\text{-}7)$$

The value of θ will lie between the relative volatilities of the light and heavy key components, which must be adjacent.

After finding θ, the minimum reflux ratio is determined from:

$$R_m + 1 = \sum_{1}^{n}\frac{\alpha_i\,x_{iD}}{\alpha_i - \theta} \quad (3\text{-}8)$$

Where:

R_m = minimum reflux ratio, L/D
α_i = relative volatility of component i to the heavy key component
q = thermal condition of the feed
Bubble point liquid, $q = 1$
Dew point vapor, $q = 0$
General feed, $q = \dfrac{(L_S - L_R)}{F}$ q = $(L_S - L_R) / F$
L_S = liquid molar rate in the stripping section
L_R = liquid molar rate in the rectification section
F = Feed molar rate
x_{iD} = mole fraction of component i in the distillate
x_{iF} = mole fraction of component i in the feed
θ = Underwood minimum reflux constant

The Underwood equation is applicable when none of the non-key components are split between the distillate and bottoms. It assumes constant molar overflow and relatively constant relative volatilities.

Example

Estimate the minimum reflux ratio for the tertiary system n-hexane, n-heptane, n-octane, with the feed at the bubble point and molar compositions 0.1, 0.5, and 0.4 respectively. 95% of the n-heptane should be contained in the distillate and 99% of the n-octane in the bottoms.

1. The Underwood multicomponent minimum reflux method is used, assigning n-heptane to the light key and n-octane as the heavy key, which requires the assumption that all of the n-hexane is contained in the distillate. This leads to a material balance:

    Distillate composition:
    X_{1D} (n-octane) $= 0.007$
    X_{2D} (n-heptane) $= 0.820$
    X_{3D} (n-hexane) $= 0.173$

2. Start by determining the feed temperature using Excel and Goal Seek as shown earlier, and illustrated in Figure 3-11.

    The answer is $102\,°C$, which then permits calculation of the relative volatilities of the components as seen in Figure 3-12.

3. Theta is determined, again using Goal Seek.

    Set up the spreadsheet with a trial value for θ that is between the relative volatilities of the light and

8	System Pressure		760	mm Hg	
9	Temperature		102	deg C	
10					
11	**Molar Composition**	**Feed**	VP, mm Hg	PP	α_i
12	n-hexane	0.10	1,915.451	191.545	5.12
13	n-heptane	0.50	837.854	418.927	2.24
14	n-octane	0.40	373.820	149.528	1.00
15				760.000	
16			Difference	(0.000)	

Figure 3-12. Calculation of relative volatilities.

heavy key components. In this example, the trial value is $\theta = 2.0$. Then, create a column for the Underwood term and sum it. Use Goal Seek to compare the summation with the value $1 - q$, finding a value of θ where they are equal, as shown in Figure 3-13.

$E22 = (C22*D22)/(D22-C\$19)$ (copied down)
$E26 = (1-C18)-E25$

4. Calculate the minimum reflux by using the distillate composition computed in Step 1 and theta as computed in Step 3 (Figure 3-14). Note that constant relative volatility is assumed, so the relative volatility at the feed condition is appropriate to use here.

$R_m + 1 = 2.17$
$R_m = 1.17$

	A	B	C	D	E	F	G	H
1			Antoine Constants					
2		A	B	C				
3	n-hexane	6.76098	1,112.56081	218.02991				
4	n-heptane	6.81342	1,224.41518	212.94937				
5	n-octane	6.85805	1,316.95574	205.52354				
6								
7								
8	System Pressure		760	mm Hg				
9	Temperature			deg C				
10								
11	**Molar Composition**	**Feed**	VP, mm Hg	PP	α_i			
12	n-hexane	0.10	45.519	4.552	16.14			
13	n-heptane	0.50	11.578	5.789	4.11			
14	n-octane	0.40	2.820	1.128	1.00			
15				11.469				
16			Difference	748.531				

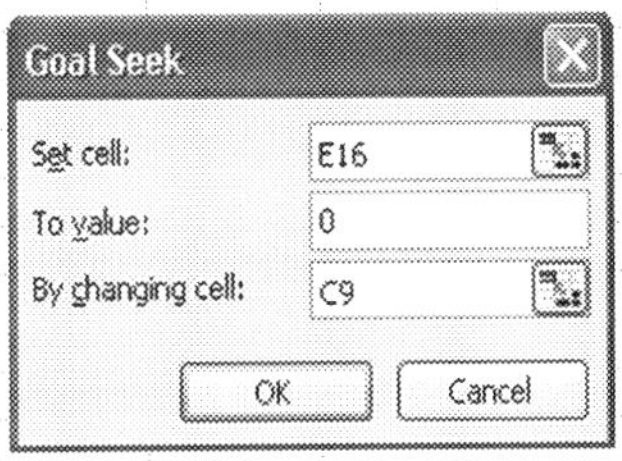

Goal Seek
Set cell: E16
To value: 0
By changing cell: C9

Figure 3-11. Setup to find feed temperature using Goal Seek.

(a)

	A	B	C	D	E	F	G	H
18	Feed Condition, q		1.00	(1 = feed at the Bubble Point)				
19	Theta		2.00					
20								
21	**Molar Composition**		**Feed**	α_i	$x_i\,\alpha_i\,/\,(\alpha_i - \theta)$			
22	n-hexane		0.10	5.12	0.16			
23	n-heptane		0.50	2.24	4.64			
24	n-octane		0.40	1.00	(0.40)			
25					4.41			
26				Difference	(4.41)			

Goal Seek
Set cell: E26
To value: 0
By changing cell: C19
OK   Cancel

(b)

	A	B	C	D	E	F
18	Feed Condition, q		1.00	(1 = feed at the Bubble Point)		
19	Theta		1.30			
20						
21	**Molar Composition**		**Feed**	α_i	$x_i\,\alpha_i\,/\,(\alpha_i - \theta)$	
22	n-hexane		0.10	5.12	0.13	
23	n-heptane		0.50	2.24	1.19	
24	n-octane		0.40	1.00	(1.33)	
25					0.00	
26				Difference	(0.00)	

Figure 3-13. Finding theta.

	A	B	C	D	E	F	G	H
18	Feed Condition, q		1.00	(1 = feed at the Bubble Point)				
19	Theta		1.30					
20								
21	**Molar Composition**		**Feed**	α_i	$x_i\,\alpha_i\,/\,(\alpha_i - \theta)$	**Distillate**	$x_{iD}\,\alpha_i\,/\,(\alpha_i - \theta)$	
22	n-hexane		0.10	5.12	0.13	0.173	0.23	
23	n-heptane		0.50	2.24	1.19	0.820	1.96	
24	n-octane		0.40	1.00	(1.33)	0.007	(0.02)	
25					0.00		2.17	= Rm + 1
26				Difference	(0.00)			

Figure 3-14. Completed calculation for minimum reflux.

Minimum Stages

The Fenske Method [9] gives a quick estimate for the minimum theoretical stages at total reflux.

$$N_m + 1 = \frac{\ln\left[\left(\dfrac{x_{LK}}{x_{HK}}\right)_D \left(\dfrac{x_{HK}}{x_{LK}}\right)_B\right]}{\ln(\alpha_{LK/HK})_{AVG}}$$ (3-9)

Where:

N_M = minimum number of theoretical stages in the column at total reflux

x_{LKD} = mole fraction light key in the distillate

x_{HKD} = mole fraction heavy key in the distillate

x_{LKB} = mole fraction light key in the bottoms

x_{HKB} = mole fraction heavy key in the bottoms

$(\alpha_{LK/HK})_{AVG}$ = geometric average of the relative volatility of light key to heavy key at the distillate, feed, and bottoms locations = $(\alpha_D\,\alpha_F\,\alpha_B)^{1/3}$. For the case where constant relative volatility is assumed this can be taken as the relative volatility at the feed condition.

The reason that this method gives Nm + 1 is that the reboiler is considered to be a theoretical stage, and it is outside the column.

Relationship of Theoretical Stages to Reflux Ratio

With values for the minimum reflux and minimum stages at total reflux in hand, the Gilliland correlation [12] charts actual reflux against actual number of stages. Several researchers have modeled the correlation numerically. The Molkanov Equation is a good one [22], which is:

$$Y = 1 - \exp\left(\frac{(1 + 54.4X)\,(X - 1)}{(11 + 117.2X)\,\sqrt{X}}\right) \qquad (3\text{-}10)$$

Where:

$$Y = \frac{(N - N_{min})}{(N + 1)}$$

$$X = \frac{(R - R_{min})}{(R + 1)}$$

The Gilliland correlation may be used when the following restrictions are satisfied. It is conservative for feeds with low values of q, and can result in erroneous results when there is a large difference in tray requirements above and below the feed [7].

- Number of components: 2 through 11
- Feed quality, q: 0.28 and 1.42
- Pressure: vacuum to 400 bar (600 psig)
- Relative volatility, α : 1.11 and 4.05
- Minimum reflux, R_{min}: 0.53 and 9.09
- Minimum number of stages, N_{min}: 3.4 to 60.3

Feed Location

The Kirkbride equation estimates the location of the feed tray [7]:

$$\frac{m}{p} = \left\{ \left(\frac{B}{D}\right) \left(\frac{x_{HK}}{x_{LK}}\right)_F \left[\frac{(x_{LK})_B}{(X_{HK})_D}\right]^2 \right\}^{0.206} \qquad (3\text{-}11)$$

Akashah et al. modified the equation with a correction factor such that [7]:

$$m = p \ (Kirkbride \ calculation) - 0.5 \log(N) \qquad (3\text{-}12)$$

Where:

$m =$ number of theoretical stages above the feed, including any partial condenser

$p =$ number of theoretical stages below the feed, including the reboiler

$m + p = N$

$N =$ total number of theoretical stages

Reflux-to-Feed Ratio

Heretofore, the reflux ratio has been defined as reflux/distillate, L/D. Another very useful molar ratio is reflux/feed, L/F. In binary systems, L/F for all practical purposes is unchanging for wide differences in feed composition, so long as the following hold:

1. The distillate and bottoms compositions, but not necessarily the quantities, are held constant.
2. The feed tray is kept matched in composition to the feed (which means the feed tray moves with changes in feed composition).

The reader can verify the above using the Underwood equations and the tower material balance. I once calculated a case where a large feed change would change L/D by 46%, whereas L/F changed only 1%. Several investigators report that the stability of L/F is well proven in the field. L/F is a good factor to use in predicting the effect of feed changes for design and in an operating plant.

Actual Trays

After actual theoretical trays are determined (see Actual Reflux and Actual Theoretical stages) one needs to estimate the actual physical number of trays required in the distillation column. This is usually done by dividing the actual theoretical trays by the overall average fractional tray efficiency. Then a few extra trays are normally added for offload conditions, such as a change in feed composition. Experience in a given service is the best guide for extra trays.

Graphical Methods – The McCabe-Thiele Diagram

Count the number of theoretical trays in a column using a McCabe-Thiele Diagram. Rather than plotting it on graph paper with pencil, use Excel to create the chart. All of the calculations and graphing can be performed with Excel using the procedures described earlier in this chapter plus additional methods given in this section; see Table 3-2.

Table 3-2
McCabe-Thiele diagram, graphical elements

Graphical Element	How to Construct the Element
Reference Line	This is a straight line with coordinates 0,0 to 1,1. Coordinates are given in terms of mole fraction liquid and mole fraction vapor.
Vapor-Liquid Equilibrium Curve	This is explained in the sections on Vapor-Liquid Equilibrium and Activity Coefficients
Operating Line – Feed	As explained in the Minimum Reflux Multicomponent section, the thermal condition of the Feed to the column is expressed by q $$q = \frac{(L_S - L_R)}{F}$$ The Feed Operating Line is a straight line. The slope is $$Slope_F = \frac{q}{(1 - q)}$$ If the Feed is at its Bubble Point temperature, $q = 1$, and the slope equation does not work. For mathematical purposes, use a very high value such as 10^6 in this case. The line crosses the Reference Line at the feed concentration, regardless of whether the feed is liquid, vapor, or partially liquid. Therefore, the intercept is $Intercept_F = x_F - Slope_F\, x_F$
Operating Line – Rectification Section	A straight line. The slope in terms of the Reflux Ratio, R, is $$Slope_R = \frac{1}{\left(1 + \frac{1}{R}\right)}$$ The line crosses the Reference Line at the distillate concentration, x_D. Therefore, the intercept is $Intercept_R = x_D - Slope_R\, x_D$
Operating Line – Stripping Section	A straight line. The slope that crosses the Reference Line at the bottoms concentration, x_B, and intersects the Feed Line at the same point that the Rectification Operating Line intersects the Feed Line. Let $A = (Intercept_F - Intercept_R)$ $B = (Slope_R - Slope_F)$ The lines intersect at $x = \frac{A}{B}$ and $y = Slope_R\, x + Intercept_R$ Therefore, the slope is $$Slope_S = \frac{(y - y_B)}{(x - x_B)}, \text{ and}$$ $Intercept_S = x_B - Slope_S\, x_B$
Step-by-Step Trays	Traditionally marked off with pencil and paper, the example problem shows how to construct the steps using Excel.

The McCabe-Thiele Diagram consists of the vapor-liquid equilibrium curve, operating lines for the stripping and rectifying sections, a feed line, and the characteristic steps that correspond to theoretical trays. The chart created in Excel is very accurate, even if the steps are close together to the point of being visually indistinguishable. Since the actual trays are inefficient, resulting in more trays being required than are calculated, the accuracy of the McCabe-Thiele Diagram is sufficient for design work for binary systems for which good VLE data is available.

The method assumes constant molal (equimolal) overflow, meaning:

- The molar heats of vaporization of the components are roughly similar.
- Heat effects are negligible due to losses from the column, heats of mixing or reaction, etc.
- For every mole of chemical that vaporizes, a mole condenses.

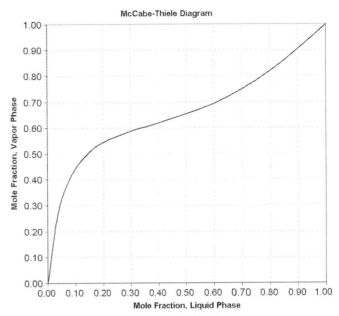

Figure 3-15. VLE chart.

Example

Construct a McCabe-Thiele diagram for the ethanol-water system. The feed composition is 40 mole% ethanol. The distillate should be 89 mole% ethanol and the bottoms 98% water. Assume the feed is at its bubble point temperature and that the column operates at 1 atmosphere pressure. Use the Margules model for activity coefficients.

1. Start with the example problem in the Activity Coefficients section to construct a table with VLE data for the ethanol-water system. Highlight (i.e., Select) the X-Y data range with the VLE data. Build a XY (Scatter) Chart in Excel; choose the subtype, "Scatter with data points connected by smoothed lines without markers." After the chart appears, format both the X and Y axes, scaling them from 0.1 to 1.0 and establishing major and minor units at 0.1

and 0.05 respectively. Resize the chart so it appears square (see Figure 3-15).

2. Add the Reference Line and Operating Lines to the spreadsheet, using the formulae given in Table 3-1. To accomplish this, data cells are required for the molar compositions of the Feed, Distillate, and Bottoms; the Reflux Ratio; and the Feed Quality (q) – see Figure 3-16.

Key formulae are:
F34 = IF(B31=1,10^6,B31/(1−B31))
F35 = E27−F34*E27
H34 = 1/(1+1/B30)
H35 = G27−H34*G27
G38 = (F35−H35)/(H34−F34)
H38 = H34*G38+H35
C38 = E38 = G38
D38 = F38 = H38
D34 = (D38−D37)/(C38−C37)
D35 = C37−D34*C37

	A	B	C	D	E	F	G	H
26	Molar Composition		Bottoms		Feed		Distillate	
27	Ethyl Alcohol		0.02		0.40		0.88	
28	Water		0.98		0.60		0.12	
29								
30	Reflux Ratio	8						
31	Feed, q	0.999						
32								
33	Reference Line		Stripping		Feed		Rectification	
34	Slope	1.00	Slope	1.14	Slope	999.00	Slope	0.89
35	Intercept	0.00	Intercept	(0.00)	Intercept	(399.20)	Intercept	0.10
36	X	Y	X	Y	X	Y	X	Y
37	0.00	0.00	0.02	0.02	0.40	0.40	0.88	0.88
38	1.00	1.00	0.40	0.45	0.40	0.45	0.40	0.45

Figure 3-16. Reference and operating lines.

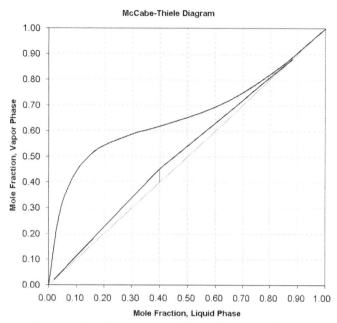

Figure 3-17. Reference and Operating Lines charted.

3. Bring the Reference and Operating Lines into the Scatter Chart, as shown in Figure 3-17.
4. Beginning at the Bottoms concentration, construct the steps. Each step has two parts: a vertical segment to the VLE line and a horizontal segment to the Operating Line. The steps start at the coordinates of the bottoms liquid mole fraction, x_B. The first segment is a vertical line to the VLE curve, the trick being determining the y value, which requires the use of Goal Seek as shown previously. The second segment is a horizontal line that intersects either the Stripping or Rectification Operating Line; this one is easy to compute. Repeat until the desired distillate concentration, x_D, is equaled or exceeded.

In Excel, create a table with the rows representing theoretical trays, beginning with the reboiler and building up to the distillate. There are two x-y values in each row, representing the beginning of the vertical and horizontal segments of the step. Then put in the formulae for the VLE curve as shown in the Activity Coefficients section of this chapter.

Reference x_B to obtain the x and y values for Segment 1 of the reboiler. Segment 2 shares the same x value with Segment 1. Enter the formula for Segment 2's y value as in the VLE table.

In the row for Step 1, use an IF statement to determine whether the horizontal Segment 2 from the reboiler intersects the Stripping or the Rectification Operating Line, recognizing that Segment 2's destination is the same point as the next step's Segment 1 origin (Figure 3-18).

Key formulae are:

M34 = C27 (the concentration of ethyl alcohol in the bottoms)

N34 = O34 = M34

P34 = O34*T34*R34/760 (same as the VLE table)

R through X are the same as the VLE table:

M35 = IF((N35−$D$35)/
$D$34<$E$38,(N35−$D$35)/
$D$34,(N35−$H$35)/$H$34)

This formula decides which Operating Line the X value intersects.

Thus, each line is the same and copied down the page except for M34 which starts off from the bottoms composition.

Next, the temperature at each step is determined using Goal Seek. This can be automated in a macro. One way of doing this is to create a cell that sums the differences (X34+X35+...) in the table. Whenever a cell is changed on the worksheet, a macro looks at the sum and if it isn't zero (within a tolerance) it executes the code to update all of the temperatures in the table. Here's a listing of that macro:

```
Private Sub Worksheet_Change(ByVal Target As
Range)
' Recalculates the Bubble Point temperature
' as needed for the McCabe-Thiele Chart
Dim Temperature As Range
Dim Difference As Range
Set Temperature = Range("Step_Temp") ' name of
cell Q33
Set Difference = Range("Step_Diff") ' name of cell
X33
Tolerance = 0.001
```

	L	M	N	O	P	Q	R	S	T	U	V	W	X
32		Segment 1 Origin		Segment 2 Origin			VP, mm Hg		Margules		Partial Pressure		
33	Step	X	Y	X	Y	Temp, deg C	Ethyl Alcohol	Water	Y_1	Y_2	Ethyl Alcohol	Water	Difference
34	Reboiler	0.02	0.02	0.02	0.00		11.75	4.59	4.516	1.001	1.06	4.50	(754.44)
35	1	0.00	0.00	0.00	0.00		11.75	4.59	4.877	1.000	0.21	4.57	(755.22)
36	2												
37	3												

Figure 3-18. First theoretical tray.

```
If Abs(Range("Step_Tot").Value) > Tolerance Then       Exit Do
i = 0                                                  End If
Do                                                     Loop
i = i + 1                                              End If
' Only executes if the row has data                    End Sub
If Range("Step_Temp").Offset(i, -1).Value <> ""
Then
Difference.Offset(i, 0).GoalSeek goal:=0, _
changingcell:=Temperature.Offset(i, 0)
Else
```

The result is shown in Figure 3-19 below:

	L	M	N	O	P	Q	R	S	T	U	V	W	X
32		Segment 1 Origin		Segment 2 Origin			VP, mm Hg		Margules		Partial Pressure		
33	Step	X	Y	X	Y	Temp, deg C	Ethyl Alcohol	Water	γ_1	γ_2	Ethyl Alcohol	Water	Difference
34	Reboiler	0.02	0.02	0.02	0.17	95.4181	1,443.10	641.88	4.516	1.001	130.35	629.65	(0.00)
35	1	0.15	0.17	0.15	0.51	84.2025	955.71	419.26	2.645	1.052	386.42	373.58	0.00
36	2	0.46	0.51	0.46	0.64	80.2304	819.98	357.92	1.281	1.426	485.35	274.65	0.00
37	3	0.61	0.64	0.61	0.70	79.3697	792.79	345.68	1.100	1.696	530.44	229.56	0.00
38	4	0.68	0.70	0.68	0.73	79.0448	782.73	341.15	1.055	1.825	557.81	202.19	0.00
39	5	0.72	0.73	0.72	0.76	78.8755	777.52	338.80	1.037	1.901	576.91	183.09	0.00
40	6	0.74	0.76	0.74	0.78	78.7712	774.33	337.37	1.027	1.952	591.42	168.58	0.00
41	7	0.77	0.78	0.77	0.79	78.7001	772.16	336.39	1.020	1.989	603.08	156.92	0.00
42	8	0.78	0.79	0.78	0.81	78.6479	770.58	335.68	1.016	2.017	612.85	147.15	0.00
43	9	0.80	0.81	0.80	0.82	78.6077	769.35	335.13	1.013	2.040	621.31	138.69	0.00
44	10	0.81	0.82	0.81	0.83	78.5754	768.37	334.69	1.011	2.060	628.82	131.18	0.00
45	11	0.82	0.83	0.82	0.84	78.5485	767.56	334.32	1.009	2.076	635.64	124.36	0.00
46	12	0.83	0.84	0.83	0.84	78.5257	766.86	334.01	1.007	2.090	641.96	118.04	0.00
47	13	0.84	0.84	0.84	0.85	78.5057	766.26	333.74	1.006	2.103	647.91	112.09	0.00
48	14	0.85	0.85	0.85	0.86	78.4881	765.73	333.50	1.005	2.114	653.59	106.41	0.00
49	15	0.86	0.86	0.86	0.87	78.4721	765.24	333.28	1.004	2.125	659.09	100.91	0.00
50	16	0.87	0.87	0.87	0.87	78.4575	764.80	333.08	1.004	2.134	664.48	95.52	0.00
51	17	0.87	0.87	0.87	0.88	78.444	764.39	332.90	1.003	2.143	669.82	90.18	0.00
52	18	0.88	0.88	0.88	0.89	78.4312	764.01	332.73	1.003	2.152	675.17	84.83	0.00

Figure 3-19. Table with results for all trays.

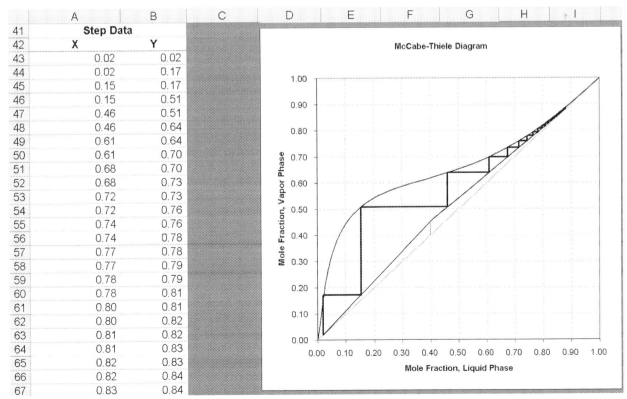

	A	B
41	Step Data	
42	X	Y
43	0.02	0.02
44	0.02	0.17
45	0.15	0.17
46	0.15	0.51
47	0.46	0.51
48	0.46	0.64
49	0.61	0.64
50	0.61	0.70
51	0.68	0.70
52	0.68	0.73
53	0.72	0.73
54	0.72	0.76
55	0.74	0.76
56	0.74	0.78
57	0.77	0.78
58	0.77	0.79
59	0.78	0.79
60	0.78	0.81
61	0.80	0.81
62	0.80	0.82
63	0.81	0.82
64	0.81	0.83
65	0.82	0.83
66	0.82	0.84
67	0.83	0.84

Figure 3-20. Completed McCabe-Thiele diagram.

5. Before completing the McCabe-Thiele chart, put the step data into a single list of segments, rather than the two lists shown before. Then, add the x-y data to the Scatter Chart as a new Series, but uncheck the "Smoothed Line" option in the "Format Data Series" dialog box. See Figure 3-20.

Plotting With Murphree Tray Efficiency

The Murphree Tray Efficiency is equal to the actual change in concentration leaving a tray divided by the theoretical change in concentration. In the example earlier, the Y value in column P is adjusted according to an assumed Murphree Tray Efficiency, E_m. Thus, for $E_m = 0.6$, the Y value in cell P35 would be $0.6 (0.51 - 0.17) + 0.17$, or 0.34.

Figure 3-21 is the entire McCabe-Thiele diagram for the example, with Murphree Tray Efficiency = 0.7. To plot the dotted Murphree Efficiency line, add the x-y data representing the points of each stage (cells O34:P64) as a new data series, and format it as a dotted and smoothed line. In this example (Figure 3-21), 20 theoretical trays are needed for the separation, but 30 actual trays at $E_m = 0.7$. However, 20 / 0.7 = 29; the overall column efficiency is not exactly the same as the tray efficiency.

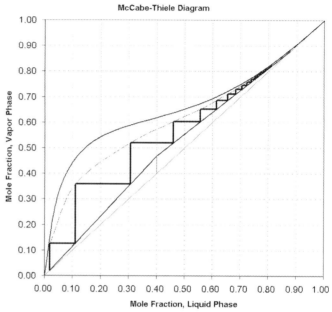

Figure 3-21. McCabe-Thiele diagram with Murphree Efficiency.

Tray Efficiency Rules of Thumb

Kolmetz et. al. [17] list these "rules of thumb" that apply to stage efficiency:

- Increased pressure increases tray efficiency
- Decreased pressure increases packing efficiency
- Increased viscosity decreases tray and packing efficiency
- Increased relative volatility decreases tray efficiency

They report that, generally, high capacity trays and packings exhibit lower efficiency. The reason is that the contact time between the liquid and vapor is reduced at higher throughput.

Using the McCabe-Thiele Plot

Detect pinched regions. Pinching occurs when the operating lines approach the equilibrium curve. Minimum reflux can draw the operating lines to the equilibrium curve. A feed too high in the column (not at the intersection of the q-line and the equilibrium curve for a binary distillation) requires extending the stripping section operating line beyond the optimum q-line intersection point. This gives a pinch at the mislocated feed point. A similar type of pinch can occur with a feed too low in the column. A "tangent pinch" can occur when an equilibrium curve doesn't have the smooth concave downward shape that we picture in a "well behaved" system. Such a curve might dip to almost touch an operating line. This "tangent pinch" probably will not occur at the feed point for binary systems like the other pinch types.

Identify excessive traffic. Over-refluxing is indicated by too large a gap between both operating lines and the equilibrium curve.

Find the chances for heat optimization. When only one operating line has a wide gap, intermediate heating or cooling may be helpful. A large gap for the bottom section could indicate potential for feed preheat or an inter-reboiler. A large gap for the top section could indicate that a precooler or inter-condenser would be attractive.

Aid optimization. The graphics can show the effects of parameter changes such as compositions, feed thermal condition, and multiple feed or draws.

Identify mislocated feed. For binary distillation, the feed point should be where the q-line intersects the equilibrium curve. For multicomponent distillation this may or may not be the case.

Column Internals – Trays and Packing

This section compares characteristics of trays and packings. For new towers the selection of a specific tray or packing type is dependent on the process requirements (including perturbations and expectations for future changes), the company's experience with particular types of internals, and economics.

Bravo generalized the selection of tower internals according to operating pressure, modified by factors such as fouling [5]. See Tables 3-3 and 3-4.

Table 3-3
Generalized recommendations for distillation tower internals [5]

Operating Pressure	Recommended Internals	Design Issues
Vacuum (<1 atm, absolute pressure)	Structured packings followed by random packing; consider trays if fouling cannot be mitigated or if ineffective packing wetting may be a problem	— Liquid distributor design and installation — High velocity vapor inlets and reboilers returns; vapor distribution — Entrainment — Distributor and collector open area and pressure drop — Flashing feed inlet design — Packing support and hold downs open area — Packing size criteria (pressure drop as well as capacity)
Medium Pressure (1 to 4 atm)	Wide range of trays or packing may be used, with primary considerations being fouling, corrosion, foaming, and cost.	— Pressure drop
High Pressure (above 4 atm)	High performance trays unless pressure drop is a prime concern (in cryogenic separations, for example) where structured packing is attractive	— Liquid feed distributor for high liquid load applications — Pressure drop on tray and its effect on downcomer backup — Efficient liquid distribution on tray — Prevention of bridging the downcomer inlet — Downcomer inlet geometry for low surface tension systems — Potential compromise between reduced efficiency and additional stages offered by increased capacity

Table 3-4
Some characteristics of trays and packings

Characteristic	Valve Trays	Sieve Trays	Bubble-Cap Trays	Packings
General application: Trayed columns generally cheaper than packed towers		Most commonly used tray type	Used infrequently, usually only for extreme turndown [1]	Usually preferred for column diameters less than 1 m (3 ft)
Pressure drop	Typically about 3 in w.c. per tray			

(Continued)

Table 3-4
Some characteristics of trays and packings—cont'd

Characteristic	Valve Trays	Sieve Trays	Bubble-Cap Trays	Packings
Corrosion resistance				Available in a wide variety of corrosion-resistant materials including ceramics, plastics, and glass
Fouling resistance	Moveable valves are less resistant to fouling	Better for fouling service		Structured favored over random packing. Use trough type distributor, not pan
Vacuum systems				Structured packing often specified for vacuum service due to its low pressure drop
Operating range factor when tray is designed to maximize the range	Floating valves have a range of about three with lower efficiency than sieve trays; small size fixed valves have greater maximum range, but not as good turndown [1]	Two or more if larger tray spacing and higher pressure drop at peak rates are acceptable [1]	Small-size bubble caps can give a range greater than five if designed to prevent liquid/vapor bypass at extremes turndown [1]	Generally between sieve trays and valve trays. Design of internals for a stable operating range of more than two or three is much more difficult for packings than for trays [1]
Cost	Lower cost than packed columns because less surface is needed for trays than for packings, and trays require far lower cost internals than packings			Up to two to three times higher cost than trays

Column Tray Spacing

Mukherjee [23] reported that tray spacing in columns in the chemical process industries generally fall between 450 mm and 900 mm (between 18 in. and 36 in.). See Table 3-5.

For bubble cap trays with downcomers, the tray spacing should be at least twice the height of the liquid downcomer. Sieve tray (with downcomer) spacing can usually be about 150 mm (6 in.) less than for a corresponding bubble tray; for perforated plate trays without downcomers Ludwig recommends that the spacing should be twice the maximum design height of the liquid/froth mixture on the tray [7].

Table 3-5
Guidelines for selection of tray spacing [23]

Condition	Tray Spacing	Comments
Column diameter >3,000 mm (10 ft)	>600 mm (>24 in.)	Large tray spacing required because tray support beams restrict access
Column diameter from 1,200 mm to 3,000 mm (4 ft to 10 ft)	600 mm (24 in.)	This spacing is sufficiently wide to allow a worker to freely crawl between trays
Column diameter from 750 mm to 1,200 mm (2.5 ft to 4 ft)	450 mm (18 in.)	Crawling between trays is seldom required
Fouling and corrosive service	>600 mm (>24 in.)	Frequent maintenance expected
Systems with a high foaming tendency	At least 450 mm (18 in.) but preferably 600 mm (24 in.) or higher	Required to avoid premature flooding
Columns operating in spray regime	At least 450 mm (18 in.) but preferably 600 mm (24 in.) or higher	Required to avoid excessive entrainment
Columns operating in froth regime	<450 mm (<18 in.)	Lower tray spacing restricts allowable vapor velocity, thereby promoting froth-regime operation

Tower Diameter – Souders-Brown Correlation

For checking designs, roughly relate tower diameter to reboiler duty by using the values in Table 3-6 [4].

Table 3-6
Rough relationship of tower diameter to reboiler duty

Situation	Reboiler Duty, million Btu/h
Pressure distillation	$0.5\ D^2$
Atmospheric pressure distillation	$0.3\ D^2$
Vacuum distillation	$0.15\ D^2$
$D =$ Tower diameter, ft	

Table 3-7
Coefficients for the Souders-Brown correlation

Tray Spacing		m	b
mm	in.		
25	10	46.1	14.7
30	12	74.1	53.2
38	15	93.3	133.5
46	18	106.6	197.2
51	20	112.6	229.1
61	24	118.8	284
76	30	121.6	334
91	36	124.3	359.8

For bubble cap trays, Souders-Brown correlated the maximum allowable mass velocity in columns with vapor and liquid density and with liquid surface tension [28]. The result is a set of curves representing tray spacing from 25 cm (10 inches) to 90 cm (36 inches).

The Souders-Brown correlation can also be used as a first approximation for sieve trays, however it is conservative since it is based on no liquid entrainment occurring between trays. Other approaches are proprietary or require knowledge or guesses of the amount of liquid entrainment [7].

These curves are expressed with the formula:

$$C = m \ln(\sigma) + b \qquad (3\text{-}13)$$

Where:

C is the "C" Factor for fractional distillation. For absorbers, multiply C by 0.55. For fractionating section of absorber oil stripper, multiply C by 0.80. For petroleum column, multiply C by 0.95. For stabilizer or stripper, multiply C by 1.15.

$\sigma =$ liquid surface tension, dynes/cm

m and b are coefficients from Table 3-7

Calculate the maximum liquid rate with:

$$W_{max} = C\left[\rho_v\left(\rho_L - \rho_v\right)\right]^{1/2} \qquad (3\text{-}14)$$

$W_{max} =$ maximum allowable mass velocity using bubble cap trays, lb/ft^2-h

$\rho_v =$ vapor density, lb/ft^3

$\rho_L =$ liquid density, lb/ft^3

From Ludwig [7]: Calculate the "C" factor at the top and bottom of the column to evaluate the point of maximum required diameter. Apply a safety factor of 1.10 to 1.25 (divide W by 1.1 or 1.25) if operating variances are anticipated. Exercising judgment and caution, increase W by 5% to 15% for columns operating in the 5 psig to 250 psig range. Round the shell diameter to nearest standard diameter (usually at 6-inch intervals starting at 24 inches).

$$D_{min} = \sqrt{\frac{4\ L}{\pi\ W_{max}}} \qquad (3\text{-}15)$$

$D_{min} =$ minimum tower diameter (round up to the nearest standard size), ft

$L =$ liquid rate, lb/h

Pressure Drop – Trays

Two factors are summed to estimate the pressure drop across a tray: 1) the *hydraulic* head of foamy liquid on top of the tray (which the vapor must pass through), and 2) the *dry* pressure drop of vapor flowing through the tray's holes. For sieve and valve trays, the two components should be approximately equal.

$$\Delta P_{total} = \Delta P_{hyd} + \Delta P_{dry}$$

$$\Delta P_{hyd} \approx \Delta P_{dry}$$

For valve and sieve trays, Lieberman derived this empirical formula, which checks for the efficient operation of the column. It can be used as a rule of thumb to back-calculate the desired pressure drop range. Use Table 3-8 to interpret the result [18].

$$K = \frac{28 \, \Delta P}{(NT)\,(TS)\,(SG)} \qquad (3\text{-}16)$$

Where:

ΔP = pressure drop across a section of trays, psi
NT = number of trays in the section
TS = tray spacing, inches

Table 3-8
Interpretation of "K" in equation 3-15

K	Interpretation
0.18 to 0.25	Tray operation is close to its best efficiency point
0.35 to 0.40	Tray suffering from entrainment
>= 0.50	Tray is in fully developed flood; opening a vent on the overhead vapor line will blow out liquid with the vapor
0.10 to 0.12	Tray deck is suffering from low tray efficiency due to liquid leakage through the sieve holes
0.00	There is no liquid level on the tray

SG = specific gravity of clear liquid at average flowing temperature

Control Schemes

Fractionators produce two results: 1) stream splitting, with so much mass going out one "end" and all other feed mass going out the other, and 2) component segregation toward one or the other of the product streams, characterized by the Fenske ratio:

$$\frac{moles \ light \ in \ overhead}{moles \ light \ in \ bottoms} \times \frac{moles \ heavy \ in \ bottom}{moles \ heavy \ in \ overhead}$$

The first result is achieved by controlling the product flow from one end of the fractionator. The second result is accomplished by controlling the heat load in the fractionator. Nearly all control schemes permit control of these two "handles." The selection of a particular scheme is influenced by secondary factors, such as sidestreams, expected process variability, and configuration of the components (e.g., receivers, pumps) in the system. It's important to provide smooth control response to limit surges; upset conditions can quickly undo a separation.

If the "end" of interest is the distillate stream, the purity of the distillate is inversely proportional to the flow rate – as the flow decreases its purity increases. At the same time the purity of the bottoms stream decreases.

Heat input determines the vapor rate through the column. As heat input increases, the separation of the light and heavy components also usually increases. If the distillate and bottoms rates are held constant while increasing heat input, more condensate is returned to the column. Thus, the reflux ratio increases, which usually results in a better separation.

In summary, heat input to a column determines the degree of separation that can be achieved, while the flow rates from the column determine how the separation is allocated between the products.

Riggs provides an excellent overview of distillation control, available online [24], with important information about column disturbances and response times. A good rule of thumb is to multiply the liquid holdup of the tower (including the overhead accumulator and reboiler) in mass units by the reflux ratio and divide this by the mass flowrate of the column feed. This will establish the minimum amount of time required for the tower to reestablish steady state after an operational change is made [30].

Modern large industrial columns are likely to employ sophisticated computerized controls, often using Model-Predictive Control (MPC) algorithms that proactively respond to disturbances rather than reacting after the effect from the disturbance is evident. MPC is outside the scope of this book. However, most fractionators will incorporate features described under the following headings. A "typical" control scheme is depicted in Figure 3-23.

Flow

The feed flow is often not controlled but is rather on level control from another column or vessel. The liquid product flows (distillate and bottoms) are often on level rather than flow control. Top vapor product is, however, usually on pressure control. The reflux is frequently on flow-ration control (FRC), but also may be on column temperature-ratio control (TRC) or accumulator level.

Column Pressure

Pressure fluctuations make column performance control difficult, and reduce unit performance. Pressure directly affects the relative volatility of key components, so changes in pressure can significantly affect product compositions. A variety of approaches may be used to control column pressure. The amount of material in the vapor phase of the overhead stream may be directly controlled, or the rate of condensation can be controlled to indirectly affect column pressure.

Sloley reviewed the major process factors involved in selecting pressure control schemes, and summarized nineteen types of pressure control used for vacuum and pressure systems. For each type, he provided a loop diagram with descriptions of the method, advantages, disadvantages, variants, and configuration notes [25].

There are three major groups of pressure control scheme:

1. Vapor product is *always* present
2. Vapor product is either absent or present at steady-state conditions, with negative vapor flow rate transients possible
3. A total condenser eliminates vapor product at steady-state; negative vapor flow rate transients possible

The fastest-responding control scheme adjusts the amount of vapor in the column overheads. This can be done by venting vapor from the accumulator, or by injecting inert gas into the accumulator. The rate of condensation can be controlled by changing the heat flux to the condenser (flow rate or temperature of the cooling medium), or by changing the heat transfer area of the condenser (by partially flooding the tubes which reduces the area available for condensation).

Level in the Overhead Receiver

For a total condenser, the accumulator level is typically set by varying distillate draw. For a partial condenser, it can be controlled with a condenser hot gas bypass.

Level in the Bottom of the Tower

The column bottom level is sometimes controlled by bottoms draw. Varying reboiler heating medium is another

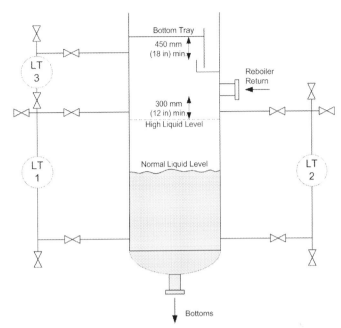

Figure 3-22. Recommended good practice for level instrumentation to ensure the reboiler's return is not submerged. (after [15]).

possibility. For some cases, bottoms draw level control works better and, for others, heating medium level control.

The reboiler return inlet should always enter the column above the liquid level. Due to varying levels of froth and foam, conventional delta P level sensors can be fooled, resulting in the liquid level climbing higher than desired in the column. Liquid levels rising above the reboiler's return inlet is one of the most common causes of damage to trays and packing.

Figure 3-22 illustrates Kister's recommendation that the column has redundant level elements. For services with a history of level problems, install a separate level sensor (LT-3) that normally indicates no level because it is located in an area that is intended to be vapor space. If the level climbs above the lower tap of LT-3, an alarm (not shown) alerts operators to investigate the column.

Composition Control

Composition can be measured directly with an on-line analyzer, or alternatively by taking samples to a lab for off-line analysis. The response time for direct composition measurements is relatively long, even with on-line instruments; therefore it is best suited to stable fractionations that may drift slowly.

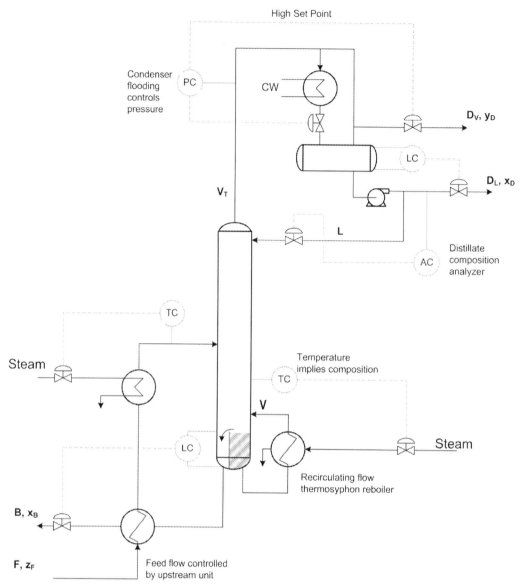

Figure 3-23. Typical distillation column control scheme.

Temperature provides an indirect indication of composition that is often sufficiently accurate and very cost effective. For a binary separation, temperature is an excellent choice if the relative volatility is greater than 2 [24]. The response time is very fast. Measure temperature on a tray near the top of the column, and use it to control the reflux rate (preferred), reflux ratio, or distillate flow rate.

Composition inferred from temperature is pressure sensitive. Therefore, a correction factor for pressure is required for accurate control. For most systems, a simple linear correction applies [24]:

$$T_{PC} = T_{meas} - K_{PR} (P - P_0) \qquad (3-17)$$

Where:

T_{PC} = pressure-compensated temperature to use for feedback control

T_{meas} = measured tray temperature

K_{PR} = pressure correction factor

P, P_0 = operating and reference pressures, respectively

K_{PR} is estimated by applying a steady-state column simulator for two different pressures within the normal operating range, and the following equation:

$$K_{PR} = \frac{T_{P1} - T_{P2}}{P_1 - P_2} \qquad (3-18)$$

Where:

T_{P1}, T_{P2} = tray temperatures predicted by the simulator at pressures P_1 and P_2

The steady-state column model can also be used to find the optimum tray for temperature measurement. Refer to Riggs for a procedure [24].

Constraints

Performance may be limited by the following factors [24]:

1. Maximum reboiler duty. This constraint may be reached by a) an increase in the column pressure resulting in reduced temperature difference between the heat transfer medium and the column bottoms, b) fouled or plugged heat exchanger tubes, c) improperly sized steam trap resulting in condensate backing up into the heat exchanger shell, d) improperly sized control valve on the steam supply to the reboiler, or e) an increase in the column feed rate resulting in boiling demand exceeding the maximum duty of the exchanger.

2. Maximum condenser duty. The constraint may be reached by a) an increase in the ambient air temperature that results in a higher temperature cooling water supply, b) fouled or plugged heat exchanger tubes, c) improperly sized coolant supply control valve, d) an increase in coolant temperature, e) an increase in column feed rate resulting in condensing demand exceeding the maximum duty of the condenser.

3. Flooding. This is caused by excessive vapor and/or liquid traffic in the column.

4. Weeping. Weeping results when the vapor flow rate is too low to keep the liquid from draining through a tray to the tray below; it is primarily applicable to sieve trays.

5. Maximum reboiler temperature. For certain systems, elevated temperature in the reboiler promotes polymerization reactions to the level that excessive fouling in the reboiler results.

Reboilers

The *GPSA Engineering Data Book*₁ has an excellent section on reboilers. The most common types are the following:

- Forced circulation (Figure 3-24)
- Natural circulation (Figure 3-25)
    - Once-through
    - Recirculating
- Vertical thermosyphon (Figure 3-26)
- Horizontal thermosyphon (Figure 3-27)
- Flooded bundle (Kettle) (Figure 3-28)
- Recirculating – Baffled bottom (Figure 3-23)

Thermosyphon reboilers are popular due to their relatively low cost and simple operation. Sloley states that nearly all services are suitable applications for thermosyphon reboilers *except* those that involve extreme fouling, viscous fluids, batch operation of the column or product draws, uncertainty in heating medium temperatures, and unstable processes [26]. Thermosyphon reboilers are horizontal with the process fluid on the shell side, or vertical with the process on the tube (most frequently) or the shell side. Generally, large duty dirty services favor horizontal thermosyphons while small duty clean services

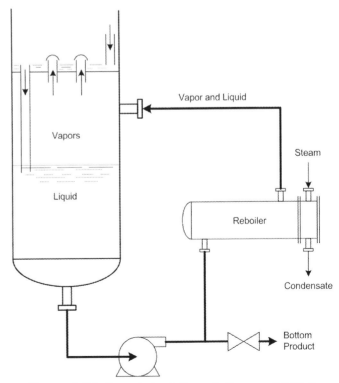

Figure 3-24. Forced circulation reboiler arrangement.

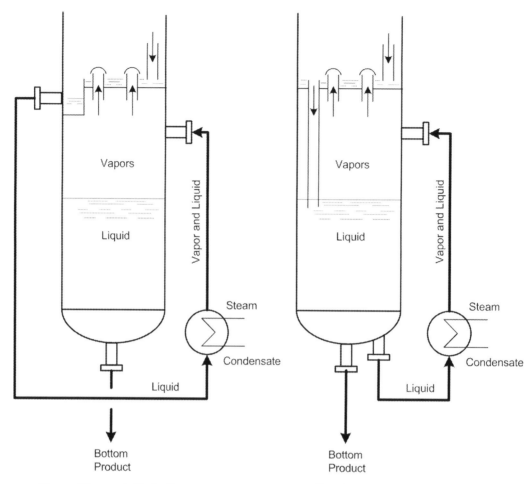

Once-Through Reboiler   Recirculating Reboiler

Figure 3-25. Natural circulation reboiler arrangements.

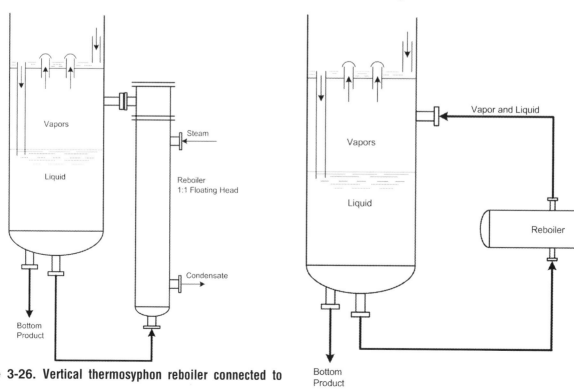

Figure 3-26. Vertical thermosyphon reboiler connected to tower.

Figure 3-27. Horizontal thermosyphon reboiler.

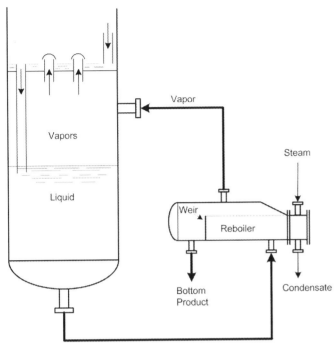

Figure 3-28. Kettle reboiler arrangement.

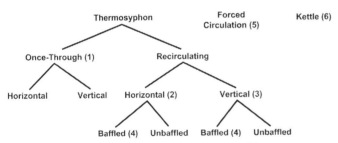

Figure 3-29. Quick selection guide.

Notes for Figure 3-29:
1. Preferable to recirculating where acceptable vaporization rates can be maintained (less than 25–30%). This type is chosen when there is a need to minimize exposure of degradable and/or fouling substances to high temperatures.
2. Used for large duties, dirty processes, and when frequent cleaning is required. The process is usually in the shell side. This type is used in 95% of oil refinery thermosyphon applications.
3. Used for small duties, clean processes, and when only infrequent cleaning is required. Vaporization is usually less than 30%, but less than 15% if the fractionators pressure is below 50 psig. The viscosity of the reboiler feed should be less than 0.5 cP. Install a butterfly valve in the reboiler inlet piping. This type is used in nearly 100% of chemical plant thermosyphon applications (70% of petrochemical).
4. Greater stability than unbaffled.
5. Usually used where piping pressure drop is high and therefore natural circulation is impractical.
6. Very stable and easy to control. Has single-phase flow. Allows low tower skirt height. This type is expensive, however.

Table 3-9
Reboiler comparison [19]

Reboiler Type	Advantages	Disadvantages
Kettle	One theoretical tray Ease of maintenance Vapor disengaging Low skirt height Handles viscosity greater than 0.5 cP Ease of control No limit on vapor load	Extra piping and space High cost Fouls with dirty fluids High residence time in heat zone for degradation tendency of some fluids Low residence time surge section of reboiler
Vertical once-through	One theoretical tray Simple piping and compact Not easily fouled Less cost than kettle	Difficult maintenance High skirt height No control of circulation Moderate controllability
Vertical natural circulation	Good controllability Simple piping and compact Less cost than kettle	No theoretical tray Accumulation of high boiling point components in feed line, i.e., temperature may be slightly higher than tower bottom Too high liquid level above design could cause reboiler to have less capacity Fouls easier Difficult maintenance High skirt height
Horizontal once-through	One theoretical tray Simple piping and compact Not easily fouled Lower skirt height than vertical Less pressure drop than vertical Longer tubes possible Ease of maintenance Less cost than kettle	No control of circulation Moderate controllability High skirt height
Horizontal natural circulation	Ease of maintenance Lower skirt height than vertical Less pressure drop than vertical Longer tubes possible Less cost than kettle	No theoretical tray Extra space and piping as compared to vertical Fouls easier as compared to vertical Accumulation of higher boiling point components in feed line, i.e., temperature may be slightly higher than tower bottom
Forced circulation	One theoretical tray Handles high viscous solids-containing liquids Circulation controlled Higher transfer coefficient	Highest cost with additional piping and pumps Higher operating cost Requires additional plant area

favor vertical units. In petroleum refining about 95% of thermosyphon reboilers are horizontal; 70% of thermosyphons in petrochemical service are vertical and nearly 100% of units in chemical plants are vertical [26].

Figure 3-29 provides an overview of reboiler selection choices. The accompanying notes provide information for a quick or "first cut" estimate of the appropriate type for a given application. Table 3-9 and Table 3-10 provide additional, more detailed, selection data. Table 3-9 gives advantages and disadvantages for all the major reboiler types. Table 3-10 is limited to thermosyphon types.

For reboilers, especially thermosyphon types, "the devil is in the details." The information presented herein is intended for preliminary work. Final design is performed by experienced engineers using detailed design techniques.

Table 3-10
Thermosyphon selection criteria [26]

Factor	Vertical	Vertical	Horizontal	Horizontal
Process side	Tube side	Tube side	Shell side	Shell side
Process flow	Once-through	Circulating	Once-through	Circulating
Heat transfer coefficient	High	High	Moderately high	Moderately high
Residence time in heated zone	Low	Medium	Low	Medium
Delta T required	High	High	Medium	Medium
Design data	Available	Available	Some available	Some available
Capital cost (total)	Low	Low	Medium	Medium
Plot requirements	Small	Small	Large	Large
Piping needed	Low cost	Low cost	High cost	High cost
Size possible (1)	Small	Small	Large	Large
Shells	3 maximum	3 maximum	As needed	As needed
Skirt height	High	High	Lower	Lower
Distillation stages	One	Less than one	One	Less than one
Maintenance	Difficult	Difficult	Easy	Easy
Circulation control	None	Possible	None	Possible
Controllability	Moderate	Moderate	Moderate	Moderate
Fouling suitability (process)	Good	Moderate	Good	Moderate
Vaporization range, minimum	5%+	10%+	10%+	15%+
Vaporization range (2)	25%	25%	25%	25%
Vaporization range, maximum (3)	35%	35%	35%	35%

1. Small = less than 8,000 ft^2/shell; large = more than 8,000 ft^2/shell
2. Normal upper limit of standard design range; design for vaporization above this level should be handled with caution
3. Maximum if field data are available

Packed Columns

Columns with random or structured packing, instead of conventional trays, are advantageous for many applications. They are commonly used for scrubbers and absorbers (see Chapter 4), and are increasingly popular for distillations. The towers are easy to specify on a turnkey basis (the vendors take care of designing the internal components), and should performance requirements change it is usually straightforward to substitute a different packing.

Unlike trays, the performance of packings is determined empirically with much of the data being proprietary to the packing manufacturers. Final design should be performed by experts, but the general engineer can provide preliminary designs using information found in open literature.

For studies using random dumped packings, one needs to estimate the column diameter and height. Diameter is determined from generalized pressure drop correlations. Column height consists of the space occupied by internals, such as liquid distributors and support plates, and the height of the packing.

Evaluate packings after performing the tray-to-tray calculations described earlier in this chapter. The key parameters from the earlier calculations are the liquid and gas loadings (L and V), liquid and gas densities (based on operating pressure), liquid viscosity, and number of theoretical trays.

There are three steps in the preliminary evaluation.

1. Select a packing

   Use the liquid and gas loadings and properties to calculate the flow parameter (Equation 3-20). Then, Figure 3-31 can be used to indicate whether to consider trays or structured packing. Consult Table 3-12 to help decide which packing to use for the evaluation. Find further information from manufacturers' websites (Table 3-11).

Table 3-11
Global manufacturers and suppliers of trays and packings

Company	Brands	Products
Artisan Industries www.artisanind.com	Dualflo®	Trays for solids-containing and fouling solutions (metallic)
Compagnie de Saint-Gobain Saint-Gobain NorPro www.norpro.saint-gobain.com	Norton™ WavePak™	Saddles (ceramic) Super Saddles (ceramic) Raschig Rings (carbon) Cross-partition rings (ceramic) WavePak™ packing for sulfuric acid towers
FinePac Structures Pvt, Ltd. www.finepacindia.in	Finepac™	Structured Packing (metal) Trays (conventional) Pall Rings and Saddles (metal)
HAT International www.hatltd.com	Alpha™	Structured packing (metal, plastic) Trays (conventional) Standard and High Performance random packings (metal, ceramic, plastic, glass)
Jaeger Products, Inc. www.jaeger.com	Super-Ring® Tri-Packs® Max-Pack®	Structured packing (metal) Trays (conventional) (metal) Rings (plastic) Saddles (plastic, ceramic) Tri-Pack® (plastic) Raschig Super-Ring® (plastic, metal)
Koch Chemical Technology Group LLC Koch-Glitsch www.koch-glitsch.com Koch Knight LLC www.kochknight.com	Intalox® IMTP® Mini-Rings® CMR™ Hy-Pak® Flexiring® Snowflake® Beta Ring™	KG-TOWER® design software (free download) Structured packing (metal) Trays (all types) Random packings (plastic, metal)

Table 3-11
Global manufacturers and suppliers of trays and packings—cont'd

Company	Brands	Products
Lantec Products, Inc. www.lantecp.com	Etapak® Flexeramic® Flexitray® GOODLOE® Q-Pac® Lanpac® Lanpac-XL® HD Q-Pac® Nupac®	High performance random packing (primarily plastic, some metal) Saddles (porcelain, ceramic)
Montz www.montz.de	Montz	Structured packing Mass transfer trays Tower internals
Pall Ring Company www.pallrings.co.uk	astraPAK®	Pall Rings (metal, plastic) astraPAK® (plastic) C-Rings, I-Rings, Raschig Rings
Raschig GmbH www.raschig.de www.raschig.com		Trays (all types) Raschig Super Rings (metal)
Sulzer Ltd. Sulzer Chemtech www.sulzerchemtech.com (purchased Nutter Engineering in 2010)	Rombopak® Mellapak™ Katapak™	SULCOL Sulzer design program for structured and random packings, and trays (free download) Structured packing (metal) Trays (all types) (metal) Nutter Ring (metal) I-Ring (similar to IMTP®) (metal) C-Ring (similar to CMR™) (metal) P-Ring (similar to Pall Ring®) (metal) R-Ring (similar to Raschig Ring) (metal)
Verantis www.verantis.com (acquired Ceilcote Air Pollution Control)	Tellerette®	Tellerettes® (plastic)
Vereinigte Fuellkoerper-Fabriken GmbH & Go. www.vff.de	NetBall® VSP® Novalox® Igel®	Packing Software (free download for hydraulics; fee for full version) Structured Packing (plastic) Pall Rings (metal, plastic, ceramic) Novalox® Saddles (metal, plastic, ceramic) VFF NetBall® (plastic) Berl Saddles (ceramic)
Beijing Zehua Chemical Engineering Co., Ltd Zehua www.zehua-chem.com		Structured Packing (metal, plastic) Random Packing (metal, ceramic, plastic)

Table 3-12
Packing type application [7]

Packing	Application Features	Packing	Application Features
Raschig Rings	Earliest type of packing. Usually cheaper, but less efficient than others. Available in widest variety of materials. Very sound structurally. Usually packed by dumping wet or dry, with larger 4- to 6- inch sizes sometimes hand stacked. Wall thickness varies between manufacturers, also some dimensions; available surface (and packing factor) changes with wall thickness. Produces considerable side thrust on tower. Usually has more internal liquid channeling, and directs more liquid to walls of tower.	Intalox Saddles and other saddle designs	One of the most efficient packings, but more costly. Very little tendency or ability to nest and block areas of bed. Gives fairly uniform bed. Higher flooding limits and lower pressure drop than Raschig rings or Berl saddles; lower HTU values for most common systems. Ceramic saddles easier to break in bed than Raschig rings.
		Metal Intalox	High efficiency, low pressure drop, reportedly good for distillations
Lessing Rings	Similar to Raschig rings. Not much performance data available, but in general slightly better than Raschig rings, pressure drop slightly higher. High side wall thrust.	Tellerette	Available in plastic, lower pressure drop and HTU values, higher flooding limits than Raschig rings or Berl saddles. Very low unit weight, low side thrust.
Pall Rings	Lower pressure drop (less than half) than Raschig rings, also lower HTU (in some systems also lower than Berl saddles), higher flooding limit. Good liquid distribution, high capacity. Considerable side thrust on column wall. Available in metal, plastic, and ceramic.	Sulzer, Flexipac, and similar	High efficiency, generally low pressure drop, well suited for distillation of clean systems, very low HETP.
Cross-Partition Rings	Usually used stacked, and as first layers on support grids for smaller packing above. Pressure drop relatively low, channeling reduced for comparative stacked packings. No side wall thrust.	GOODLOE® Packing; wire mesh packing	Available in metal and plastic, used in large and small towers for distillation, absorption, scrubbing, liquid extraction. High efficiency, low HETP, low pressure drop. Limited data available. Consider for multi-purpose batch distillation systems, systems with many theoretical stages, and corrosive applications.
Spiral Rings	Usually installed as stacked, taking advantage of internal whirl of gas-liquid and offering extra contact surface over Raschig rings, Lessing rings or cross-partition rings. Available in single, double, and triple internal spiral designs. Higher pressure drop. Wide variety of performance data not available.	Grid Tile	Available with plain side and bottom or serrated sides and drip-point bottom. Used stacked only. Also used as support layer for dumped packings. Self supporting, no side thrust. Pressure drop lower than most dumped packings and some stacked, lower than some ¼-inch x 1-inch and ¼-inch x 2-inch wood grids, but greater than larger wood grids. Some HTU values compare with those using 1-inch Raschig rings.
Berl Saddles	More efficient than Raschig rings in most applications, but more costly. Packing nests together and creates "tight" spots in bed which promotes channeling but not as much as Raschig rings. Do not produce as much side thrust, have lower HTU and unit pressure drops with higher flooding point than Raschig rings. Easier to break in bed than Raschig rings.	Wood grids	Very low pressure drop, low efficiency of contact, high HETP or HTU, best used in atmospheric towers of square or rectangular shape. Very low cost.

2. Determine the column diameter
Make a preliminary selection of the packing size, find the packing factor from Table 3-16 [16], or the manufacturer; then estimate the pressure drop at flooding (Equation 3-21). Larger packings provide greater capacity, and less pressure drop, but also a higher HETP than smaller ones. Assign a pressure drop that is 60% to 80% of the flooding pressure drop; compare with Table 3-15 and adjust the pressure drop assignment if necessary. Use the flow parameter, Figure 3-30, and Equation 3-19 to find the superficial vapor velocity. Calculate the column

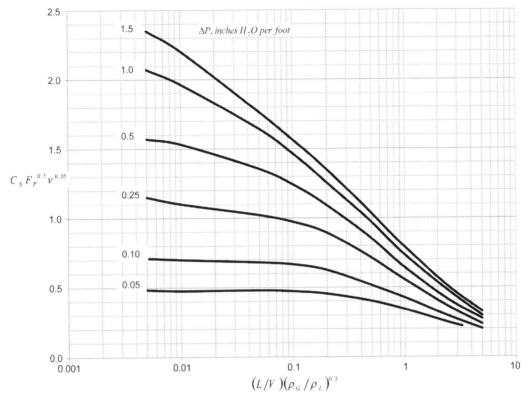

Figure 3-30. Generalized pressure drop correlation for random packings. (after [29]).

Table 3-13
Maximum packing size [10]

Nominal Packing Size (inches)	Minimum Column ID (in)
1	12
1½	18
2	24
3½	42

diameter from the velocity. Compare the recommendations in Table 3-13 and Table 3-14 with the calculated diameter and selected packing size. Minimum wetting rates are 0.1 to 0.2 gpm/ft^2 for structured packing, and 0.5 to 2 gpm/ft^2 for random packings [14]. If necessary, adjust the packing size and repeat this step.

Table 3-15
Design pressure drop [4]

Service	Pressure Drop (in. water per ft. packed depth)
Absorbers and regenerators (non-foaming systems)	0.25 to 0.40
Absorbers and regenerators (moderate foaming systems)	0.15 to 0.25
Fume scrubbers (water absorbent)	0.40 to 0.60
Fume scrubbers (chemical absorbent)	0.25 to 0.40
Atmospheric or pressure fractionators	0.40 to 0.80
Vacuum fractionators	0.15 to 0.40

Table 3-14
Maximum recommended liquid loading for random packings [10]

Nominal Packing Size (inches)	Liquid Rate (gpm/ft^2)
³/₄	25
1	40
1½	55
2	70
3½	125

Table 3-16
Packing Factors (F_P) for random dumped packings

	mm	12	16	19	25	30	38	50	80
	inch	0.5	0.625	0.75	1	1.25	1.5	2	3 or 3.5
Raschig Rings	ceramic	580	380	255	179	125	93	65	37
Raschig Rings	$\frac{1}{32}$" metal	300	170	155	115				
Raschig Rings	$\frac{1}{16}$" metal	410	300	220	144	110	83	57	32
Pall Rings	plastic		96		54		33	24	17
Pall Rings	metal		81		56		40	27	18
Cascade Mini-Rings	ceramic				75	69	55	37	32
Super Cascade Rings	ceramic				75	69	55	37	32
Berl Saddles	ceramic	240		170	110		65	45	
Intalox® Saddles	ceramic	200		145	92		52	40	22
Super Intalox® Saddles	ceramic				60			30	
Super Intalox® Saddles	plastic				40			28	18
IMTP®	metal		51		41		24	18	12
Hy-Pak®	metal				45		32	26	16
Intalox Snowflake®	plastic							13	
Tellerettes	plastic				35			24	17

3. Determine the height

Using the theoretical number stages and liquid viscosity, determine the total packing depth from Table 3-17 and Table 3-18. Packed beds have limited height; metal and ceramic packings are typically up to 30 feet per bed, plastic up to 24 feet. However, no more than 10 to 14 theoretical stages per packed section are recommended to avoid liquid distribution problems such as channeling [14]. Packing manufacturers will provide specific guidelines.

Packed columns operate over a continuum through the depth of the packing, not in discrete stages like a trayed column. Subtle differences in packing geometry, materials, and packing method affect performance; this is the reason that experts with access to detailed data should perform final designs. The number of theoretical trays is related to an empirical "Height Equivalent to a Theoretical Plate" or HETP. An alternative way to calculate the packing depth, which is used for mass transfer applications, is to compute the number of transfer units (NTU) required for the mass transfer and the corresponding height of a transfer unit (HTU). This is described in Chapter 4.

The current primary suppliers of packings and tower internals are listed in Table 3-11. In addition to these companies, a large number of suppliers of "generic" random packings exist, primarily in Asia.

Table 3-12 compares packings as an aid to initial selection. For simple applications use Pall rings for studies as a "tried and true" packing. This will give conservative results when compared with more recently engineered, random dumped packings.

Strigle's version of the venerable generalized pressure drop correlation (GPDC), considered the "best and latest"

Table 3-17
Separation efficiency in standard distillation systems [4]

Nominal Packing Size (inches)	HETP (inches)
³/₄	11 to 16
1	14 to 20
1½	18 to 27
2	22 to 34
3 or 3½	31 to 45

Table 3-18
Effect of liquid viscosity on packing efficiency [10]

Liquid Viscosity (cP)	Relative HETP (%)
0.22	100
0.35	110
0.75	130
1.5	150
3.0	175

[16], is shown in Figure 3-30. The chart correlates the column's liquid and vapor flow rates, properties, and an empirical packing factor with pressure drop through the packing.

Kister, et. al. described how to use the Stigle chart [16]. The ordinate, called the capacity parameter (CP), is given by:

$$CP = C_S F_P^{0.05} \nu^{0.05} \tag{3-19}$$

Where:

$$C_S = U_S \left[\frac{\rho_G}{(\rho_L - \rho_G)} \right]^{0.5}$$

U_S = superficial vapor velocity, ft/s
ρ = gas and liquid density, lb/ft^3
F_P = packing factor specific to the random packing, dimensionless
ν = kinematic viscosity, cSt (= dynamic viscosity, cP, divided by liquid density, g/cm^3)

The abscissa, called the flow parameter, is given by:

$$F_{lv} = \left(\frac{L}{V} \right) \left(\frac{\rho_G}{\rho_L} \right)^{0.5} \tag{3-20}$$

The pressure drop at the flooding point is correlated to the packing factor with:

$$\Delta P_{flood} = 0.12 \, F_P \tag{3-21}$$

ΔP_{flood} = pressure drop at flooding, inches H$_2$O per foot of packing

Once this pressure drop is known, the flood velocity can be found from the GPDC.

The same technique applies to structured packing using Figure 3-31.

The GPDC should not be relied upon for critical design work, but it is well suited for preliminary designs. For more accurate predictions, experimental data should be used. When experimental data is superimposed on the GPDC, the combined chart provides an accurate tool for

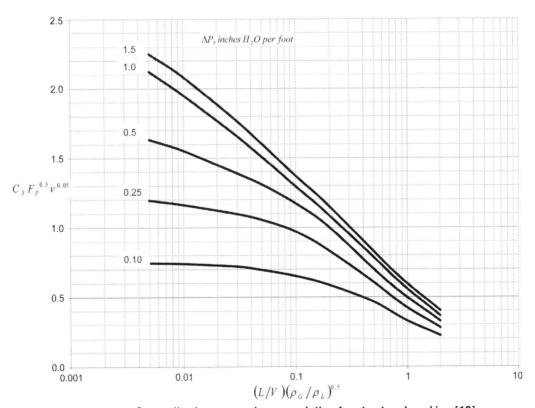

Figure 3-31. Generalized pressure drop correlation for structured packing [16].

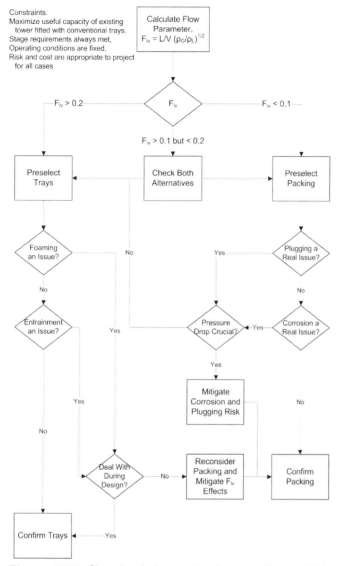

Figure 3-32. Choosing between structured packing or high performance trays for distillation retrofits [6].

final design. The combined charts are called "interpolation charts." Kister, et. al. have published an atlas of interpolation charts, along with a detailed procedure for using them. The atlas has individual charts for each packing, and includes packing factors. See [16] for further information. A decision tree comparing trays to structured packing is shown in Figure 3-32.

Nomenclature

D_T	=	column diameter
E_m	=	Murphree tray efficiency
F	=	Feed molar rate
F_P	=	packing factor specific to the random packing, dimensionless
L_{mass}	=	liquid mass rate
L_S	=	liquid molar rate in the stripping section
L_R	=	liquid molar rate in the rectification section
N_m	=	minimum number of theoretical stages in the column
P_1	=	vapor pressure or partial pressure of pure component 1
P_2	=	vapor pressure or partial pressure of pure component 2
q	=	thermal condition of the feed
R_m	=	minimum reflux ratio, L/D
S_F	=	separation factor
t	=	temperature, $°C$ or $°F$ R_m = minimum reflux ratio, L/D
U_S	=	superficial vapor velocity, ft/s
V_{mass}	=	vapor mass rate
V_S	=	vapor molar rate in the stripping section
V_R	=	vapor molar rate in the rectification section
x_{LD}	=	mole fraction of the light component in the distillate
x_{LF}	=	mole fraction of the light component in the feed
α	=	relative volatility
γ_i	=	the activity coefficient specific to the particular system
ρ	=	gas and liquid density, lb/ft^3
θ	=	Underwood minimum reflux constant
μ	=	dynamic viscosity, cP
ν	=	kinematic viscosity, cSt (= dynamic viscosity, cP, divided by liquid density, g/cm^3)

References

[1] Bennett D, Kovak K. Optimize Distillation Columns Part 1: Trayed Columns. *Chemical Engineering Progress*, May 2000.

[2] Branan CR. *Pocket Guide to Chemical Engineering.* Gulf Professional Publishing; 1999.

[3] Branan CR. *The Fractionator Analysis Pocket Handbook.* Gulf Publishing Co; 1978.

[4] Branan CR. In: *The Process Engineer's Pocket Handbook*, vol. 2. Gulf Publishing Co; 1983.

[5] Bravo J. Column Internals. *Chemical Engineering*, February 1, 1998.

[6] Bravo J. Select Structured Packing or Trays? *Chemical Engineering Progress*, July 1997.

[7] Coker AK. *Ludwig's Applied Process Design for Chemical and Petrochemical Plants*. 4th ed. vol. 2. Gulf Professional Publishing; 2010.

[8] Colburn AP. *Trans A.I.Ch.E*, 1941;37:805.

[9] Fenske M. *Ind Eng Chem*, 1932;24:482.

[10] Frank O. Shortcuts for Distillation Design. *Chemical Engineering*, March 14, 1977.

[11] Fruehauf P, Mahoney D. Distillation Column Control Design Using Steady State Models: Usefulness and Limitations. *ISA Transactions*, July 1993;32(2).

[12] Gilliland ER. *Ind Eng Chem*, 1940;32:1101–6.

[13] Gas Processors Suppliers Association (GPSA). *Engineering Data Book, SI Version*, vol. 2. 12th ed. 2004.

[14] Kister HZ. *Distillation-Design*. McGraw-Hill, Inc; 1992.

[15] Kister HZ. Ask the Experts: Introducing Reboiler Return. *Chemical Engineering Progress*, March 2006.

[16] Kister HZ, Scherffius J, Afshar K, Abkar E. Realistically Predict Capacity and Pressure Drop for Packed Columns. *Chemical Engineering Progress*, July 2007.

[17] Kolmetz K, Ng W, Lee S, Lim T, Summers D, Soyza C. Optimize Distillation Column Design for Improved Reliability in Operation and Maintenance, 2nd Best Practices in Process Plant Management. Kuala Lumpur, Malaysia: March 14–15, 2005. Downloaded from www.kolmetz.com.

[18] Lieberman N, Lieberman E. *A Working Guide to Process Equipment*. McGraw-Hill; 1997.

[19] Love D. No Hassle Reboiler Selection. *Hydrocarbon Processing*, October 1992.

[20] McCabe WL, Thiele EW. *Ind Eng Chem*, 1925;17:605.

[21] Martin HZ, Brown GG. *Trans A.I.Ch.E*, 1939;35:679.

[22] Molokanov YK, Korablina TP, Mazurina NI, Nikiforov GA. *Int Chem Eng*, 1972;12(2):209–12.

[23] Mukherjee S. Tray Column Design: Keep Control of the Details. *Chemical Engineering*, September 2005.

[24] Riggs J. Distillation: Introduction to Control. at, www.controlguru.com/wp/p60.html. Downloaded August, 2011.

[25] Sloley A. Efficiently Control Column Pressure. *Chemical Engineering Progress*, January 2001.

[26] Sloley A. Properly Design Thermosyphon Reboilers. *Chemical Engineering Progress*, March, 1997.

[27] Smith BD. *Design of Equilibrium Stage Processes*. McGraw-Hill; 1963.

[28] Souders Jr M, Brown G. Design of Fractionating Columns. *Industrial Engineering Chemistry*, 1934;26:98.

[29] Strigle Jr RF. *Packed Tower Design and Applications*. 2nd ed. Houston, Texas: Gulf Publishing; 1994.

[30] Summers D. Evaluating and Documenting Tower Performance. *Chemical Engineering Progress*, February 2010.

[31] Underwood AJV. Fractional Distillation of Multicomponent Mixtures. *Chemical Engineering Progress*, 1948;44:603.

[32] Van Winkle, Matthew. Distillation. McGraw-Hill; 1967.

4

Absorbers

Introduction

There are two major types of absorption. In each case, a gas stream contacts the liquid absorbent, transfers components into the liquid, and therefore cleans the gas.

Hydrocarbon absorption is very similar to distillation, with the vapor-liquid equilibrium driving the process, as discussed in Chapter 3. The absorber uses a lean oil in which the hydrocarbon components are much heavier than the component absorbed from the gas stream. There may or may not be a reboiler. The columns are often fitted with trays rather than packing. Canned computer distillation programs usually include hydrocarbon absorber options.

Inorganic components are absorbed into aqueous solutions. When it is strictly a physical process, mass transfer coefficients determine the column design. Packed towers are almost always utilized. Reactive absorption is when chemical reactions accompany the absorption of gases into liquid solutions. The design of reactive absorption processes should also use mass transfer film models, taking into account the chemical reactions [1].

In these processes, we define "solute" to mean the component in the gas stream that is absorbed into a liquid "solvent."

An Excel workbook, containing worked examples, accompanies this chapter.

Hydrocarbon Absorber Design

Absorbers are used to remove hydrocarbons from natural gas. The "rich gas" enters the bottom of a column with trays or packing, flowing countercurrent to a "lean oil" with a molecular weight of about 100 to 200. For ambient temperature absorbers, a heavy lean oil is used with molecular weight of 180 to 200. Refrigerated absorbers use a lighter lean oil of molecular weight 120 to 140. The circulation rate of the oil depends on its molecular weight, so the lighter oil will have a lower circulation rate. However, the lighter oil will have higher vaporization losses [2].

For all detailed absorber designs, tray-to-tray calculations should be done using an appropriate computer program. Vendors will perform the calculations if the engineer lacks access to a program. However, preliminary designs can be done with the method given in this section.

The 1947 Edmister short-cut method uses "absorption and stripping factors" to predict the performance of absorption into lean oils. Others have published Edmister's chart (see Refs [1] or [2]) which gives curves for columns with 0.2 to infinite theoretical trays, and an abscissa with partially compressed scale. Because this graphical method is now of interest primarily for preliminary work, the chart is reworked and simplified in Figure 4-1.

The primary assumptions for the method are:

- Relative volatility between the key component and the lean oil is constant throughout the column.

- The average temperature of the column is representative of the overall column.
- Vapor and liquid traffic are constant through the column.

1. Given:
   Composition of rich gas stream, mole fractions
   Feed rate of rich gas stream, moles/h
   Relative volatility of each component of rich gas with lean oil
   Mole fraction of each hydrocarbon in lean oil
   Desired recovery of key component from gas into the oil fraction
   Number of theoretical trays in the column
2. Use the chart (Figure 4-1) to find the absorption factor. For example, if the desired recovery of the key component is 75%, with six theoretical trays, the absorption factor is 0.80.
3. Calculate the required feed rate of the lean oil stream:

$$L_o = A_i \, \alpha_{i,o} \, V \qquad (4\text{-}1)$$

Where

L_o = feed rate, lean oil, mol/h
A_i = Absorption factor from chart, for component i
$\alpha_{i,o}$ = average relative volatility, hydrocarbon component i and lean oil
V = feed rate, rich gas, mol/h

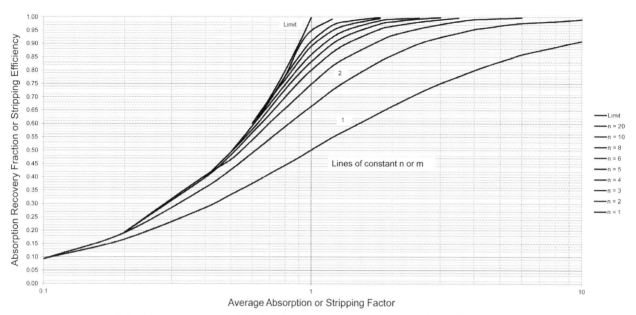

Figure 4-1. Absorption and stripping factors vs. stripping functions. (Adapted from [1].)

4. Calculate the absorption factor for each of the remaining components in the rich gas stream, using Equation 4-1 rearranged.
5. Use the chart to find the recovery fraction for each of the remaining components. For example, if the absorption factor is 0.33, the recovery fraction with six theoretical trays is 0.36.
6. For each component, calculate the mole fraction in the lean gas stream that leaves the column with:

$$Y_{1,i} = \frac{V\,Y_{n+1,i} - E_i\,(V\,Y_{n+1,i} - L_o\,Y_{o,i})}{V} \quad (4\text{-}2)$$

Where
  $Y_{1,i}$ = component i in lean gas stream, mole fraction
  $Y_{n+1,i}$ = component i in rich gas stream, mole fraction
  E_i = recovery fraction for component i
  $Y_{o,i} = \alpha_{i,o}\,X_{o,i}$ = equilibrium vapor concentration of component i in lean oil stream
  $X_{o,i}$ = component i in liquid phase of lean oil stream, mole fraction

Inorganic Absorbers

Film theory is widely used to model absorption units. It assumes that the gas and liquid phases are well mixed, but separated from each other by a thin boundary layer. The boundary layer consists of a gas film and a liquid film. In each case the films are assumed to have little motion (no mixing) and mass transfer occurs by molecular diffusion. The diffusion rate is defined by absorption coefficients that depend on the concentration of the solute in the gas, at the interface, and in the liquid.

The absorption coefficients require knowledge of the surface area of the boundary layer. This is difficult to define. Instead, "volume coefficients" are determined experimentally for specific packings. When the solute is very soluble in the solvent, the gas film controls the mass transfer and K_{GA} is used for calculations. When the solute has low solubility, the liquid film commands and K_{LA} is preferred. However, the mass transfer calculations can use either value regardless of which side is controlling.

For *physical absorption*, where there is no chemical reaction, the vapor-liquid equilibrium defines the driving force. Since the solvent usually leaves the column containing a dilute concentration of solute, use Henry's Law to model the system if the value of the coefficient is known. Henry's Law is applicable for gaseous solutes (i.e., the solute has a vapor pressure well above the system pressure at the operating temperature), with a maximum 0.01 mole fraction in the solvent and pressure below 2 atmospheres. For some systems Henry's Law is applicable outside this envelope; plot experimental data with the Henry's Law expression to compare [7].

Rules of thumb for packed column absorbers:

* Select liquid rates at least 25% to 100% greater than the theoretically calculated minimum rate.
* Typical liquid rates are between 12 and 125 (m³/h)/m² (5 and 50 gpm/ft²) [7]. See Table 4-1.
* Typical gas flow rates are between 40 and 70% of the calculated flooding rate.
* Tower diameter should be at least eight times the packing size.
* For packing factors from 10 to 60 (ft⁻¹), an empirical equation for the limiting pressure drop at flooding is [4]:

$$\Delta P_{flood} = 0.115 \, F_p^{0.7} \qquad (4\text{-}3)$$

Table 4-1
Maximum recommended liquid loading for random packings [9]

Packing Size		Liquid Rate (gpm/ft²)
mm	in	
19	³/₄	25
25	1	40
38	1½	55
50	2	70
90	3½	125

Where:

ΔP_{flood} = pressure drop at flooding, inches H_2O per foot of packing

* For higher values of the packing factor, assume 2.0 inches H_2O per foot [4]. However, Strigle recommends that after determining the tower diameter and height based on mass transfer equations (next section), calculate the pressure drop. The design parameters should be iterated until a maximum pressure drop of 0.60 in H_2O/ft with water as the solvent, or 0.40 in H_2O/ft with other liquids, is found. For foaming systems, the maximum pressure drop should be 0.25 in H_2O/ft [9].
* Choose a solvent: a) with high gas solubility, b) with low volatility, c) as non-corrosive as possible, d) with low cost and high availability, e) with relatively low viscosity, and f) with positive safety and toxicity profile.
* Consider structured packing, rather than random dumped packing, for applications requiring very low pressure drop or when an existing column capacity must be increased (see Table 4-2).
* For absorbers with chemical reaction, provide at least 33% excess of the reactant in the solvent [7].

Packing Height for Mass Transfer in Packed Columns

The following procedure gives a reasonable estimate for the height of packing. As written it neglects temperature change in the column; the temperature may change due to the heat of solution, heat of reaction, and heat of vaporization. This procedure is applicable to physical

Table 4-2
Comparison of random packing and trays for acid-gas absorption columns [2]

Attribute	Random Packing	Trays
Pressure drop	Typically about ¹/₃ that of trays due to larger open area and lack of liquid head	Each tray has a liquid head (typically 50 mm or 2 in. per tray) contributing to higher pressure drop through the column
Typical pressure drop for tower with 25 transfer stages	7 kPa (1 psi)	21 kPa (3 psi)
Foaming	Excellent performance due to low gas and liquid velocities and large open area	Comparable to random packing if downcomers are well designed. However, prediction of downcomer choke and aeration factors are uncertain.
Column diameter	No restriction if the size of the packing is small compared with the column diameter. Recommended for columns with diameter less than 1 m (3 ft)	

(Continued)

Table 4-2
Comparison of random packing and trays for acid-gas absorption columns [2]—cont'd

Table 4-2
Comparison of random packing and trays for acid-gas absorption columns [2]—cont'd

Attribute	Random Packing	Trays
Solids	Solids tend to accumulate in packing voids.	Handle solids much better than packing due to much higher gas and liquid velocities, and fewer locations on trays where solids can be deposited. Use trays with large sieve holes or large fixed valves when plugging and fouling are primary considerations.
Maldistribution	Most severe in large-diameter towers, where there are low liquid flowrates, and smaller packing. Maldistribution can be remedied with good distributor design, water-testing, and inspection.	Inherently more robust
Hydrogen-rich systems	Fast movement of hydrogen molecule can result in reverse diffusion in packing	Much less influence on turbulent contact on trays
Turndown	Distributor turndown often restricts packing turndown to about 2 on the liquid flowrate. If steady liquid flowrate is used (e.g., pumparounds), turndown performance of packing can match or exceed trays.	Moving valve trays typically achieve turndown of 4 to 5. Large diameter sieve holes or fixed valve trays typically achieve turndown of 2 to 2.5, which is comparable to packings.
Resistance to corrosion	Ceramic and plastic packings are highly corrosion-resistant at low cost	Corrosion-resistant alloys are expensive
Flexibility	Greater flexibility because packing is relatively easy to change to modify column characteristics	Fixed trays are relatively difficult and expensive to change

absorption. See Table 4-3 for a list of data required for the calculations. The accompanying Excel workbook has a worked example.

For systems with a fast chemical reaction, the concentration of solute in the solvent may be assumed to be nil throughout the column; this assumption decreases the calculated packing height. However, if there is a slow reaction the required packed height may be more than that calculated [1].

1. Define the equilibrium curve. Use Henry's Law, if applicable, to establish the concentration of solute in solvent that is in equilibrium with the gas feed stream concentration (y_1) at the bottom of the column (x_1^*).

$$x_1^* = \frac{y_1}{k_H}$$

2. Calculate the slope of the equilibrium curve. This is the minimum ratio of molar flow, L/G:

$$\left(\frac{L}{G}\right)_{min} = \frac{y_2 - y_1}{x_2 - x_1^*}$$

Table 4-3
Data required for absorber calculations

Category	Data Required
Absorber Packed Column and System	Operating pressure, P Column diameter, D Packing type and size Absorption volume coefficient, K_{GA} Henry's Law coefficient, k_H, or vapor-liquid equilibrium data, evaluated at the average column temperature
Gas Stream	Molar flow rate, G_1 Molecular weight of inert gas and solute, M_G and M_S Temperature Mole fraction of solute, feed stream, y_1 Mole fraction of solute, exit stream, y_2, or percent of solute to remove in the column
Liquid Stream	Molecular weight of solvent, M_L Temperature Density, ρ_L Molar concentration of solute in the feed stream, x_2 (important if the solvent is recirculated)

3. Choose a liquid rate that is about 20% to 100% higher than minimum [7]. For example, with 50% excess liquid flow:

$$L_2 = 1.5\,G_1 \left(\frac{L}{G}\right)_{min}$$

4. Assume that none of the solvent is vaporized. Determine the molar gas flow out of the column.

$$G_2 = \frac{G_1\,(1 - y_1)}{(1 - y_2)}$$

5. Close the material balance to find the actual concentration of solute in solvent at the bottom of the column. First compute the moles of solute in the liquid discharge. Then calculate the concentration and molar flow rate.

$$Moles = L_2\,x_2 + G_1\,y_1 - G_2\,y_2$$

$$x_1 = \frac{Moles}{Moles\,L_2\,(1 - x_2)}$$

$$L_1 = G_1 + L_2 - G_2$$

6. Calculate the molal gas velocity using the average gas rate through the column.

$$G_m = \frac{(G_1 + G_2)/2}{\pi\,D^2/4}$$

7. Find the vapor phase concentration of solute that would be in equilibrium with the liquid concentration at the bottom and top of the column. If there is an irreversible chemical reaction the equilibrium concentration is zero ($y^* = 0$).

$$y_1^* = k_H\,x_1$$
$$y_2^* = k_H\,x_2$$

8. Obtain the log-mean concentration driving forces with the following two expressions:

$$(y - y^*)_{LM} = \frac{(y_1 - y_1^*) - (y_2 - y_2^*)}{\ln\left[\frac{(y_1 - y_1^*)}{(y_2 - y_2^*)}\right]}$$

$$(1 - y)_{LM}^* = \frac{(1 - y_1) - (1 - y_1^*)}{\ln\left[\frac{(1 - y_1)}{(1 - y_1^*)}\right]}$$

9. Calculate the number of transfer units required for absorption:

$$N_{OG} = \frac{(y_1 - y_2)}{(y - y^*)_{LM}}$$

10. Calculate the height of each transfer unit:

$$H_{OG} = \frac{G_m}{K_{GA}\,(1 - y)_{LM}^*}$$

11. The overall packed height is found by multiplying the number of transfer units by the height of each transfer unit:

$$Z = N_{OG}\,H_{OG}$$

The dimensions of a packed column are shown in Figure 4-2.

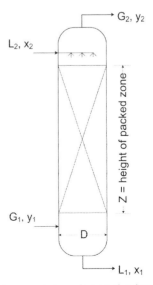

Figure 4-2. Nomenclature for packed tower absorber.

Overall Mass Transfer Coefficient

Absorption coefficient data may be difficult to obtain. The values are specific to the gas-liquid system of interest, and change with both gas and liquid rate. Packing manufacturers often publish K_{GA} values for the CO_2-caustic system, but these are of little use in predicting coefficients for other systems.

The coefficient is a function of gas flow rate and liquid flow rate according to the relationship [7]. However, the K_{GA} value reaches a maximum as the liquid rate is increased at constant gas rate, when a pressure drop of about 0.75 in H_2O/ft is achieved for 2-inch and smaller sizes of random packings [9].

$$K_{GA} \propto L^b G^c \tag{4-4}$$

For liquid-film controlled systems, the value of exponent b lies between 0.22 and 0.34 depending on the characteristics of the packing. If data is not available assume that $b = 0.30$. Assume exponent $c = 0.06$ to 0.08.

For gas-film controlled systems, the value of exponent b also lies between 0.22 and 0.34 (assume 0.30 if unknown). However, exponent c ranges from 0.67 to 0.80 and should be assumed to be 0.75 if unknown. In any event, the sum of b and c should be greater than 1.0.

Example

Absorption of hydrogen chloride into water is gas-film controlled. The K_{GA} at a liquid rate of 4 gpm/ft^2 and gas rate of 3.5 ft/s is reported to be 14 lb-mol/h-ft^3-atm. What is the K_{GA} at a liquid rate of 10 gpm/ft^2 and gas rate of 5.0 ft/s?

Solution:

$$K_{GA} = (14) \left(\frac{10}{4}\right)^{0.3} \left(\frac{5}{3.5}\right)^{0.75} = 24$$

If K_{GA} values are available for a known system, those of an unknown system can be approximated by:

$$K_{GA}(unknown) = K_{GA}(known) \left(\frac{D_v(unknown)}{D_v(known)}\right)^{0.56} \tag{4-5}$$

Where:

K_{GA} = gas film overall mass transfer coefficient, kg-mol/s-m^3-atm or lb-mol/h-ft^3-atm
D_v = diffusivity of solute in gas, m^2/s or ft^2/h [5]

The simplest gas diffusivity relationship is the Gilliland relationship:

$$D_v = 0.0069 \frac{T^{3/2} \left(1/M_A + 1/M_B\right)^{0.5}}{P \left(V_A^{1/3} + V_B^{1/3}\right)^2} \tag{4-6}$$

Where:

T = absolute temperature, R
M_A and M_B = molecular weights of the two gases, A and B
P = total pressure, atm
V_A and V_B = molecular volumes of gases, cc/g-mol

It is convenient that packing manufacturers have largely standardized the reporting of K_{GA} values, based on a system of 1% carbon dioxide absorbed into a solution of 4% NaOH with 25% conversion to carbonate. Comparison of K_{GA} values at the same liquid loading provides the overall mass transfer relationship among the packings. Using Table 4-4, convert the reported, or known, value of K_{GA} for a specific packing to the value to use for a different packing type or size.

Some commercial systems are listed in Table 4-5.

Table 4-4
Relative overall mass transfer coefficient for packings

| Packing | Material | Packing Size | | | | |
		16 mm (⅝")	25 mm (1")	38 mm (1.5")	50 mm (2")	80 mm (3" or 3.5")
βeta-Ring®	Metal		2.22		1.41	1.13
Cascade Mini-Ring®	Metal		1.99	1.80	1.68	1.30
Flexiring®	Metal	1.68	1.54	1.32	1.09	
Hy-Pak®	Metal		1.5	1.26	1.05	0.69
IMTP®	Metal		2.00	1.66	1.37	0.87

Table 4-4
Relative overall mass transfer coefficient for packings—cont'd

Packing	Material	16 mm ($^5/_8$")	25 mm (1")	38 mm (1.5")	50 mm (2")	80 mm (3" or 3.5")
I-Ring	Metal	3.12	2.68	1.98	1.56	0.98
Nutter Ring	Metal		2.05	1.71	1.51	1.07
Pall Rings	Metal		1.51	1.26	1.06	0.62
Intalox® Saddles	Ceramic		1.38	1.11	0.92	0.54
Raschig Rings	Ceramic		1.13	0.94	0.80	0.50
Super Intalox® Saddles	Ceramic		1.63		1.00	0.56
βeta-Ring®	Plastic		1.64		1.21	0.84
Cascade Mini-Ring®	Plastic				1.29	1.07
Flexiring®	Plastic	1.61	1.19	1.16	0.95	0.60
Intalox Snowflake®	Plastic				1.16	
Pall Rings	Plastic		1.29	1.10	1.02	0.60
Jaeger Ring	Plastic	1.63	1.20	1.09	0.98	0.59
Tri-Pack®	Plastic		1.80		1.37	1.20
Jaeger Saddle	Plastic		1.54		0.97	0.59
Jaeger Low-Profile Ring	Plastic		1.17		1.17	1.02
Tellerette	Plastic				1.22	1.07

Comparison of K_{GA} published by packing manufacturers for the CO_2/NaOH system at 25 °C at a liquid rate of 10 gpm/ft^2 and gas rate of between 400 and 970 lb/ft^2-h

Table 4-5
Commercially important absorption systems

Solute	Solvent	Type of Absorption	Typical K_{GA} (lb-mol/h-ft^3-atm)	Henry's Law Constant $k_H = p / x$ [8]	
				$k_{H,px}^*$	C
Acetone	Water	Physical		1.95	4600
Acrylonitrile	Water	Physical		5.0	2800
Ammonia	Dilute Acid	Physical	13 (Note 4)	0.94	4100
Ethanol	Water	Physical		0.30	6500
Formaldehyde	Water	Physical	4.4 (Note 3)	0.017	6800
Hydrogen Chloride	Water	Physical	14 (Note 4)	0.03	0
Hydrogen Fluoride	Water	Physical	6.0 (Note 4)	–	–
Sulfur Dioxide	Water	Physical	2.2 (Note 3)	46.1	3100
Sulfur Trioxide	Water	Physical	20 (Note 1)	–	–
Benzene and Toluene	Hydrocarbon Oil	Physical		–	–
Butadiene	Hydrocarbon Oil	Physical		–	–
Butanes and Propane	Hydrocarbon Oil	Physical		–	–
Naphthalene	Hydrocarbon Oil	Physical		–	–
Carbon Dioxide	Aqueous Sodium Hydroxide	Irreversible Chemical	1.5 (Note 3)	–	–
Hydrochloric Acid	Aqueous Sodium Hydroxide	Irreversible Chemical		–	–
Hydrocyanic Acid	Aqueous Sodium Hydroxide	Irreversible Chemical	4.4 (Note 3)	–	–
Hydrofluoric Acid	Aqueous Sodium Hydroxide	Irreversible Chemical		–	–
Hydrogen Sulfide	Aqueous Sodium Hydroxide	Irreversible Chemical	4.4 (Note 3)	–	–
Chlorine	Water	Reversible Chemical	3.4 (Note 3)	608	2500
Carbon Monoxide	Aqueous Cuprous Ammonium Salts	Reversible Chemical		–	–
CO_2 and H_2S	MEA or DEA	Reversible Chemical	(Note 2)	–	–
CO_2 and H_2S	DEG or TEG	Reversible Chemical		–	–
Nitrogen Oxides	Water	Reversible Chemical		18000	2000

The Henry's Law constant (units: atm) listed here is the reciprocal, representing volatility, at 298.15 K. Use this expression to adjust for temperature:

$$k_H = k_H^* \exp\left(-C\left(\frac{1}{T} - \frac{1}{298.15}\right)\right) \text{ (atm)}$$

Table 4-5
Commercially important absorption systems—cont'd

Solute	Solvent	Type of Absorption	Typical K_{GA} (lb-mol/h-ft³-atm)	Henry's Law Constant $k_H = p / x$ [8]	
				$k_{H,px}^{\cdot}$	C

1. 2-inch ceramic Intalox saddles at a liquid rate of 7.5 gpm/ft² and gas rate of 1200 lb/ft²-h [7]
2. Refer to [7] for a discussion of this system
3. #2 plastic Super Intalox® packing at a liquid rate of 4 gpm/ft² and gas velocity of 3.5 ft/s; liquid film controlled [7]
4. #2 plastic Super Intalox® packing at a liquid rate of 4 gpm/ft² and gas velocity of 3.5 ft/s; gas film controlled [7]

Nomenclature

A_i = Absorption factor from chart, for component i

G = vapor rate, moles/time

D = tower diameter, m or ft

D_v = diffusivity of solute in gas, m²/s or ft²/h

E_i = recovery fraction for component i

F_P = packing factor

G = gas rate, moles/time

G_m = molal gas velocity, moles/time/m² or moles/time/ft²

H_{OG} = height of a transfer unit with gas-film resistance, m or ft

K_{GA} = gas film overall mass transfer coefficient, kg-mol/s-m³-atm or lb-mol/h-ft³-atm

k_H = Henry's law constant

L = liquid rate, moles/time

L_o = feed rate, lean oil, mol/h

$M_A \text{ and } M_B$ = molecular weights of the two gases, A and B

N_{OG} = number of transfer units with gas-film resistance

P = total pressure, atm

T = absolute temperature, R

V = feed rate, rich gas, mol/h

$V_A \text{ and } V_B$ = molecular volumes of gases, cc/g-mol

x = mole fraction, liquid phase

y = mole fraction, vapor phase

$X_{o,i}$ = component i in liquid phase of lean oil stream, mole fraction

$Y_{1,i}$ = component i in lean gas stream, mole fraction

$Y_{n+1,i}$

$Y_{o,i} = \alpha_{i,o}$, $X_{o,i}$ = equilibrium vapor concentration of component i in lean oil stream

Z = height of packed section of column, m or ft

$\alpha_{i,o}$ = average relative volatility, hydrocarbon component i and lean oil

References

[1] Coker AK. *Ludwig's Applied Process Design for Chemical and Petrochemical Plants*. 4th ed, vol. 2. Gulf Professional Publishing; 2010.

[2] Gas Processors Suppliers Association (GPSA). *Engineering Data Book, SI Version*. 12th ed, vol. 2; 2004.

[3] Kenig E, Seferlis P. Modeling Reactive Absorption. *Chemical Engineering Progress*, January, 2009:65–73.

[4] Kister H. Ask the Experts: Acid-Gas Absorption. *Chemical Engineering Progress*, June, 2006:16–7.

[5] Marreo T, Mason E. Gaseous Diffusion Coefficients. *Journal of Physical and Chemical Reference Data. American Institute of Physics*, 1972; 1:3.

[6] McCabe W, Smith J, Harriott P. *Unit Operations of Chemical Engineering*. 7th ed. (New York): McGraw-Hill, Inc; 2004.

[7] McNulty K. Effective Design for Absorption and Stripping. *Chemical Engineering*, November, 1994.

[8] Sander R. *Compilation of Henry's Law Constants for Inorganic and Organic Species of Potential Importance in Environmental Chemistry* (Version 3), www.henrys-law.org; 1999.

[9] Strigle Jr R. *Packed Tower Design and Applications*. 2 ed. Gulf Publishing Co; 1994.

Introduction

Pumps are crucial for the operation of almost every installation in the chemical process industry. This chapter gives some guidance in choosing from the many different types of pumps available. System curves, primarily used in conjunction with centrifugal pumps, are explained, and a dissertation on flow control is included. Finally, centrifugal pump equations are given.

Pumped systems are either closed-loop or open-loop. In a closed-loop system, such as recirculation of chilled water through a heat exchanger, frictional losses predominate. In an open-loop system, such as pumping from one tank to another, static head may be significant due to elevation differences and tank pressurization requirements. These differences are important when selecting a pump and deciding how to control the flow rate.

An Excel workbook with worked examples accompanies this chapter.

Pump Types and Selection

Pumps fall into two broad categories: kinetic (centrifugal or rotodynamic) and positive displacement. Approximately 70% of pump sales are kinetic with the remainder being positive displacement [10].

Centrifugal Pumps

Centrifugal pumps are more commonly used than positive displacement pumps for several reasons [12]:

- Low cost
- Simple and safe to operate
- Require minimal maintenance
- Long service life
- Operate under broad range of conditions
- Risk of catastrophic failure due to deadheading is low

There is a tendency to oversize pumps. Engineers often select larger than necessary pumps to try to accommodate uncertainties in system design, fouling effects, or future capacity increases. They also tend to oversize pumps to prevent being responsible for inadequate system performance. Oversizing increases the cost of operating and maintaining a pump, and creates problems including excess flow noise, inefficient operation, and pipe vibrations.

The next sections in this chapter show how to match pump performance with system curves, but it is up to the engineer to critically analyze available design and pump data, and then make appropriate design decisions.

Positive Displacement Pumps

PD (positive displacement) pumps have a fixed displacement volume and their flow rates are directly proportional to their speed. They generally need more safeguards than a centrifugal pump, such as relief valves to prevent over-pressurization. They also typically have more wear parts than a centrifugal pump, and require more frequent maintenance. Positive displacement pumps are typically used for:

- Highly viscous working fluids, especially when viscosity exceeds 150 cP
- High pressure, low flow conditions
- Self-priming
- Shear sensitive or non-Newtonian working fluids
- Precise metering
- High efficiency, especially at flow below $25 \, m^3/h$ (100 gpm) [10]

Recommendations from Vandell and Foerg are summarized in Table 5-1 [13], supplemented by Petersen and Jacoby [10].

Pump Specification Checklist

Here is a list of considerations to bear in mind when specifying a pump (see Table 5-2).

Good Design Practice

For all pumps, the following design practices are recommended:

Table 5-1
Selection and design considerations for PD pumps

PD Pump Type	Application Notes
Reciprocating, all	Maximum fluid lift is about 7 m (22 ft) for cold water at sea level (less at higher altitudes or with fluids with higher vapor pressure).
	Pulsating flow. This characteristic dictates that the pressure drop analysis use peak flow conditions. This is especially important for NPSH calculations; with large pulsations the NPSH required by the pump may not be met.
	Consider using a suction stabilizer to smooth pulsations in the suction piping and improve NPSH. Larger diameter piping on the suction side also improves NPSH.
	Adequately sized suction stabilizers can reduce the flow variation to +/−5% of the velocity at rated pump flow.
	Use a pulsation dampener to reduce high pressure fluctuations, attenuate the frequency of pressure pulsations, eliminate relief valve chatter, provide relatively steady flow, reduce pump power, and minimize check valve wear.
	Install a pressure relief valve between the pump and the discharge check and block valves. At a minimum, the set pressure should be 115% of the pump's rated discharge pressure, with full flow relief at 10% above the setpoint.
	Consider installing a bypass valve for use in start-up and capacity control.
	Limit the discharge velocity to 1.5 to 5 m/s (5 to 15 ft/s).
Reciprocating, piston	Maximum differential pressure about 135 bar (2,000 psi)
	Maximum fluid viscosity about 8,000 SSU (1,700 cP).
Reciprocating, plunger	Maximum differential pressure about 700 bar (10,000 psi)
	Maximum fluid viscosity about 8,000 SSU (1,700 cP).
Reciprocating, diaphragm	Maximum fluid viscosity about 3,500 SSU (750 cP).
	Specify the maximum speed (strokes per minute, SPM) to put bidders on the same plane. Best practice is to limit the speed to 100 SPM.
	Most commonly used in metering applications (from mL/h to a few gal/m). Capacity is controlled with a variable speed drive or by varying the stroke length.
	Commonly used for slurries and as sump pumps.
	Hydraulically driven units are capable of very high pressure (to 10,000 bar, 50,000 psi).
	Sealless design is good for hazardous fluids.
	Usually have an internal relief valve, but external valves are often installed in addition to the internal valve for ease of adjustment and maintenance.
	Consider air-operated units for low flow applications in hazardous areas.
Rotary, all	Relatively smooth flow characteristics
	Specify the amount and nature of any entrained gases so that the proper pump design and speed are selected.
	Suction stabilizers and pulsation dampeners are not required
	Install a pressure relief valve between the pump and the discharge check and block valves. At a minimum, the set pressure should be 115% of the pump's rated discharge pressure, with full flow relief at 10% above the setpoint.
	The discharge line should be fully ventable to prevent vapor lock during start-up.
	There are always paths of leakage around the rotors or from suction to discharge, called "slip." Slip is generally reduced with increased fluid viscosity, and the pump's efficiency increases.
	Consider using a rotary pump instead of centrifugal when large variations in pressure are anticipated.
	For non-Newtonian fluids, specify the apparent viscosity which is the viscosity at the pump given the shear rate of the pump; this must often be determined empirically.
	If solids or abrasives are expected in the process fluid, consult the pump manufacturer for proper application. Consider installing a strainer in the suction line to remove solids.
	Due to the close tolerances, rotary pumps can require frequent maintenance. However, they are relatively easy to service.
Rotary, screw	Maximum fluid viscosity about 4 million SSU (1 million cP).
	3-screw pumps are limited to non-abrasive fluids with good lubricity.
	Flow is axial through the pump resulting in lower velocity through the pump compared with gear or vane pumps. Some screw pumps operate at 5000 rpm, but most are at 1800 or 3600 rpm (for 60 Hz power supplies).
Rotary, progressing cavity	Single screw rotates in a stator that is often made of an elastomer. The rotor is a single helix while the stator is a double helix.
	Compatibility between the stator material and process fluid is essential.
	Handles shear-sensitive fluids well, but heads and flows are limited in comparison with screw pumps.
	Some abrasives can be handled if the rotor and stator are both metallic. However, as the hardness of the abrasive approaches that of the rotor's abrasion resistance, deterioration can rapidly occur. "Cut less" elastomers deform to accommodate abrasive or large solids, thus solving the problem.
Rotary, internal gear	Fluid is circulated around the outside of the gears, so high angular velocities are achieved. Maximum shaft speed is about 1800 rpm, with most gear pumps operating below 1000 rpm.

(Continued)

Table 5-1
Selection and design considerations for PD pumps—cont'd

PD Pump Type	Application Notes
	The practical pressure limit for internal gear pumps is determined by the rotor-shaft support bearings and is usually limited to a differential of 14 bar (200 psi).
	High viscosity fluids (1,000,000 cSt) can be pumped with a gear pump
	Flow rates range from 0.1 m^3/h (0.5 gpm) to 350 m^3/h (1,500 gpm).
Rotary, external gear	Used for high pressure applications, such as hydraulics, where pumping pressure can be several thousand psi.
	Smaller pumps operate up to 3600 rpm, while larger ones are usually less than 1000 rpm.
	Flow rates range from drops per minute to 350 m^3/h (1,500 gpm).
Rotary, lobe	External timing gears ensure the lobes do not actually touch
	Because there is no metal-to-metal contact, good choice for pumping fluids that lack lubricity.
	Pressure rise is usually limited to about 15 bar (200 psi).
	Two shaft seals are required; material selection is critical
	Lobe pumps are often used for hygienic applications such as food or pharmaceutical processing.
	Higher viscosity results in diminished performance and is usually limited to 100,000 cSt. At higher viscosities the pump speed must be reduced.
	Maximum flow rate is usually about 680 m^3/h (3000 gpm).
Rotary, vane	Rigid vane design: centrifugal force allows the vanes to seal against the casing; the vanes slide in and out of slots. Rigid vane pumps are self-adjusting to wear. On some models, variable displacement is produced by altering the eccentricity of the rotor within the casing.
	Flexible vane design: also self-adjusting but not available in a variable volume configuration. Elastomeric vanes "wipe" along the inside of the casing. Care must be taken to ensure the vanes material of construction is compatible with the process fluid and temperature.
	Fluid is circulated around the outside of the vanes, so high angular velocities are achieved. Maximum shaft speed is about 1800 rpm, with most vane pumps operating below 1000 rpm
	Maximum viscosity is about 25,000 cSt.
	Pressure rise is usually limited to about 8 bar (120 psi).
	Operating speed is usually between 1000 rpm and 1800 rpm; flow rates range to 450 m^3/h (2000 gpm).

Table 5-2
Pump specification checklist

Consideration	Commentary
Service requirements	Determine the maximum and minimum flow and/or pressure requirements for the pump. What control accuracy is required? How variable are the flow or pressures?
	Provide a System Curve if possible.
	Detail the flow or pressure control method (*i.e.,* throttling control valve, fixed orifice, bypass valve, or variable speed).
	Calculate NPSH available at operating and extreme conditions.
Liquid properties	Specify the flowing liquid and system pressures and temperatures. Include normal operating conditions and also extreme conditions such as start-up, shutdown, and process upsets.
	Provide physical properties for unpublished materials.
	Characterize impurities, especially solids and gases.
	Define hazards including corrosivity, flammability, and toxicity.
	Define sensitivities to shear, temperature, pressure, or stagnation, if applicable.
Safety and environmental	Identify Code, government, or industry compliance requirements.
	Specify label requirements (e.g., UL, CE)
	List protections required for installation in a hazardous rated area such as type of motor, maximum surface temperature, or electrical device specifications.
	Specify maximum emissions from seals or connections.
	Identify relief valves and specify the fate of relieving fluids.
	List any guarding requirements, including coupling guards or flange covers.
Pump construction and features	Specify materials of construction for all components (e.g., casing, impeller, shaft, bearings, coupling, housings, seals, baseplates).
	Provide design limits such as maximum impeller size relative to casing size, minimum motor horsepower, efficiencies, speed, and minimum flow.

Table 5-2
Pump specification checklist—cont'd

Consideration	Commentary
Operations, maintenance, and reliability	Indicate preference for shaft seal (e.g., packing, single mechanical, double mechanical, sealless). Specify seal materials, flushing fluid (if applicable), and cooling.
	Specify if bearings or casing must be heated, cooled, or insulated.
	Specify casing drain and valve, if applicable.
	Specify lubricants and lubrication requirements
	Define baseplate requirements (e.g., size, material, containment, drainage, methods to mount pump and secure baseplate).
	Consider pump design as it relates to installation and replacement. For example, a back pull-out pump can be removed from a piping system without disturbing the piping, whereas an in-line pump may be more challenging.
	Is technical assistance required for installation, start-up, or training of operators or maintenance technicians?

- Minimize turbulence and keep gas bubbles from entering suction lines. One way to accomplish this is with a weir plate and vortex breaker installed in the suction tank (if applicable).
- Keep suction lines short and velocity low (0.3 to 1 m/s or 1 to 3 ft/s), but use NPSH calculations (page 116) for definitive determination. Provide 5 to 10 pipe diameters of straight piping at the suction of the pump. Suction pipes are usually at least one size larger than the suction flange on the pump.
- Install pressure gages on the suction and discharge sides of the pump.
- Use eccentric reducers, installed to prevent a trapped vapor pocket from forming (flat on bottom when the piping turns up, flat on top when the piping is flat or turns down). Eccentric reducers installed at the pump suction are normally flat on top. If there is an increaser on the discharge, install it between the pump and the check valve.
- If the discharge path of a centrifugal pump can be blocked, install a recirculation line to accommodate the pump's minimum flow requirement. Smaller pumps often use a continuous flow recirculation line, by using a restriction orifice sized to pass the minimum flow at the rated shutoff head of the pump. Controlled flow is usually used for larger pumps, using a self-contained pressure control valve or a control valve with flow meter and controller. Recirculation lines should not feed back to the suction of the pump to avoid undesired temperature rise.
- Install a low level switch in the pump feed tank to automatically stop the pump if the tank runs dry.
- An installed spare is customarily provided for major pumps.

Pump Testing

AIChE's Equipment Testing Procedures Committee recommends that positive displacement pumps be tested soon after installation to obtain a performance baseline that future tests can be compared to. Testing is described in their publication [2], summarized in *CEP* [1].

The two recommended tests are:

- Standardized Performance Test. This test charts flow rate vs. differential pressure, and input power vs. differential pressure, at a constant liquid viscosity, suction pressure, and shaft speed.
- Standardized MRSP Test. This test determines the effect of changing the suction pressure on the performance of the pump, and determines the minimum required suction pressure (MRSP).

Other tests are described by AIChE, with the following being performed most often:

- Sound pressure level measures the sound at normal operating conditions and across a range of conditions.
- Vibration measurements are taken for maintenance reasons, used to predict an impending failure so the problem can be resolved early.
- Temperature measurements are taken at the pump surfaces, and inside if instruments are installed at bearings, rotors, and shaft seals.

System Curves

This section illustrates how to create a system curve, using the example in Chapter 1 (see page 3.). Then, pump characteristics and control valve settings are charted.

The system consists of a feed tank, suction pipe, pump, discharge pipe, and destination tank. A system curve plots the flow rate through the system (m^3/h or gpm) against the pressure change through the pump (m or ft of fluid). Chapter 1 presented calculations for pressure drop due to pipe friction and fittings. The system curve adds liquid head (due to gravity).

Data for the example are listed in Table 5-3. Line 215 is the pump suction pipe. The liquid level in the suction is 12 feet above the datum elevation. Line 216 is the discharge pipe; it discharges 23 feet above the datum elevation.

The procedure for creating a system curve is as follows:

1. Use a spreadsheet to model the suction and discharge pipes. The primary input variable is flow rate. US units are used in the example. Start by listing all of the inputs needed to calculate the head pressures and pressure drop due to friction (see Figure 5-1).

2. Calculate the Reynolds Number, Friction Factor, and Pressure Drop due to friction as described in Chapter 1. Also, obtain physical property data for vapor pressure, viscosity, and density based on the fluid temperature (see Figure 5-2).

3. For the suction pipe, calculate the Net Suction Pressure by adding the pressure above the liquid (converted to units of feet of fluid, described below) to the fluid pressure. Subtract the pressure loss due to friction, also converted to units of feet of fluid. Formulae are:
   $$D48 = (D11+14.7)*2.31/(D37/62.4)$$
   $$D49 = (D13-D14)$$
   $$D50 = -D28*2.31/(D37/62.4)$$
   Where the second term, $2.31/(Dxx/62.4)$ corrects for the specific gravity of the fluid. There are 2.31 feet of water in 1 lb/in.^{2}.

4. For the discharge pipe, calculate the Net Discharge Pressure by summing the three terms (pressure at the discharge, elevation change, and pressure drop due to friction). Note that pressure drop is subtracted on the suction side, but added on the discharge side of the pump.

Table 5-3
The example is illustrated in Chapter 1, page 3

Line	Physical Length	Contingency	Subtotal	Fittings	Total Equiv L
215	4.5 m (14 ft)	1 m (3 ft)	5.5 m (17 ft)	3 m (10 ft)	8.5 m (27 ft)
216	26 m (85 ft)	5 m (15 ft)	30 m (100 ft)	15 m (50 ft)	45 m (150 ft)

	A	B	C	D	E	F
7	**Inputs**			US Units		
8		Pumped liquid		Water		
9		Temperature		86.0	F	
10						
11		Head pressure		0.0	psig	
12		Discharge pressure		0.0	psig	
13		Liquid height above datum		12.0	ft	Primary Input Variable
14		Pump suction height above datum		1.5	ft	
15		Pump discharge height above datum		23.0		
16		**Flow rate**		**193,200.0**	lb/h	
17		Suction pipe inside diameter		4.0	in	
18		Equivalent length of suction pipe		27.0	ft	
19		Discharge pipe inside diameter		3.0	in	
20		Equivalent length of discharge pipe		150.0	ft	
21		Roughness		0.00015	ft	

Figure 5-1. Input data for generating system curve.

	A	B	C	D	E
24	**Output**				
25		*Suction Pipe - Line 215*			
26		Reynolds Number		375,275	
27		Friction Factor		0.0176	
28		Pressure Drop due to friction		0.90	psi
29					
30		*Discharge Pipe - Line 216*			
31		Reynolds Number		503,620	
32		Friction Factor		0.0183	
33		Pressure Drop due to friction		22.58	psi
34					
35		Vapor Pressure		0.62	psia
36		Viscosity		0.81	cP
37		Density		62.56	lb/ft3

Figure 5-2. Pressure drop due to friction calculated with non-compressible flow equation and VBA function (see Chapter 1).

	A	B	C	D	E
46		Total Dynamic Head			
47		*Suction Pressure*			
48		Head pressure		33.9	ft fluid
49		Fluid pressure		10.5	ft fluid
50		Friction loss		(2.1)	ft fluid
51		Net Suction Pressure		42.3	ft fluid
52					
53		*Discharge*			
54		Head pressure		33.9	ft fluid
55		Fluid pressure		21.5	ft fluid
56		Friction loss		52.0	ft fluid
57		Net Discharge Pressure		107.4	ft fluid
58					
59		Differential Pressure (TDH)		65.11	ft fluid

Figure 5-3. Calculate Total Dynamic Head in three steps.

5. Next, calculate the Total Dynamic Head (TDH). Subtract the Net Suction Pressure from the Net Discharge Pressure (see Figure 5-3).
6. Create a table that relates flow rate to TDH. Substitute a range of values in the Flow Rate input cell (D16) and record the resultant TDH. Convert the flow rate to gallons per minute units. Use a scatter diagram to chart the system curve. See Figure 5-4 for table of results, Figure 5-5 for chart.

Sensitivity

Check the system curve for sensitivity to assumptions. Rules of thumb were used to estimate the equivalent length of the piping segments for the example problem. The fluid level in the feed tank affects the calculation for Net Suction Head. Vary the parameters

and recalculate TDH. For the example, at 180,000 lb/h (about 360 gpm) the TDH ranges from 48 ft to 64 ft. The Base Case (above) computes to 58 ft so the variance is +10% to −17%.

	G	H	I
68	Flow Rate	TDH	Flow Rate
69	lb/h	ft fluid	gpm
70	10,000	11	20
71	20,000	12	40
72	40,000	14	80
73	60,000	17	120
74	80,000	21	159
75	100,000	26	199
76	120,000	32	239
77	140,000	40	279
78	160,000	48	319
79	180,000	58	359
80	200,000	69	399
81	220,000	81	438
82	240,000	94	478
83	260,000	108	518

Figure 5-4. Convert flow rate to gpm (US Units) before charting the system curve.

System Curve

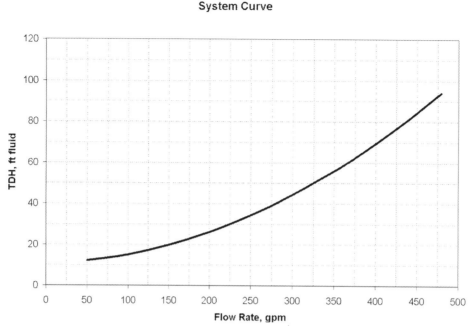

Figure 5-5. Completed system curve relates TDH (through the pump) to flow rate.

Operating Point and Pump Selection

There are many factors to consider when selecting a pump. This section discusses hydraulic characteristics. Important hydraulic considerations include:

- Best Efficiency Point (BEP). Strive to select a pump that operates near its best efficiency. This minimizes energy consumption, and also decreases loads on the pump (from vibration, for instance) and maintenance requirements [12].
- Impeller size relative to pump housing. Centrifugal pumps usually accommodate a range of impeller sizes. Select a pump with an impeller that is smaller than the maximum permissible; this provides contingency for increasing pumping capacity in the future by substituting a lager impeller size.
- Speed. Pump wear increases with speed to seventh power [7].
- Slope of the operating line. Flow control for pumps operating on a flat portion of their operating line is more difficult; consider variable speed control in this case (page 112).

The example problem is used to illustrate these points. If it is stated that the flow rate should normally be controlled at a constant 360 gpm, and occasionally at 250 gpm, select a pump. Furthermore, assume that flow is controlled with a modulating valve.

1. Tentatively assign a pressure drop to the control valve. At 360 gpm, the TDH is 58 ft. Using the rule of thumb in Chapter 1, the control valve pressure drop should be 10% to 15% of the total, or 10 psi, whichever is greater. Since in this example the fluid is water, 10 psi = 23 ft fluid. This would give a controlled TDH of 58 + 23 = 81 ft. Checking, 23/81 = 28%. Therefore, a tentative pressure drop for the valve is 23 ft, and the normal operating point is 360 gpm at 81 ft TDH.

2. Use manufacturers' published data to find pumps where the normal operating point falls near the BEP. Chart the System Curve and Pump Curves together as illustrated in Figure 5-6. The selected pump has a 10-inch impeller, trimmed to about 9.7 inches. When controlled at 360 gpm, the pressure drop through the control valve is about 20 ft to 30 ft, depending on how well the actual piping system adheres to the input assumptions. When controlled

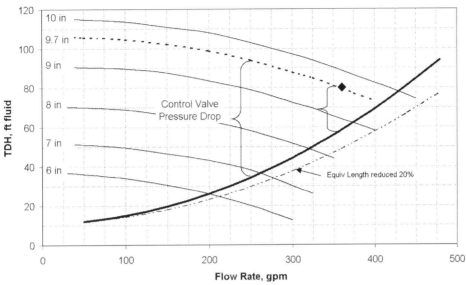

Figure 5-6. Chart system and pump curves together. Indicate the operating point (in this case with a diamond) and show the pressure drop taken by the control valve.

at 250 gpm, the pressure drop through the valve is about 60 ft.

3. Add the efficiency curves to the plot to confirm that the selected pump operates near its BEP. Figure 5-7 indicates an efficiency of about 68%.

4. Pump curves may also indicate the power requirement (charted for water, specific gravity = 1).

Select a motor that would not be overloaded if the pump operates at the extreme maximum flow rate, termed "run out." In the example, this is at a flow rate of about 430 gpm. This could be achieved if, for instance, a branch line was installed between the pump discharge and the control valve, and the branch had a much lower pressure drop and static

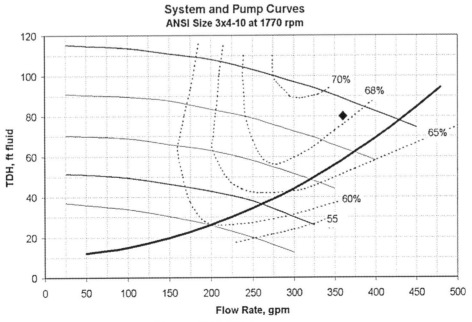

Figure 5-7. Efficiency curves.

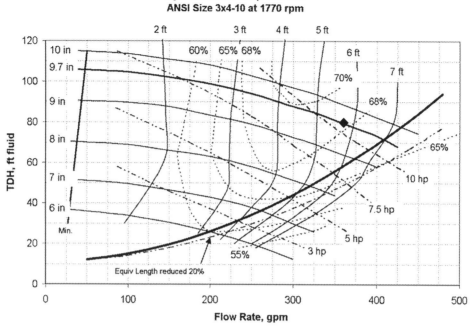

Figure 5-8. Complete pump curve chart. The selected pump operates on the "9.7-in." impeller line; the control valve absorbs the difference in head between the pump and the system curves.

head than the primary discharge pipe. If the control valve is completely open, the operating point would be at about 410 gpm and 70 ft TDH. See Figure 5-8.

5. Pump curves also show the Net Positive Suction Head (NPSH) the pump requires (see page 116) and, often, the minimum recommended flow rate through the pump.

Flow Control with Throttling Valve

Control the pumped flow with a modulating control valve, or by varying the speed of the pump. Control valves are traditional technology and applicable to most installations containing centrifugal pumps. Speed modulation, discussed in the next section, saves pumping energy.

Control Valve

This section continues the discussion in Chapter 1 (pages 20–21) using the example from page 108. Recall that the valve normally controls to 360 gpm with a pressure drop of 20 ft to 30 ft, and may control to 250 gpm with a pressure drop of 60 ft.

1. Calculate the valve coefficient with Equation 1-24, first converting units ft fluid to psi. Results for the example problem are given in Table 5-4.

$$\Delta P = \frac{ft\,fluid\left(\dfrac{density}{62.4}\right)}{2.31} \quad (5\text{-}1)$$

2. Tentatively choose a valve. At the control point, the valve should be no more than 80% open (page 21). In this case, the full-open coefficient should be

Table 5-4
Valve flow coefficients

Flow Rate gpm	Pressure Drop		Valve Coefficient C_v
	ft fluid	psi	
360	21.6	9.4	118
360	30	13	100
250	60	26	49

greater than 118/0.8 = 147. For example, a 3-inch linear flow globe valve has a C_v of 148 [1].
3. Check for the possibility that cavitation or flashing will occur in the valve. Assume the worst case

conditions. Consult with the valve manufacturer to determine if cavitation might be a problem.
4. Chart the control characteristics to help choose between an equal percentage or linear flow valve, if applicable. See the next section for details.

Charting Control Characteristics

Chart the valve lift against flow rate to visualize controllability. Tighter control is achieved when the curve is flatter, because a larger change in valve lift results in a given change in flow rate.

The slope of the valve curve is called "gain." Gain is a measure of the change in flow for a given change in valve opening. It is easier to tune the control loop if the gain is relatively constant over the controlled flow range.

The same example problem is used to explain the procedure.

Refer to Figure 5-6 on page 109. For a series of flow rates ranging from 0 to the maximum control point (360 gpm in the example), determine the pressure drop through the control valve. Convert this to a valve coefficient using Equation 1-24. Then, make a column containing the ratio of C_v to the coefficient calculated for the maximum flow. An example result is shown in Figure 5-9.

The valve curves are created by estimating the percent open (or "lift") required to achieve the desired coefficient for each flow rate in the table.

For the linear valve, which has a full open coefficient of 148 in the example, the lift is the ratio of desired coefficient to full open coefficient. At 250 gpm, the linear valve is 53.8/148 = 36% open.

Use Equation 5-2 to calculate the lift for an equal percentage valve. The valve manufacturer provides the flow coefficient and controllable flow ratio.

$$Lift = \frac{\ln\left(C_v \dfrac{FlowRatio}{ValveCoefficient}\right)}{\ln(FlowRatio)} \quad (5\text{-}2)$$

Where:

$Lift$ = valve lift, %
C_v = flow coefficient required at flow rate
$Flow\ Ratio$ = the controllable ratio of the valve, often 50 for globe valves
$ValueCoefficient$ = the full open valve coefficient

The lift for a globe valve with control ratio of 50, and C_v of 136 at 250 gpm, is:

$$Lift = \frac{\ln\left(53.8 \dfrac{50}{136}\right)}{\ln(50)} = 0.76$$

The completed chart compares the response of the valves with the installation requirement. In this example, the desired control range is from 250 to 360 gpm, so the linear valve is preferred because it has a flatter curve. If the control point was less than about 100 gpm then the equal percentage valve would be a better choice. See Figure 5-10.

Other factors that influence controllability include:

- Hysteresis and deadband. These characteristics are usually treated together, and the uncertainty they

	B	C	D	E	F	G	H	I
30	Installation Curve							
31	Flow Rate	Pump TDH	System	Valve Pressure Drop		Coefficient	Full Open	Ratio of
32	gpm	ft fluid	ft fluid	ft fluid	psi	Cv	Coefficient	Coefficients
33	0	107	12	-	-	-	117.66	0.00
34	50	106	15	90.75	39.39	7.98	117.66	0.07
35	100	105	20	85.05	36.91	16.48	117.66	0.14
36	150	102	26	76.16	33.05	26.12	117.66	0.22
37	200	99	34	64.70	28.08	37.79	117.66	0.32
38	250	94	44	49.85	21.64	53.81	117.66	0.46
39	300	88	56	32.02	13.90	80.58	117.66	0.68
40	360	80	58	21.63	9.39	117.66	117.66	1.00
41								

Figure 5-9. The installation curve is calculated from the pump curve and system curve.

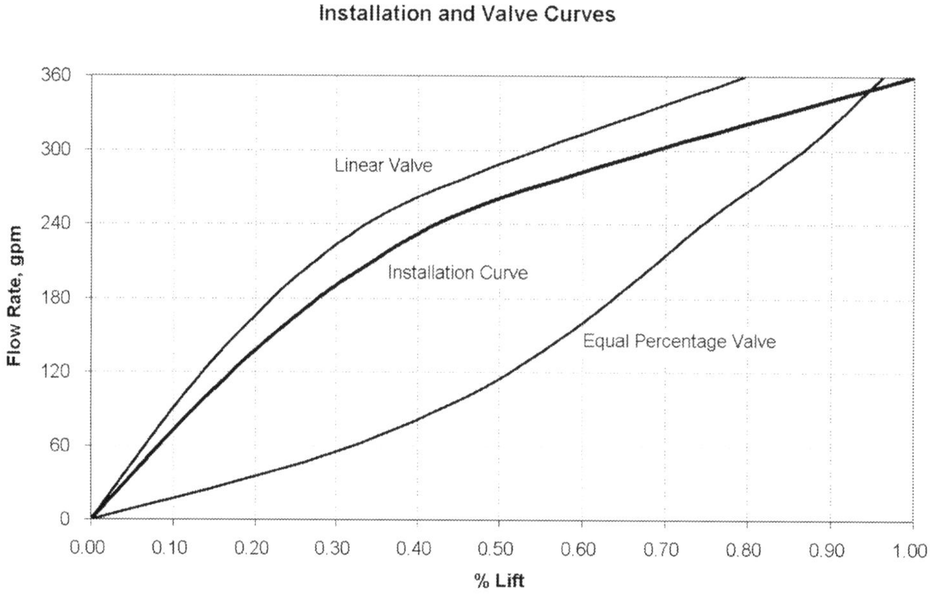

Installation and Valve Curves

Figure 5-10. Installation and valve curves chart.

cause in achieving a specific setpoint can be signifi-cant (up to 10%, [5]). Hysteresis is the difference in position when opening the valve compared with closing the valve to a certain desired point. It is more pronounced in rotary motion (e.g., butterfly and plug valves) and is caused by gear backlash and other "loose" actuation components. Deadband is the amount of change that must be commanded before the valve actually moves.

• Process side error. There is more to the control loop than the valve. If the sensing element responds poorly to changes in flow, then the flow control valve will be equally unresponsive.

• Stroking speed and overshoot. The speed with which a valve strokes to the set position must be commen-surate with the process requirement. For instance, an air actuated open-close ball valve may take 5 to 10 seconds to fully respond to a command.

• Positioning resolution. This is a measure of how well the valve position can track very small adjustments, and also how repeatable it is. The larger the control gain, the more important this criterion becomes.

• Load sensitivity. The actuator must exert force to maintain the valve in a desired position. If the flowing fluid overcomes that force then control is lost. Globe valves, in particular, are specified as "flow to open" or "flow to close" which indicates the direction the flowing fluid takes through the valve body and plug.

• Nonlinearity. This is a measure of how the actual valve behaves compared with an ideal, or theoretical, valve.

Flow Control by Changing the Pump Speed

Pump speed can be changed in fixed increments by using multiple windings in the motor, with a mechanical mechanism such as the belt and pulley system, or by changing the voltage and frequency of the power that feeds the pump's motor. This section discusses the use of the VFD (variable frequency drive) systems, which is the predominant and most recommended speed control method.

VFDs offer many advantages, including:

• Low start-up current, usually only about 1.5 times the normal operating current
• Energy savings
• Integrated process control and diagnostic software
• Reduced maintenance / improved reliability
• Lowered installation cost
• Better control

Table 5-5
Partial data for pump curves at various operating speeds

GPM 1770 rpm	TDH 1770 rpm	GPM 1500 rpm	TDH 1500 rpm	GPM 1250 rpm	TDH 1250 rpm
200	75	169	54	141	38
250	71	212	51	177	35
300	64	254	46	212	32

The first two columns are obtained from the pump manufacturer. The remaining columns are calculated using affinity laws.

Use the *affinity laws* (page 114) to predict the performance of a centrifugal pump at different operating speeds. First, select a pump based on the maximum design flow rate. If there's a choice, favor a pump where the operating point is slightly to the right of the Best Efficiency Point. List the TDH for a series of flow rates on the pump's operating curve. Then, for various pump speeds, use equations 5-3 and 5-4 to construct additional operating curves.

With the same example (page 106), a size 3x4 inch pump is selected, based on the system curve, but no throttling control valve. The operating point is at 360 gpm and 58 ft TDH. Points from the pump curve, obtained from the pump manufacturer, are tabulated. See Table 5-5.

Similarly, predict the BEP for the new pump speeds by applying equations 5-1 and 5-2 to the BEP of the pump at its published speed. The results for the example are shown in Figure 5-11. Notice that, for this example, the same pump is selected for control with a throttling valve or VFD, but the impeller is smaller with VFD control as is the pump motor (15 hp vs. 10 hp).

The maximum speed reduction in this example is from 1770 rpm to about 750 rpm. Further reduction would move the pump performance curve below the system curve, resulting in zero flow. This is because of the fixed head in the system that is associated with static lift. It's typical for centrifugal pumps to operate between 50% and 100% of the motor speed, so the VFD will output 30 Hz to 60 Hz (U.S. power).

Strongly consider purchasing pump-specific VFDs that incorporate control algorithms and allow for adjustment of many parameters such as speed limits, starting frequency, starting time, motor braking, etc. See Table 5-6.

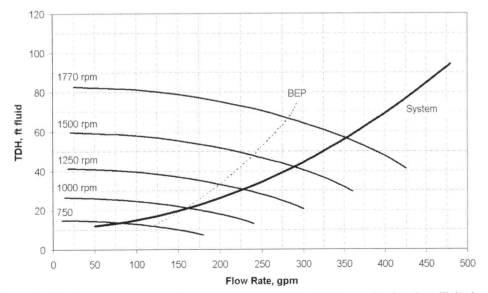

System and Pump Curves
ANSI Size 3x4–8.6 inch-impeller at Indicated Pump Speeds

Figure 5-11. Pump curves for various operating speeds are constructed using the affinity laws.

Table 5-6
Typical features for a pump-specific VFD controller

Category	Features
Mounting location	Attached directly to the pump motor, mounted nearby (e.g., on the wall by the pump), in a cabinet, or on a control panel
Settings	Minimum/maximum speed during normal operation
	Minimum speed at start
	Ramp-up and ramp-down time
	Restart delay (prevent pump from quickly cycling on and off)
	Critical speed windows (speed ranges the pump will not operate at)
	Measurement units
Pump Control	Control the speed based on set point of: flow, pressure, level, temperature, or other measurements. The controller functions as a PID device.
	Start/stop the pump locally, from a remote switch, manually from a panel (HMI), or automatically from a plant control system
Multiple Pump Control	Control several pumps that are piped in parallel or series, automatically starting and stopping them in accordance with a user-determined sequence.
Set Point	The control set point can originate from: the VFD controller's interface, an analog signal from a remote device (e.g., HMI) or a digital signal using various protocols such as Modbus, Profibus, etc. Some controllers can store two set points and the drive will toggle between them based on a digital input such as a low flow switch or high level switch.
Diagnostics	Automatic detection and pump shutdown if conditions that indicate block suction (dry running) or uncontrolled flow (pump moves to run-out).
	Current limiting; motor thermal protection
	Alarms
	Jog the motor (used for maintenance and diagnostic field tests)

Affinity Laws

Dynamic type pumps (e.g., centrifugal pumps) obey these affinity laws:

1. Capacity varies directly with impeller diameter and speed.
2. Head varies directly with the square of impeller diameter and speed.
3. Horsepower varies directly with the cube of impeller diameter and speed.

Volume capacity:

$$\frac{q_1}{q_2} = \left(\frac{n_1}{n_2}\right)\left(\frac{d_1}{d_2}\right) \tag{5-3}$$

q = volume flow capacity, m^3/s, gpm, etc.
n = impeller rotational speed, rpm
d = impeller diameter

Pressure (or head) relationship:

$$\frac{\Delta P_1}{\Delta P_2} = \left(\frac{n_1}{n_2}\right)^2\left(\frac{d_1}{d_2}\right)^2 \tag{5-4}$$

ΔP = the pressure increase imparted by the pump, Pa, psi, m of fluid, ft of fluid

Power consumption relationship:

$$P_{shaft} = \left(\frac{n_1}{n_2}\right)^3\left(\frac{d_1}{d_2}\right)^3 \tag{5-5}$$

P_{shaft} = the shaft power of the pump (hydraulic power divided by pump efficiency), assuming constant pump efficiency (true for speed changes and diameter changes up to 10%)

The following relationships are not exact, but are good for approximate work, as reported by William McNally [9].

$NPSH_{required}$ at the best efficiency point (BEP) varies by the speed (or impeller diameter) to the 1.5 power.

Shaft deflection varies by the impeller speed (or impeller diameter) squared.

Component wear varies by the impeller speed (or impeller diameter) cubed.

Power

Estimate hydraulic power with the formula:

$$P_{hydraulic} = q H \rho g \qquad (5\text{-}6)$$

In SI Units,

Watts = (m³/s)(meters of pumped fluid head)(kg/m³)(9.81 m/s²)

In US Units, modify the formula slightly by substituting the specific gravity for the density and embedding conversion factors:

$$P_{hydraulic} = \frac{q H SG}{3960} \qquad (5\text{-}7)$$

Hp = (gpm)(feet of pumped fluid head)(specific gravity, water = 1)/3960
Shaft power = hydraulic power divided by the pump efficiency
Motor power = shaft power divided by the motor efficiency

Minimum Flow

The Hydraulic Institute [7] offers guidance for determining the minimum flow through a centrifugal pump:

- Temperature rise of the liquid. This is usually established at 10 °C and results in a very low minimum flow limit.
- Radial hydraulic thrust on impellers. This is most serious with single volute pumps and, even at a flow rate that is 50% of the BEP, could cause reduced

bearing life, excessive shaft deflection, seal failure, impeller rubbing, and shaft breakage.
- Flow recirculation in the pump impeller. This can also occur below 50% of the BEP causing noise, vibration, cavitation, and mechanical damage.
- Total head characteristic curve. Some curves droop toward shutoff. Operation in such a region should be avoided.

Temperature Rise

As a rule of thumb, shaft power is divided between pumping the fluid and heating the fluid. The pump's efficiency relates shaft power to hydraulic power. Efficiency is dependent on the flow and head, and is obtained from the pump manufacturer (usually indicated on the pump curves).

Use this equation to estimate the temperature rise through the pump:

$$\Delta t = C P_{shaft} \frac{(1-e)}{c_p q \rho} \qquad (5\text{-}8)$$

Δt = temperature rise through the pump, °C or °F

C = conversion factor: C = 1 for SI units; C = 317 for US units
P_{shaft} = brake power to the pump, kw or hp
e = pump efficiency, evaluated at the flow and head where the pump is operating
c_p = heat capacity of pumped fluid, kJ/kg °C or btu/lb °F
q = volumetric flow, m³/s or gal/m
ρ = fluid density, kg/m³ or lb/ft³

The absolute minimum flow through the pump is such that the fluid does not vaporize in the pump head. Use the equation to check that the boiling point (at suction pressure) is not reached.

Suction System NPSH Available

Calculate the available Net Positive Suction Head (NPSH$_{available}$), using compatible units, as follows: Absolute pressure at surface of liquid reservoir + Pressure exerted by depth of liquid (elevation difference between the liquid reservoir and the pump entrance) − Vapor pressure of the liquid − Line loss due to friction = NPSH$_{available}$

Vapor pressure of water at 30 °C is 4.2 kPa (0.615 psia).

Density of water at 30 °C is 995 kg/m^3 (62.1 lb/ft^3)

Vapor pressure of n-butane at 38 °C is 356 kPa (51.6 psia).

Specific gravity of n-butane at 38 °C is 0.59.

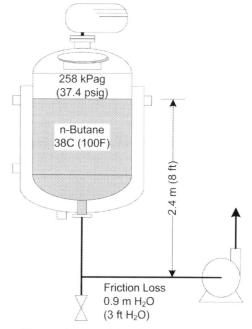

Example 5-2. Liquid at boiling point.

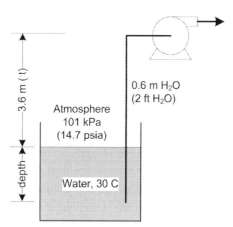

Example 5-1. Suction lift.

Pressure at surface of reservoir	101 kPa	14.7 psia
Convert to water pressure	10.3 m	34.1 ft
Depth of reservoir	irrelevant	
Elevation difference between reservoir and pump	−3.6 m	−12.0 ft
Friction loss in suction pipe	−0.6 m	−2.0 ft
Vapor pressure	−0.4 m	−1.4 ft
NPSH$_{available}$	5.7 m	18.7 ft

Pressure at liquid surface in the tank	359 kPa	52.1 psia
Convert to liquid pressure (52.1 psi * 2.31 ft H$_2$O/ psi / 0.56 SG)	65.5 m	214.9 ft
Elevation between liquid surface and pump	+2.4 m	+8 ft
Friction loss in suction pipe (0.9 m H$_2$O / 0.56 SG)	−1.6 m	−5.4 ft
Vapor pressure converted to liquid pressure	−64.9 m	−212.8 ft
NPSH$_{available}$	1.4 m	4.7 ft

Suction System NPSH for Studies

For studies or initial design, estimate the $NPSH_{required}$ with the general formula [4]:

$$\frac{n\,q^{0.5}}{NPSH^{0.75}} = C \qquad (5\text{-}9)$$

Where:

n = pump speed, rpm
q = flow rate, gpm
C = a constant between 7,000 and 10,000

Multiple Centrifugal Pumps in Parallel or in Series

When two or more centrifugal pumps are piped in series, the heads are additive and the flow is roughly equal to the pump with the smallest capacity. This assumes that the pumps are of approximately equal size.

For pumps operated in parallel, again choose units of approximately equal size. Then, the capacities of the pumps are added and the system head is found at the rate where the flows are recombined.

As a rule of thumb, when two identical pumps are installed in parallel, the flow rate when one pump is operating is about two-thirds the flow when both are operating.

Nomenclature

c	=	acoustic velocity, m/s or ft/s
c_p	=	heat capacity of pumped fluid, kJ/kg °C or btu/lb °F
C_v	=	flow coefficient (gpm at 1 psi pressure drop with water at 60 °F)
d	=	impeller diameter
e	=	pump efficiency, evaluated at the flow and head where the pump is operating
g_c	=	conversion factor, 1 m/s² or 32.17 ft/s²
H	=	pressure, m or ft of fluid head
K_{bulk}	=	bulk modulus, Pa or lb$_f$/in.²
n	=	impeller rotational speed, rpm
P_{shaft}	=	brake power to the pump, kW or hp
q	=	volumetric flow, m³/s or gal/m
SG	=	specific gravity (water = 1)
Δt	=	temperature rise through the pump, °C or °F
U	=	velocity, m/s or ft/s
ρ	=	fluid density, kg/m³ or lb/ft³

References

[1] AIChE Equipment Testing Procedures Committee. Assess the Performance of Positive-Displacement Pumps. *Chemical Engineering Progress*, December 2007:32.

[2] AIChE Equipment Testing Procedures Committee. *Positive Displacement Pumps: A Guide to Performance Evaluation*. 1st ed. Wiley; 2007.

[3] Emerson Process Management. *Control Valve Handbook*. 4th ed. Downloaded from, www.emersonprocess.com; 2005.

[4] Evans FL. In: *Equipment Design Handbook for Refineries and Chemical Plants*. 22nd ed., vol. 1. Gulf Publishing Co; 1979.

[5] Fitzgerald B, Linden C. *The Control Valve's Hidden Impact on the Bottom Line, Part 1*. Maintenance World; 2003.

[6] Hicks TG. *Standard Handbook of Engineering Calculations*. 2nd ed. McGraw-Hill; 1985.

[7] Hydraulic Institute, www.pumps.org/public/pump_resources/faq.htm

[8] Hydraulic Institute. Europump, and U.S. Department of Energy, Variable Speed Pumping: A Guide to Successful Applications. DOE/GO-102004–1913; May, 2004.

[9] McNally Institute. Technical Information for Pumps and Seals, http://www.mcnallyinstitute.com/. Downloaded July, 2010.

[10] Petersen J, Jacoby R. Selecting a Positive Displacement Pump. *Chemical Engineering*, August, 2007:42.

[11] Sarco-Spirax, Ltd. Steam Engineering Tutorials. at, http://www.spiraxsarco.com/resources/steam-engineering-tutorials.asp. Downloaded September 2010.

[12] U.S. Department of Energy. *Hydraulic Institute, Improving Pumping System Performance*. 2nd ed. May, 2006.

[13] Vandell C, Foerg W. The Pluses of Positive Displacement. *Chemical Engineering*, January, 1993:74.

6

Fans, Blowers, and Compressors

Introduction

Fans, blowers, and compressors are used to pressurize and move gases, usually through ducts or pipes. Like pumps, the gas movers are categorized as kinematic or positive displacement. However, gas compression is accompanied by density and temperature increases which change the calculations and the selection factors.

This chapter provides narrative and tabular comparisons and application notes for many different types of fans, blowers, and compressors. It then describes calculations and procedures used to predict performance characteristics, especially head and power.

An Excel workbook with worked examples and VBA functions accompanies this chapter.

Rules of Thumb

Rules of Thumb (Compressors)

- 1 kW = 7 m^3/h at 7 bar *or* 1 hp = 4 cfm at 100 psi.
- 1 to 40 kW (1 to 50 hp), choose a piston (i.e., reciprocating) compressor.
- 75 kW (100 hp) and above, choose a rotary screw or centrifugal compressor.
- A typical compressed air system leaks 20% to 30% of its compressed air capacity, or as little as 10% if a proactive leak detection and repair program is in place.
- 80% to 90% of the electrical energy used by a compressor is converted to heat.
- Compression ratios range from about 1.05 to 7 per stage; use 3.5 to 4 for first pass design.
- Compressor sealing systems are usually limited to a gas temperature of 150 °C (300 °F).

Good Engineering Practice

- Monitor compressed air filters, such as moisture coalescing filters, to ensure they are cleaned or changed when dirty. A typical coalescing filter has a pressure drop of 2 psi. A pressure drop of 6 psi adds 2% to the energy cost for running the compressor.
- Install differential pressure gauges on compressed air inlet filters. A rule of thumb is that a 2 psi pressure drop reduces capacity by 1%.
- Provide pressure sensing points at the compressor discharge, before and after dryers and filters, and throughout the distribution system to each point of use (compressed air systems).

Compressed Air Systems

The Compressed Air Challenge®, Inc. provides this summary of best practices [9]. The first three are called "critical concepts":

- Deliver air at the lowest practical pressure.
- Use storage and automatic system controls to anticipate peak demands. Operate only the number of compressors required to meet the demand at any given time. Operate only one compressor in "trim" control mode.
- Identify leaks and understand cost of leakage. Repair all leaks, beginning with the most significant.
- Make sure that compressed air is the best alternative for the application.
- Use a blower rather than a compressor, if appropriate.
- For applications that do not require air 100% of the time, shut off the air supply when not needed. Turn off the compressed air supply at a process when it is not running.
- All parts of a process may not need air simultaneously. Analyze the peak and average rates of flow to determine actual needs and whether local secondary storage may be advantageous.
- Check the appropriateness of equipment used to control and deliver compressed air, including air compressor controls; primary and secondary receiver sizes; distribution piping size; and in-line filters, regulators, and lubricators.
- Determine the cost of compressed air for each machine or process.
- Follow application of the preceding recommendations with a review of the number of compressors in

operation and their control settings so that a corresponding reduction in energy is realized.

- Make sure that the compressed air supply side personnel are involved in process- and end use-related decisions.

Reference [9], with over 300 pages, has far more useful information than can be provided here, and is highly recommended for design, operating, and maintenance engineers who work with compressed air.

Fans

Fan types and applicability. Fans are normally rated at a standard air density of 1.20 kg/m^3 (0.075 lb/ft^3) at zero elevation (sea level). Size fans carefully; oversized fans create problems that can increase operating costs and decrease reliability. See Tables 6-1 to 6-3 for more information.

Table 6-1
There are two primary types of fan

Fan Type	Application
Axial	0.5 to 0.75 kPa (2 to 3 inches w.c.)
	All static pressure increase is derived from the change in velocity
	Sub categories: propeller, tubeaxial (generate higher static pressure than propeller), and vaneaxial (basically tubeaxial with guide vanes to improve efficiency)
	Although sometimes used interchangeably with centrifugal fans, axial fans are commonly used in "clean air", low-pressure, high-volume applications. Axial fans have less rotating mass and are more compact than centrifugal fans of comparable capacity. Additionally, axial fans tend to have higher rotational speeds and are somewhat noisier than in-line centrifugal fans of the same capacity; however, this noise tends to be dominated by high frequencies which tend to be easier to attenuate [8].
Centrifugal	5 kPa (20 inches w.c.) to 25 kPa (100 inches w.c.)
	Fans with forward curved blades depend more on velocity pressure conversion than on centrifugal force, and are generally less efficient than fans with backward curved blades
	Centrifugal fans are capable of generating relatively high pressures. They are frequently used in airstreams with high moisture and particulate content, in material handling applications, and in systems at higher temperatures [8].

Table 6-2
Centrifugal fan application notes

Attribute	Forward-Curved Blades	Radial-Blade	Radial-Tip	Backward-Inclined
Typical use	Low to medium air volumes at low pressure Limited to clean applications Residential HVAC	Low to medium air volumes at high pressure High-particulate air streams including dust, wood chips, and metal scrap Corrosive gases	Airstreams with small particulates at moderate concentration and airstreams with high moisture content	Clean applications Forced draft service
Performance	Performance curve dips at low flow creating a stall region	Can operate at low flows without vibration problems that usually accompany operating in stall		Three blade shapes available: flat, curved, and airfoil Highly susceptible to unstable operation due to stall (airfoil)
Typical efficiency	55% to 65%		Up to 75%	85% (airfoil)
Control	Difficult to accurately control flow Particularly vulnerable to stall			Particularly vulnerable to stall
Noise	Low when operating at low speeds			
Size	Small compared with other types			
Driver	Careful driver selection required to avoid overload at runout			Power drops off at high flow rates providing a non-overloading characteristic

Summarized from [8].

Table 6-3
Axial fan application notes

Attribute	Propeller	Tubeaxial	Vaneaxial
Typical use	High airflow at low pressure Usually not combined with extensive ductwork Rooftop ventilation systems	Medium pressure, high airflow applications Ducted HVAC Exhausts	Medium to high pressure such as induced draft for boiler exhaust Exhaust
Performance	Can operate in reverse	Uneven airflow profile downstream of fan Low rotating mass so they quickly accelerate to rated speed Can operate in reverse	Outlet vanes provide uniform airflow profile Low rotating mass so they quickly accelerate to rated speed Can operate in reverse
Typical efficiency	Low		85% (airfoil blades)
Control	Particularly vulnerable to stall. Not recommended in systems with widely varying flow requirements unless a means for keeping air flow rates above the stall point, such as a bleed line or recirculation is provided		Can be equipped with variable pitched blades for effective and efficient control
Noise	Comparatively noisy	Moderate	
Size		Relatively space efficient	
Driver	Highest efficiency near free delivery so motors are non-overloading	Power decreases with increased flow so motors are non-overloading Most use belt drives for speeds below 1100 rpm	Frequently driven directly by the motor

Summarized from [8].

Fan Noise

If manufacturer's data is unavailable, the noise generated by a fan can be predicted by using [5]:

$$PWL = 56 + 30 \log\left(\frac{MPM}{304.8}\right) + \log HP \qquad (6\text{-}1)$$

Where:

PWL = sound power level, dB(A)
MPM = tip speed, m/min = fan diameter x pi x fan speed
HP = horsepower (motor, at operating conditions)

Noise attenuates with distance by:

$$SPL_R = PWL - 20 \log(3.28\,R) \qquad (6\text{-}2)$$

Where:

SPL_R = the sound power level at distance R, dB(A)
R = distance from the center of the sound source, line of sight, m

Blower Types

Blower types and applicability [3] are summarized in Table 6-4 below.

Table 6-4
Blower types and application

Blower Type	Application
Regenerative	3.5 to 55 kPa (0.5 to 8 psig) duty, up to 1700 m^3/h (1000 cfm) Rotary centrifugal blower with non-contacting impeller, providing oil-free air or gas Advantages: compact, relatively quiet, cost competitive Disadvantages: pressure limitation, fluctuations in flow and efficiency with pressure changes when compared with PD blowers

(Continued)

Table 6-4
Blower types and application—cont'd

Blower Type	Application
Liquid Ring	0 to 250 kPa (0 to 35 psig) duty, 12 to 20,000 m^3/h (7 to 12,000 cfm)
	Rotary positive displacement blower, used for pressure and vacuum service, sealed with a low viscosity liquid (usually water, but can be any liquid that is compatible with the blower and process). Discharged gas is usually the same temperature as the incoming sealing liquid, and is saturated with the liquid.
	Advantages: excellent for use in severe operating conditions, extremely simple and low maintenance, delivered air is cool and free of dust or oil, quiet
	Disadvantages: higher power consumption, large quantity of cooling water needed, cost of exotic materials to withstand corrosion from sealing liquid (in some cases)
Rotary Lobe	Single stage 14 to 100 kPa (2 to 15 psig); multi-stage to 275 kPa (40 psig) with interstage cooling; to 85,000 m^3/h (50,000 cfm) Vacuum to 50 kPa (15 inches Hg)
	PD blower, used for pressure and vacuum service, up to 2:1 compression ratio. Non-contacting impellers operate dry, producing oil-free air or gas.
	Advantages: minor variations in flow compared to differential pressure, reliable, competitive cost
	Disadvantages: noisy, small internal clearances are unforgiving if solids or dusts are present
Multistage Centrifugal	14 to 310 kPa (2 to 45 psig); 85 to > 170,000 m^3/h (50 to >100,000 cfm)
	Pressure lubricated. Horizontally split machines provide access to rotors, diaphragms, and bearings without disturbing piping. Vertically split machines are modular, with ability to change the number of compression stages to meet operating requirements.
	Advantages: gas flow can be controlled with a suction throttling valve at constant speed, inlet guide vanes, or speed variation
Helical Lobe	To 275 kPa (40 psig); 170 to 14,000 m^3/h (100 to 8,000 cfm)
	PD blower with external timing gears to precisely control the position of the rotors. Non-contacting impellers operate dry, producing oil-free air or gas.
	Advantages: low maintenance
Single Stage Centrifugal	35 to 240 kPa (5 to 35 psig); 1,700 to 500,000 m^3/h (1,000 to 300,000 cfm)
	Variable flow machines commonly used to move air or gas up to a 3:1 compression ratio for either pressure or vacuum duty. Flow control with suction throttling valve, inlet guide vanes, discharge diffusion vanes, or speed variation.
	Advantages: High efficiency over a wide operating range, oil and pulsation free, flow and power consumption can be reduced up to 50% without using a recycle or blow-off line
	Disadvantages: Lower capacity machines more expensive than PD alternatives

Compressor Types

Gallick, Phillippi, and Williams have written an excellent comparison of reciprocating and centrifugal compressor types [4]. The main points are summarized in Table 6-5.

Ohama, et. al. [11] and Bruce [2] make the case for using screw compressors instead of reciprocating or centrifugal machines in oil and gas fields, in petroleum refineries, and for petrochemical production. There are significant advantages, including higher reliability, lower maintenance costs, lower consumed power at unloaded condition, and suitability for process fluctuation such as gas composition and pressure. See Tables 6-6 and 6-7.

Table 6-5
Comparison of reciprocating and centrifugal compressor types [4]

Attribute	Reciprocating Compressors	Centrifugal Compressors
Maximum discharge pressure	828 bar (12,000 psi); special compressors for low density polyethylene production discharge to 3500 bar (50,000 psi)	100 bar (1450 psi) for horizontally split compressors 1034 bar (15000 psi) for radially split compressors
Maximum suction pressure	Atmospheric or slight vacuum	Atmospheric or below

Table 6-5
Comparison of reciprocating and centrifugal compressor types [4]—cont'd

Attribute	Reciprocating Compressors	Centrifugal Compressors
Maximum discharge temperature	Hydrogen rich service limited to 135 °C (275 °F) by API 618 (1995) Natural gas service usually limited to 150 °C (300 °F); up to 175 °C (350 °F) permissible Air compressors discharge in excess of 200 °C (400 °F)	Compressor design limits discharge temperature to 204 °C to 232 °C (400 °F to 450 °F) Process conditions may limit temperature due to fouling, downstream components, and process efficiency
Minimum suction temperature	−40 °C (−40 °F) due to common cylinder materials Lower temperature applications require very special materials	−19 °C to −46 °C (−20 °F to −50 °F) due to standard centrifugal compressor materials Special materials such as stainless steels used for lower temperature applications
Maximum flow	Limited by cylinder size and number	680,000 m³/h (400,000 ft³/min) in a single body, limited by the compressor's choke point which is the point where the velocity through the some part of the compressor nears Mach 1
Minimum flow Flow range	Very small reciprocating compressors are available 20% (or lower) to 100% through speed change, clearance pockets, cylinder end deactivation, or system recycle	A few hundred cfm, limited by the surge point 70% to 100% (fixed speed); 50% to 100% (variable speed or with inlet guide vanes)
Weight	Heavier than centrifugal on a mass per power basis	Driver, baseplate, and auxiliary systems contribute significantly to the overall weight
Size	In general, the higher the speed, the smaller the size	Generally a function of flow capacity (sets the diameter) and number of stages (sets the length). Diameters range from 500 mm (20 inches) to 3800 mm (150 inches)
Reliability	Less reliable due to large number of parts and more rubbing seals that wear and require more frequent replacement than any seal or part in a centrifugal machine. Liquid or solid debris in the gas significantly increases wear	Availability is typically 98% to 99%
Typical Maintenance Intervals	Vary significantly with application. Valve and seat element intervals range from a few months to three-to-five years. Major overhaul every 10 years or longer	Per API 617 (2002) a centrifugal compressor must be designed for at least five years of uninterrupted service; they can operate continuously for 10 years or longer
Compressed Gas Molecular Weight	No limit	Compression ratio is highly dependent on molecular weight
Compression Ratio	Limited by discharge temperature. Typically 1.2 to 4.0 per stage	For a specific gas, compression ratio is limited by speed and number of stages in a single body. Temperature can be controlled with intercooling
Materials	Common materials such as iron, carbon steel, alloy steel, and stainless steel	Carbon, alloy, and stainless steel
Multiservice Capability	Easy, limited only by the number of "throws" available (up to 12 per frame)	Not typical
Efficiency	Adiabatic efficiency curve. Efficiency drops with compression ratio and also decreases with increased molecular weight	Polytropic efficiency usually used for centrifugal compressors (70% to 85%), except air compressors which use adiabatic efficiency.
Cost: Capital and Operating	Lower capital but higher operating costs compared with centrifugal	Higher capital but lower operating costs compared with reciprocal
Minimum/Maximum Power	Under 7.5 kW (10 hp) to 9 MW (12,000 hp)	75 kW (100 hp) to 97 MW (130,000 hp) or more
Lead Time	14 to 40 weeks for a bare compressor; critical path may be the motor, depending on power rating	35 to 75 weeks, affected by shop loading, availability of special materials/parts, and special or unique design requirements
Installation Time and Complexity	Varies significantly with size and whether or not the compressor is packaged. Packaged units are available to 3.4 MW (5,000 hp) and can be installed in less than 2 weeks. Machines assembled on site may require 3 to 4 weeks to install	Similar to reciprocal

Table 6-6
Application in the natural gas compression, petroleum and petrochemical industries [11]

Attribute	Reciprocating Compressor		Screw Compressor		Centrifugal Compressor
	Lube	Non-Lube	Oil Flooded	Oil Free	
Maximum Discharge Pressure	300 barG (4500 psig)	100 barG (1500 psig)	100 barG (1500 psig)*	40 barG (600 psig)	200 barG (3000 psig)
Maximum single-stage pressure ratio	3:1	3:1	> 50:1	4:1 to 7:1	1.5:1 to 3:1
Maximum Inlet Flow	15000 m³/h (8800 cfm)	1500 m³/h (8800 cfm)	25000 m³/h (15000 cfm)	70000 m³/h (41000 cfm)	400,000 m³/h+ (240,000 cfm+)
Turndown accomplished by:	Suction valve unloaders (step and stepless) Clearance pockets Bypass	Suction valve unloaders (step and stepless) Clearance pockets Bypass	Slide valve (15% to 100%) stepless Bypass	(None) Bypass	Inlet guide vane Speed control (70% to 100%) Bypass
Polymer Gas	Difficult	Difficult	Difficult	Possible	Difficult
Dirty Gas	Possible	Difficult	Possible	Possible	Difficult

*Bruce gives a maximum discharge pressure of 23 barG (350 psig) for screw compressors, stating that "There are some screw machines available capable of operating at higher pressures by using cast steel casings but these are not yet commonly used in the natural gas industry due to capital cost and availability" [2].

Table 6-7
Compressed air applications. 100 cfm = 170 m³/h

Compressor Type	Application Range	Advantages (air service)	Disadvantages (air service)	Operating Cost
Centrifugal (dynamic)	Pressure to 300 psig	— lubricant-free air delivery — generally well packaged with no need for special foundations — relatively smooth air delivery — relative first cost per unit of power improves with size	— limited constant discharge pressure capacity control range — requires unloading for reduced capacity — need for specialized bearings for high rotational speeds and monitoring of vibrations and internal clearances — specialized maintenance considerations	16 to 20 kW/100 cfm
Rotary Screw, lubricant injected (PD)	Single-stage: 8 to 4,000 cfm 50 to 250 psig Two-stage: Decreases power by 12% to 15% 3 to 700 hp	— complete compact package — relatively low first cost — vibration-free; no need for special foundation — part-load capacity control systems can match system demand — routine maintenance includes lubricant and filter changes	— less efficient at full and part load than water-cooled reciprocating type — potential problem of oil carryover; requires proper maintenance of air/lubricant separator element — periodic lubricant changes	Single-stage: 18 to 19 kW/100 cfm Two-stage: 16 to 17 kW/100 cfm
Rotary Screw, lubricant free (PD)	Dry type range from 80 to 4,000 cfm; single stage to 50 psig, two-stage to 150 psig Water injected type single stage to 150 psig	— completely packaged — designed to deliver lubricant-free air — no need for special foundations	— cost premium over lubricant-injected type — less efficient than lubricant-injected type — limited to load/unload capacity control unless variable displacement or variable speed control is available	18 to 22 kW/100 cfm

Table 6-7
Compressed air applications. 100 cfm = 170 m³/h—cont'd

Compressor Type	Application Range	Advantages (air service)	Disadvantages (air service)	Operating Cost
Sliding Vane, lubricant injected (PD)	40 to 800 acfm 80 to 125 psig 10 to 200 hp	— complete compact package — relatively low first cost — vibration-free; no need for special foundation — routine maintenance includes lubricant and filter changes	— higher maintenance costs, with recommended periodic air end replacement — capacity control limitations — less efficient than rotary screw type	21 to 23 kW/100 cfm
Single Screw, lubricant injected (PD)	30 to 450 acfm Versions for refrigeration from 200 to 2,500 acfm 5 to 100 hp	— complete compact package — relatively low first cost — vibration-free; no need for special foundation — routine maintenance includes lubricant and filter changes	— less efficient at full and part load than water-cooled reciprocating type — capacity control limitations — potential problem of oil carryover; requires proper maintenance of air/lubricant separator element — periodic lubricant changes	Single-stage: 18 to 19 kW/100 cfm Two-stage: 16 to 17 kW/100 cfm
Rotary Scroll, (PD)	6 to 14 acfm up to 145 psig 2 to 5 hp	— completely packaged — relatively efficient operation — can be lubricant-free — low noise levels — air cooled	— limited range of sizes in the lower capacity range	20 to 22 kW/100 cfm
Single-Acting Reciprocating (PD)	Single-stage from 25 to 125 psig Two-stage from 125 psig to 175 psig More stages above 175 psig Up to 150 hp, but are much less common above 25 hp	— small size and weight — generally can be located close to point of use, avoiding lengthy piping runs and pressure drops — integral cooling system — simple maintenance procedures	— lubricant carryover when rings wear — relatively high noise — relatively high cost of compression — generally designed to run not more than 50% of the time (some models have duty cycle of 70% to 90%) — generally compress and store air in a receiver at a pressure higher than required at point of use	22 to 24 kW/100 cfm
Double-Acting Reciprocating (PD)	Discharge pressure to several thousand psi Single-stage common for 100 psig, but two-stages with intercooling gives better efficiency Range from about 10 hp to 1,000 hp; not often used for compressed air applications	— heavy duty, continuous service — efficient compression, particularly with multi-stage compressors — three-step (0–50–100%) or five-step (0–25–50–75–100%) capacity controls — relatively routine maintenance procedures	— relatively high first cost compared with rotary compressors — relatively high space requirements — relatively high vibrations require high foundation costs, especially for single-cylinder types — lubricant carryover when rings wear — larger compressors seldom sold as complete packages — requires flywheel mass to overcome torque and current pulsations in motor driver — repair procedures require some training and skills	15 to 16 kW/100 cfm

Summarized from [8].

Surge Control

Centrifugal compressors have a low flow limitation, usually at about 50% to 70% of rated flow. Whether controlled with a variable-speed drive or by using suction or discharge vanes or valves, if the flow through the compressor is reduced to the low flow limitation point, an unstable condition called "surge" occurs. At the surge point, the gas flows alternately forward and backward through the compressor at a frequency of roughly 2 seconds. This is accompanied by increased noise, vibration, and heat. Prolonged operation at the surge point can damage the compressor.

Control systems prevent surge, usually by recycling flow from the compressor discharge to its suction, or venting flow to atmosphere (especially for air compressors). Both methods result in increasing the flow through the compressor while delivering lower flow to the process. Good practice requires cooling of recycled flow.

Performance Calculations

Engineers frequently need to estimate the power required to compress a gas stream. For a new installation with well-defined requirements, the compressor manufacturers are best prepared to evaluate the data and provide a suitably sized machine. However, for conceptual work, retrofits, and proposed changes to existing installations, the plant engineer often does the analysis.

Compressor calculations and specifications are almost always expressed in terms of the air volume at the compressor's inlet conditions. Unless explicitly defined, assume that volume units "cfm," "acfm," and "icfm" are interchangeable and identical units that reference the temperature and pressure at the inlet to the compressor.

"Free Air Delivery" (FAD) is sometimes used to rate air compressors. The actual discharge from the compressor is converted to standard conditions, defined either as 1 bar (abs) pressure and 20 °C *or* simply the pressure and temperature at the compressor inlet. A way to approximately estimate FAD for an existing air compressor is to measure the pump-up time of the receiver. Then,

$$FAD = (P_2 - P_1)\frac{V_R\, t}{P_{atm}} \qquad (6\text{-}3)$$

Where:

FAD = free air delivery, m^3/h or ft^3/m
P_2 = final pressure in the receiver, kPa or psig
P_1 = initial pressure in the receiver, kPa or psig
P_{atm} = pressure at the inlet to the compressor, kPa or psig
V_R = receiver volume, m^3 or ft^3
t = time, h or min

Positive displacement compressors generally follow an adiabatic, or isentropic, compression path. If multi-stage, interstage cooling is often provided. The GPSA equation for reciprocating compressors considers these facts.

Dynamic compressors (e.g., centrifugal) generally follow a polytropic compression path, where the relationship PV^n remains constant. Equations are given for estimating the power requirements of dynamic compressors.

Isothermal compression, while treated in thermodynamic texts, is not found in practice.

Fan calculations often assume the gas is incompressible, since pressure and temperature changes through a fan are very low.

Definitions

Head. Force to compress a unit mass of gas. Units are N-m/kg or ft-lb$_f$/lb$_m$. When a performance curve for a dynamic pressure is labeled "ft" on the ordinate, it means ft-lb$_f$/lb$_m$.

Work. Force over time to compress a constant mass flow of gas. Units are kilowatts or horsepower.

Power. Force over time that an actual compressor needs to compress a constant mass flow of gas. Divide Work by Efficiency. Units are kilowatts or horsepower.

Efficiency. Compression efficiency is the ratio of theoretical power to the power actually imparted by the compressor. See Table 6-8. Compressor efficiency accounts for mechanical losses in the compressor gears, bearings, etc. Additional efficiencies include the motor (or driver) efficiency and variable speed controller efficiency (if applicable).

Table 6-8
Centrifugal compressor flow range

Nominal Flow Range (inlet m³/h)	Nominal Flow Range (inlet ft³/m)	Average Polytropic Efficiency	Average Isentropic Efficiency	Speed to Develop 30,000 N-m/kg (50,000 ft-lb$_f$/lb$_m$)head per Wheel
170 – 850	100 – 500	0.70	0.67	20,500
850 – 12,700	500 – 7,500	0.80	0.78	10,500
12,700 – 34,000	7,500 – 20,000	0.86	0.83	8,200
34,000 – 56,000	20,000 – 33,000	0.86	0.83	6,500
56,000 – 94,000	33,000 – 55,000	0.86	0.83	4,900
94,000 – 136,000	55,000 – 80,000	0.86	0.83	4,300
136,000 – 195,000	80,000 – 115,000	0.86	0.83	3,600
195,000 – 245,000	115,000 – 145,000	0.86	0.83	2,800
245,000 – 340,000	145,000 – 200,000	0.86	0.83	2,500

If available obtain efficiency values from the compressor manufacturer rather than from this table. (GPSA [5])

Affinity Laws

Dynamic type air movers (e.g., centrifugal fans, centrifugal compressors) obey these affinity laws: The affinity laws only apply to single stages or multi-stages with very low compression ratios or very low Mach numbers [1].

Volume capacity:

$$\frac{q_1}{q_2} = \left(\frac{n_1}{n_2}\right)\left(\frac{d_1}{d_2}\right)^3 \tag{6-4}$$

q = volume flow capacity, m³/s, ft³/min, etc.
n = wheel rotational speed, rpm
d = wheel diameter

Pressure (or head) relationship:

$$\frac{\Delta P_1}{\Delta P_2} = \left(\frac{n_1}{n_2}\right)^2\left(\frac{d_1}{d_2}\right)^2\left(\frac{\rho_1}{\rho_2}\right) \tag{6-5}$$

ΔP = the pressure increase imparted by the fan, Pa, psi, m of fluid, ft of fluid
ρ = gas density at inlet conditions

Power consumption relationship [12]:

$$P_{shaft} = \left(\frac{n_1}{n_2}\right)^3\left(\frac{d_1}{d_2}\right)^5\left(\frac{\rho_1}{\rho_2}\right) \tag{6-6}$$

P_{shaft} = the shaft power of the fan (compression power divided by compressor efficiency), assuming constant efficiency (true for speed changes and diameter changes up to 10%)

At constant flow, horsepower and pressure vary inversely with absolute temperature. Similarly, horsepower and pressure vary directly with density.

Power – Fan

Use these equations for estimating the shaft power for a fan:

$$Power(kW) = \frac{Q\Delta P}{3,600,000\,\eta_f} \tag{6-7}$$

Q = fan volume, m³/h
ΔP = pressure rise, Pa
η_f = fan efficiency

$$Power(hp) = \frac{Q\,P}{6350\,\eta_f} \tag{6-8}$$

Q = fan volume, ft³/min
ΔP = pressure rise, inches H$_2$O

Power – Reciprocating Compressor

GPSA gives this equation for obtaining a "quick and reasonable estimate for [reciprocating] compressor horsepower". It was developed for large, slow-speed (300 to 450 rpm) compressors handling gases with a relative density of 0.65, and having stage compression ratios above 2.5. For high-speed compressors, add up to 20% additional power, but consult with the compressor manufacturer [5].

$$Brake\ Power = C\ F\ R\ N\ Q \qquad (6\text{-}9)$$

$Brake\ Power$ = compressor shaft power, kW or hp

$C =$

in SI units

0.010 to 0.012 for compression ratios from 1.5 to 2.0
0.013 for gas with specific gravity 0.8 to 1.0
0.014 for compression ratio above 2.0 and gas with specific gravity < 0.8

in US units

0.023 to 0.027 for compression ratios from 1.5 to 2.0
0.030 for gas with specific gravity 0.8 to 1.0
0.032 for compression ratio above 2.0 and gas with specific gravity < 0.8

$F =$

1.0 for single-stage compression
1.08 for two-stage compression
1.10 for three-stage compression

$R =$ compression ratio per stage

$N =$ number of stages
$Q =$ gas flow rate evaluated at 100 kPa (14.5 psia) and intake temperature, m^3/h or ft^3/m

Example

What shaft power is needed to drive a reciprocating compressor that compresses 500 m^3/h (295 ft^3/m) of air at 25 °C from 101.3 kPa (14.7 psia) to 900 kPa (130 psia)?

1. The overall compression ratio is $900/101.3 = 8.88$. A two-stage compressor has a compression ratio of $8.88^{(1/2)} = 2.98$ per stage which seems reasonable.
2. The ratio of specific heats for air is 1.4. Estimate the discharge temperature from the first stage using Equation 6-10, setting n = 1.4. T2 = 134 °C. Since this is less than 150 °C use a two-stage compressor (page 119).
3. Correct the volumetric flow to a pressure of 100 kPa. Q = 500 (101.3/100) = 506.5.
4. Calculate shaft power with C = 0.013, F = 1.08, R = 2.98, N = 2, and Q = 506.5. Assume 10% additional because this is a small compressor operating at speed above 500 rpm. Answer: Shaft Power = 47 kW.
5. If the motor efficiency is assumed to be 94%, the motor power is 50 kW (67 hp), or 16.9 kW/100 cfm.

Power – Centrifugal Compressor

When a pressure-enthalpy diagram is not available, the following procedure may be used. Calculations are performed *per compressor stage* and then summed. Multiple wheels within one casing are treated as a single stage unless intercooling is provided. Use the polytropic equation unless the compressor manufacturer specifies otherwise. Be sure to account for interstage cooling. Obtain the polytropic efficiency from the compressor manufacturer or from Table 6-8.

1. Obtain the following process data: volumetric flow rate at inlet conditions, inlet pressure, discharge pressure, inlet temperature, molecular weight, heat capacity ratio, critical pressure, and critical temperature. The critical properties are used to

calculate the compressibility factor and may be omitted for applications where perfect gas behavior is assumed.

2. Calculate the compression ratio per stage.

$$Ratio_{stage} = \left(\frac{P_2}{P_1}\right)^{1/Number\ of\ Stages} \qquad (6\text{-}10)$$

Where:

$P_2 =$ Compressor discharge pressure, absolute
$P_1 =$ Compressor inlet pressure, absolute

3. Calculate the polytropic exponent (for isentropic compression, the exponent $= \gamma$)

$$n = \frac{\eta_p\ \gamma}{1 + \eta_p\ \gamma - \gamma} \qquad (6\text{-}11)$$

Where:

η_p = polytropic efficiency (provided by the compressor manufacturer)

γ = ratio of specific heats

4. Calculate the discharge temperature from the first stage

$$T_2 = T_1(Ratio_{stage})^{\frac{n-1}{n}} \tag{6-12}$$

Where:

T_2 = Stage discharge temperature, absolute

T_1 = Stage inlet temperature, absolute

5. If the discharge temperature exceeds $200\,°C$ ($390\,°F$), check the assumptions and calculations. Consider increasing the number of stages, or consult with the compressor manufacturer.

6. Calculate the stage inlet density

$$\rho_1 = \frac{P_1 M_w}{R T_1} \tag{6-13}$$

Where:

ρ_1 = density, kg/m^3 or lb/ft^3

P_1 = stage inlet pressure, kPa or $lb/in.^2$, absolute

M_w = molecular weight, kg/kg-mole or lb/lb-mole

R = gas constant, $8314.5\ \frac{m^3\ kPa}{K\ gmol}$ or $10.732\ \frac{ft^3\ psi}{R\ lbmol}$

T_1 = stage inlet temperature, K or R

7. Calculate the stage polytropic or isentropic head. Note the change in US pressure units

$$W_p = \frac{P_1 Z n}{\rho_1(n-1)}\left[\left(\frac{P_2}{P_1}\right)^{\frac{n-1}{n}} - 1\right] \tag{6-14}$$

$$W_{\Delta S=0} = \frac{P_1 Z \gamma}{\rho_1(\gamma-1)}\left[\left(\frac{P_2}{P_1}\right)^{\frac{\gamma-1}{\gamma}} - 1\right] \tag{6-15}$$

Where:

W_p = polytropic head, kN-m/kg or ft-lb_f/lb_m

$W_{\Delta S=0}$ = isentropic head, kN-m/kg or ft-lb_f/lb_m

P_1 = stage inlet pressure, kPa or lb/ft^2

P_1 = stage discharge pressure, kPa or lb/ft^2

Z = average of inlet and discharge compressibility factors (see Chapter 27)

8. Repeat for each compression stage. Base the inlet temperature to subsequent stages on the temperature of cooling water, with an approach of $10\,°C$ ($18\,°F$) or less. For a more rigorous calculation, assume a pressure loss of 35 kPa (5 psi) in the interstage cooler; this requires a corresponding increase in compression ratio to obtain the desired stage discharge pressure.

9. Sum the heads from the stage calculations.

10. Convert head to gas power using the polytropic or isentropic efficiency value.

For SI Units:

$$Power_{gas} = W\frac{Q\rho}{3600\ \eta} \tag{6-16}$$

For U.S. Units:

$$Power_{gas} = W\frac{Q\rho}{33000\ \eta} \tag{6-17}$$

Where:

$Power_{gas}$ = power to compress the gas, kW or hp

$W = W_p$ or $W_{\Delta S=0}$

Q = flow rate at inlet conditions, m^3/h or ft^3/min

ρ = density at inlet conditions, kg/m^3 or lb/ft^3

η = polytropic or isentropic efficiency

11. Use this formula to account for mechanical losses in the compressor (friction, gears, etc.).

$$Power_{losses} = Power_{gas}{}^{0.4} \tag{6-18}$$

12. Calculate the shaft power and total power.

$$Power_{shaft} = Power_{gas} + Power_{losses} \tag{6-19}$$

$$Power_{total} = \frac{Power_{shaft}}{\eta_{drive}\ \eta_{motor}} \tag{6-20}$$

Where:

η_{drive} = efficiency of the drive train, such as a VFD

η_{motor} = motor efficiency

Comparison with Manufacturer's Data

Manufacturers publish measured power ratings for air compressors in accordance with the standardized method published by the Compressed Air and Gas Institute (www.cagi.org).

Figure 6-1 shows the specific power (kW/100 cfm) for a variety of rotary screw and rotary vane compressors, all with 75 kW motors. The two-stage units in the lower trend (e.g., 16 to 17 kW/100 cfm at a compression ratio of 7.8)

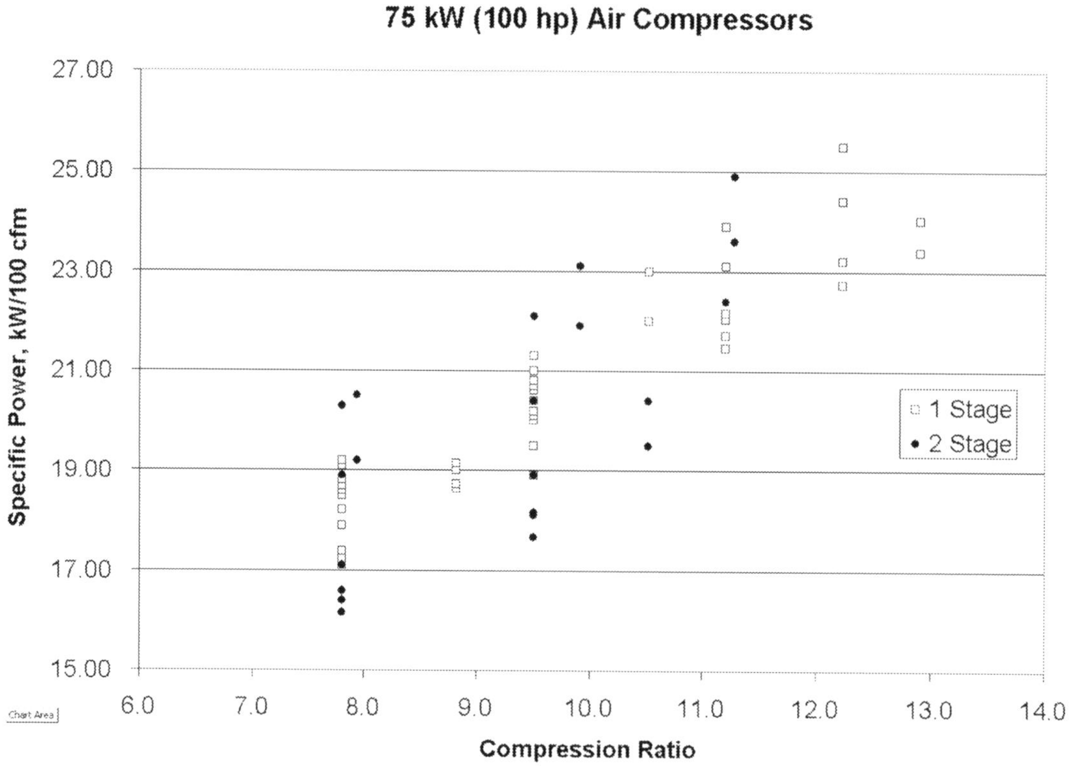

75 kW (100 hp) Air Compressors

Figure 6-1. Power consumption for 62 compressors reported by the compressor manufacturers in accordance with CAGI guidelines. (Sources: Atlas Copco, Comp Air, Ingersoll Rand, Kaeser Compressors, Mattei Compressors, Inc., Quincy Compressor, and Sullair Corp.)

are oil injected compressors, while those at the upper range are non-injected. Being positive displacement machines, the reciprocating compressor formula (Equation 6-9, page 128) might be expected to provide the best power estimate. It returns answers that are about 30% too high for the single-stage compressors, 7% too high for the two-stage lubricated machines, and 10% too low for the two-stage non-lubricated units.

Using the formulae from the centrifugal compressor section (pages 128–129), assuming isentropic compression with an efficiency of 90%, the results are better. The answers are about 11% too high for the single-stage compressors, 8% too high for the two-stage lubricated

units, and 11% too low for the two-stage non-lubricated machines. Assuming 80% efficiency, the agreement with the two-stage non-lubricated results is within about +/−5%.

Oil injection explains the discrepancy between calculated and published data. The injected oil removes much of the heat of compression, achieving near isothermal operation [13].

Engineers who need to estimate compressor power should attempt to obtain data for actual compressors from manufacturers, and use the calculations in this chapter for sanity checking, rough order-of-magnitude work, or as a last resort.

Power with P-E Diagram

Another way to estimate the power consumption for a blower or compressor is to use a P-E (pressure-enthalpy) diagram to find the work needed to compress the gas, and divide by an assumed efficiency. This method eliminates the need to find the compressibility factor because the P-E

diagram is based on experimental measurements or a tested Equation of State.

The method described is from GPSA [1]. Although explicitly stated for centrifugal compressors, it is a valid method for any type of compressor if the compressor's efficiency and compression path are known.

P-E data is often published in tabular, rather than graphical, form. An Excel worksheet that interpolates tabular data accompanies this chapter; the example is solved in the worksheet, and the chart is created by Excel.

Example

What shaft power is needed to drive a reciprocating compressor that compresses 500 m³/h (295 ft³/min) of air at 25 °C from 101.3 kPa (14.7 psia) to 900 kPa (130 psia)? Assume that the adiabatic efficiency is 80% and that intercoolers reduce the temperature of the compressed gas to 30 °C.

1. The overall compression ratio is 900/101.3 = 8.88. A two-stage compressor has a compression ratio of $8.88^{(1/2)} = 2.98$ per stage which seems reasonable.
2. The ratio of specific heats for air is 1.4. Estimate the discharge temperature from the first stage using Equation 6-10, setting n = 1.4. T2 = 134 °C. Since this is less than 150 °C use a two-stage compressor (page 119).

3. Locate the coordinate for pressure (1 bar) and temperature (298 K) on a Pressure-Enthalpy diagram for air. Read the enthalpy from the x-axis to be 298 kJ/kg. See Figure 6-2, Point 1.
4. Follow the line of constant entropy (S) to the stage discharge pressure (2.98 bar). Read the enthalpy from the x-axis to be 418 kJ/kg. See Figure 6-2, Point 2.
5. Apply the adiabatic efficiency to calculate the discharge enthalpy at Point 3, H_3.

$$H_3 = \frac{H_2 - H_1}{\eta} + H_1 = \frac{418 - 298}{0.80} + 298$$
$$= 448$$

6. The work of compression for the first stage is $H_3 - H_1$, or 150 kJ/kg.
7. Interpolate the chart to determine that the discharge temperature is 450 K (177 °C).
8. For the second stage, locate the coordinate for pressure (2.98 bar) and temperature (30 °C or 303 K) on the chart and repeat steps 3 through 6. The result (not shown) is 142 kJ/kg.

Enthalpy-Pressure Diagram for Air

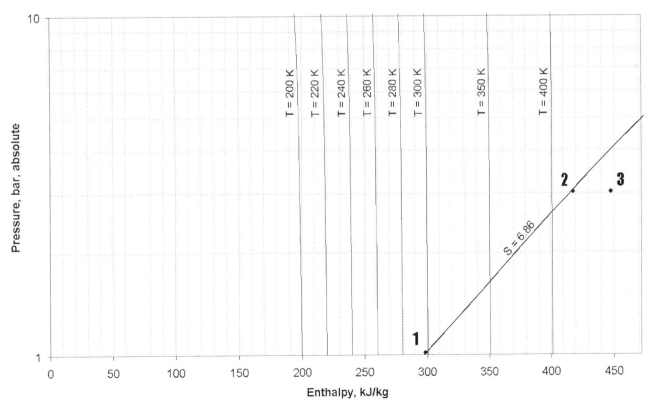

Figure 6-2. Isentropic compression path for one stage of compression. It is coincidental that the temperature values are approximately equal to the enthalpy values.

9. Add the results from the two stages to get 292 kJ/kg.
10. Calculate the mass flow rate using the gas temperature, pressure, and molecular weight at inlet conditions, assuming perfect gas behavior:

$$w = Q\,\rho = Q\frac{P\,M_w}{R\,T} = \frac{(500)(101.3)(29)}{(8314)(25+273.15)}$$

$$= 594\ kg/h$$

11. Calculate the gas power:

$$Power_{gas} = \Delta H\,w = 292\frac{kJ}{kg}\ 594\frac{kg}{h}\ \frac{h}{3600\ s}$$

$$= 48.2\ kW$$

12. Add for mechanical losses in the compressor (equations 6-16 and 6-17):

$$P_{shaft} = Power_{gas} + Power_{gas}{}^{0.4}$$

$$= 48.2 + 48.2^{0.4} = 52.8\ kW$$

This answer is 12% higher than the shortcut formula (Equation 6-9, page 128). The gap is easily eliminated by adjusting assumptions for efficiency, mechanical losses, and intercooled temperature proving once again the validity of using shortcut methods while reinforcing the need to exercise caution when applying the results.

Cost of Compressed Air

Use this formula to estimate the cost of compressed air:

$$Annual\ Cost = BHP \times 0.746\frac{kW}{BHP}\ \frac{Electric\ Rate \times Annual\ Hours \times Load\ Factor}{Motor\ Efficiency} \tag{6-21}$$

If the load factor is unknown, use 80%.
If the motor efficiency is unknown, use 93%.

A more accurate treatment would also consider electricity demand charges and on/off peak electric rates.

Receiver Volume

The minimum size of the receiver for a compressed air system is:

$$V_{min} = D\frac{P_i}{P_d} \tag{6-22}$$

Where:

V_{min} = minimum volume of receiver
D = compressor displacement, volume free air per minute (only the first stage if a multi-stage compressor)

P_i = inlet pressure, absolute
P_d = discharge pressure, absolute

A safety factor of 50% to 100% is usually added.

Time to Pump System to Pressure

Use the following formula to estimate the time (in minutes) to pump a system from an initial pressure, P_1, to a final pressure, P_2.

$$t = V_{system}\frac{(P_2 - P_1)}{P_{atm}\,D} \tag{6-23}$$

V_{system} = volume of pressurized system (receiver and piping)
P_{atm} = absolute pressure of 1 atmosphere in compatible units

Nomenclature

t	=	time, minutes
D	=	compressor displacement, volume free air per minute (only the first stage if a multi-stage compressor)
H	=	enthalpy, kJ/kg or Btu/lb
M_w	=	molecular weight
P_{atm}	=	absolute pressure of 1 atmosphere (101 kPa or 14.7 psia)
P_i	=	compressor inlet pressure, absolute
P_d	=	compressor discharge pressure, absolute
P_{shaft}	=	shaft power (brake power), kW or hp
Q	=	volume flow, m^3/h or ft^3/min
S	=	entropy, kJ/kg or Btu/lb
T	=	absolute temperature, K or R
t	=	time, h or min
V_{min}	=	minimum volume of receiver
V_{system}	=	volume of pressurized system (receiver and piping)
W_p	=	polytropic head, kN-m/kg or ft-lb$_f$/lb$_m$
$W_{\Delta S=0}$	=	isentropic head, kN-m/kg or ft-lb$_f$/lb$_m$
w	=	mass flow, kg/h or lb/h
Z	=	average of inlet and discharge compressibility factors (see Chapter 27)
γ	=	ratio of specific heats
η	=	efficiency
ρ	=	density, kg/m^3 or lb/ft^3

References

[1] Ask the Experts. Best Practices for Compressor Operation. *Chemical Engineering Progress*, September, 2006;14.

[2] Bruce J. *Screw Compressors: A Comparison of Applications and Features to Conventional Types of Machines*. Calgary, AB, Canada: Toromont Process Systems; 2001.

[3] Compressed Air and Gas Institute. Blower Selection Program. www.cagi.org

[4] Gallick P, Phillippi G, Williams B. What's Correct for my Application – A Centrifugal or Reciprocating Compressor? *Proceedings of the Thirty-Fifth Turbomachinery Symposium*. Texas A&M University; 2006.

[5] Gas Processors Suppliers Association (GPSA). *Engineering Data Book, SI Version*, vol. 1, 12th ed. 2004.

[6] Golden S, Fulton S, Hanson D. Understanding Centrifugal Compressor Performance in a Connected Process System. *Petroleum Technology Quarterly*, Spring; 2002.

[7] Green D, Perry R. *Perry's Chemical Engineers' Handbook*. 8th ed. McGraw-Hill; 2008.

[8] Lawrence Berkeley National Laboratory. *Resource Dynamics Corporation, Improving Fan System Performance: A Sourcebook for Industry*. U.S. Department of Energy and Air Movement and Control Association International (AMCA); 1989.

[9] McCulloh D, Scales W. *Best Practices for Compressed Air Systems. The Compressed Air Challenge®*, 2nd ed. 2009.

[10] McCulloh D. Don't err with Air Compressors. *Chemical Processing*, www.chemicalprocessing.com/articles/2008/050.html; 2008.

[11] Ohama T, Kuroka Y, Tanaka H, Koga T. Process Gas Applications Where API 619 Screw Compressors Replaced Reciprocating and Centrifugal Compressors. *Proceedings of the Thirty-Fifth Turbomachinery Symposium*. Texas A&M University; 2006.

[12] Pontyak Resource Centre, www.pontyak.com

[13] Renz H. Design and Application of Small Screw Compressors – Part 1. *Air Conditioning and Refrigeration Journal*, April-June 2000. Indian Society of Heating, Refrigerating, and Air Conditioning Engineering.

7

Drivers

Introduction

Several technologies are used to drive machines in the petroleum and chemical process industries. This chapter provides information about electric motors, air motors, hydraulic motors, steam turbines, gas expanders, and internal combustion engines.

Electric motors are by far the most commonly used driver. Reasons why a *different* driving technology is chosen include:

- Electricity not available (e.g., remote location)
- Extremely large load (torque and power)
- Backup required during power outages (often used for fire water pumps)
- Hazardous location
- Portability needed
- Unique features of alternative driver are compelling

Electric Motors

In the US, most motors are the induction type supplied with three-phase, 240- or 480-volt power. Power supplies are typically rated at slightly higher voltages than motors because of anticipated voltage drops in the distribution systems, so motors are typically rated at 230 or 460 volts.

In Europe, where there is 50 Hz power, voltages vary but are now being standardized at 230V single-phase and 400V three-phase. Other popular voltages are 220/380 and 240/415. Motors rated for 400 volts, three-phase, normally have sufficient tolerance to run on 380-volt or 415-volt systems without damage.

A characteristic of induction motors is that their torque is directly related to slip, or the difference between the speed of the magnetic field and the speed of the motor shaft. Consequently, actual operating speeds are usually around 2% less than the motor's nominal speed. For example, a theoretical four-pole induction motor with no slip would rotate at 1,800 rpm with a 60 Hz power supply; however, rated operating speeds for this motor are usually around 1,750 rpm, indicating that slip rates are a little over 2.7% at rated load.

Another component is the motor controller. This is the switch mechanism that receives a signal from a low power circuit, such as an on/off switch, and then energizes or de-energizes the motor by connecting or disconnecting the motor windings to the power line voltage. Soft starters are electrical devices that are often installed with a motor controller to reduce the electrical stresses associated with the start-up of large motors. Soft starters gradually ramp up the voltage applied to the motor, reducing the magnitude of the start-up current.

A key advantage of variable frequency drives (VFDs) is that they are often equipped with soft starting features that decrease motor starting current to about 1.5 to 2 times the operating current. Although VFDs are primarily used to reduce operating costs, they can significantly reduce the impact of motor starts on an electrical system.

MotorMaster+ is a free, online, National Electrical Manufacturers Association (NEMA), Premium® efficiency motor selection and management tool that simplifies motor and motor systems planning by identifying the most efficient action for a given repair or motor purchase decision. The tool includes a catalog of more than 20,000 low-voltage induction motors, and features motor inventory management tools, maintenance log tracking, efficiency analysis, savings evaluation, energy accounting, and environmental reporting capabilities. Download it from the US Department of Energy website at: http://www1.eere.energy.gov/industry/bestpractices/software.html.

Table 7-1 lists design considerations for various motor characteristics.

Motor Voltages

See Table 7-2 for a listing of common 60 Hz supply voltages and configurations, with ratings for compatible motors. See Table 7-3 for a listing of common 50 Hz supply voltages and configurations, with ratings for compatible motors.

Table 7-1
Consider each of these characteristics to select the best motor for an application [8]

Characteristic	Considerations
Full-load efficiency	The annual cost for operating a motor is usually much greater than the purchase cost of a motor. The most efficient available motor is often the most cost effective.
Part-load efficiency	Study part-load efficiency data if the motor will operate at reduced load for long periods. Manufacturers may not provide the data; use the MotorMaster+ program to obtain part-load performance information.
Power rating	Serious oversizing of the motor causes it to operate at reduced efficiency. However, it rarely pays to replace a motor simply because it is oversized. Larger motors are more efficient overall.
Service factor	Service factor provides a margin for error in calculating the size of motor that is needed for an application. A service factor higher than 1.0 is useful in applications where the motor operates for a large fraction of the time at low loads. This allows selection of the smallest possible motor, thereby achieving greater efficiency at low loads without fear of overheating the motor during occasional periods of peak load. A higher service factor also allows operation at higher ambient temperature or higher altitude. Operating a motor above its rated load reduces its service life and usually reduces its efficiency.
Torque characteristics	Full-load torque, measured at the motor's rated load. This rating is especially important for loads that develop more resistance as the speed increases, such as pumps and fans.
	Locked-rotor torque, measured when the shaft is held stationary. This is the motor's available starting torque and is most important that has large starting resistance such as a loaded conveyor belt.
	Pull-up torque is the minimum torque that the motor can produce, typically occurring at a fraction of full speed. This rating is important when driving loads that quickly develop resistance when starting, such as reciprocating compressors.
	Breakdown torque is the maximum torque that the motor can produce, developed at a speed slower than the rated speed with the motor in an overload condition. It is most important in applications that have transient requirements for extreme torque, such as a rock crusher.
Full-load speed	All other things being equal, a more efficient motor has less slip at a given load, so its speed is slightly higher. The amount of slip has a very large effect on torque.
Operating temperature	The most common motor failures are insulation failure and bearing failure, both related to operating temperature. A 10 °C increase in winding temperature reduces insulation life by about half, and bearing life by about one-quarter. If a motor must operate in a high-temperature environment, either get a motor with insulation that is rated for higher temperatures, or get a motor with a larger power rating, so that the motor operates at a lower percentage of full load.
	Factors that may raise motor temperature include: frequent starting, variable-frequency drives, poor power quality, phase balance, dirty environment, and high altitude.
High voltage vs. low voltage	Lower operating voltages allow higher motor efficiencies. However, the site's transformer efficiency may offset the savings.
Small voltage differences	Select motors for efficiency based on the actual voltage used. Motors that are designed to operate at several voltages operate most efficiently at the highest rated voltage. If possible, adjust the taps in the power supply transformers to match the supply voltage to the motor rating.
Power factor	The effective way to deal with power factor is to increase the power factor of the facility's electrical system by using power factor correction capacitors, or other devices intended for this purpose.
Locked-rotor and starting amperage	In retrofits, the starting current of a high-efficiency replacement motor may be too high for the existing fuses, circuit breakers, or wiring. If this problem occurs, a fairly inexpensive solution is to use a staged ("soft") motor starter.
Enclosure type	Select on the basis of the motor's environment. The type of casing has little effect on efficiency.
Frame and face type	Frame and face configuration do not strongly affect efficiency, except perhaps with some specialized motors.

Table 7-2
Typical commercial and industrial power system voltages for 60 Hz [1]

| Supply Voltage | System Configuration | Utilization Equipment Voltage Ratings | | Classification |
		Single-Phase	Three-Phase	
120/208	3-phase, 4-wire, grounded wye	115 208–230	200 208–230	Low Voltage
240	3-phase, 3-wire delta connected, normally ungrounded (Note 1)	230 208–230	230 208–230	
120/240/240	3-phase, 4-wire tapped delta, neutral grounded	115 230 208–230	230 208–230	
277/480	3-phase, 4-wire, grounded wye	277 265 (Note 2)	460	
480	3-phase, 3-wire, delta connected, normally ungrounded (Note 1)	460	460	
600	3-phase, 3-wire, delta connected, normally ungrounded (Note 1)	575	575	
2400	3-phase, 3-wire, delta connected	2300	2300 2300/4160	Medium Voltage
4160	3-phase, 4-wire, grounded wye	2300 4000 4160	4000 2300/4160	

1. On some systems grounding of one leg may be utilized.
2. Some single-phase equipment may be rated for 265 volts.

Motors: Efficiency

Standard and high efficiency motors are compared in Table 7-4 and Table 7-5. Synchronous and induction motors are compared in Table 7-6. Table 7-7 shows the effect of a large range of speeds on efficiency.

The power factor is a measure of the motor's reactive load. For most facilities, the power factor for individual motors is inconsequential. Overall motor efficiency measures the true power consumption relative to the work performed by the motor. Increasing a motor's efficiency will reduce the kilowatt-hour consumption and power cost for all classes of power users, regardless of their particular rate structure or power factor situation [1].

The efficiency of motors operating a partial load can be much less than at full load, especially when operated at less than 50% of full load. This is an important reason for avoiding oversizing a motor. See Figure 7-1.

Table 7-3
Typical commercial and industrial power system voltages for 50 Hz [1]

| Supply Voltage | System Configuration | Utilization Equipment Voltage Ratings | |
		Single-Phase	Three-Phase
115/200	3-phase, 4-wire, grounded wye	115 200	200
127/220	3-phase, 4-wire, grounded wye	127 220	220
220/380	3-phase, 4-wire, grounded wye	220 380	380 400 (Note 1)
230/400	3-phase, 4-wire, grounded wye	230 400	400
240/415	3-phase, 4-wire, grounded wye	240 415	415 400 (Note 1)
250/440	3-phase, 4-wire, grounded wye	250 440	440
220	3-phase, 3-wire, delta connected	220	220
440	3-phase, 3-wire, delta connected	440	440

1. Alternate rating.

Table 7-4
Energy evaluation chart; IEC frame size induction motors, 380 volt, 50 Hz [4]

kW	Approximate Ratings at Full Load			
	rpm	amps	Power Factor	Efficiency, %
0.55	1410	1.45	0.80	74.0
	910	1.70	0.74	65.0
0.75	1410	1.90	0.81	76.0
	890	2.30	0.75	68.3
1.1	1410	2.75	0.81	77.0
	910	3.40	0.72	69.8
1.5	1415	3.60	0.82	78.0
	940	4.40	0.70	76.4
2.2	1405	5.10	0.82	81.0
	945	5.80	0.75	80.0
3.0	1400	7.20	0.79	82.5
	960	6.80	0.80	85.3
4.0	1415	8.60	0.85	85.0
	955	9.10	0.80	85.0
5.5	1440	11.40	0.85	87.0
	955	12.40	0.80	85.0
7.5	1445	15.50	0.84	88.5
	965	16.00	0.82	87.9
11.0	1460	22.50	0.84	89.7
	965	23.50	0.82	89.0
15.0	1450	29.50	0.86	90.5
	965	31.00	0.83	90.0
18.5	1470	36.00	0.86	91.0
	970	37.50	0.83	90.8
22.0	1465	43.00	0.85	91.8
	965	45.00	0.83	91.0
30.0	1470	57.00	0.87	92.5
	975	60.00	0.83	91.7
37.0	1475	73.00	0.84	92.8
	985	76.00	0.80	92.5
45.0	1470	88.00	0.84	93.2
	985	85.00	0.87	93.3
55.00	1475	100.00	0.88	94.2
	985	102.00	0.88	93.2
75.0	1475	141.00	0.86	94.4
	985	142.00	0.85	94.5
90.0	1480	166.00	0.87	94.8
	985	170.00	0.85	94.7
110.0	1480	206.00	0.85	95.2
	990	225.00	0.86	95.2
132.0	1485	260.00	0.85	95.5
	990	245.00	0.86	95.3
160.0	1485	300.00	0.85	95.6
	990	296.00	0.86	95.5
200.0	1485	375.00	0.85	95.7
	990	370.00	0.86	95.7
250.0	1485	465.00	0.85	96.3
	990	467.00	0.86	95.9
315.0	1485	585.00	0.85	96.5
	994	580.00	0.86	96.2
355.0	1491	655.00	0.86	96.4
	994	650.00	0.86	96.4
400.0	1492	730.00	0.86	96.5
	994	730.00	0.86	96.5
450.0	1493	815.00	0.87	96.6

Table 7-4
Energy evaluation chart; IEC frame size induction motors, 380 volt, 50 Hz [4]—cont'd

kW		Approximate Ratings at Full Load		
	rpm	amps	Power Factor	Efficiency, %
	994	825.00	0.86	96.6
500.0	1493	890.00	0.88	96.6
	994	910.00	0.86	96.7
1000.0	1492	1805.00	0.87	96.5
	994	1855.00	0.85	96.4

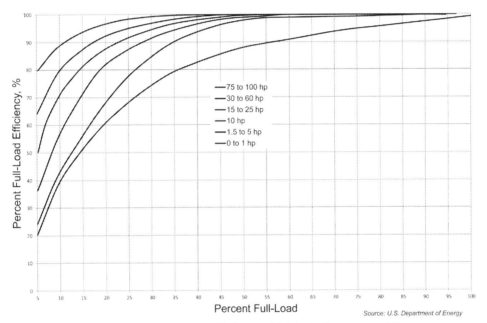

Figure 7-1. Part-load efficiency of AC induction motors [8].

Table 7-5
Energy evaluation chart; NEMA frame size induction motors, 460V, 60 Hz [1]

HP	Approx Full-Load rpm	Amperes Based on 460V		Efficiency at Full-Load, %	
		Std Eff	High Eff	Std Eff	High Eff
1	1,800	1.9	1.5	72.0	84.0
	1,200	2.0	2.0	68.0	78.5
1½	1,800	2.5	2.2	75.5	84.0
	1,200	2.8	2.6	72.0	84.0
2	1,800	2.9	3.0	75.5	84.0
	1,200	3.5	3.2	75.5	84.0
3	1,800	4.7	3.9	75.5	87.5
	1,200	5.1	4.8	75.5	86.5
5	1,800	7.1	6.3	78.5	89.5
	1,200	7.6	7.4	78.5	87.5
7½	1,800	9.7	9.4	84.0	90.2
	1,200	10.5	9.9	81.5	89.5
10	1,800	12.7	12.4	86.5	91.0

(Continued)

Table 7-5
Energy evaluation chart; NEMA frame size induction motors, 460V, 60 Hz [1]—cont'd

HP	Approx Full-Load rpm	Amperes Based on 460V		Efficiency at Full-Load, %	
		Std Eff	High Eff	Std Eff	High Eff
	1,200	13.4	13.9	84.0	89.5
15	1,800	18.8	18.6	86.5	91.0
	1,200	19.7	19.0	84.0	89.5
20	1,800	24.4	25.0	86.5	91.0
	1,200	25.0	24.9	86.5	90.2
25	1,800	31.2	29.5	88.6	91.7
	1,200	29.2	29.1	88.5	91.0
30	1,800	36.2	35.9	88.5	93.0
	1,200	34.8	34.5	88.5	91.0
40	1,800	48.9	47.8	88.5	93.0
	1,200	46.0	46.2	90.2	92.4
50	1,800	59.3	57.7	90.2	93.6
	1,200	58.1	58.0	90.2	91.7
60	1,800	71.6	68.8	90.2	93.6
	1,200	68.5	69.6	90.2	93.0
75	1,800	92.5	85.3	90.2	93.6
	1,200	86.0	86.5	90.2	93.0
100	1,800	112.0	109.0	91.7	94.5
	1,200	114.0	115.0	91.7	93.6
125	1,800	139.0	136.0	91.7	94.1
	1,200	142.0	144.0	91.7	93.6
150	1,800	167.0	164.0	91.7	95.0
	1,200	168.0	174.0	91.7	94.1
200	1,800	217.0	214.0	93.0	94.1
	1,200	222.0	214.0	93.0	95.0

Table 7-6
Synchronous vs. induction motors; 3-phase, 60 Hz, 2,300 or 4,000V [1]

HP	Speed, rpm	Synch Motor Efficiency, Full-Load, 1.0 PF	Induction Motor	
			Eff Full-Load	Power Factor
3,000	1,800	96.6	95.4	89.0
	1,200	96.7	95.2	87.0
3,500	1,800	96.6	95.5	89.0
	1,200	96.8	95.4	88.0
4,000	1,800	96.7	95.5	90.0
	1,200	97.0	95.4	88.0
5,000	1,800	96.8	95.6	89.0
	1,200	97.0	95.4	88.0
5,500	1,800	96.8	95.6	89.0
	1,200	97.0	95.5	89.0
6,000	1,800	96.9	95.6	89.0
	1,200	97.	95.5	87.0
7,000	1,800	96.9	95.6	89.0
	1,200	97.2	95.6	88.0
8,000	1,800	97.0	95.7	89.0
	1,200	97.3	95.6	89.0
9,000	1,800	97.0	95.7	89.0
	1,200	97.3	95.0	88.0

Table 7-7
Full-load efficiencies; 460V, 60 Hz

HP	3,600 rpm	1,200 rpm	600 rpm	300 rpm
5	80.0	82.5	—	—
	—	—	—	—
20	86.0	86.5	—	—
				82.7*
100	91.0	91.0	93.0	—
	—	—	91.4*	90.3*
250	91.5	92.0	91.0	—
	—	93.9*	93.4*	92.8*
1,000	94.2	93.7	93.5	92.3
	—	95.5*	95.5*	95.5*
5,000	96.0	95.2	—	—
	—	—	97.2*	97.3*

Synchronous motors, 1.0 PF.

Motors: Starter Sizes

Motor starter (controller) sizes are given in Tables 7-8 and 7-9.

Table 7-8
Single-phase motor starter sizes

NEMA Size	Maximum Horsepower, Full-Voltage Starting (two-pole contactor)	
	115 Volts	230 Volts
00	1.3	1
0	1	2
1	2	3
2	3	7.5
3	7.5	15

Table 7-9
Three-phase motor starter sizes

NEMA Size	Maximum Horsepower, Full-Voltage Starting	
	230 Volts	460–575 Volts
00	1.5	2
0	3	5
1	7.5	10
2	15	25
3	30	50
4	50	100
5	100	200
6	200	400
7	300	600

Motors: Useful Equations

Full-Load Current, Three-Phase

$$I = \frac{0.746\, hp}{1.73\, E(eff)PF}$$

Full-Load Current, Single-Phase

$$I = \frac{0.746\, hp}{E(eff)PF}$$

kVA Input, Three-Phase

$$kVA = \frac{1.73\, I\, E}{1,000}$$

kVA Input, Single-Phase

$$kVA = \frac{I\, E}{1,000}$$

kW Input, Single or Three-Phase

$$kW = kVA(PF)$$

Horsepower Output

$$hp = \frac{kW(eff)}{0.746}$$

$$hp = \frac{Torque(rpm)}{5250}$$

Full-Load Torque

$$Torque = \frac{5250\, hp}{rpm}$$

Power Factor

$$PF = \frac{kW\ input}{kVA\ input}$$

Operating Cost (Average Hp may be Lower than Nameplate Hp); Excludes Power Factor Penalty and Demand Charges

$$kWh = \frac{0.746(avg\ hp)(hrs\ of\ operation)}{eff}$$

$$Cost = kW(avg\ cost\ per\ kWh)$$

Synchronous speed (95% of synchronous motors are 2-, 4-, or 6-pole)

$$rpm = \frac{120\ Hz}{(number\ of\ poles)}$$

Where:

E = voltage, line-to-line
eff = efficiency (fraction)
hp = power, horsepower
Hz = alternating current frequency
I = current, amps
kVA = kilovolt-amperes
kW = power, kW
kWh = kilowatt-hours
PF = power factor (fraction)
rpm = motor speed, revolutions per minute

Motors: Overloading

When a pump has a motor drive, the process engineer must verify that the motor will not overload in extreme process changes. The horsepower for a centrifugal pump increases with flow. If the control valve in the discharge line opens fully, or an operator opens the control valve bypass, the pump will tend to "run out on its curve", giving more flow and requiring more horsepower. The motor must have the capacity to handle this.

Air-Driven Motors

Air motors are available with a power output of up to 22 kW (30 hp). Advantages and disadvantages of air-driven motors are listed in Table 7-10. They feature:

• Continuous variability of speed and torque, from about 20 rpm to 1,800 rpm and up to 1,000 ft-lb torque
• Fast starting, stopping, and reversing

• Inherent overload protection; air motors stall rather than burn out
• Non sparking; they are safe to use in hazardous locations
• Air consumption increases with increased supply pressure

Hydraulic Motors

Motors powered by a pressurized fluid, typically oil or water, have these features:

- High horsepower to weight ratio (e.g., they are very small compared with an electric motor with the same power output)
- Low speed torque
- Spark free making them suitable for use in highly hazardous areas such as areas with an electrical classification Group B (hydrogen)

The fluid power industry has established design and performance standards through the National Fluid Power Association (www.nfpa.com), ANSI, and ISO.

Hydraulic motors can be operated in a hazardous location while the hydraulic fluid is circulated from a pump located in a safe area. Multiple motors can be driven by a single pump. They provide high torque with variable speed.

Steam Turbines: Steam Rate

The theoretical steam rate (sometimes referred to as the water rate) for stream turbines can be determined from Keenan and Keyes or Mollier charts following a constant entropy path. Typical rates are given in Table 7-11. One word of caution – when using Keenan and Keyes, steam pressures are given in PSIG, based on sea level. For low steam pressures at high altitudes appropriate corrections must be made. The theoretical steam rate must then be divided by the efficiency to obtain the actual steam rate.

Table 7-10
Advantages and disadvantages of air-driven motors

Advantages	Disadvantages
Intrinsically safe. Can be used in just about every hazards (potentially explosive) environment.	Inefficient. It takes a 10 horsepower compressor to operate a 1 horsepower rotary vane motor.
Variable speed, controlled by throttling the air supply with a needle valve or ball valve.	High maintenance due to wearing elements in a rotary vane motor design.
Generally less expensive than an explosion-proof electric motor with comparable power	Lubricated air is usually required. Dry running air motors are available, but they are much more expensive and they have a longer lead time to procure.
Inherent overload protection due to stalling	Air supply requires a filter, regulator, and lubricator. Lubricant steadily discharges from the outlet port.
	Noisy.

Steam Turbines: Efficiency

Evans provides a chart of steam turbine efficiencies [2], shown in Figure 7-2. Smaller turbines can vary widely in efficiency depending greatly on speed, horsepower, and pressure conditions. Very rough efficiencies to use for initial planning below 500 horsepower are listed in Table 7-12. Some designers limit the speed of the cheaper small steam turbines to 3,500 rpm.

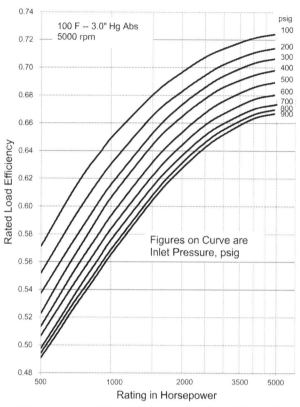

Figure 7-2. Typical efficiencies for mechanical drive turbines [7].

Table 7-11
Typical theoretical steam rates (lb/kWh) for determining actual steam combustion of turbines [6]

Exhaust Pressure	Steam Inlet Conditions (psig / °F)				
	1,500 / 925	900 / 825	600 / 750	150 / 500	50 / 400
600 psig	29.99	68.2			
150 psig	13.97	18.18	23.83		
50 psig	10.7	12.90	15.36	39.9	
1 atm	8.09	9.25	10.40	17.51	29.10
4.0 in Hg abs	6.368	7.03	7.65	10.78	14.00
1.5 in Hg abs	5.845	6.388	6.888	9.30	11.52

Multiply by 0.746 to obtain units of lb/h/hp.

Table 7-12
Small turbine efficiencies at 3500 rpm

Horsepower	Efficiency, %
1 to 10	15
10 to 50	20
50 to 300	25
300 to 350	30
350 to 500	40

Gas Turbines: Performance

The performance of gas turbines is expressed in terms of power and heat rate. Power is defined as the net power available at the output shaft after all losses have been subtracted. Heat rate is the power input to the turbine; heat rate and power are related by the turbine's thermal efficiency.

Gas turbine fuel rates (heat rates) vary considerably, however, Evans provides the following fuel rate graph (Figure 7-3) for initial estimations. The heat rate is usually expressed in Btu/hp/h or kJ/(kW-h), and is based on the lower heating value of the fuel gas.

Turbines are usually rated at standard conditions of:

Ambient temperature $= 15\ ^{\circ}\text{C}$
Altitude $=$ sea level
Ambient pressure $= 101.325$ kPa (absolute) or 14.7 psia
Relative humidity $= 60\%$

Use Table 7-13 to correct for site-specific conditions. Relative humidity has an insignificant effect on gas turbine power or heat rate.

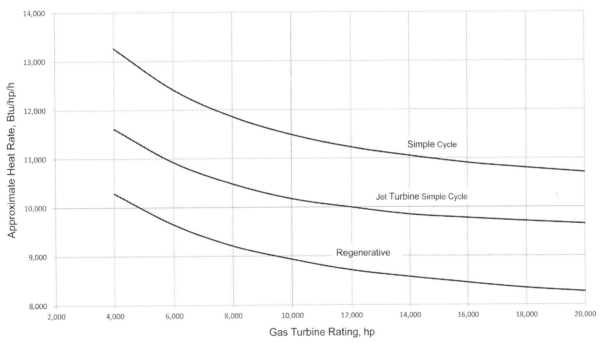

Figure 7-3. Approximate gas turbine fuel rates [2].

Table 7-13
Correction factors for site-specific conditions

Site Condition	Correction Factor Equations	
	Power Correction Factor	Heat Rate Correction Factor
Altitude, ft	1 − Altitude / 30000	Not affected by altitude
Inlet pressure loss, inches of water	1 − Inlet Loss / 250	1 + Inlet Loss / 640
Exhaust pressure loss, inches of water	1 − Exhaust Loss / 560	1 + Exhaust Loss / 560
Temperature, °F	1.25 − (0.0041) (Temp.)	0.88 + (0.002) (Temp)

Adapted from graphs in [3].

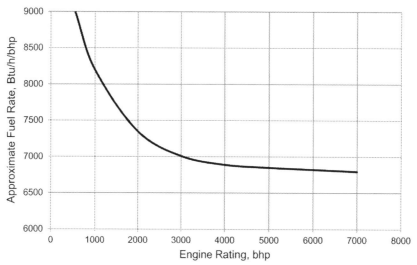

Figure 7-4. Approximate gas engine fuel rates [2].

Internal Combustion Engines: Fuel Rates

Fuel rates for initial estimates of gas engines are shown in Figure 7-4 below.

Gas Expanders: Available Energy

As energy costs increase, expanders will be used more than ever. A quick rough estimate of actual expander available energy is:

$$\Delta H = 0.5 \, C_P \, T_1 \left[1 - \left(\frac{P_2}{P_1} \right)^{(K-1)/K} \right] \qquad (7\text{-}1)$$

Where:

ΔH = actual available energy, Btu/lb
C_P = heat capacity at constant pressure, Btu/lb-°F
T_1 = inlet temperature, R
$P_1 \; and \; P_2$ = inlet and outlet pressures, psia
K = ratio of heat capacities, C_P/C_V

To get lb/h/hp divide as follows:

$$\frac{2545}{\Delta H}$$

A rough outlet temperature can be estimated with:

$$T_2 = T_1 \left(\frac{P_2}{P_1} \right)^{(K-1)/K} + \left(\frac{\Delta H}{C_P} \right) \qquad (7\text{-}2)$$

For large expanders, Equation 7-1 may be conservative. A full rating using vendor data is required for accurate results. Equation 7-1 can be used to see if a more accurate rating is worthwhile. For comparison, the outlet temperature for gas at critical flow across an orifice is given by:

$$T_2 = T_1 \left(\frac{P_2}{P_1} \right)^{(K-1)/K} = T_1 \left(\frac{2}{K+1} \right) \qquad (7\text{-}3)$$

The proposed expander may cool the working fluid below the dew point. Be sure to check for this.

The expander equation (Equation 7-1) is generated from the standard compressor head calculation (see Compressors, Horsepower Calculation) by:

1. Turning $[(P_2/P_1)^{(K-1)K} - 1]$ around (since work = $-\Delta h$)
2. Substituting $C_P = (1.9865/Mol.Wt.)[K/(K-1)] = (1.9865/1544)R[K/(K-1)]$
3. Cancelling 1.9865/1544 with 779 ft-lb/Btu
4. Assuming Z = 1
5. Using a roundhouse 50% efficiency

References

[1] Cowen Papers, Amps, Watts, Power Factor, and Efficiency: What Do You Really Pay For? Motors and Drives, LLC, www.motorsanddrives.com.

[2] Evans F. *Equipment Design Handbook for Refineries and Chemical Plants*, vol. 1, 2nd ed. Gulf Publishing Co., 1979.

[3] Gas Processors Suppliers Association (GPSA). *Engineering Data Book, IPS Version*, vol. 2. 12th ed. 2004.

[4] Gas Processors Suppliers Association (GPSA). *Engineering Data Book, SI Version*, vol. 2, 12th ed. 2004.

[5] Keenan JH, Keyes FG. Theoretical Steam Rate Tables. *Transactions ASME;* 1938.

[6] Lawrence Berkeley National Laboratory. *Resource Dynamics Corporation, Improving Fan System Performance: A Sourcebook for Industry*, U.S. Department of Energy and Air Movement and Control Association International (AMCA); 1989.

[7] Peterson J, Mann W. Steam System Design. In: McKetta J, editor. *Encyclopedia of Chemical Processing and Design*. New York: Marcel Dekker, Inc., 1996.

[8] Wulfinghoff D. *Energy Efficient Manual*. Wheaton, MD: Energy Institute Press; 1999.

8

Vessels

Introduction

Vessels provide many functions in a chemical plant. They have many forms and sizes, and are fabricated from nearly any structural material. Vessels contain valuable inventory that can be hazardous to life and property if released, so it's important for process engineers to pay close attention to the use, sizing, and design of all of the vessels in the plant. This chapter presents some useful rules of thumb, equations, and design procedures for vessel design.

Engineers should strive to minimize the size of vessels, while maintaining the desired plant functionality. The primary reasons to minimize vessel size (and number) fall into two categories: cost and safety.

Cost Considerations

- Vessel size has a ripple effect on the size of piping, pumps, heat exchangers, agitators, and other related vessels.
- Ancillary costs such as structural supports, skirts, ladders, insulation, and painting are directly affected.
- Vessel size, and especially the number of vessels, are directly related to the plant size, both in area and elevation.

- Inventory held in the tanks, whether raw material, intermediate, or final product, is an operating cost that can significantly affect the cost of goods sold.
- Other operating costs are affected, including operator hours, cooling water, steam, and power.

Safety Considerations

- If a vessel inadvertently discharges its contents, the quantity of the release is directly related to the potential harm that can be done, especially for hazardous materials.
- Codes, such as NFPA 30 and OSHA 1910.119, peg certain requirements to vessel size or plant inventory.

Balance plant operating requirements with the economic considerations, but never sacrifice safety to save money. Operating requirements are derived from the material and energy balances, anticipated throughput variations (e.g., weekend shutdowns, discontinuous flow through unit operations), physical size of the facility (e.g., ceiling heights, door widths), size of trucks or rail cars delivering raw materials, desired inventory level (e.g., one week, one month), hold times for inspection and release of product, etc.

General Rules of Thumb for Vessels

Atmospheric Storage Tanks [2]

- Defined as having a design pressure less than 2.5 psig.
- Use fixed-roof tank if the true vapor pressure (TVP, defined as the equilibrium partial pressure for a liquid at 100 °F) is less than 1.5 psia.
- Use floating-roof tanks for TVPs from 1.5 psia to 11 psia.
- The bottom outlet nozzle is usually at least 10 inches above the tank bottom, and the tank volume beneath the nozzle is excluded from any calculation of the working capacity.
- Provide an overfill protection section, above the normal high level, to accommodate upsets. This is usually 10% to 15% of the total tank volume, and may be called "freeboard".

- Use an economic calculation to determine the working volume for a large storage tank. Consider factors such as bulk transportation costs, size and frequency of shipments, and risks of plant shutdown. Many plants specify 30 days of capacity for both raw materials and products. Tanks are often sized at about 1.5 times the size of the railcar or truck that delivers the material to the plant.
- Mild carbon steel (A-36, A-328) is the most commonly used material for large storage tanks. Corrosion is mitigated with a corrosion allowance, a tank lining such as rubber or plastic, or switching to a corrosion-resistant alloy such as stainless steel.
- The optimal height-to-diameter ratio is determined from wind and seismic loadings, available space, and soil-bearing capacity. For field-erected tanks, the cost

is roughly proportional to the surface area of the shell and roof.

- The largest shop-fabricated tanks are approximately 50,000 gallons (12 ft diameter by 50 ft long), based on their ability to be transported by truck. Most tanks storing flammable or combustible liquids are under 15,000 gallons.
- Typical upset conditions include: overpressure, overflow, boil-over, over-temperature, water ingress, floating-roof failure, unexpected phase separation, lightening, static-charge buildup, steam coil failure, fires, and implosion caused by a vacuum condition in the tank. Consider these when conducting a Process Hazards Analysis.

Pressure Vessels and Reactors

- Must be designed, fabricated, and tested in accordance with applicable Codes.
- The user is responsible for specifying loadings that are used to calculate the vessel wall thicknesses and reinforcements. Factors include: internal/external pressure; ambient and operational temperatures; static pressure and mass of contents in operating and test conditions; wind and earthquake conditions; reaction forces and moments resulting from supports, attachments, piping, agitators, thermal expansion, etc.; corrosion; fatigue; and decomposition of unstable fluids.
- The aspect ratio (vertical straight-side height divided by diameter) is usually between 1:1 and 1.5:1. Taller vessels, with aspect ratios ranging to about 4:1, are used when necessary to maximize heat transfer through a jacket, to maximize contact time of a sparged gas, or for other process requirements.
- ASME F&D heads (torispherical) are usually specified for pressures to 20 bar. Ellipsoidal (2:1) heads are used for pressures from 20 to 100 bar. Very high pressure applications, above 100 bar, utilize hemispherical heads. Conical bottoms are used for some crystallizers when it is desired that precipitates flow freely to the bottom outlet nozzle.
- The working volume of an agitated vessel should be about 80% of the volume measured at the top tangent line.
- Determine heating and cooling duty using factors including: control of process exotherms, heat-up and cool-down loads (time-based), boiling, thermal losses to the environment, and heat input from agitators and pumps.

Tank Dimensions

Tank dimensional nomenclature is illustrated in this section. See Figures 8-1 to 8-5.

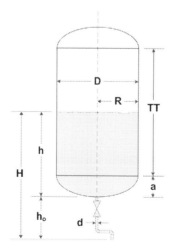

Figure 8-1. General vessel nomenclature, with discharge pipe outlet below the bottom of the tank that is filled to an arbitrary level, h. Because the heads include a short straight section, dimension "TT" is greater than the distance between the welds.

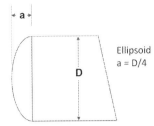

Figure 8-2. The 2:1 semi-ellipsoidal head is more expensive than the torispherical due to its greater depth. The straight flange is roughly 10 mm to 30 mm (4 in. to 12 in.) depending on diameter and thickness.

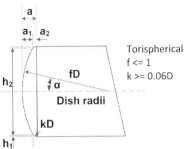

Figure 8-4. The torispherical head is also known as "flanged and dished" or "F&D". It is the most common type of head used on pressure vessels because it is the least expensive to form and fabricate. The standard ASME F&D head has a dish radius no greater than the vessel diameter, and a knuckle radius that is no less than 6% of the diameter or three times the metal thickness, whichever is greater. The straight flange is roughly 10 mm to 30 mm (4 in. to 12 in.) depending on diameter and thickness.

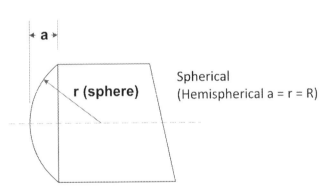

Figure 8-3. Dished heads with shallow depth are generally used for atmospheric tanks. The shallow dish may be inverted, sometimes used for the bottom of silos. Deep dishes, up to full hemispherical, have the highest pressure rating of all type of heads but are also the most expensive to form.

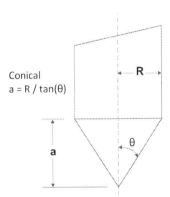

Figure 8-5. The cone angle is variable and at the discretion of the customer. Often used for vessels handling powders, the angle is normally steeper than the angle of repose. For pressure vessels, the maximum angle (two times theta) is 120 °. Note that theta must be expressed in radians for the formula. Radians = Degrees x pi / 180. The straight flange is roughly 10 mm to 30 mm (4 in. to 12 in.) depending on diameter and thickness.

Liquid Volume at Any Fill Height

Jones presented formulae for calculating the tank volume corresponding to a given level for standard vertical and horizontal tanks [5] (see Tables 8-1, 8-2). Use a math program (e.g., MathCAD) to solve the integral equations; the author's spreadsheet, TANKVOL, which solves all of the equations, is available at www.chemengsoftware.com.

Table 8-1
Vertical tank volume

Tank	Formula
Vertical flat bottom	$V_f = \pi R^2 h$
Conical bottom head	For $h >= a$,

$$V_f = \pi R^2 \left(h - \frac{2a}{3} \right)$$

For $h < a$,

$$V_f = \pi \left(\frac{Rh}{a} \right)^2 \left(\frac{h}{3} \right)$$

Ellipsoidal bottom head

For $h >= a$

$$V_f = \pi R^2 \left(h - \frac{a}{3} \right)$$

For $h < a$

$$V_f = \pi \left(\frac{Rh}{a} \right)^2 \left(a - \frac{h}{3} \right)$$

Spherical bottom head

For $h >= a$

$$V_f = \pi \left(\frac{a^3}{6} - \frac{a R^2}{2} + h R^2 \right)$$

For $h < a$

$$V_f = \pi h^2 \left(\frac{a}{2} + \frac{R^2}{2a} - \frac{h}{3} \right)$$

Torispherical bottom head

For $h >= a$

$$V_f = \pi \left(\frac{a_1^3}{6} + \frac{a_1 R_1^2}{2} \right) + \frac{\pi t}{2} \left[(R - 2kR)^2 + s \right] + \frac{\pi t^3}{12}$$
$$+ \pi R(1 - 2k) \left[\frac{t\sqrt{s}}{2} + 4kR^2 \sin^{-1}(\cos \alpha) \right] + \pi R^2 (h - a)$$

For $h <= a_1$

$$V_f = \pi h^2 \left(\frac{a_1}{2} + \frac{R_1^2}{2a_1} - \frac{h}{3} \right)$$

For $a > h > a_1$

$$V_f = \pi \left(\frac{a_1^3}{6} + \frac{a_1 R_1^2}{2} \right) + \pi u \left[(R - 2kR)^2 + s \right] + \frac{\pi t u^2}{2} - \frac{\pi u^3}{3}$$
$$+ 2\pi R(1 - 2k) \left[\frac{2u - t}{4} \sqrt{s + tu - u^2} + \frac{t\sqrt{s}}{4} + 2k^2 R^2 \left(\cos^{-1} \frac{t - 2u}{4kR} - \alpha \right) \right]$$

Where

$$\alpha = \sin^{-1} \frac{1 - 2k}{2(f - k)} = \cos^{-1} \frac{\sqrt{4f^2 - 8fk + 4k - 1}}{2(f - k)}$$
$$a_1 = 2fR(1 - \cos \alpha)$$
$$a_2 = 2kR \cos \alpha$$
$$R_1 = 2fR \sin \alpha$$
$$s = (2kR \sin \alpha)^2$$
$$t = 4kR \cos \alpha = 2a_2$$
$$u = h - 2fR(1 - \cos \alpha)$$

Table 8-2
Horizontal tank volume with both heads the same (in these formulae, $L = TT$)

Tank	Formula

The cross sectional area of the cylindrical section of the vessel is given by:

$$A_f = R^2 \cos^{-1}\left(\frac{R - h}{R}\right) - (R - h)\sqrt{2Rh - h^2}$$

Horizontal with flat heads
Conical heads

$V_f = A_f L$

For $0 <= h < R$

$$V_f = A_f L + \frac{2 a R^2 K}{3}$$

For $h = R$

$$V_f = A_f L + \frac{\pi a R^2}{3}$$

For $R < h <= 2R$

$$V_f = A_f L + \frac{2 a R^2 (\pi - K)}{3}$$

Where

$$K = \cos^{-1} M + M^3 \cosh^{-1}\frac{1}{M} - 2M\sqrt{1 - M^2}$$

$$M = \left|\frac{R - h}{R}\right|$$

Ellipsoidal heads

$$V_f = A_f L + \pi a h^2 \left(1 - \frac{h}{3R}\right)$$

Spherical heads

For the case where the tank is exactly half full and $a <= R$. Double this for a completely full vessel.

$$V_f = A_f L + \frac{\pi a}{6}(3R^2 + a^2)$$

When the tank is *not* half full or completely full, and $a >= 0.01 D$

$$V_f = A_f L + \frac{a}{|a|}\left\{\frac{2 t^3}{3}A - B\right\}$$

Where

$$A = \cos^{-1}\frac{R^2 - r w}{R(w - r)} + \cos^{-1}\frac{R^2 + r w}{R(w + r)} - \frac{z}{r}\left(2 + \left(\frac{R}{r}\right)^2\right)\cos^{-1}\frac{w}{R}$$

$$B = 2\left(w r^2 - \frac{w^3}{3}\right)\tan^{-1}\frac{y}{z} + \frac{4 w y z}{3}$$

When $a < 0.01 D$ and the tank is *not* half full or completely full

$$V_f = A_f L + \frac{a}{|a|}\left[2\int_{w}^{R}(r^2 - x^2)\tan^{-1}\sqrt{\frac{R^2 - x^2}{r^2 - R^2}}\,dx - A_f z\right]$$

Where

$$r = \frac{a^2 + R^2}{2|a|}$$

$a = \pm(r - \sqrt{r^2 - R^2})$ $\quad + (-)$ for convex (concave) heads

$w = R - h$

$y = \sqrt{2Rh - h^2}$

$z = \sqrt{r^2 - R^2}$

Torispherical heads

If $0 <= h <= h_1$

$V_f = A_f L + 2 v_1$

For $h_1 < h < h_2$

$V_f = A_f L + 2(v_{1,\max} + v_2 + v_3)$

If $h > h_2$

$V_f = A_f L + 2(2 v_{1,\max} + v_{2,\max} + v_{3,\max} - v_1 (with\ h = D - h))$

$$v_1 = \int_{0}^{\sqrt{2kDh - h^2}}\left[n^2\sin^{-1}\frac{\sqrt{n^2 - w^2}}{n} - w\sqrt{n^2 - w^2}\right]dx$$

(Continued)

Table 8-2
Horizontal tank volume with both heads the same (in these formulae, $L = TT$)—cont'd

Tank	Formula
	$$v_2 = \int_0^{kD\cos\alpha} \left[n^2 \left(\cos^{-1}\frac{w}{n} - \cos^{-1}\frac{g}{n} \right) - w\sqrt{n^2 - w^2} + g\sqrt{n^2 - g^2} \right] dx$$

For $0.5 < f <= 10$

$$v_3 = \frac{r^3}{3}\left[\cos^{-1}\frac{g^2 - r\,w}{g\,(w - r)} + \cos^{-1}\frac{g^2 + r\,w}{g(w + r)} - \frac{z}{r}\left(2 + \left(\frac{g}{r}\right)^2 \right) \cos^{-1}\frac{w}{g} \right] - \left(w\,r^2 - \frac{w^3}{3} \right) \tan^{-1}\frac{\sqrt{g^2 - w^2}}{z}$$

$$+ \frac{w\,z\sqrt{g^2 - w^2}}{6} + \frac{wz}{2}\sqrt{2\,g\,(h - h_1) - (h - h_1)^2}$$

For $10 < f <= 10000$

$$v_3 = \int_w^g (r^2 - x^2)\tan^{-1}\frac{\sqrt{g^2 - x^2}}{z}\,dx - \frac{z}{2}\left(g^2\cos^{-1}\frac{w}{g} - w\sqrt{2g(h - h_1) - (h - h_1)^2} \right)$$

$v_{1.\max} = v_1$ with $h = h_1$

$v_{2.\max} = v_2$ with $h = h_2$

$v_{3.\max} = v_3$ with $h = h_2$, or $v_{3.\max} = \frac{\pi\,a_1}{6}(3\,g^2 + a_1{}^2)$ with $a_1 = r(1 - \cos\alpha)$

$$\alpha = \sin^{-1}\frac{1 - 2\,k}{2(f - k)} = \cos^{-1}\frac{\sqrt{4\,f^2 - 8\,f\,k + 4\,k - 1}}{2(f - k)}$$

$r = f\,D$

$h_1 = k\,D(1 - \sin\alpha)$

$h_2 = D - h_1$

$n = R - k\,D + \sqrt{k^2 D^2 - x^2}$

$g = f\,D\sin\alpha = r\sin\alpha$

$w = R - h$

$z = \sqrt{r^2 - g^2} = f\,D\cos\alpha = r\cos\alpha$

Time to Drain a Tank

Calculate the time to drain a tank in two parts: 1) the cylindrical body, and 2) the bottom head (in the case of vertical tanks with heads). Sum the times for the two parts to estimate the total.

The integrated expressions are based on explicit formulae that give the cross-sectional area of the tank section as a function of liquid depth. Because torispherical heads are complex, especially in the case of horizontal tanks, use the formulae for tanks with flat or ellipsoidal heads instead; the accuracy is within a few percent in most cases which should be more than adequate for general work.

The procedure for deriving the formulae is presented by Schwarzhoff and Sommerfeld [9]. This section only gives the resultant formulae.

Figure 8-1 defines the nomenclature for the tank geometry. The total liquid depth, H, provides the driving force that is balanced by frictional pressure drop through the discharge pipe. H decreases with the tank level as does the flow rate out of the tank.

To account for the differential pressure from the vapor space in the tank to the outlet pipe discharge point, if any, express the differential pressure in consistent units (m or ft of liquid) and add the result to both H and h_o. If the vapor space is higher pressure, H increases; it decreases if the tank is under vacuum. See Table 8-3 for formulae.

Assumptions:

1. Constant differential pressure between the tank vapor space and the outlet pipe discharge.
2. Constant friction factor.

Table 8-3
Formulae for time to drain a tank through a discharge pipe

Tank	Formula
Vertical cylindrical section	Apply this formula from the initial liquid level, H_o, to the final level if $H_f > (h_o + a)$, or to the bottom of the cylindrical section where $H_f = h_o + a$. $$t = \frac{D^2}{d^2}\sqrt{\frac{2}{g}\left(1 + \frac{fL}{d}\right)}(\sqrt{H_o} - \sqrt{H_f}) \quad \text{(8-1)}$$
Elliptical bottom head	If the initial liquid level is within the cylindrical section, $H_o = (h_o + a)$ $$t = C\left[\left(\frac{2}{5}H_f{}^2 - \frac{4B}{3}H_f + 2E^2\right)\sqrt{H_f} - \left(\frac{2}{5}H_o{}^2 - \frac{4B}{3}H_o + 2E^2\right)\sqrt{H_o}\right]$$ Where $B = h_o + a$ $$C = \left(\frac{D}{d\,a}\right)^2\sqrt{\frac{1}{2g}\left(1 + \frac{fL}{d}\right)}$$ $E^2 = h_o{}^2 + 2\,a\,h_o$
Spherical bottom head	If the initial liquid level is within the cylindrical section, $H_o = (h_o + R)$ $$t = \frac{4}{d^2}\sqrt{\frac{1}{2g}\left(1 + \frac{fL}{d}\right)}\left[\left(\frac{2}{5}H_f{}^2 - \frac{4b}{3}H_f + 2e^2\right)\sqrt{H_f} - \left(\frac{2}{5}H_o{}^2 - \frac{4b}{3}H_o + 2e^2\right)\sqrt{H_o}\right]$$ Where $b = h_o + R$ $e^2 = h_o{}^2 + 2R\,h_o$
Conical bottom head	$$t = \frac{4}{d^2}\sqrt{\frac{1}{2g}\left(1 + \frac{fL}{d}\right)}\tan(\theta)^2\left[\begin{array}{l}\left(\frac{4}{3}H_f{}^{(3/2)}h_o - \frac{2}{5}H_f{}^{(5/2)} - 2H_f{}^{0.5}h_o{}^2\right) - \\ \left(\frac{4}{3}H_o{}^{(3/2)}h_o - \frac{2}{5}H_o{}^{(5/2)} - 2H_o{}^{0.5}h_o{}^2\right)\end{array}\right]$$
Spherical tank	$$t = \frac{4}{d^2}\sqrt{\frac{1}{2g}\left(1 + \frac{fL}{d}\right)}\left[\left(\frac{2}{5}H_f{}^2 - \frac{4b}{3}H_f + 2e^2\right)\sqrt{H_f} - \left(\frac{2}{5}H_o{}^2 - \frac{4b}{3}H_o + 2e^2\right)\sqrt{H_o}\right]$$ Where $b = h_o + R$ $e^2 = h_o{}^2 + 2R\,h_o$

t = time to drain tank or section from initial level, H_o, to final level, H_f.
g = gravitational constant, 9.8 m/s^2 or 32.17 ft/s^2
f = Darcy friction factor, dimensionless
L = pipe equivalent length, m or ft
Other nomenclature in units of m or ft as defined in Figure 8-1

Emissions from Storage Tanks

AP-42, Chapter 7, presents the generally accepted method for estimating uncontrolled fugitive emissions from a storage tank [1]. API-42 gives formulae for standing and working losses from vertical and horizontal tanks with fixed or floating roofs. It also provides chemical and US geographic data to localize the calculations.

The US EPA also publishes a Windows-based computer software program called *TANKS* that is based on the procedures from AP-42. *TANKS* is free to download from www.epa.gov/ttn/chief/software/tanks/index.html. Quoting from AP-42:

Emissions from organic liquids in storage occur because of evaporative loss of the liquid during its storage and as a result of changes in the liquid level. The emission sources vary with tank design, as does the relative contribution of each type of emission source. Emissions from fixed roof tanks are a result of evaporative losses during storage (known as breathing losses or standing storage losses) and evaporative losses during filling and emptying operations (known as working losses). External and internal floating roof tanks are emission sources because of evaporative losses that occur during standing storage and withdrawal of liquid from the tank. Standing storage losses are a result of evaporative losses through rim seals, deck fittings, and/or deck seams. The loss mechanisms for fixed roof and external and internal floating roof tanks are described in more detail in this section [i.e., AP-42, Chapter 7]. Variable vapor space tanks are also emission sources because of evaporative losses that result during filling operations. The loss mechanism for variable vapor space tanks is also described in this section. Emissions occur from pressure tanks, as well. However, loss mechanisms from these sources are not described in this section.

Load Cells

Weight is often used for process monitoring and control, especially for storage tanks, bins, silos, and batch reactors. Typically, three or four load cells are mounted beneath the vessel and a summing instrument transmits the combined measured weight to the process control system. The load cells are installed under the tank legs or between side-mounted lugs and the steel support structure.

Best results are obtained when extraneous forces are eliminated. In addition to the weight of the vessel system, forces may be generated from lateral loads, thermal expansion or contraction, and transmission from attached components and piping.

Here are some tips for load cell design and installation:

- Provide a rigid support structure. The load cells and vessel system should be tightly connected to eliminate any flexing that would affect the readings. If the tank sits on long legs, the load cells should be installed near the top of the legs, not at the bottom. Provide adequate stiffening and cross-bracing. If four or more load cells are used, ensure they carry approximately equal load.
- Isolate the vessel system from its surroundings. Ensure that appurtenances float with the tank; don't give them "hard" connections to external structures. For instance, ladders should be attached to the tank and not the structure, or attached to the structure with a small gap between the ladder and tank.
- Piping and conduit connections must be flexible. Although many engineers routinely specify flex connections, such as hoses or bellows, for this purpose, adequate flexibility is usually obtainable with careful routing and support of the piping. The goal is to eliminate extra horizontal or vertical forces from the piping. Taking the potential thermal expansion and contraction of the vessel into account, piping bending moments can be calculated between the connection point and the closest rigid pipe support. The bending moments translate into vertical and horizontal forces imposed on the tank; their magnitude determines the potential affect on the load cell reading, and may be inconsequential when taken as a percentage of the total mass of the vessel system.

Metler-Toledo publishes an excellent handbook that shows how to calculate reaction forces from piping and environmental effects (e.g., wind, seismic events) [7]. If the total piping force (all pipe connections combined) is F then:

$$F \leq 0.1(Accuracy)(Capacity) \qquad (8\text{-}1)$$

Where:

Accuracy = the required system accuracy stated as a % of system capacity
Capacity = the maximum live load that is expected to be weighed

For example, if the accuracy is 0.25% and the system capacity is 25,000 lb, then:

$$F = 0.1(0.25)(25000) = 625 \text{ lb}$$

System calibration accounts for piping forces; the magnitude of the allowed force should be confirmed with the manufacturer of the specific load cell system that will be used.

Calculate the force from each piping connection using:

$$F_p = \frac{0.59\left(D^4 - d^4\right)\Delta h\, E}{L^3} \qquad (8\text{-}2)$$

Where:

F_p = vertical force exerted by pipe, N or lb
D = outside diameter of pipe, mm or in.
d = inside diameter of pipe, mm or in.
Δh = tank deflection when the system capacity is added to the tank, mm or in.
E = Young's modulus, N/mm^2 or lb/in.^{2}
L = horizontal piping distance from connection to nearest piping support point, mm or in.

This assumes that the pipe exerts no extra force if the connection point moves as a result of thermal expansion or contraction. In other words, the piping support point moves with the tank nozzle. If this is not true, use engineering judgment to decide whether the differential should be added to the tank deflection value. Base the judgment on thermal conditions during processing and when the load cells are calibrated; if calibration occurs at the same temperature as operation then no adjustment is needed. The calculation is conservative because it assumes a horizontal straight pipe (no elbows or vertical sections,

which are typical for connections to tanks) and a rigid pipe hanger.

Assuming that the pipe support is fixed and not moving with the tank's support structure, calculate tank deflection with:

$$\Delta h = \frac{(Load\ Cell\ Deflection)(Capacity)}{(Rated\ Capacity)N} + (Structural\ Deflection) \qquad (8\text{-}3)$$

Where:

Load Cell Deflection = the deflection at the load cell's rated capacity, available from the manufacturer's datasheet, mm or in.
Capacity = the maximum live load that is expected to be weighed, kg or lb
Rated Capacity = from the manufacturer's datasheet, kg or lb
N = number of load cells supporting the vessel
Structural Deflection = deflection of the support structure under the vessel at Capacity, mm or in.

Liquid-Liquid Separators

The separation of immiscible liquids constitutes one of the important chemical engineering operations. Barton has provided an empirical method for sizing liquid-liquid separators [3]. Do not use this procedure for emulsions.

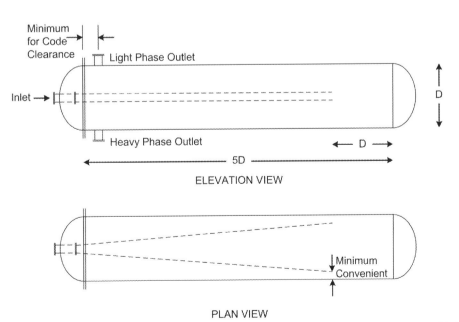

Figure 8-6. Recommended design for a liquid-liquid separator. For large vessels, a manway and internally dismantled cone may be used. Locate gauge glass and level instruments at the inlet-outlet end.

1. Calculate the total required residence time with:

$$T = 0.1\left[\frac{\mu}{SG_b - SG_t}\right] \qquad (8\text{-}4)$$

Where:
T = residence time, hours
μ = viscosity of the continuous phase, mPa-s or cP
SG_b = specific gravity of the bottom (heavy) phase, dimensionless
SG_t = specific gravity of the top (light) phase, dimensionless

2. Assign a length-to-diameter ratio of 5, and size a tank to accommodate the calculated residence time

3. Provide inlet and outlet nozzles at one end, and an internal flat cone to distribute the continuous phase. See Figure 8-6.

While this design procedure is empirical, there is a rationale behind it. The relationship between viscosity and specific-gravity difference of the phases corresponds to those of the equations for terminal settling velocity in the Stokes-law region and free-settling velocity of isometric particles. Also, the dimensions of the tank and cone recognize that the shape of turbulence created by nozzles discharging into liquids spreads at an angle whose slope is about 1 to 5.

Vapor-Liquid Separators

Optimal sizing and configuration of vapor-liquid separators remains an art, however, there are several guidelines that will provide a serviceable design. Obtain the liquid and vapor flow rates and physical properties from the process flow diagrams, being sure to note flow variations. Discussed in the next three sections, the separator size is based on liquid surge volume and vapor velocity in the disengagement section. Orientation, horizontal or vertical, depends primarily on the liquid volume, but is also influenced by the presence of an immiscible water phase. The overall aspect ratio (length divided by diameter) is usually kept between 3 and 5.

Liquid Surge Volume

Liquid surge volume is often expressed in terms of minutes of residence time when the separator is half full. Half full is defined by the total vessel volume (divided by 2) or the maximum permissible liquid inventory in the tank if there is a high level shutoff (also divided by 2). Rules of thumb for surge volume:

1. 5 to 10 minutes half-full for most applications, such as a distillation column reflux drum.
2. 10 to 30 minutes half-full for a fired heater surge drum.

3. 10 to 20 minutes full for a compressor feed liquid knockout drum (but minimum volume sized to 10 minutes of gas flow).

The reflux drum surge volume may be reduced if there is good process control implying lower variation in the drum level and fast response when there is an upset. Watkins developed an expression that relates process control to surge volume [12]; modern control systems are likely to compute to a volume at the low end of Watkins's range, about 2 to 3 minutes half-full. Relevant information is shown in Tables 8-4 to 8-6.

Table 8-4
Instrument and labor factors for sizing reflux drums. The labor factor is based on the perceived quality of the operators and is influenced by staffing levels, training, and experience

Control Scheme	Instrument Factor, F_1		Labor Factor, F_2		
	With Alarm	No Alarm	Good	Fair	Poor
Flow Ratio Control (FRC)	0.5	1	1	1.5	2
Level Ratio Control (LRC)	1	1.5	1	1.5	2
Temperature Ratio Control (TRC)	1.5	2	1	1.5	2

Table 8-5
Factor for overhead product flow to external equipment

Operating Characteristics	Factor F_3
Under good control	2.0
Under fair control	3.0
Under poor control	4.0
Feed to or from storage	1.25

Table 8-6
Factor for operator's ability to monitor level in the drum

Drum Level Visibility	Factor F_4
Board-mounted level recorder	1.0
Level indicator on board	1.5
Gage glass at equipment only	2.0

$$V_d = 2\,F_4(F_1 + F_2)(L + F_3\,D) \qquad (8\text{-}5)$$

Where:

V_d = volume of the reflux drum full, gal

L = reflux to the column, gal/min
D = distillate product to next unit operation, gal/min
F_i = factors listed in the tables

Vapor Velocity in the Disengagement Section

A separator uses gravity to disengage entrained liquid droplets from the vapor stream. For many applications, carryover of about 5% of the liquid in the vapor is acceptable. The calculation presented in this section is based on 5% carryover. If better separation is required, demister pads are used, which reduce carryover to less than 1% because liquid droplets impinge and coalesce on the metal or plastic wire mesh demister elements. Even better separation is possible through the use of centrifugal devices such as cyclone separators.

Watkins has charted a factor, K, which is used to calculate the maximum vapor velocity in the separator [12]. Branan regressed this plot, giving the following expression for a vertical separator, K_v. For a horizontal separator, $K_h = 1.25\,K_v$.

$$K_V = \exp\big(-1.94 - 0.815\,X - 0.179\,X^2$$
$$- 0.0124\,X^3 + 0.00039\,X^4 + 0.00026\,X^5\big)$$
$$(8\text{-}6)$$

$$X = \ln\left[\left(\frac{W_L}{W_V}\sqrt{\frac{\rho_v}{\rho_L}}\right)\right] \qquad (8\text{-}7)$$

Where:

W_L = liquid flow rate, lb/s
W_V = vapor flow rate, lb/s
ρ_L = density of liquid, lb/ft^3
ρ_v = density of vapor, lb/ft^3

Then;

$$U_{vapour,\max} = K\left(\frac{\rho_L - \rho_v}{\rho_v}\right)^{0.5} \qquad (8\text{-}8)$$

Where:

$U_{vapor,\max}$ = maximum vapor velocity in disengagement section, ft/s
K = factor for vertical or horizontal vessel

Vapor-Liquid Separator Size

A vessel handling large amounts of liquid or a large liquid surge volume will usually be horizontal. Where water must be separated from hydrocarbon liquid, the vessel will also be horizontal. A vessel with small surge volume such as a compressor knockout drum will usually be vertical.

Vertical Separator

1. Calculate the maximum vapor velocity by using the procedure in the previous section.
2. Calculate the minimum vessel cross-sectional area and diameter with:

$$A_{min} = \frac{W_V}{\rho_v \, U_{vapor,max}} \tag{8-9}$$

$$D_{min} = \sqrt{\frac{4 \, A_{min}}{\pi}} \tag{8-10}$$

3. Adjust the diameter to the next higher standard increment. In the US, vessels are commonly fabricated with diameters in 6-inch increments.
4. Estimate the average density of the vapor-liquid mixture. Find the approximate range of velocities for the inlet nozzle and calculate the corresponding nozzle sizes; pick a standard size within the range.

$$\rho_{mix} = \frac{W_V + W_L}{\left(\dfrac{W_V}{\rho_v}\right) + \left(\dfrac{W_L}{\rho_L}\right)} \tag{8-11}$$

$$U_{nozzle,max} = \frac{100}{\sqrt{\rho_{mix}}} \tag{8-12}$$

$$U_{nozzle,min} = \frac{60}{\sqrt{\rho_{mix}}} \tag{8-13}$$

$$A_{nozzle} = \frac{W_V + W_L}{\rho_{mix} U} \tag{8-14}$$

5. Sketch the vessel. For height above the inlet nozzle to the top tangent, use 36 inches plus half the nozzle diameter, or 48 inches minimum. For distance below the inlet nozzle to the high liquid level, use 12 inches plus half the nozzle diameter, or 18 inches minimum. Base the high liquid level on the recommendations given in the Liquid Surge Volume section on page 158.
6. Check the aspect ratio, total height (tangent-to-tangent) divided by diameter. If the aspect ratio is greater than 5, use a horizontal vessel instead of a vertical one. If the ratio is less than 3, it may be necessary to include excess surge volume to bring the aspect ratio to 3.

Horizontal Separator

1. Follow steps 1 and 2 under *Vertical Separator*, above. In this case, A_{min} is the cross-sectional area of the vapor space of the tank.

2. Use the rule of thumb that the vapor volume must be at least 20% of the vessel volume. Therefore, ignoring the vessel heads, the minimum diameter is

$$D_{min} = \sqrt{\frac{4 \, A_{min}}{0.2 \, \pi}} \tag{8-15}$$

3. Base the liquid inventory on the recommendations given in the Liquid Surge Volume section on page 158. The total vessel volume is the Liquid Surge Volume divided by 0.8.
4. Use the next standard tank diameter increment (from D_{min}) to calculate the length, L, with

$$L = \frac{Total \ Vessel \ Volume}{\left(\pi/4\right)D^2} \tag{8-16}$$

5. Calculate L/D and check that it is between 3 and 5. If too short, L can be increased (increases the liquid surge volume), or increase the vapor volume. If too long, increase the diameter which will also decrease the vapor velocity. Dimensions are shown in Figure 8-7.

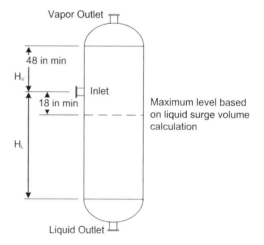

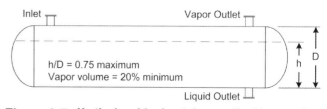

Figure 8-7. Vertical and horizontal vapor-liquid separators.

Horizontal Drum with Water Separation

If there is water mixed with an oil liquid phase, the water's settling time requirement must be checked. The water settling requirement, rather than other process considerations, might set the liquid surge volume.

1. Estimate the water terminal settling velocity from:

$$U_T = 44.7x10^{-8}(\rho_w - \rho_o)F_S/\mu_o \qquad (8\text{-}17)$$

Where:

U_T = terminal settling velocity of water, ft/s
ρ_w, ρ_o = density of water and oil, lb/ft^3
F_S = correction factor for hindered settling
μ_o = absolute viscosity of oil phase, lb/ft-s
This assumes a water droplet diameter of 0.0005 ft. Setting X = volume fraction of oil in the liquid phase, F_s is determined from:

$$F_S = \frac{X^2}{10^{1.82(1-X)}} \qquad (8\text{-}18)$$

2. Calculate the modified Reynolds Number from:

$$N_{RE} = \frac{0.0005\,\rho_o\,U_T}{\mu_o} \qquad (8\text{-}19)$$

3. Calculate the actual settling velocity:

$$U_S = U_T[A + B \ln N_{RE} + C \ln(N_{RE}^2)$$
$$+ D \ln(N_{RE}^3) + E \ln (N_{RE}^4)] \qquad (8\text{-}20)$$

Where:
U_S = actual settling velocity, ft/s
$A = 0.9198$
$B = -0.09135$
$C = -0.01716$
$D = 0.002926$
$E = -0.0001159$

4. Calculate the minimum length of the settling section

$$L_S = \frac{h\,Q}{A\,U_S} \qquad (8\text{-}21)$$

Where:
L_S = length of settling zone, ft
h = height of liquid phase, ft
Q = liquid flow rate, ft^3/s
A = cross-sectional area of the settling zone, ft^2
This allows the water to fall out and be drawn off at the bootleg before leaving the settling zone. Dimensions are shown in Figure 8-8.

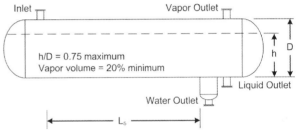

Figure 8-8. Horizontal separator with water collection bootleg.

Vessel Specifications

Although the vessel manufacturer is subject to strict Code requirements, the purchaser is responsible for specifying many aspects, as described in this section. Most engineers use data sheets to document the requirements; data sheets can be readily found online or obtained from vessel fabricators. Written specifications often accompany the data sheets, providing additional requirements that aren't easily filled into a form.

Figure 8-9 illustrates a typical vessel data sheet. This particular one was created in Excel, and is included in the online materials that support this book.

An important feature is a vessel sketch, where critical features can be shown and annotated. The sketch often resembles the P&ID, and is sometimes copied directly

from it. It is the first thing that fabricators and others who use the data sheet will look at. Nozzles are normally shown and identified on the sketch, cross-referenced to the nozzle schedule, or list, that also appears on the data sheet.

The nozzles (called "openings" in the example) are called out with their size and type. Size means the nominal pipe size of the nozzle. Type refers to the connection method. Typical types used in the US are listed in Table 8-7 (this isn't a complete list). Use the "Service" column to describe the purpose of the opening, or give additional details such as "with blind flange", or "sight-glass with light". Use the Notes section, or an accompanying specification, to specify the pipe to be used for nozzles (e.g.,

				PRESSURE VESSEL		
				CLIENT	*EQUIP. NO*	*PAGE*
REV	PREPARED BY	DATE	APPROVAL	*W.O.*	*REQUISITION NO.*	*SPECIFICATION NO.*
0						
1				*UNIT AREA*	*PROCURED BY*	*INSTALLED BY*
2						

#	Item			Sketch
1	Reference Drawing No.			
2	Manufacturer			
3	Manufacturer's Drawing No.			
4	Code		Stamp?	
5				
6		Shell	Jkt.	
7	Operating Pressure, psig			
8	Design Pressure, psig			
9	Operating Temperature, deg F			
10	Design Temperature, deg F			
11	Relief Valve set pressure, psig			
12	Corrosion Allowance, in			
13	Position (horizontal/vertical)			
14	Location (indoor/outdoor)			
15	Height or Length (T-T), ft-in			
16	Inside Diameter, ft-in			
17	Head type			
18	Jacket Type			
19	Nominal Volume, gal			
20	Specific Gravity of Fluid			
21	**Materials of Construction**	Shell	Jkt.	
22	Shell			
23	Top Head			
24	Bottom Head			
25	Internals			
26	Lining			
27	Gaskets			
28	Bolts			
29	**Finishes**			
30	Process wetted parts			
31	Jacket wetted parts			
32	Exterior			
33				
34	Supports			
35	Insulation			
36	Earthquake Zone			
37	Design wind speed, mph			
38	Internals			
39	Nameplate			
Notes				

	Openings		
Item	Number	Size/Type	Service
A			
B			
C			
D			
E			
F			
G			
H			
I			
J			
K			
L			
M			
N			
O			
P			
Q			
R			
Openings Key			

Figure 8-9. Typical vessel data sheet.

Table 8-7
Typical nozzle types used in the US

Abbreviation	Description
FF	ANSI flat face flange
RF	ANSI raised-face flange
FNPT	3000 lb half-pipe coupling, threaded
MNPT	Pipe nipple, threaded
TC	Tri-Clamp sanitary flange (for hygienic applications)
3A	Sanitary threaded connection
MW	Manway (provide details in the Notes)
SG	Sight-glass

welded or seamless), gaskets, and special flanges (e.g., lined, loose).

The numbered lines on the example data sheet are described in Table 8-8.

The "Notes" section is used for additional information that has no other place on the data sheet. The requirements listed in the Notes are usually concise and should be easily comprehended. An accompanying specification can be used for complex requirements.

Table 8-8
Vessel data sheet entries

Data Sheet Lines	Description
1–3	Use to cross-reference the data sheet to other drawings, such as a P&ID or vendor drawing
4	Enter the Code requirements such as ASME or TÜV
7–12	For the vessel and jacket (if applicable), enter the normal operating and design pressures and temperatures. The design pressure is usually the larger of 10%, or 10 psi (0.7 bar), greater than the maximum normal operating pressure. Pressure is measured in the vapor space and does not include static pressure from the vessel's liquid contents. Specify maximum and minimum operating temperatures; the design temperature is usually at least 20 °F (10 °C) greater than the maximum operating temperature. Vacuum condition can be stated as actual pressure or with the abbreviation "FV." Example: 50 psig/FV.
	Relief valve set pressure is usually the same as the design pressure
	Corrosion allowance is used by the vessel designer by adding to the metal thickness after calculating required thickness. Then the thickness is rounded up to the nearest standard thickness.
13–16	These are parameters that describe design requirements, and might duplicate information found in the sketch. Use the Notes section to elaborate if necessary, such as stating a maximum or minimum diameter or height. "T-T" is the tangent-to-tangent distance, or the distance from the beginning of curvature of the heads. It is greater than the distance between the weld seams at the heads since the heads have a straight portion already included.
17	Head types are described in the "Tank Dimensions" section beginning on page 150. In the U.S., pressure vessels are most commonly specified with "ASME F&D" heads which are torispherical heads with dimensions meeting the ASME Code.
18	Jacket types include:
	Half-pipe coil (specify 180 ° or 120 ° of the available 360 ° circumference of the pipe that is included in the coil); also specify the spacing of the coils (vessel manufacturer will provide their "standard" for these parameters if not specified)
	Dimple jacket
	Conventional (annular) jacket, which is not recommended except for glass-lined vessels. Conventional jackets require internal baffles or agitating nozzles to be effective for heat transfer.
	The number of heat transfer zones should also be specified
19–20	Nominal volume is roughly calculated to the top tangent line, and is normally greater than the working volume. The fabricator needs the worst-case specific gravity of the vessel content for calculations.
21–28	Initial specifications may use generic terms for materials of construction (e.g., "CS" for carbon steel), but specific grades must be determined prior to fabrication (e.g., "SA-285 C")
29–32	There are several grades of "mill" finish, which refers to the condition of plate as made by the manufacturer. Grinding and polishing can be specified using various industry terms, with the ASME BPE surface finish designations recommended for vessels used in the biopharmaceutical industry (see Chapter 13).
34–39	Appurtenances must be specified by the user. Use the Notes section to elaborate, and to provide additional information needed to calculate and price the vessel.

Good Engineering Practices for Storage Tank and Vent Relief Design

The following checklists are intended to guide engineers to eliminate potential hazards during the design phase of a project [4]:

Storage tank design

- Question the need for all intermediate hazardous material storage and minimize quantities where storage is really needed, with greater attention to "just in time" supply.
- Minimize the inventory in transfer lines by careful attention to pipe routing and using the minimum practicable pipe size (while maintaining sufficient size, generally 25 mm (1 inch) minimum to withstand physical abuse). Consider the distance to an area of concern (such as a fence line or office building).
- Consider storing materials under less hazardous conditions. Two examples are dilution, to reduce the storage pressure and lower the initial atmospheric concentration in the event of a release, and refrigeration, which reduces the vapor pressure and driving force for a leak.
- It may be possible to reduce the dust explosion hazard of handling solid materials by using large particle size materials or by handling solids as a wet paste or a slurry in water. Consider the possibility of particle attrition, which can result in the production of small particles that could increase dust explosion hazards.
- Include dikes in the design, which will not allow flammable or combustible materials to accumulate around the bottom of tanks or equipment in case of a spill.

Storage tank vents

- Design tanks with as high a pressure as practicable. Consider 103.4 kPa gage (15 psig) API tank design limits for low pressure tanks to make it easier and more economical to design suitable vent relief systems.
- Special concern should be given to large, low-pressure tanks to avoid high- and low-pressure conditions, since these tanks are fragile and sensitive to internal pressure changes.
- Do not permit filling of top cones of low-pressure tanks.
- Install level detectors and have remote redundant high level switches, or redundant level transmitters and alarms for tanks containing hazardous materials.

- Define and document the worst-case venting and relief scenario.
- Be sure vents are large enough for the worst case; include allowances for headers, etc. Use properly designed pressure relief systems consistent with the design pressure of the tank.
- Have no traps in the vent line. Be especially careful with headers.
- All possibilities for blockage of vents must be considered and minimized, such as polymer formation, ice, solids, coatings, insects, birds, flying vegetation, flamer arresters, corrosion, etc. To avoid the possible plugging of vents by ice, insulate and heat with tracing if necessary.
- Design a system for the safe inspection and maintenance of vents and relief devices. Develop and use good operating procedures for inspecting and maintaining vents and relief systems. Inspect vents and relief devices periodically – at least once a year.
- Install drains in the low point of relief vent lines to drain any water that can collect. A loose-fitting weather cap should be installed over the stack openings, but this does not eliminate the need for a drain due to condensation and possible relief valve leakage.
- Design the outlets of vents and vent drains so that localized tank shell heating is avoided if flammable vapors are ignited. Avoid flame impingement on any part of the tank if vapors from vents are ignited.
- Consider the environmental consequences of tank venting.
- Have the vents from tanks directed to contained areas (preferably remote from the tank) so liquid overflow is contained.
- Supply vacuum relief if a vacuum can occur.
- Have no traps in a nitrogen line used for inerting tanks.

Nomenclature

A = cross-sectional area of the settling zone, ft^2

a = head extension, from tangent to end or bottom of head

D = distillate product to next unit operation, gal/min

D = tank inside diameter, m or ft

D = outside diameter of pipe, mm or in.

d = inside diameter of pipe, mm or in.

E = Young's modulus, N/mm^2 or lb/in.^2

F_p = vertical force exerted by pipe, N or lb

F_S = correction factor for hindered settling

f = Darcy friction factor, dimensionless

g = gravitational constant, $9.8\ \text{m/s}^2$ or $32.17\ \text{ft/s}^2$

H = height of liquid above point where drain pipe discharges, m or ft

h = height of liquid from bottom of tank, m or ft

h = height of liquid phase, ft

h_o = height of liquid from bottom of tank to point where drain pipe discharges, m or ft

Δh = tank deflection when the system capacity is added to the tank, mm or in.

L = reflux to the column, gal/min

L = horizontal piping distance from connection to nearest piping support point, mm or in.

L = pipe equivalent length, m or ft

L_S = length of settling zone, ft

Q = liquid flow rate, ft^3/s

R = tank inside radius, m or ft

T = residence time, hours

t = time to drain tank or section from initial level, H_o, to final level, H_f.

TT = tangent-to-tangent dimension, m or ft

U_S = actual settling velocity, ft/s

U_T = terminal settling velocity of water, ft/s

V_d = volume of the reflux drum full, gal

W_L = liquid flow rate, lb/s

W_V = vapor flow rate, lb/s

$\rho_w \rho_o$ = density of water and oil, lb/ft^3

ρ_L = density of liquid, lb/ft^3

ρ_v = density of vapor, lb/ft^3

μ_o = absolute viscosity of oil phase, lb/ft-s

μ = viscosity of the continuous phase, mPa-s or cP

References

[1] AP-42. Compilation of Air Pollution Emission Factors, Volume 1: Stationary Point and Area Sources, US EPA, 5th ed., with Supplements, http://www.epa.gov/ttn/chief/ap42/index.html; 1995.

[2] Amrouche Y, Dave C, Gursahani K, Lee R, Montemayor L. Aboveground Storage Tank Design and Operation. *Chemical Engineering Progress*, December, 2002.

[3] Barton R. Sizing Liquid-Liquid Phase Separators. *Chemical Engineering*, July 8, 1974.

[4] Englund S. Inherently Safer Plants: Practical Applications, paper presented at the AIChE Summer National Meeting. Colorado: Denver; August 16, 1994.

[5] Jones D. Calculating Fluid Tank Volumes. *Chemical Processing*, November, 2002:46–50.

[6] Kachelhofer K. Decoding Pressure Vessel Design. *Chemical Engineering*, June, 2010:28–35.

[7] Mettler-Toledo Inc. *Weigh Module Systems Handbook*. Switzerland: Metler-Toledo AG; 2010.

[8] Osage D, Straub M, Buchheim M, Amos D, Chiasson T, Samodell D. Precision Equations and Enhanced Diagrams for Local Stresses in Spherical and Cylindrical Shells Due to External Loadings for Implementation of WRC Bulletin 107. Welding Research Council, www.forengineers.org, December, 2010.

[9] Schwarzhoff J, Sommerfeld J. How Fast Do Spheres Drain? *Chemical Engineering*, June 20, 1988:158.

[10] Shoaei M, Sommerfeld J. Draining Tanks: How Long Does It Really Take? *Chemical Engineering*, January, 1989:154.

[11] Sommerfeld J. Tank Draining Revisited. *Chemical Engineering*, May, 1990:171.

[12] Watkins R. Sizing Separators and Accumulators. *Hydrocarbon Processing*, November, 1967:253.

9
Boilers

Introduction

Boilers are essentially shell-and-tube heat exchangers, with water being boiled to steam or transformed to high pressure hot water. Heat is generated by burning a fossil fuel; the combustion gas can be ducted around tubes containing the water (water-tube boiler), or the hot gas can pass inside the tubes with the water being in the shell (fire-tube boiler).

Stationary boilers, the very large boilers used in power generating stations, operate at high pressures and are invariably of the water-tube type. Water-tube designs are also popular for packaged boilers because they can be designed for higher pressure. Fire-tube boilers are limited to about 2.4 Mpa (350 psig) pressure and are used primarily for heating and process steam applications.

This chapter provides information on a variety of topics that are useful for process engineers who are involved with the design, installation, or troubleshooting of a boiler system.

Boiler Horsepower

Boiler horsepower is a measure of boiler energy output. It is different to motor horsepower.

Boiler Horsepower x 34.5 = steam production rate, lb/h

Boiler Horsepower x 0.069 = water-to-steam evaporation rate, gpm
Boiler Horsepower x 33,472 = energy output, Btu/h
Boiler Horsepower x 9.8 = energy output, kW

Types of Packaged Boilers

Boilers that are factory assembled and tested are called "packaged" boilers. They are fully piped and instrumented at the shop. Some of their features are given in Table 9-1.

There are two types of fire-tube boilers:

1. Dryback: a refractory-lined chamber outside the vessel is used to direct the combustion gases from

Table 9-1
Comparison of common packaged boiler types [3]

Characteristic	Fire-tube (Scotch Marine)	Membrane Water-tube	Industrial Water-tube	Electric
Efficiency	High	Medium	Medium	High
Floor space required	Medium to High	Very Low	High	Very Low
Maintenance	Low	Medium	High	Medium to High
Initial cost	Medium to High	Low to Medium	High	High
No. of options available	High	Medium	High	Medium
Pressure range	Hot Water	Hot Water	High Temperature	Hot Water
	Low Pressure Steam	Low Pressure Steam	Hot Water	Low Pressure Steam
	HPS to 350 psig	HPS to 600 psig	HPS to 900 psig	HPS to 600 psig
Typical sizes	To 1500 hp	To 250 hp	To 150,000 lb/h saturated or superheated steam	To 300 hp 9 kW to 3375 kW
Typical applications	Heating Process	Heating Process	Process	Heating Process
Best for	Most applications with steam pressures less than 350 psig	Building heat or modest process applications	Superheated steam, large or fluctuating steam loads, or pressures greater than 350 psig	Emission regulations prevail; areas where the cost of electricity is low

The pressure of "Low Pressure Steam" is below 15 psig.
"HPS" = "High Pressure Steam".

the furnace to the tube banks. Easy access to all internal areas of the boiler, including tubes, burner, furnace, and refractory from either end makes maintenance easier and reduces associated costs

2. Wetback: water-cooled turnaround chamber used to direct the flue gases from the furnace to the tube

banks. Requires less refractory maintenance, but internal pressure vessel maintenance is more difficult and costly. More prone to water-side sludge buildup because of restricted flow areas near the turnaround chamber

Power Plants

Battelle has provided a well-written report that discusses power plant coal utilization in great detail [13]. It gives a thermal efficiency of 80–83% for steam generation plants and 37–38% thermal efficiency for power generating plants at base load (about 70%). A base load plant designed for about 400MW and above will run at steam pressures of 2,400 or 3,600 psi and 1,000 °F with reheat to 1,000 °F and regenerative heating of feedwater by steam extracted from the turbine. A thermal efficiency of 40% can be obtained

from such a plant at full load and 38% at high annual load factor. The 3,600 psi case is supercritical and is called a once-through boiler, because it has no steam drum. Plants designed for about 100–350MW run around 1,800 psi and 1,000 °F with reheat to 1,000 °F. Below 100 MW a typical condition would be about 1,350 psi and 950 °F with no reheat. Below 60% load factor, efficiency falls off rapidly. The average efficiency for all steam power plants on an annual basis is about 33%.

Controls

This section discusses three basic parts of boiler controls:

1. Level control
2. Firing control (also applies to heaters)
3. Master control

For steam drum level control, select the modern three-element system – steam flow, feedwater flow, and drum level. Steam and feedwater flows are compared, with feedwater being requested accordingly and trimmed by the drum level signal. This system is better than having the drum level directly control the feedwater, because foaming or changing steam drum conditions can cause a misleading level indication. Also, the three-element controller responds faster to changes in demand.

The firing controls must be designed to ensure an air-rich mixture at all times, especially during upward or downward load changes. Steam header pressure signals the firing controls for a boiler. The signal to the firing controls comes from a master controller fed by the steam header pressure signal, if multiple boilers are operating in parallel. The firing controls that best ensure an air-rich mixture are often referred to as metering type controls, because gas flow and air flow are metered, thus the fuel-air ratio is controlled. The fuel-air ratio is the most important

factor for safe, economical firing, so it is better to control it directly. Do not settle for low-budget controllers that attempt to perform this basic job indirectly (such as by controlling fuel-steam or other ratios).

The idea for safe control is to have the air lead the fuel on increases in demand, and the fuel lead the air on decreases in demand. On load increases, the air is increased ahead of the fuel. On load decreases, the fuel is decreased ahead of the air. This is accomplished with high- and low-signal selectors.

A high-signal selector inputs to the air flow controller, which often adjusts forced draft fan inlet vanes. The high-signal selector compares the steam pressure and fuel flow signals, selects the highest, and passes it on to the air flow controller. For a load decrease, the steam line pressure tends to rise and its controller reduces the amount of firing air to the boiler, but not immediately. The high-signal selector picks the fuel flow signal instead of the steam pressure signal, which has decreased. The air flow will therefore wait until the fuel has decreased. On load increases, the steam pressure signal exceeds the fuel signal and the air flow is immediately increased.

A flue gas oxygen analyzer should be installed to continuously monitor, or even trim, the fuel-air ratio. The optimum percentage of oxygen in the flue gas depends on the type of fuel and varies with load. Therefore, the oxygen setpoint is modified with steam flow.

A master controller is necessary to control a single steam header pressure from multiple parallel boilers.

Depending on the load and the performance of the individual boilers, the most efficient operation may be achieved with some boilers shut down, some boilers base loaded at a constant firing rate, and the remaining boilers allowed to swing with the load.

Thermal Efficiency

This graph (Figure 9-1) shows how thermal efficiency can be determined from excess air and stack gas temperature. The difference between gross thermal efficiency and combustion efficiency is the heat lost through the walls of the boiler system to the surroundings.

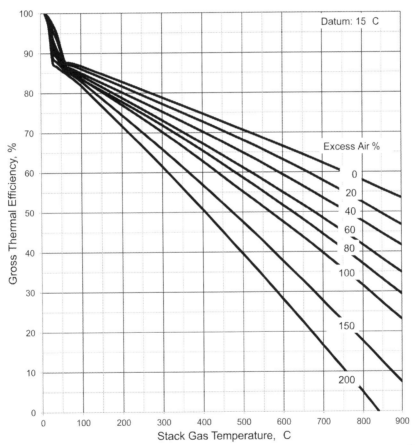

Figure 9-1. Gross thermal efficiency for a fuel with Higher Heating Value (HHV) = 37.3 kJ/Sm^3 (1000 Btu/scf) [8].

Stack Gas Enthalpy

The typical enthalpy of the combustion (or stack) gases when natural gas is burned with 20% excess dry air is shown in Figure 9-2.

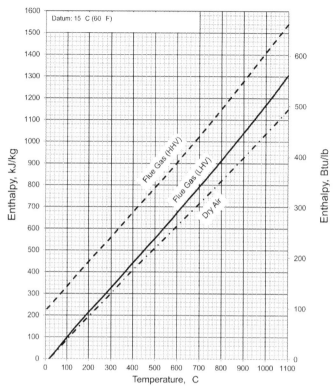

Figure 9-2. Typical enthalpy of combustion gases for a dry natural gas fuel and 20% excess dry air [8].

Steam Drum Stability

Steam drums are located at the top of a water-tube boiler, to provide an upper reservoir for the water that covers the tube bank and to separate the generated steam from entrained water. Steam drums also include safety relief valves and air eliminators.

The internal fittings in the steam drum help to distribute the water evenly throughout the drum, separate the generated steam from the water and remove moisture from the steam before it leaves the boiler [17] (refer to Figure 9-3).

1. **Lower baffle plates or apron plates.** Separate the incoming feedwater and generated steam and direct the steam to the separators.
2. **Primary separators (cyclone separators).** Separate most of the water from the steam by giving it a cyclone or rotary motion so that the water particles are expelled from the steam by the centrifugal forces. These separators are vertically mounted in the steam drum so that the steam rises out the top and the water falls back into the steam drum.

3. **Secondary separators (chevron dryers).** Remove additional moisture from the steam by changing the direction of steam flow several times. The steam passes on but the moisture cannot make the direction change with the steam. These separators are mounted above the primary separators and direct steam to the dry box which collects the steam at the top of the steam drum, directing it to the steam outlet piping to the superheater.
4. Feedwater leaves the economizer and enters the boiler through the internal feed pipe and becomes boiler water. Perforations along the side of the feed pipe allow water to be distributed evenly throughout the steam drum.
5. Since suspended solids may accumulate on the surface of the water in the steam drum, there must be a means of removing them. The surface blow pipe is used to remove these light suspended solids from the surface of the water and to reduce the total dissolved

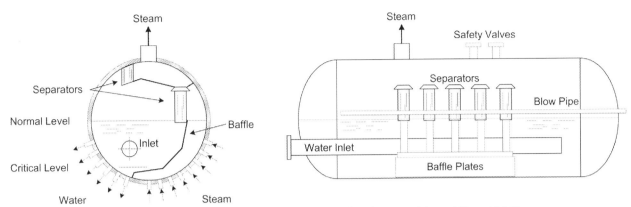

Figure 9-3. General steam drum configuration (adapted from [4] and [17]).

solid content of the boiler water. Suspended solids usually consist of oil, salt contaminants, or excessive treatment chemicals, which can cause foaming on the water surface. Dissolved solids usually consist of salt contaminants and treatment chemicals that are in solution.

Ellison has published an extremely important factor for steam drum design called the Drum-Level-Stability Factor [4]. As manufacturers have learned how to increase boiler design ratings, the criteria for steam drum design have lagged behind. The three historical steam drum design criteria have been:

1. The drum is sized for the required steaming rate.
2. The drum must be large enough to contain the baffling and separators required to maintain separation and steam purity.
3. The drum must extend some minimum distance past the furnace to mechanically install the tubes.
   A fourth criterion needs to be:
4. The drum must have a water holding capacity with enough reserve in the drum itself, such that all the steam in the risers, at full load, can be replaced by water from the drum without exposing critical tube areas.

This fourth criterion can be met by spending less than 1% of the cost of the boiler on the steam drum. It is met by requiring that the Drum-Level-Stability Factor (DLSF) be equal to a minimum of 1.0. When this exists the steam drum level will be stable for wide and sudden operational changes.

The DLSF is defined as follows:

$$DLSF = \frac{V_a}{V_m} \qquad (9\text{-}1)$$

Where:

V_a = The actual water holding capacity of the drum between the normal water level and the level at which tubes would be critically exposed, gal

V_m = The minimum water holding capacity required to replace all of the steam bubbles in the risers, gal.

V_a is calculated based on Figure 9-3, which uses, as the "critical level", a height of one inch above the lower end of the drum baffle separating the risers from the downcomers. Ellison derived equations to simplify the calculations of V_m.

$$V_m = \frac{(\%SBV)G(HR)}{600,000} \qquad (9\text{-}2)$$

Where: (the constants are based on test data)

G = volume of water required to fill the entire boiler to the normal water level, gal

HR = furnace heat release per area of effected projected radiant surface, Btu/ft^2

$$\%SBV = \frac{V_s}{(C-1)V_w + V_s}$$

C = average boiler circulation ratio at full load, lb water-steam mixture circulated in a circuit per lb of steam leaving that circuit (C is the average of all circuits in the boiler)

V_s = specific volume of steam at saturation temperature and pressure at operating conditions

V_w = specific volume of water at saturation temperature and pressure at operating conditions

Ellison gives the following example:
Given

Steam drum pressure = 925 psig
C (from manufacturer) = 18.5
G (from manufacturer) = 6,000 gal
HR (from manufacturer) = 160,000 Btu/ft^2
V_a = 1,500 gal

Calculations

V_w = 0.0214ft^3/lb (from steam tables)
V_s = 0.4772ft^3/lb (from steam tables)
%SBV = 0.4772/[(18.5 − 1)   (0.0214) + 0.4772] = 0.56 (or 56% steam by volume)
V_m = [(0.56) (6,000) (160,000)]/600,000 = 896 gal
DLSF = V_a/V_m = 1,500 / 896 = 1.67

This steam drum level would be very stable (DLSF well above 1.0).

Deaerator Venting

Boiler feedwater, consisting of fresh water plus returned condensate, is heated in the deaerator to full saturation temperature corresponding to the steam pressure. The purpose is to scrub out and carry away dissolved gases, especially oxygen.

Most of the steam provided to the deaerator condenses, but a small fraction (usually 5% to 14%) must be vented to accommodate the stripping requirements. Normal design practice is to calculate the steam required for heating, and then make sure that the flow is sufficient for stripping as well. If the condensate return rate is high (>80%) and the condensate pressure is high in comparison with the deaerator pressure, then very little steam is needed for heating, and provisions may be made for condensing the surplus flash steam [16].

Steam consumption is equal to the steam required to heat the incoming water to its saturation temperature, plus the amount vented with the non-condensable gases, less

any flashed steam from hot condensate or steam losses through failed traps. Calculate the heat balance with the incoming water at its lowest expected temperature. The vent rate is a function of deaerator type, size (rated feedwater capacity), and the amount of makeup water.

Knox has published graphs for estimating the steam vent requirement from boiler feedwater deaerators [12]. These formulas replicate the graphs. The input variables are the fraction of the water that is fresh water (F), and the deaerator rating in thousands of kg or lb per h (R). The result is the amount of steam vented in kg or lb per h (V), which must be added to the steam required to heat the water to saturation temperature to obtain the total steam flow to the deaerator.

For a spray-type deaerator:

$$V = (0.66\,F + 0.68)R \tag{9-3}$$

For a tray-type deaerator:

$$V = (0.49\,F + 0.5)R \tag{9-4}$$

Water Alkalinity

Most water analysis results are rather easily interpreted. However, two simple and useful tests need explanation. These are the P and M alkalinity.

M alkalinity is a measure of the *total* alkalinity of the water, expressed as ppm $CaCO_3$. It includes alkalinity caused by one or all of: 1) hydroxide (OH^-), 2) carbonate (CO_3^{2-}), or 3) bicarbonate (HCO_3^-). It is found by titration with sulfuric acid to a methyl orange end point range of pH 4.2 to 4.4.

P alkalinity is defined as that alkalinity which exists at pH values above the phenolphthalein endpoint range of pH 8.2 to 8.4. It measures just the hydroxide content.

In natural waters, the alkalinity is usually caused by bicarbonate. Carbonate or hydroxide are rarely encountered in untreated water. Due to chemical equilibria, the contribution from each type of alkaline is determined from the M and P measurements according to Table 9-2.

Table 9-2
Alkalinity relationship based on P and M tests

Situation	Hydroxyl	Carbonate	Bicarbonate
P = M	M	0	0
P > ½ M	2P − M	2 (M − P)	0
P = ½ M	0	M	0
P < ½ M	0	2 P	M − 2P
P = 0	0	0	M

Alkali ions found in wastewaters, such as bisulfides, sulfides, and phosphates, interfere with the M and P titrations. In such cases other tests may be warranted, such as acid evolution of carbon dioxide.

Example: If P = 78 ppm as $CaCO_3$ and M = 131 ppm as $CaCO_3$, then the situation in Table 9-2 is P > ½ M, or P is greater than ½ of M.

Alkalinity contributed by each species is then:

Hydroxyl = 2P − M = (2 × 78) − 131 = 25 ppm as $CaCO_3$

Carbonate = 2 (M − P) = 2 × (131 − 78) = 106 ppm as $CaCO_3$

Bicarbonate = 0 ppm as $CaCO_3$

Check Total = 25 + 106 + 0 = 131 ppm = M

Source Water Impurities

Source water is normally either ground water or surface water. Ground waters contain dissolved inorganic impurities that came from the rock and sand strata through which the water has passed. Surface waters often contain silt particles in suspension (suspended solids) and dissolved organic matter, in addition to dissolved inorganic impurities (dissolved solids). GPSA published this table (Table 9-3) showing major water impurities, the difficulties they cause, and their treatment [8].

Table 9-3
Water impurities and characteristic treatment [8]

Constituent	Chemical Formula	Difficulties Caused	Means of Treatment
Turbidity	None, usually expressed in Jackson Turbidity Units	Imparts unsightly appearance to water; deposits in water lines, process equipment, boilers, etc.; interferes with most process uses	Coagulation, settling, and filtration
Color	None	Decaying organic material and metallic ions causing color may cause foaming in boilers; hinders precipitation methods such as iron removal; hot phosphate softening can stain product in process use	Coagulation, filtration, chlorination, adsorption by activated carbon
Hardness	Calcium, magnesium, barium, and strontium salts expressed as $CaCO_3$	Chief source of scale in heat exchange equipment, boilers, pipelines, etc.; forms curds with soap; interferes with dyeing, etc.	Softening, distillation, internal boiler water treatment, surface active agents, reverse osmosis, electrodialysis
Alkalinity	Bicarbonate (HCO_3^{1-}), carbonate (CO_3^{2-}), and hydroxyl (OH^{1-}), expressed as $CaCO_3$	Foaming and carryover of solids with steam; embrittlement of boiler steel; bicarbonate and carbonate produce CO_2 in steam, a source of corrosion	Lime and lime-soda softening, acid treatment, hydrogen zeolite softening, demineralization, dealkalization by anion exchange, distillation, degasifying

(Continued)

Table 9-3
Water impurities and characteristic treatment [8]—cont'd

Constituent	Chemical Formula	Difficulties Caused	Means of Treatment
Free Mineral Acid	H_2SO_4, HCl, etc. expressed as $CaCO_3$, titrated to methyl orange end-point	Corrosion	Neutralization with alkalis
Carbon Dioxide	CO_2	Corrosion in water lines and particularly steam and condensate lines	Aeration, deaeration, neutralization with alkalis, filming and neutralizing amines
pH	Hydrogen Ion concentration defined as $pH = \log \dfrac{1}{(H^{1+})}$	pH varies according to acidic or alkaline solids in water; most natural waters have a pH of 6.0 to 8.0	pH can be increased by alkalies and decreased by acids
Sulfate	$(SO_4)^{2-}$	Adds to solids content of water, but, in itself, is not usually significant; combines with calcium to form calcium sulfate scale	Demineralization, distillation, reverse osmosis, electrodialysis
Chloride	Cl^{1-}	Adds to solids content and increases corrosive character of water	Demineralization, distillation, reverse osmosis, electrodialysis
Nitrate	$(NO_2)^{1-}$	Adds to solid content, but is not usually significant industrially; useful for control of boiler metal embrittlement	Demineralization, distillation, reverse osmosis, electrodialysis
Fluoride	F^{1-}	Not usually significant industrially	Adsorption with magnesium hydroxide, calcium phosphate, or bone black; alum coagulation; reverse osmosis; electrodialysis
Silica	SiO_2	Scale in boilers and cooling water systems; insoluble turbine blade deposits due to silica vaporization	Hot process removal with magnesium salts; adsorption by highly basic anion exchange resins, in conjunction with demineralization; distillation
Iron	Fe^{2+} (ferrous) Fe^{3+} (ferric)	Discolors water on precipitation; source of deposits in water lines, boilers, etc.; interferes with dyeing, tanning, paper mfr, etc.	Aeration, coagulation, and filtration, lime softening, cation exchange, contact filtration, surface active agents for iron retention
Manganese	Mn^{2+}	Same as iron	Same as iron
Oil	Expressed as oil or chloroform extractable matter, ppmw	Scale, sludge, and foaming in boilers; impedes heat exchange; undesirable in most processes	Baffle separators, strainers, coagulation and filtration, diatomaceous earth filtration
Oxygen	O_2	Corrosion of water lines, heat exchange equipment, boilers, return lines, etc.	Deaeration, sodium sulfite, corrosion inhibitors, hydrazine or suitable substitutes
Hydrogen Sulfide	H_2S	Cause of "rotten egg" odor; corrosion	Aeration, chlorination, highly basic anion exchange
Ammonia	NH_3	Corrosion of copper and zinc alloys by formation of complex soluble ion	Cation exchange with hydrogen zeolite, chlorination, deaeration, mixed-bed demineralization
Conductivity	Expressed as µS/cm, specific conductance	Conductivity is the result of ionizable solids in solution; high conductivity can increase the corrosive characteristics of a water	Any process that decreases dissolved solids content will decrease conductivity; examples are demineralization, lime softening
Dissolved Solids	None	"Dissolved solids" is measure of total amount of dissolved matter, determined by evaporation; high concentrations of dissolved solids are objectionable because of process interference and as a cause of foaming in boilers	Various softening processes, such as lime softening and cation exchange by hydrogen zeolite, will reduce dissolved solids; demineralization; distillation; reverse osmosis; electrodialysis
Suspended Solids	None	"Suspended solids" is the measure of undissolved matter, determined gravimetrically; suspended solids plug lines, cause deposits in heat exchange equipment, boilers, etc.	Subsidence, filtration, usually preceded by coagulation and settling
Total Solids	None	"Total solids" is the sum of dissolved and suspended solids, determined gravimetrically	See "Dissolved Solids" and "Suspended Solids"

Conductivity versus Dissolved Solids

For a quick estimate of total dissolved solids (TDS) in water one can run a conductivity measurement. The unit for the measurement is mhos/cm. A mho is the reciprocal of an ohm. The mho has been renamed the Sieman (S) by ISO. Both mhos/cm and S/cm are accepted as correct terms. In water supplies (surface, ground, etc.) conductivity will run about 10^{-6} S/cm or 1 µS/cm.

Without any data available the factor conductivity to TDS is:

$$TDS(ppm) = Conductivity(\mu S/cm)/2$$

However, the local water supplier will often supply TDS and conductivity data. Table 9-4 gives conductivity factors for common ions found in water supplies.

Table 9-4
Water quality – conductivity factors of ions commonly found in water [14]

Ion	µS/cm per ppm
Bicarbonate	0.715
Calcium	2.60
Carbonate	2.82
Chloride	2.14
Magnesium	3.82
Nitrate	1.15
Potassium	1.84
Sodium	2.13
Sulfate	1.54

Silica in Steam

Silica in steam used to drive turbines should be limited to 0.01 to 0.02 ppm [15]. Excessive silica precipitates when the steam pressure is reduced in the turbine; the precipitate can deposit on the turbine blades and severely degrade the turbine efficiency. These glasslike deposits are very difficult to remove.

Caustic Embrittlement

Whenever free caustic (NaOH) is present in a boiler, failure from caustic embrittlement may occur. Embrittlement is a form of stress corrosion cracking. It occurs in areas where there are specific stress conditions and free NaOH in the boiler water. Boiler tubes usually fail from caustic embrittlement at points where tubes are rolled into sheets, drums, or headers. However, it also occurs at weld cracks and if there is tube-end leakage.

Sodium nitrate is usually used to inhibit embrittlement in boilers operating at low pressures. The US Bureau of Mines recommends that the ratio of sodium nitrate to

sodium hydroxide be maintained in accordance with Table 9-5. At pressures above 7000 kPa (1000 psig), coordinated phosphate/pH control is usually employed which precludes the development of high concentrations of caustic.

Calculate the ratio using the following formula:

$$\frac{NaNO_3}{NaOH}$$

$$= \frac{ppm\ nitrate(as\ NO_3)}{ppm\ M\ alkalinity(as\ CaCO_3) - ppm\ phosphate(as\ PO_4^{3-})}$$

Table 9-5
Recommended ratio of sodium nitrate to sodium hydroxide
for boilers

Boiler Pressure, kPa	Boiler Pressure, psig	Ratio
0 to 1725	0 to 250	0.20
1725 to 2750	250 to 400	0.25
2750 to 4825	400 to 700	0.40
4825 to 7000	700 to 1000	0.50

Waste Heat

Fire-tube boilers are widely used to recover energy from waste gas streams commonly found in chemical plants, refineries, and power plants. Typical examples are exhaust gases from gas turbines and diesel engines, and effluents from sulfuric acid, nitric acid, and hydrogen plants. Generally, they are used for low-pressure steam generation. Typical arrangement of a fire-tube boiler is shown in Figure 9-4. Sizing of waste heat boilers is quite an involved procedure. However, using the method described here one can estimate the performance of the boiler at various load conditions, in addition to designing the heat transfer surface for a given duty.

The design method is similar to the evaluation of any heat exchanger, but since a common application is to recover heat from a flue gas typical properties for flue gas are used to create the nomograph (Figure 9-5, [6]).

The energy transferred from the waste gas stream is evaluated using the equation:

$$Q = W_g \, c_p \, (T_{out} - T_{in}) \tag{9-5}$$

Where:

Q = enthalpy change, kJ/h (multiply by 0.278 for W) or Btu/h
W_g = mass flow rate, kg/h or lb/h
c_p = specific heat, kJ/kg-°C or Btu/lb-°F
T_{out} = temperature at waste boiler outlet, °C or °F
T_{in} = temperature at waste boiler inlet, °C or °F

The waste heat boiler heat transfer depends on the temperature difference between the waste gas stream flowing inside the tubes and saturated steam at the

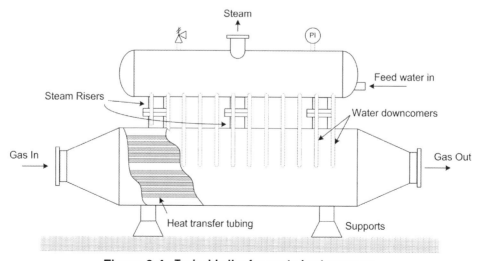

Figure 9-4. Typical boiler for waste heat recovery.

generation pressure, the overall heat transfer coefficient, and the inside area of the tubes:

$$A = \frac{Q}{U\Delta T_{mean}} \qquad (2\text{-}2)$$

A = heat transfer area, usually calculated at the inside tube diameter, m^2 or ft^2
Q = heat transferred, W or Btu/h
U = overall heat transfer coefficient, W/m^2-°C or Btu/h-ft^2-°F
ΔT_{mean} = mean temperature difference (MTD) between generated steam and waste gas streams, °C or °F

Assume:
$$U = 0.95 \, h_i$$

With the inside film coefficient calculated from:

$$N_{Nu} = \frac{h_i \, d_i}{k} = 0.023 \, N_{Re}^{0.8} \, N_{Pr}^{0.4} \qquad (9\text{-}6)$$

N_{Nu} = Nusselt number
h_i = inside film coefficient, W/m^2-°C or Btu/ft^2-°F
d_i = inside diameter of one tube, m or ft
k = gas thermal conductivity, W/m-h-°C or Btu/ft-h-°F

$$N_{Pr} = \text{Prandtl number} = \frac{c_p \, \mu}{k}$$

$$N_{Re} = \text{Reynolds number} = \frac{d_i \, \rho \, u}{\mu}$$

ρ = density of the waste gas stream, kg/m^3 or lb/ft^3
μ = dynamic viscosity of the waste gas stream, kg/m-s or lb/ft-h
u = velocity in tubes, m/s or ft/s

For the mean temperature, use:

$$\Delta T_{mean} = \frac{(T_{in} - t_s) - (T_{out} - t_s)}{\ln \dfrac{(T_{in} - t_s)}{(T_{out} - t_s)}} \qquad (9\text{-}7)$$

t_s = saturation temperature of steam at the generation pressure, °C or °F

With some rearrangement and substitution ($A = \pi \, d_i \, N \, L$ with N = number of tubes and L = length of tube bank; $w_g = W_g/N$),

$$\ln \frac{(T_{in} - t_s)}{(T_{out} - t_s)} = \frac{U \, A}{W_g \, c_p} = \frac{0.62 \, L \, k^{0.6}}{d_i^{0.8} \, w_g^{0.2} \, c_p^{0.6} \, \mu^{0.4}}$$

$$= \frac{0.62 \, L \, F(t)}{d_i^{0.8} \, w_g^{0.2}} \qquad (9\text{-}8)$$

Where:

$$F(t) = \frac{k^{0.6}}{c_p^{0.6} \, \mu^{0.4}}$$

Equation 9-8 is charted in Figure 9-5 using properties for commonly found flue gas streams. Features of the nomograph are:

- The surface area, A, required to transfer a given Q may be found (design)
- The exit gas temperature, T_{out}, and Q at any off design condition may be found (performance)
- If there are space limitations for the boiler (L or shell dia. may be limited in some cases) we can use the chart to obtain a different shell diameter given by $D_s = 1.3 \, d_i \, \sqrt{N}$
- Heat transfer coefficient does not need to be computed
- Reference to flue gas properties is not necessary

Thus, a considerable amount of time may be saved by using this chart. Evaluate gas properties at the average gas temperature. In the case of design, the inlet and exit gas temperatures are known and hence the average can be found. In the case of performance evaluation, the inlet gas temperature alone will be known and, hence, a good estimate of average gas temperature is $0.6 \, (T_{in} + t_s)$. Also, a good starting value of w for design is 80 to 150 lb/h.

Example 1. A waste heat boiler is to be designed to cool 75,000 lb/h of flue gases from 1,625 °F to 535 °F, generating steam at 70 psia. Tubes of size 2-in. OD and 1.8-in. ID are to be used. Estimate the surface area and geometry using 120 lb/h of flow per tube. Find Q and W_s if feed water enters at 180 °F.

Solution. Calculate $(T_{in} - t_s)/(T_{out} - t_s) = (1,625 - 303)/(535 - 303) = 5.7$. Connect w = 120 with $d_i = 1.8$ and extend to cut line 1 at "A". Connect t = $(1,625 + 535)/2 = 1,080$ with $(T_{in} - t_s)/(T_{out} - t_s) = 5.7$ and extend to cut line 2 at "B". Join "A" and "B" to cut the L scale at 16. Hence, tube length is 16 ft. Number of tubes, $N = 75,000/120 = 625$. Surface area, $A = 1.8 (16) (625) (\pi/12) = 4,710$ ft^2. Note that this is only one design and several alternatives are possible by changing w or L. If L has to be limited, we can work the chart in reverse to calculate w.

$$Q = 75,000 \, (0.3) \, (1,625 - 535) = 24.52 \text{ MMBtu/h}$$

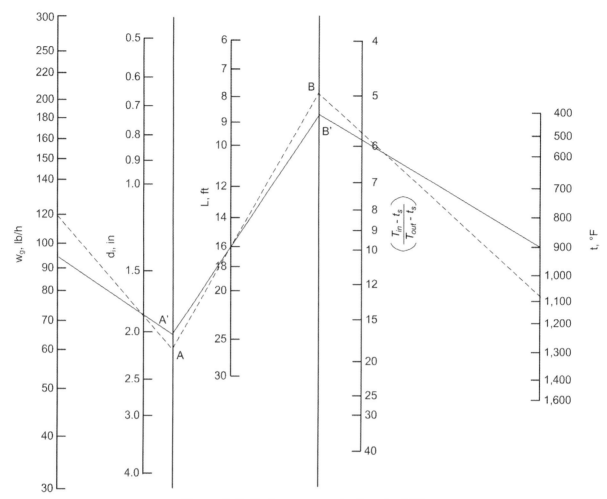

Figure 9-5. Performance evaluation chart [6].

$$W_s = \frac{24.52(10^6)}{(1,181-148)} = 23,765 \text{ lb/h}$$

(1,181 and 148 are the enthalpies of saturated steam and feed water.)

Example 2. What are the duty and exit gas temperature in the boiler if 60,000 lb/h of flue gases enter the boiler at 1,200 °F? Steam pressure is the same as in example 1.

Solution. Connect w = (60,000/625) = 97 with d_i = 1.8 and extend to cut line 1 at "A". Connect with

$L = 16$ and extend to cut line 2 at "B". The average gas temperature is 0.6 (1,200 + 303) = 900 °F. Connect "B" with t = 900 to cut $(T_{in} - t_s)/(T_{out} - t_s)$ scale at 5.9. Hence $t = 455$ °F. Assuming a c_p of 0.285, $Q = 60,000$ (0.285) (1,200 − 455) = 12.74 MMBtu/h. $W_s = 12.74$ (10^6)/ (1,181 − 148) = 12,340 lb/h [c_p can be approximated by (0.22 + 7.2 (10^{-5}) t) in the temperature range 600 to 1,500 °F].

Boiler Feedwater

ASME published their consensus recommendations for water chemistry of feedwater and boiler water for industrial water-tube, fire-tube, coil type, marine propulsion, and specialty (electrode, forced circulation, jet) steam

boilers. These recommendations supersede the widely published American Boiler Manufacturers Association (ABMA) table. Table 9-6 lists the recommended limits for water-tube and fire-tube boilers *without* superheaters or

Table 9-6
Suggested water chemistry limits for industrial water-tube and fire-tube boilers, high duty, fuel-fired with up to 100% feedwater makeup percentage and 1.0 ppm maximum TDS in the steam (Note 7)

Characteristic	Water-tube Boilers		Fire-tube Boilers
	0 to 2.07 MPa (0 to 300 psig)	2.08 to 4.14 MPa (301 to 600 psig)	0 to 2.07 MPa (0 to 300 psig)
Feedwater (Note 3)			
Dissolved oxygen, ppm; measured before chemical oxygen scavenger addition (Notes 1 and 2)	< 0.007	< 0.007	< 0.007
Total iron, ppm	< 0.1	< 0.05	< 0.1
Total copper, ppm	< 0.05	< 0.025	< 0.05
Total hardness, ppm as $CaCO_3$	< 0.5	< 0.3	< 1.0
pH @ 25 °C	8.3 to 10.5	8.3 to 10.5	8.3 to 10.5
Nonvolatile TOC, ppm (Note 6)	< 1	< 1	< 10
Oily matter, ppm	< 1	< 1	< 1
Boiler Water			
Silica, ppm SiO_2	< 150	< 90	< 150
Total alkalinity, ppm as $CaCO_3$	< 1000	< 850	< 700
Free OH alkalinity, ppm (Note 4)	Not Specified	Not Specified	Not Specified
Specific conductance, µS/cm @ 25 °C without neutralization	< 7000	< 5500	< 7000

1. Values in the table assume existence of a deaerator
2. Chemical deaeration should be provided in all cases, especially if mechanical deaeration is nonexistent or inefficient
3. Water-tube boilers with relatively large furnaces, large steam release space and internal chelant, polymer, and/or antifoam treatment can often tolerate higher levels of feedwater impurities than those in the table and still achieve adequate deposition control and steam purity. Removal of these impurities by external pretreatment is always a more positive solution. Alternatives must be evaluated as to practicality and economics in each individual case. The use of some dispersant and antifoam internal treatment is typical in this type of boiler operation. Fire-tube boilers of conservative design, with internal chelant, polymer, and/or antifoam treatment can often tolerate higher levels of feedwater impurities than those in the table (<0.5 ppm Fe, <0.2 ppm Cu, <10 ppm total hardness) and still achieve adequate deposition control and steam purity.
4. Minimum and maximum hydroxide alkalinities must be individually specified by a qualified water treatment consultant with regard to silica solubility and other components of internal treatment. Consensus could not be reached on these values.
5. Alkalinity and conductance values are consistent with steam purity limits. Practical limits above or below tabulated values should be individually established by careful steam purity measurements.
6. Nonvolatile TOC is that organic carbon not intentionally added as part of the water treatment program. See Section 6.4 in [5].
7. This limit represents steam purity that should be achievable if other tabulated water quality values are maintained. The limit is not intended to be nor should it be construed to represent a boiler performance guarantee.

turbines. Refer to the ASME source document for other types of boilers [5].

Blowdown rates typically range between 4% and 8% of feedwater flow [11], but should never be less than 1%, even if the feedwater is high purity, demineralized or evaporated water [5].

The suggested limits for boiler water solids are based upon one or more of the following factors [8]:

• Sludge and Total Suspended Solids – these result from the precipitation in the boiler of feedwater hardness constituents due to heat and interaction of treatment chemicals, and from corrosion products in the feedwater. They can contribute to boiler tube deposits and enhance foaming characteristics, leading to increased carryover.
• Total Dissolved Solids – these consist of all salts naturally present in the feedwater, of soluble silica,

and of any chemical treatment added. Dissolved solids do not normally contribute to scale formation but excessively high concentrations can cause foaming and carryover or can enhance "under deposit" boiler tube corrosion.

• Silica – this may be the blowdown controlling factor in softened waters containing high silica. High boiler water silica content can result in silica vaporization with the steam, and, under certain circumstances, siliceous scale. Silica content of the boiler water is not as critical for steam systems without steam turbines.
• Iron – occasionally in high pressure boilers where the iron content is high in relation to total solids, blowdown may be based upon controlling iron concentrations. High concentrations of suspended iron in boiler water can produce serious boiler deposit problems and are often indications of potentially serious corrosion in the steam or steam condensate systems.

Blowdown Control

Blowdown from industrial steam boilers is typically controlled automatically using a continuous system. Automatic systems usually maintain the specific conductance of the boiler water at a specified level, especially if sodium zeolite softened makeup water is used. Other boiler water constituents such as chlorides, sodium, and silica are also used as a means of controlling blowdown. Continuous blowdown is supplemented by manual blowdown to discharge suspended solids from the low points in the boiler system.

Here are some guidelines for blowdown control [9]:

- Dissolved solids – specific conductance gives an indirect measure of dissolved solids and can usually be used for blowdown control. Conductance is caused by the ionization of the various salts that are present; in dilute solutions there is almost complete ionization; however, in concentrated solutions ionization is suppressed. Therefore, it is best to correlate specific conductance with dissolved solids for each system. For a "rule of thumb" estimate, very dilute solutions such as condensate may be calculated with a factor of 0.5 to 0.6 ppm of dissolved solids per microsiemens of specific conductance. For a more concentrated solution such as boiler water, the factor varies between 0.55 and 0.90 ppm of dissolved solids per microsiemens of specific conductance. It is common practice to neutralize any hydroxide ions with gallic acid or boric acid prior to measuring conductivity.
- Chloride – if the chloride concentration in the feedwater is high enough to measure accurately, it can be used to control blowdown and to calculate the rate of blowdown. Chlorides do not precipitate in the boiler water, therefore the relative chloride concentrations provide an accurate basis for calculating the rate of blowdown. The chloride test is unsuitable when the feedwater chloride is too low for accurate determination.
- Silica – on-line methods of silica monitoring are primarily applicable to boilers operating at 600 psig or greater, utilizing a demineralized makeup and where the steam is used in turbines. Silica in the steam must be limited to 0.01 to 0.02 ppm; to do this the silica in the boiler water must be limited to a specified maximum that is dependent on boiler pressure and system design.

Best Practices – Burner System Piping

Axon gives these tips [2]:

1. Use black iron (not galvanized) pipe with malleable iron fittings for combustion systems.
2. Install drip legs on gas lines to trap dirt and moisture and prevent fouling of regulators, valves, burners, etc.
3. Leak-test all piping connections with a soapy water solution.
4. When pressure-testing new equipment, be careful not to exceed the allowable maximum pressure for included components.
5. Where possible, avoid reducing bushings and street ell fittings (i.e., 45-deg. Or 90-deg. fittings with a male pipe thread on one end and a female pipe thread on the other).
6. Always provide flexible connections for the burner and pilot. Do not use the burner as a pipe support. Include unions and flanges for easy access and removal.
7. Use a good quality anti-seize joint compound on pipe connections that will be exposed to elevated temperatures. Do not use soldered joints in fuel lines.
8. Center-fed manifolds with constant diameter are preferred over end-fed or decreasing-diameter configurations.
9. Check with the equipment manufacturer regarding the required lengths of straight pipe runs upstream and downstream of regulators, valves, orifices, etc., to ensure good control and accurate tracking.
10. If the combustion system will be modulating, select a combustion air blower that has a relatively flat pressure curve over the entire operating range.
11. Install a blower inlet filter to help keep both the burner air and pilot air passageways clean and dirt-free. Filter media can be disposable or washable. Blower noise silencers or combination filter/silencers are also available.

References

[1] American Boiler Manufacturers Association (ABMA). Arlington, Virginia. www.abma.com

[2] Axon B. Improve Combustion System Efficiency. *Chemical Engineering Progress*, August, 2009.

[3] Cleaver-Brooks, Inc. *Boiler Book*. Milwaukee, WI. www.cleaverbrooks.com, 2011.

[4] Ellison G. Steam Drum Level Stability Factor. *Hydrocarbon Processing*, May, 1971.

[5] Feedwater Quality Task Group. *Consensus on Operating Practices for the Control of Feedwater and Boiler Water Chemistry in Modern Industrial Boilers, Industrial Subcommittee of the ASME Research and Technology Committee on Water and Steam in Thermal Systems*. American Society of Mechanical Engineers; 1994.

[6] Ganapathy V. Size or Check Waste Heat Boilers Quickly. *Hydrocarbon Processing*, September, 1984.

[7] Ganapathy V. *Applied Heat Transfer*. PennWell Books; 1982.

[8] Gas Processors Suppliers Association (GPSA). *Engineering Data Book, SI Version*. 12th ed. 2004.

[9] GE Power & Water. *Handbook of Industrial Water Treatment*. Chapter 13, downloaded from www.gewater.com/handbook/index.jsp; January, 2011.

[10] Hoehenberger L. Scaling, Corrosion and Water Treatment on Steam Generator Systems < 30 bar. Hung Yen, Vietnam: presented at an Industrial Seminar, http://www.sme-gtz.org.vn; 2007.

[11] Jaber D, McCoy G, Hart F. Follow These Best Practices in Steam System Management. *Chemical Engineering Progress*, December, 2001.

[12] Knox A. Venting Requirements for Deaerating Heaters. *Chemical Engineering*, January 23, 1984.

[13] Locklin D, Hazard H, Bloom S, Nack H. Power Plant Utilization of Coal, A Battelle Energy Program Report. Columbus, Ohio: Battelle Memorial Institute; September, 1974.

[14] McPherson L. How Good Are Your Values for Total Dissolved Solids? *Chemical Engineering Progress*, November, 1995.

[15] Selby K. On-Line Silica Measurement for Steam Generating Systems. Orlando, Florida: presented at Corrosion 2000, NACE International; March 26–31, 2000.

[16] U.S. Department of Energy. Steam Tip Sheet #18, DOE/GO-102006-2263. January, 2006.

[17] U.S. Navy. Boiler Construction, Information Sheet 62B-203 downloaded from http://www.fas.org/man/dod-101/navy/docs/swos/eng/62B-203I.html, 2011.

10

Cooling Towers

Introduction

Every plant requires cooling, usually using a cooling medium (water, air, or heat transfer fluid) that is circulated through heat exchangers, coils, and equipment jackets. Cooling is obtained from several sources, including:

- Evaporative cooling towers, where the latent heat of evaporation cools the circulating water to near the wet-bulb temperature.
- Closed-loop cooling towers, which rely on air to cool the circulating fluid to near the dry-bulb temperature.
- Mechanical refrigeration machines and heat pumps, where fluids are compressed and liquefied, then expanded and evaporated. Electrical or steam energy is used to drive the compressors; the latent heat of evaporation cools a circulating heat transfer fluid while the energy is dissipated into the atmosphere, surface water, or the earth.
- Direct cooling with river water or air; useful when the required cooling temperature is high.

This chapter discusses evaporative cooling towers. Water cooled by towers is often circulated through refrigeration equipment, or it may be used in the plant to cool process streams flowing through heat exchangers.

Packaged cooling towers are built at a factory and shipped to the site, ready for installation. The units usually consist of one or more "cells", designed to be connected together. Each cell has one or more fans, internal components to ensure efficient contact between air and water, and piping connections. Foundations and, often, a water sump are provided by the purchaser. Appurtenances such as ladders and platforms are supplied with the towers. Each of the parts is limited in size to enable shipment to the site; this usually means a maximum envelope of about 4 m x 4 m x 12 m (12 ft x 12 ft x 40 ft).

Field-erected towers are built at the site, using engineered components such as structure, cladding, fill, water deck, fans unit, and drift eliminator. They are available in fiberglass, concrete, metal, and wood.

Hyperbolic towers, typically found in power plants, get their motive force from natural draft. They are usually constructed of concrete, and can be up to 200 m (650 ft) tall and 100 m (325 ft) in diameter.

Mechanical Draft Cooling Towers

Most packaged cooling towers use fans to force or draw air through the tower. Countercurrent-induced draft towers are the most prevalent in the process industries, capable of cooling water within just over 1 °C (2 °F) of the wet-bulb air temperature.

Induced draft towers have fans at the discharge that pull the air through. This produces low entering and high exit air velocities, reducing the possibility of recirculation, in which discharged air flows back into the air intake. Location of the fan in the warm air stream provides excellent protection against icing of the mechanical components. Towers range from 3 to 160,000 m^3/h (15 to 700,000 gpm).

Forced draft towers have fans at the air intake to the tower. This creates high entering and low exiting air velocities. The low exit velocity is more prone to recirculation. Forced draft designs typically require more motor power than equivalent induced draft towers because they are usually equipped with centrifugal blower type fans. The primary benefit is that high pressure drops through the tower are possible, enabling a more compact design and indoor installations.

Counterflow towers are similar to packed bed strippers. The air flows up through the packed fill material while the water flows down, countercurrently. Circulated water is distributed across the fill through spray nozzles. See Figure 10-1. Most new towers are of the counterflow design, with low-clog film type fill material.

Crossflow towers are those in which the air flows perpendicularly to the water. Water is introduced through a pan with holes or nozzles in the bottom to evenly distribute the water across the fill material. This is shown in Figure 10-2, which also identifies the streams that are used for a water material balance. Crossflow towers are specified when the composition of the cooling water would lead to fouling of film type fill material (Figure 10-3).

Figure 10-1. Field-erected cooling towers.

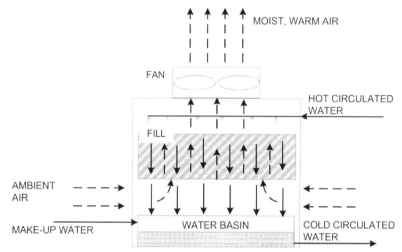

Figure 10-2. Counterflow induced draft cooling tower.

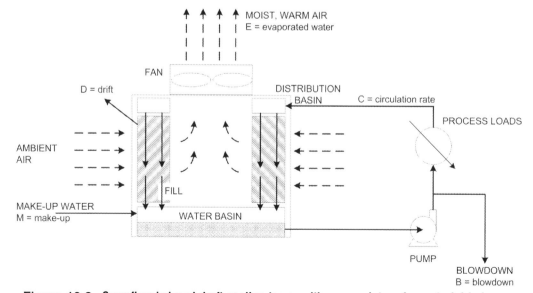

Figure 10-3. Crossflow induced draft cooling tower with nomenclature for material balance.

Material and Energy Balance Calculations

Refer to Figure 10-2 for a simplified flow diagram with nomenclature.

Defined:

M = make-up water from all sources. It includes intentional make-up that has (usually) been treated prior to adding to the tower, and unintentional flow from sources such as rainwater or condensate streams.

B = blowdown and other losses from leaks, filter backwash, etc.

C = water that returns to the tower after being circulated through the water distribution system.

D = drift, which is defined as water that is lost from the tower as liquid droplets entrained in the exhaust air. Drift is independent of evaporative loss. It typically ranges from 0.001% to 0.1% of the recirculation rate; the loss depends on system-specific factors such as the age and design of the tower, the presence of drift eliminators, and on condition-dependent factors such as the direction and velocity of the wind. For a new induced draft tower with drift eliminators, assume D = 0.005% of C for crossflow or 0.001% of C for counterflow. For towers manufactured before about 1990 assume D = 0.02% of C [1].

E = water that evaporates in the tower – the primary force that reduces the temperature of the remaining liquid water.

C_H = hydraulic concentration factor which is a measure of the efficiency of water use. It is also called the "cycles of concentration" or just "cycles". It also represents the accumulation of dissolved minerals in the recirculating cooling water, with a higher value meaning a higher concentration of minerals. Because evaporated water contains no chloride salts, use the chloride concentration in the make-up and circulated waters to calculate the cycles. Towers typically operate within a C_H range of 3 to 7.

$$C_H = \frac{X_C}{X_M} = \frac{M}{B + D} = \frac{M}{M - E} = 1 + \frac{E}{B + D} \quad (10\text{-}1)$$

X_C and X_M are mass-concentration of chlorides in the C and M streams. Use consistent flow units (kg/h, lb/h, gpm) for M, B, D, C, and E.

Hoots, et. al. report that, for systems operating above about seven to eight cycles, normal water treatment dosages need to be greatly increased, particularly for corrosion inhibitors and, more importantly, dispersants [4].

Use this expression to estimate the evaporation rate using a simple heat balance:

$$E = \frac{C \, \Delta T \, c_p}{H_v} \quad (10\text{-}2)$$

Where:

E = evaporation rate, kg/h or lb/h
C = circulation rate, kg/h or lb/h
ΔT = water temperature difference from the return water to the basin, °C or °F
c_p = specific heat of water, 4.184 kJ/kg-°C or 1 Btu/lb-°F
H_v = latent heat of vaporization of water, 2260 kJ/kg or 1045 Btu/lb

Cooling tower efficiency, typically between 70% and 75%, is defined as:

$$\eta = \frac{t_i - t_o}{t_i - t_{wb}} \times 100 \quad (10\text{-}3)$$

Where:

η = tower efficiency, %
t_i = temperature of circulation water at the tower inlet, °C or °F
t_o = temperature of the cooled water in the basin, °C or °F
t_{wb} = inlet air wet-bulb temperature, °C or °F

The *holding time index* (HTI) is defined as the half-life of an ion in the system. Most industrial cooling systems operate with a HTI of 48 hours or less. Increasing the concentration factor also increases the half-life and puts stress on the organic portions of the treatment chemicals.

$$HTI = 0.693 \frac{V_T}{B + D} \quad (10\text{-}4)$$

Where:

HTI = holding time index, hours
V_T = total volume of the system (tower, basin, circulation loop), kg or lb of water
B and D in kg/h or lb/h

Rating Nomenclature

To size a tower, critical parameters include: circulating water flow rate, return temperature, supply temperature, wet-bulb temperature, and dry-bulb temperature (or relative humidity).

The wet-bulb temperature design criterion must be based on the location of the tower, and should consider the time of peak load demand. According to SPX, the manufacturer of Marley cooling towers, most industrial installations that are based on the wet-bulb temperatures that are exceeded no more than 5% of the total hours during a normal summer have given satisfactory results. This is because the hours during which the wet-bulb temperature exceeds the upper 5% level are seldom consecutive, and usually occur in periods of relatively short duration. The "flywheel" effect of the total water system is sufficient to carry through the above-average periods without detrimental results [3].

Cooling towers may be defined by their approach temperature (the difference between the cooled water and entering air web-bulb temperatures) and circulating flow rate. Typical approach temperatures are 5.5 °C to 8.5 °C (10 °F to 15 °F). Figure 10-4 shows how the approach temperature specification affects the tower size.

Towers that are rated in "tons" of cooling capacity use one of two conversion factors. Generally, for process cooling applications, 1 ton = 12,000 Btu/h. However, for towers that are part of an HVAC system, where the cooling water is used by a refrigeration unit, 1 ton = 15,000 Btu/h – which accounts for the inefficiency of the cascaded system. Engineers should clarify with the manufacturer which conversion factor is used for a particular tower.

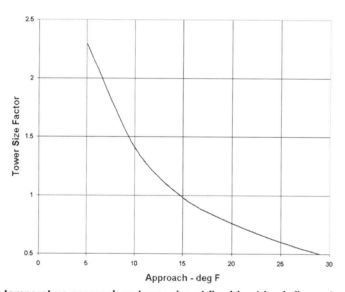

Figure 10-4. Effect of chosen temperature approach on tower size at fixed heat load, flow rate, and wet-bulb temperature [3].

Performance Calculations

Cooling tower manufacturers use computer programs to predict tower performance given the duty requirements and environmental factors. The programs utilize test data from the specific fill materials and tower geometries of the manufacturer's products.

To analyze the performance of an existing tower, or to design a new installation, manufacturers should be

consulted. If this is not feasible or desirable, the Cooling Technology Institute [2] offers a comprehensive computer program that performs thermal design calculations for a new tower, or performance evaluation calculations for an existing unit.

Operations and Accessories

Cooling water should be filtered. This is usually done at the circulation pump. For small towers, the entire circulation stream is filtered. Larger towers use a side-stream filter, with 1% to 10% of the circulation water diverted to the filter. Filters are generally rated to remove 25 micron particles. Solid contaminants enter cooling towers from three primary sources:

- Ambient air dirt load. Most contaminants in the air end up in the tower basin.

- Circulation water build-up. This includes calcium carbonate scale that forms in the tower and flakes off, treatment chemical residues, and biological growth.
- Make-up water.

Huchler lists a number of operating and monitoring practices, recommended to maximize cooling tower efficiency (Table 10-1).

Table 10-1
Recommended operating and monitoring practices [5]

Objective	Measure	Comments
Minimize water loss due to flooding of upper deck. Water is lost over the side of the tower, lowering efficiency.	Manually adjust valves on the distribution header to ensure even distribution.	Exercise the manual flow valve through its entire range prior to setting final position
Operate tower at the designed temperature approach	Utilize variable-speed fan drives to reduce fan speed when loads or air temperature are lower	
Eliminate ice formation on drift eliminators	Consider removing a single row of nozzles at the perimeter of the deck to form a "water wall" which reduces water spray that causes icing	Never run fans in reverse for de-icing unless the tower operating manual specifically allows it
Maximize fan life	Mount vibration sensors on fan drive shafts (not the oil supply lines) to warn of imminent fan failure	
Monitor mechanical operation and cooling water chemistry	Use a central control system supplemented with routine visual inspection. Conduct weekly inspections of the tower, checking for integrity of the structure, pump screens, appearance of the water in the basin, and performance of rotating equipment	
	Sample cooling water once or twice per day and make adjustments to water treatment chemical federates and blowdown rates	
Control blowdown	Online conductivity meters are commonly used	
Control chemical feed	The common automated feed systems are for acid (for pH control) and chlorine (for control of microbiological fouling)	Automating the specialty chemicals for deposit and corrosion control requires complex solutions, such as fluorescent-tagged chemicals or feed-forward flow control algorithms
Monitor bacteria levels	Paddle Test is most commonly used	Rapid methods such as Biological Activity Reaction Tests (BARTs) and Adenosine triphosphate (ATP) methods may supplement paddle test results
Monitor system reliability	Install corrosion coupons in a bypass rack to evaluate the risk of corrosion. Coupons fabricated from alloys identical to the process heat exchangers provide the best results.	
Detect process leaks	Hand-held and online instruments are used to detect light hydrocarbon contaminants in cooling water	US EPA has imposed strict time limits for correction of uncontrolled emissions from cooling towers
Protect integrity of rotating equipment	Pumps, fan bearings, and fan gearboxes require routine lubrication and immediate corrective action should vibration sensors record high readings	Provide automatic cut-out switches on fan motors tied to high vibration sensor
Protect pump impellers against damage from debris	Install two sets of screens in the supply pump inlet bays. Routinely clean the screens. Install lifts to simplify the cleaning and replacement of screens.	

Water Quality Guidelines

General guidelines for water quality in the circulation water are listed in Table 10-2.

Table 10-2
General guidelines for water quality limits in circulation water as it returns to a cooling tower [6]

Characteristic	Analysis
Temperature	<50 °C (<120 °F) up to 65 °C (150 °F) for high temperature PVC decking in counterflow towers
pH	6.5 to 9.0
Chlorides	Galvanized steel: <750 ppm
	300 Series Stainless Steel: <1500 ppm
Calcium (as $CaCO_3$)	<800 ppm should not result in calcium sulfate scale
Sulfates	If calcium > 800 ppm, sulfates < 800 ppm (or less in arid climates) to limit scale
	If calcium < 800 ppm, sulfates < 5000 ppm
Silica (as SiO_2)	<150 ppm to prevent silica scale
Iron	<3 ppm
Manganese	<0.1 ppm
Total Suspended Solids	<25 ppm for film type fill
Total Dissolved Solids	<5000 for thermal performance
Oil and Grease	Avoid with film type fill (contributes to fill plugging, prevents proper seasoning)
Ammonia	Limit to 50 ppm if copper alloys are present

Scale Prevention

The solubility of calcium carbonate is dependent on temperature and pH and, therefore, varies through a cooling water system. There are two indices that indicate the directional tendency of water to either deposit a coating or be corrosive. Water treatment companies often use these equations to arrive at a method of treatment; examination of corrosion coupons can confirm that the treatment is in balance.

Both indices are derived from a computed saturation pH [3]:

$$pH_S = (9.3 + A + B) - (C + D) \qquad (10\text{-}5)$$

Where:

pH_S = pH value at which water is in equilibrium with solid $CaCO_3$

A = value reflecting total solids (Table 10-3)
B = value reflecting temperature (Table 10-4)
C = value reflecting calcium hardness (Table 10-5)
D = value reflecting total alkalinity (Table 10-5)

Table 10-3
Data for total solids, coefficient A

Total Solids, ppm	A
50 to 300	0.1
400 to 1000	0.2

Table 10-4
Data for temperature, coefficient B

Temperature, °F	B
32 to 34	2.6
36 to 42	2.5
44 to 48	2.4
50 to 56	2.3
58 to 62	2.2
64 to 70	2.1
72 to 80	2.0
82 to 88	1.9
90 to 98	1.8
100 to 110	1.7
112 to 122	1.6
124 to 132	1.5
134 to 146	1.4
148 to 160	1.3
162 to 178	1.2

Table 10-5
Data for calcium hardness and M-alkalinity, coefficients C and D

Calcium Hardness or M-Alkalinity, ppm	C	D
10 to 11	0.6	1.0
12 to 13	0.7	1.1
14 to 17	0.8	1.2
18 to 22	0.9	1.3
23 to 27	1.0	1.4
29 to 34	1.1	1.5
35 to 43	1.2	1.6
44 to 55	1.3	1.7
56 to 69	1.4	1.8
70 to 87	1.5	1.9
88 to 110	1.6	2.0
111 to 138	1.7	2.1
139 to 174	1.8	2.2
175 to 220	1.9	2.3
230 to 270	2.0	2.4
280 to 340	2.1	2.5
350 to 430	2.2	2.6
440 to 550	2.3	2.7
560 to 690	2.4	2.8
700 to 870	2.5	2.9
880 to 1000	2.6	3.0

$$\text{Saturation Index} = \text{pH(actual)} - \text{pH}_S$$

$$\text{Stability Index} = (2\,\text{pH}_S) - \text{pH(actual)}$$

When the Saturation Index is zero, the water is in equilibrium with solid $CaCO_3$ at that temperature; when it is positive, the water is supersaturated and may deposit a coating or scale; when it is negative, the water will dissolve $CaCO_3$ and may be corrosive.

A value of 6 or 7 of the Stability Index indicates water which is the most balanced, while values above 8 indicate increasing corrosion tendencies.

Nomenclature

B	=	blowdown rate
C	=	circulation rate, kg/h or lb/h
C_H	=	hydraulic concentration factor
c_p	=	specific heat of water, 4.184 kJ/kg-°C or 1 Btu/lb-°F
D	=	drift rate
E	=	evaporation rate, kg/h or lb/h
H_v	=	latent heat of vaporization of water, 2260 kJ/kg or 1045 Btu/lb
HTI	=	holding time index, hours
M	=	make-up water flow rate
ΔT	=	water temperature difference from the return water to the basin, °C or °F
t_i	=	temperature of circulation water at the tower inlet, °C or °F
t_o	=	temperature of the cooled water in the basin, °C or °F
t_{wb}	=	inlet air wet-bulb temperature, °C or °F
V_T	=	total volume of the system (tower, basin, circulation loop), kg or lb of water
η	=	tower efficiency, %

References

[1] Bugler T, Fields B, Miller R. Cooling Towers, Drift, and Legionellosis. *CTI Journal*. Houston, TX: Cooling Technology Institute; 2010:33(1), Winter.

[2] CTI ToolKit Software. Cooling Technology Institute, Houston, Tx, www.cti.org.

[3] Hensley J, editor. *Cooling tower fundamentals*. 2nd ed. Overland Park, KS: SPX Cooling Technologies, Inc.; 2009.

[4] Hoots J, Johnson D, Lammering J, Meier D. Operate Cooling Towers Correctly at High Cycles of Concentration. *Chemical Engineering Progress*, March, 2001.

[5] Huchler L. Cooling Towers, Part 2: Operating, Monitoring, and Maintaining. *Chemical Engineering Progress*, October, 2009.

[6] Lindahl P. Operating Strategies and Water Treatment. downloaded from Cooling Technology Insitute, www.cti.org, January, 2011.

11
Refrigeration

Introduction

Refrigeration systems in the chemical industry use, for the most part, standard equipment that has application in many industries. However, the chemical industry demands construction and performance that differ from commercial facilities, due to factors such as temperature range requirements, capacity needs, fluctuating loads, hazardous locations, and the presence of corrosive chemicals. The systems are normally engineered by specialists; process engineers are responsible for providing design criteria.

Three technologies are used: steam jet ejectors, absorption chillers, and mechanical refrigeration systems. See Table 11-1.

A "ton of refrigeration" is defined by the amount of heat equivalent to melting 2,000 lb (one ton) of ice in 24 hours. One ton equals 12,000 Btu/h, 200 Btu/min, or 3.5 kW. To be comparative, refrigeration equipment must have the refrigerant level (or evaporation temperature) specified [4].

Table 11-1
Types of refrigeration systems

Technology	Approximate Temperature Range, °C (°F)	Refrigerant
Steam-jet ejector	2 °C to 20 °C (35 °F to 70 °F)	Water
Absorption	4 °C to 20 °C (40 °F to 70 °F)	Lithium Bromide solution
Absorption	−40 °C to −1 °C (−40 °F to 30 °F)	Ammonia
Mechanical compression	−130 °C to 4 °C (−200 °F to 40 °F)	Ammonia, halogenated hydrocarbons, propane, propylene, ethane, ethylene, and others

Refrigerant	Approximate Temperature Range, °C (°F)
Methane	−185 °C to −130 °C (−300 °F to −200 °F)
Ethane and ethylene	−115 °C to −60 °C (−175 °F to −75 °F)
Propane and propylene	−45 °C to 4 °C (−50 °F to 40 °F)

Specifications

Although mechanical refrigeration systems are usually furnished by companies who specialize in designing and building such systems, the plant engineer remains responsible for specifying many aspects of the system. Table 11-2 is a checklist of items to consider when writing a specification.

The American Society of Heating, Refrigerating and Air-Conditioning Engineers (ASHRAE) rants [1]:

Refrigeration facilities represent only a minor part of the total plant investment. The entire utilities installation for the chemical industry usually falls in the range of 5 to 15% of the total plant investment, with the refrigeration system only a small portion of the utilities investment. Process requirements may be overruling, but process engineers must recognize legitimate process necessities and avoid unnecessary and costly restrictions on the refrigeration system design.

The major point is that process engineers should provide sufficient process information together with design flexibility so that the refrigeration engineer can design an optimized solution, analyzing temperature levels, potential load combinations, and energy recovery potentials.

Ludwig [4] described the system design and selection process as follows; the datasheet in Figure 11-1 is adapted from Ludwig.

Basically, the system design consists of the selection of component parts to combine and operate in the most economical manner for the specified conditions. Unfortunately, the specific conditions are not only for the evaporator where the refrigerant is actually used but include all or part of the following. These conditions are identified whether the system is a separate component selection or a package furnished assembled by a manufacturer.

1. Evaporator: temperature and refrigerant
2. Compressor: centrifugal, screw or reciprocating; electric motor, steam turbine, or other driver
3. Condenser: horizontal or vertical, temperature of cooling water, water quantity limit
4. Receiver: system refrigerant volume for shutdown refrigerant storage
5. Operation: refrigeration tonnage load changes.

Table 11-2
Checklist of items to include in a specification for a refrigeration system

Design Aspect	Specification Guidance
Application	Describe the application for the refrigeration system. Include Process Flow Diagrams and heat and material balances. Include information about the plant location, intended location for the compressor and evaporator components, hours of operation, and existing capabilities for operating and maintaining the equipment (e.g., maintenance performed by employees or contractors, maintenance staff work schedules). If this is a batch process, then indicate operating schedules and discuss diversity. Provide expectations for aspects such as conformance with existing plant standards, matching existing refrigeration systems, etc. Also describe the importance of the system to plant operations in terms of production, safety, and environment. Redundancy and interlocks should be described. What, if any, periodic plant shutdowns (for maintenance) are normally planned?
Temperature range and load	Give the temperature and heat load for each use point as well as the overall (blended) load. Specify if secondary coolants are used. Loads should be expressed as maximum design condition, normal condition, and minimum condition. Consider the expected duration of peak conditions, as well as factors other than normal process loads that may impact the sizing of the refrigeration system (such as unusual start-up conditions).
Scope of supply	Clearly delineate the scope that the specification includes and excludes. Items such as interconnecting piping, insulation, controls, heat exchangers for secondary loops, etc. must be defined. For equipment connected to the refrigeration system, but out of the scope of the specification, provide enough information so that the supplier of the refrigeration system can understand the interface. Specify the scope for features like containment (of leaks).
Design principles	Specify any requirements for particular technologies, contingency factors, reliability, maintainability, and start-up/shut-down requirements.
Safety	Specify the electrical area classifications, adjacent hazards, and requirements for emergency vents and drains. Include any relevant plant standards such as emergency stops, machine guarding, noise constraints, and labeling. Potential hazards such as contact with corrosive or toxic process fluids must be identified.
Automation	Provide minimum automation requirements (degree of automation) and plant standards including interfaces to plant control systems. There will often be a stand-alone specification that dictates requirements for automation provided with "packaged" systems.
Electrical	Identify the available power sources. Specify if the system will be connected to a standby power system and if the controls require uninterruptable power (UPS). Specify the location of any switchgear or motor control centers that may be required. Be sure the scope of these items is clear.
Other requirements	Provide specifications for components such as piping, instrumentation, insulation, heat exchangers, vessels, painting, etc. Recognize and respect that the vendors already have proven design solutions, and that substitution of unfamiliar components could lead to cost increases, schedule delays, or even performance issues.
Testing, training, and spare parts	Clearly define testing requirements for factory acceptance tests (FAT), site acceptance tests (SAT), and commissioning. Specify the scope of after-sale support including training of operators and mechanics. Obtain a list of recommended spare parts and preventative maintenance procedures.

Ludwig concludes:

For final design horsepower and equipment selection, the usual practice is to submit the refrigeration load and utility conditions/requirements to a reputable refrigerant system designer/manufacturer and obtain a warranted system with equipment and instrumentation design and specifications including the important materials of construction.

	MECHANICAL REFRIGERATION SYSTEM		
	CLIENT	EQUIP. NO	PAGE
REV	PREPARED BY	DATE	APPROVAL	W.O.	REQUISITION NO.	SPECIFICATION NO.

REV	PREPARED BY	DATE	APPROVAL	W.O.	REQUISITION NO.	SPECIFICATION NO.
0						
1				UNIT AREA	PROCURED BY	INSTALLED BY
2						

1			**Service Conditions**				
2	Capacity	Tons	Refrigerant		Process Fluid		
3	Manufacturer			Model			
4	Compressor Suction	psig	°F	Discharge	psig		°F
6	Condenser Refrigerant Flow				lb/h		°F exit
7	Cooling Water Supply				°F		psig
8	Cooling Water Requirement		Maximum gpm		°F exit		
9	Evaporator Refrigerant	psig	°F		lb/h		
10	Evaporator Process Fluid	lb/h	°F inlet		°F outlet		psig
11	Compressor Driver provided by		Driver type				
12			hp		rpm		
13							
14			**Specifications**				
15	Temperature Control		Instruments & Controls		IP Code		
16			Elec. Area Classification		Class/Group/Div		
17	Level Controls (specify):			Connections (flange type):			
18	Pressure Gages:						
19			Connection Sizes, inch		Flanges		
20			Inlet Outlet		Rating, psig Face		
21	Condenser Cooling Water						
22	Refrigerant to Evaporator						
23	Suction Cross Exchanger						
24	Compressor Suction		N/A				
25	Compressor Discharge	N/A					
26	Steam Supply		psig	°F			
27	Power Supplies	Driver	V/Ph/Hz	Instruments	V/Ph/Hz		
28	Fouling Factors for Cooling Water		ft²-h-°F/Btu	Process Fluid	ft²-h-°F/Btu		
29	Condenser Tube Water Velocity Limits		ft/s	to	ft/s		
30	Refrigerant Charge		lb				
31							
32			**Materials of Construction**				
33	Condenser Tubes		OD	inch	Tube Wall	BWG	
34	Number		Length	ft	Pitch	BWG	
35	Tube Sheet		Channel				
36	Evaporator Furnished by						
37	Tubes		OD	inch	Tube Wall	BWG	
38	Number		Length	ft	Pitch	BWG	
39	Tube Sheet		Channel				
40			**Notes**				
41	1. Vendor to specify make and model for flow, temperature, and pressure control valves; hand valves;						
42	thermometers and steam traps; electrical components; and miscellaneous equipment and components.						
43							
44							
45							
46							
47							
48							
49							
50							
51							
52							
53							
54							

Figure 11-1. Datasheet for a mechanical refrigeration system. Separate datasheets are used for system components such as compressors and heat exchangers. (Adapted from [4]; included in the companion spreadsheet).

Indirect and Direct System Design

Indirect refrigeration systems use heat exchangers to cool a secondary heat transfer fluid. The application of indirect systems, with an extensive list of heat transfer fluids, is discussed in Chapter 12. With an indirect system, the refrigeration units can be grouped together in a utility area or building. The chilled heat transfer fluid is distributed to the plant like other utilities such as steam and cooling water.

Systems for very low temperature applications are commonly installed close to the load, utilizing special low-temperature heat transfer fluids. These indirect systems are chosen specifically for the isolated load and are generally not connected to a central indirect refrigeration system.

Direct systems are applied when there are relatively few loads and the piping lengths can be short. In these systems, the refrigerant is circulated directly to the process heat exchanger. By omitting an intermediate heat transfer fluid, the temperature of the coolant can be 5 °C to 10 °C (10 °F to 20 °F) colder than with an indirect system that uses the same refrigerant. The advantages and disadvantages of direct refrigeration are summarized in Table 11-3.

Table 11-3
Advantages and disadvantages of direct refrigeration [1]

Advantages	Disadvantages
Careful control of corrosion inhibitors is not necessary; secondary heat transfer fluids must sometimes be inhibited to maintain their stability.	Difficult to keep an extensive refrigeration piping system leak-free; leakage of secondary coolants is frequently less objectionable than leakage of refrigerants.
Less equipment and maintenance required; secondary coolant storage, circulation, and control not needed	Higher piping costs are often involved when all things are considered, including large and expensive vapor and liquid control valves at the individual process heat exchangers.
Lower power costs because of higher suction pressures and, in some designs, because pumps are not needed.	No system reserve capacity is available as is the case with secondary systems that include a storage tank; process upsets can suddenly and directly increase the load on the refrigeration system causing rapid cycling.
Damage due to equipment freezing is not likely, though it can occur in a secondary loop if the coolant condition or the refrigeration plant are not properly operated.	Constant temperature control is sometimes more difficult or costly to achieve.
Initial cost for refrigerants is usually much lower than the cost of secondary heat transfer fluids; in the case of system leaks, the cost for make-up of fluids is about the same.	Initial pressure testing must be done pneumatically rather than hydrostatically to prevent problems with water inadvertently left in the system.

Horsepower per Ton

The quick but accurate graph in Figure 11-2 shows how much horsepower is required for mechanical refrigeration systems, using the most practical refrigerant for the desired temperature range [2].

Example. A water-cooled unit with an evaporator temperature of $-40\,°F$ will require 3 horsepower per ton of refrigeration. Multiply horsepower by 0.746 to convert to kW. A ton of refrigeration is equal to 12,000 Btu/hr, or 3.51 kW. Therefore, in SI units, the machine will require 0.64 kW per kW of refrigeration.

The curves fit this equation, and example coefficients are given in Table 11-4:

$$hp/ton = A + B\,t + C\,t^2 + D\,t^3 + E\,t^4$$

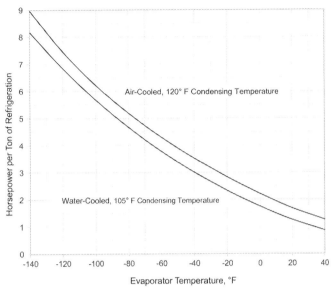

Figure 11-2. Horsepower requirement per ton of refrigeration.

Table 11-4
Coefficients for the horsepower per ton equation

Condenser Temperature, °F	A	B	C	D	D
105	1.751	−2.686e-2	1.152e-4	3.460e-8	1.320e-9
120	2.218	−2.882e-2	1.036e-4	3.029e-7	3.961e-9

Refrigerants

The tables (Tables 11-5 to 11-9, [3]) in this section identify refrigerants for interim or long-term application to comply with ozone depletion agreements (i.e., the Montreal Protocol). Hydrofluorocarbons (HFCs) are considered to be suitable for long-term use; they have no ozone depletion potential. Hydrochlorofluorocarbons (HCFCs) are approved for the transitional period ending in 2030, when their phase-out period ends.

Table 11-5
Long-term refrigerants for commercial refrigeration systems

ASHRAE #	Trade Name	Manuf.	Replaces	Type	Typical Lubricant	Applications	Comments
R-404A 125/143a/134a (44%/52%/4%)	Genetron 404A Suva HP62 Forane 404A	Honeywell DuPont Arkema	R-502 R-22 HP-80 R-408A	Blend HFC (low glide)	Synthetic (POE, PVE, etc.)	New Equipment Retrofits	Most widely used low and medium temperature replacement
R-507 125/143a (50%/50%)	Genetron AZ-50 Suva 507	Honeywell DuPont	R-502 R-22 HP-80 R-408A	Azeotrope (no glide) HFC	Synthetic (POE, PVE, etc.)	New Equipment Retrofits	Slightly higher pressures and efficiency than R-404A. Best choice for systems with flooded evaporators

(Continued)

Table 11-5
Long-term refrigerants for commercial refrigeration systems—cont'd

ASHRAE #	Trade Name	Manuf.	Replaces	Type	Typical Lubricant	Applications	Comments
R-422A 125/134a/600a (85.1%/11.5%/ 3.4%)	One Shot Isceon 79	ICOR DuPont	R-502 R-22	Blend HFC	Synthetic (POE, PVE, etc.)	New Equipment Retrofits	Similar performance to R-404A. Equipment with suction line accumulators and receivers should use synthetic oils to avoid oil return issues
R-422D 125/134a/600a (65.1%/31.5%/ 3.4%)	Genetron 422D Isceon MO29	Honeywell DuPont	R-22	Blend (moderate glide) HFC	Mineral Oil POE	New Equipment Retrofits	Lower capacity Use of POE will enhance oil return, if required
R-407C 32/125/134a (23%/25%/52%)	Genetron 407C Suva 9000 Forane 407C	Honeywell DuPont Arkema	R-22	Blend (high glide) HFC	Synthetic (POE, PVE, etc.)	New Equipment Retrofits	Reasonable performance match to R-22 in medium temperature refrigeration. Lower capacity in low temperature refrigeration system.
R-134a	Genetron 134a Suva 134a Forane 134a Klea 134a	Honeywell DuPont Arkema INEOS	R-12	Single component HFC	Synthetic (POE, PVE, etc.)	New Equipment	Performs well in small hermetic systems.

Table 11-6
Interim HCFC-based refrigerants for commercial refrigeration systems

ASHRAE #	Trade Name	Manuf.	Replaces	Type	Typical Lubricant	Applications	Comments
R-401A 22/152a/124 (53%/13%/34%)	Genetron MP39 Suva MP39	Honeywell DuPont	R-12	Blend (moderate glide) HCFC/HFC	Alkylbenzene Synthetic (POE, PVE, etc.) Mineral oil	Med temp retrofits	In most cases no oil change is needed. Best for applications with > 0°F suction
R-401B 22/152a/124 (61%/11%/28%)	Genetron MP66 Suva MP66	Honeywell DuPont	R-12 R-500	Blend (moderate glide) HCFC/HFC	Alkylbenzene Synthetic (POE, PVE, etc.) Mineral oil	Transport refrigeration Low temp retrofits Retrofits including air conditioner and dehumidifiers	In most cases no oil change is needed. Best for low temp R-12 and R-500 retrofit applications
R-409A 22/124/142b (60%/25%/15%)	Genetron 409A Suva 409A Forane FX-56	Honeywell DuPont Arkema	R-12 R-500	Blend (high glide) HCFC	Alkylbenzene Synthetic (POE, PVE, etc.) Mineral oil	Retrofits Low and med temp	In most cases no oil change is needed. Good broad range R-12 substitute

Table 11-6
Interim HCFC-based refrigerants for commercial refrigeration systems—cont'd

ASHRAE #	Trade Name	Manuf.	Replaces	Type	Typical Lubricant	Applications	Comments
R-402A 125/290/22 (60%/2%/38%)	Genetron HP80 Suva HP80	Honeywell DuPont	R-502	Blend (low glide) HFC/HC/HCFC	Alkylbenzene Synthetic (POE, PVE, etc.)	Retrofits Low and med temp	Most widely used R-502 retrofit substitute. Higher discharge pressure than R-502. Use either synthetic oil or blend of AB/MO with AB > 50%
R-402B 125/290/22 (38%/2%/60^)	Genetron HP81 Suva HP81	Honeywell DuPont	R-502	Blend (low glide) HFC/HC/HCFC	Alkylbenzene Synthetic (POE, PVE, etc.)	Ice machines	Niche refrigerant used in some ice machines
R-408A 125/143a/22 (7%/46%/47%)	Genetron 408A Suva 408A Forane FX-10	Honeywell DuPont Arkema	R-502	Blend (low glide) HCC/HCFC	Alkylbenzene Synthetic (POE, PVE, etc.)	Retrofits Low and med temp	Works well as R-502 substitute. Higher discharge temperatures than R-502. Use either synthetic oil or blend of AB/MO with AB > 50%

Table 11-7
Ultra-low temperature refrigerants

ASHRAE #	Trade Name	Manuf.	Replaces	Type	Typical Lubricant	Applications	Comments
R-508B 23/116 (46%/54%)	Genetron 508B Suva 95	Honeywell DuPont	R-13 R-503	Azeotrope HFC	Synthetic (POE, PVE, etc.)	New equipment Retrofits	Lower discharge temperatures than R-13 and R-23. Good performance match to R-503

Table 11-8
Centrifugal chiller refrigerants for air conditioning

ASHRAE #	Trade Name	Manuf.	Replaces	Type	Typical Lubricant	Applications	Comments
R-123	Genetron 123 Suva 123 Forane 123	Honeywell DuPont Arkema	R-11	Single component fluid HCFC	Alkylbenzene Mineral oil Synthetic (POE, PVE, etc.)	New equipment Retrofits	Due for phase out in 2030
R-245fa	Genetron 245fa	Honeywell	R-11	Single component fluid HFC	Synthetic (POE, PVE, etc.)	New equipment	Equipment redesign Organic Rankine cycle and as heat transfer fluid
R-134a	Genetron 134a Suva 134a Forane 134a Klea 134a	Honeywell DuPont Arkema INEOS	R-12 R-500	Single component fluid HFC	Synthetic (POE, PVE, etc.)	New equipment Retrofits	Used in many new chiller designs

Table 11-9
Long-term refrigerants for air conditioning and heat pumps

ASHRAE #	Trade Name	Manuf.	Replaces	Type	Typical Lubricant	Applications	Comments
R-407C 32/125/134a (23%/25%/52%)	Genetron 407C Suva 9000 Forane 407C	Honeywell DuPont Arkema	R-22 R-500	Blend (high glide) HFC	Synthetic (POE, PVE, etc.)	New Equipment Retrofits	Best retrofit alternative to R-22. Close performance match with slightly higher operating pressures
R-410A 32/125 (50%/50%)	Genetron AZ-20 Suva 410A Forane 410A Puron	Honeywell DuPont Arkema Carrier	R-22	Azeotropic mixture (near zero glide) HFC	Synthetic (POE, PVE, etc.)	New equipment	High pressure, high efficiency refrigerant designed for new equipment. Not for retrofitting
R-422D 125/134a/600a (65.1%/31.5%/ 3.4%)	Genetron 422D Isceon MO29	Honeywell DuPont	R-22	Blend (moderate glide) HFC	Mineral oil POE	New equipment Retrofits	Lower capacity. Use of POE will enhance oil return, if required
R-134a	Genetron 134a Suva 134a Forane 134a Klea 134a	Honeywell DuPont Arkema INEOS	R-12 R-500	Single component fluid (no glide) HFC	Synthetic (POE, PVE, etc.)	New equipment Retrofits	Used in large screw chillers

Refrigerant blend compositions are stated as mass%. Trade names are registered trademarks of the respective manufacturers. "Glide" is the difference in temperature between the evaporator outlet and inlet due to fractionation of the blend, theoretically calculated by finding the difference between the dew and bubble point temperatures at constant pressure, but actual measurements may differ slightly. Synthetic lubricants include polyolester (POE) and polyvinyl ether (PVE).

Refrigerant Containment

Since refrigerants have a significant impact on global warming and ozone depletion, their emissions to atmosphere must be reduced to the maximum practicable extent. ASHRAE provides information that is useful for manufacturers of refrigeration equipment as well as the engineers who are designing plants that use the equipment. Some of ASHRAE's tips are summarized here [1].

Six types of refrigerant emissions, or losses, are identified:

- Fugitive emissions, where the source cannot be precisely located
- Tightness degradation, caused by temperature variations, pressure cycling, and vibrations
- Component failures, mostly originating from poor construction or faulty assembly
- Losses during refrigerant handling, mainly when charging the system or opening it without first recovering the refrigerant
- Accidental losses, caused by fires, explosions, sabotage, theft, etc.
- Losses at equipment disposal, caused by intentionally venting, rather than recovering, refrigerant at the end of system life

The process engineer can help minimize refrigerant losses in several ways. See Table 11-10.

Leak detection is performed using various procedures and instruments. Engineers should ensure that appropriate checks are performed when a new system is installed, or if a system has been emptied of its charge. Monitoring practices should also be employed, using the refrigeration system control system as well as external instrumentation. See Table 11-11.

Table 11-10
Process engineers influence refrigerant losses throughout the lifecycle of the equipment

Life Cycle Aspect	Practices that Reduce Losses
System design	Specify and design systems that are leaktight for the length of their useful lives, and reliable, to minimize the need for service. Critical factors include: selection of materials, joining techniques, design for easy installation, and design for service access.
	Minimize the amount of refrigerant charge, to reduce the amount of release in the event of catastrophic loss. Select heat exchangers, piping, and components to reduce the amount of refrigerant in the system (but not at the expense of energy efficiency).
Installation	Properly clean joints before brazing, purge the system with an inert gas during brazing, and evacuate noncondensables.
	Inspect for manufacturing defects before charging the system.
	Properly charge the system per specifications to prevent overcharging.
	Perform careful system performance and leak checks.
Maintenance	Monitor to ensure the system remains well sealed, properly charged, and operating within normal parameters.
	During maintenance, ensure refrigerant is isolated or recovered by equipment capable of handling the specific refrigerant.
	Update maintenance documents every time the system is serviced, to record additions and removals of refrigerant. Track whether recharging operations are actually associated with repairs of leaks.
	When a system is decommissioned, recover the refrigerant for recycling, reuse, or disposal.
Training	Provide training for maintenance technicians. Include education on the environmental effects of refrigerants; recovery, recycling, and reclamation of refrigerants; leak checks and repairs; and introduction to new refrigerants.
	Provide refresh training to ensure new developments are understood, such as new designs, new refrigerants and their compatibility with lubricants, new low-emission purge units, retrofitting requirements, and service practices.

Summarized from [1].

Table 11-11
Leak detection practices

Leak Detection Type	Practices and Instruments
Global detection — system checking	Approaches applicable to a system that has been emptied of its charge. The tracer gas is often nitrogen or helium. • Pressurize system with tracer gas and isolate it. A pressure drop within a specified time indicates leakage. • Evacuate the system and measure the vacuum level over a specified time. A pressure rise indicates leakage. • Place the system in a chamber and charge with tracer gas. Then evacuate the chamber and monitor for system leaks by testing the chamber with a mass spectrometer or residual gas analyzer. • Evacuate the system and place it in an atmosphere with a tracer gas. Monitor for leaks with a mass spectrometer or residual gas analyzer.
Global detection — continuous monitoring during operation	Electronic leak detectors in machinery room may be efficient if: 1) They are sensitive enough to refrigerant dilution in the air, and 2) Air is circulated properly in the room.
Local detection	These methods pinpoint location of leaks and are usually used during servicing. Sensitivity varies widely; it is usually stated as ppm/volume but, for clarity, mass flow rates (g/year) are often used. • Visual checks. Look for telltale traces of oil at joints. >50 g/year.

(Continued)

Table 11-11
Leak detection practices—cont'd

Leak Detection Type	Practices and Instruments
	• Soapy water detection (bubble testing). Simple and inexpensive. >20 g/year. • Tracer color added to oil or refrigerant. Shows a leak's location. • Electronic detectors. Different techniques. Require care and training. 5 to 50 g/year depending on sensitivity. • Ultrasonic detectors. Less sensitive than electronic detectors. Easily disturbed by air circulation. • Helium and HFC mass spectrometers. With probes or hoods. Very sensitive. <1 g/year
Automated performance monitoring systems	Automated diagnostic programs incorporated into the control systems monitor parameters such as temperatures and pressures and help identify any change in the equipment. In addition to potential leak detection (due to loss of refrigerant), monitors provide data that is useful for performing diagnostics on the condition of heat exchanger surfaces, refrigerant pumping, and shortage of refrigerant charge.

Summarized from [1].

Steam Jet Ejectors

Ludwig [4] gives the following advantages for steam jet ejectors, assuming the refrigerant is water at a temperature greater than 2 °C (35 °F) and the cost of water and steam are "reasonable."

1. No moving parts except for water pumps
2. Refrigerant (water) is non-hazardous and has a low cost
3. Low system pressures
4. Outdoor installation if desired; self supporting
5. Flexible physical arrangement
6. Steam and cooling water requirements are adjustable to reasonable economical balance
7. Steam condensate is recoverable in surface type units
8. Refrigeration tonnage can be varied as to amount and temperature level
9. Simple start up and operation
10. Barometric units can use brackish, dirty, or waste water for a "once through" system. Recirculating

systems require a water quality that avoids fouling heat transfer equipment (steam condensate, with blowdown or treatment, is usually used).
11. Relatively low cost per ton of refrigeration

Steam pressure is usually between 350 kPa to 1400 kPa (50 psig to 200 psig) although pressures down to 15 kPa (2 psig) are possible. See Table 11-12.

The temperature of the chilled water can be close to 0 °C (32 °F), but it is usually not economical to operate below 4 °C (40 °F). Assume that the water temperature will rise slightly – about 1 °C (2 °F) between the ejector and the process load. For recirculated systems, assume that make-up water is approximately 1% of the circulation rate.

The system specifications should state the temperature of available cooling water (for the condenser) at its highest expected value to ensure the system will operate at full load during hot weather.

Figure 11-3 is a typical data sheet that can be used to specify a steam jet refrigeration system.

Table 11-12
Thermal efficiency of steam jet ejectors depends on the steam pressure [4]

Operating Steam Pressure, kPa (psig)	Approximate Thermal Efficiency, %
100 kPa (15 psig)	30%
700 kPa (100 psig)	60%
2800 kPa (400 psig)	80%

STEAM JET REFRIGERATION SYSTEM		
CLIENT	EQUIP. NO	PAGE
W.O.	REQUISITION NO.	SPECIFICATION NO.
UNIT AREA	PROCURED BY	INSTALLED BY

REV	PREPARED BY	DATE	APPROVAL
0			
1			
2			

#						
1	Service Conditions					
2	Type		Capacity:	tons	Manufacturer	Model:
3	Chilled Water Supply	lb/h	°F	Return	lb/h	°F
4	Make-Up Water			lb/h	°F	
6	Booster Jet Stream (a)	lb/h	psig	°F		
7	(b)	lb/h	psig	°F		
8	Secondary Jet Stream (a)	lb/h	psig	°F		
9	(b)	lb/h	psig	°F		
10	Cooling Water		Primary Condenser:	lb/h	°F exit	
11	psig	°F	Secondary Condenser:	lb/h	°F exit	
12	Fouling Factor for Water	ft²-h-°F/Btu				
13						
14	Specifications					
15	Steam Hand Valves provided by:					
16	Level Controls (specify):		Connections (flange type):			
17	Pressure Gages:					
18						
19		Connection Sizes, inch		Flanges		
20		Inlet Outlet		Rating, psig Face		
21	Flash Tank Chilled Water					
22	Make-Up Water					
23	Primary Condenser Steam					
24	Secondary Condenser Steam					
25	Primary Condenser Cooling Water					
26	Secondary Condenser Cooling Water					
27						
28	Condenser Tube Water Velocity Limits	ft/s	to	ft/s		
29						
30	Materials of Construction					
31	Flash Tank	Corr Allw	Barometric Leg		Corr Allw	
32	Booster Ejector Nozzle	Secondary Ejector Nozzles				
33	Primary Tubes	OD	inch	Tube Wall	BWG	
34	Number	Length	ft	Pitch	BWG	
35	Tube Sheet	Channel				
36	Surf Condenser Furnished by					
37	Tubes	OD	inch	Tube Wall	BWG	
38	Number	Length	ft	Pitch	BWG	
39	Barometric Primary Condenser		Barometric After Condenser			
40	Shell	Corr Allw	Shell		Corr Allw	
41	Baffles	Corr Allw	Baffles		Corr Allw	
42	Spray Noz	No.	Spray Noz		No.	
43						
44	Notes					
45	1. Vendor to specify make and model for flow, temperature, and pressure control valves; hand valves;					
46	thermometers and steam traps; electrical components; and miscellaneous equipment and components.					
47	2. Flexibility (determines number of booster jets):					
48	3. Temperature extremes and durations of cooling water:					
49	4. Closed or open circuit cooling water:					
50	5. Pumps (by Vendor or Owner; discharge heads if applicable):					
51	6. Space limitations:					
52	7. Cost of utilities (if Vendor required to do economic analysis):					
53						
54						

Figure 11-3. Datasheet for a steam jet refrigeration system. (Adapted from [4]; included in the companion spreadsheet).

Table 11-13
Steam and cooling water required for ammonia absorption refrigeration systems [4]

Single-Stage				
Evap Temp, °F	Steam Sat. Temp, °F, Required in Generators	Btu/Min Required in Generator Per Ton Refrigeration (200 Btu/Min)	Steam Rate, lb/h/ton Refrigeration	Water Rate Through Condenser (7.5 °F Temp Rise), gpm/ton
50	210	325	20.1	3.9
40	225	353	22.0	4.0
30	240	377	23.7	4.1
20	255	405	25.7	4.3
10	270	435	28.0	4.6
0	285	467	30.6	4.9
−10	300	507	33.6	5.4
−20	315	555	37.3	5.9
−30	330	621	42.5	6.6
−40	350	701	48.5	7.7
−50	370	820	57.8	9.5

Two-Stage			
Steam Sat. Temp, °F, Required in Generators	Steam Sat. Temp, °F, Required in Generators	Steam Sat. Temp, °F, Required in Generators	Steam Sat. Temp, °F, Required in Generators
175	595	35.9	4.3
180	625	37.8	4.5
190	655	40.0	4.6
195	690	42.3	4.9
205	725	44.7	5.3
210	770	47.5	5.7
220	815	50.6	6.3
230	865	54.0	6.9
240	920	58.0	7.8
250.	980	62.3	9.0
265	1050	67.5	11.0

Ammonia Absorption Utilities Requirements

Steam and cooling water requirements for ammonia absorption refrigeration systems are shown in Table 11-13 for single-stage and two-stage units. The tables are based upon cooling water to the condenser of 85 °F with 100 °F condensing temperature. Water from the condenser is used in the absorbers.

Example. For an evaporator temperature of − 10 °F, a steam rate (300 °F saturated temperature in the generators) of 33.6 lb/hr/ton refrigeration is required. Also, 5.4 gpm cooling water/ton refrigeration, assuming a 7.5 °F rise through the condenser, are required in this system.

References

[1] *ASHRAE Handbook – Refrigeration*. Atlanta, GA: ASHRAE; 2010.

[2] Ballou, Lyons, Tacquard. Mechanical Refrigeration Systems. *Hydrocarbon Processing*, June, 1967:127.

[3] Honeywell Corporation. Guide to Alternative Refrigerants. Bulletin G-525-043, 2005.

[4] Ludwig E. *Applied Process Design for Chemical and Petrochemical Plants*. vol. 3, 3rd ed. Gulf Publishing Co; 2001.

12

Closed Loop Heat Transfer Systems

Introduction

Closed loop heat transfer systems are ubiquitous in industry. They are used in refrigeration systems (as "chilled water" or "brine"), and for process heating and cooling. Closed loop systems service chemical reactors, storage tanks, condensers, reboilers, and many other unit operations.

Collaborating with the end user, the process engineer develops user requirements, including:

- A block diagram of the entire system, illustrating the extent of the closed loop system.
- Dynamic and steady state heating and cooling loads for each unit operation in the process.
- Performance requirements such as the time to heat or cool a tank.
- Diversity factors or calculations.
- Design requirements or limitations, such as environmental conditions, "boilerplate" mechanical and electrical specifications, and spatial constraints.

The engineer creates the system P&ID, designs the overall system (including calculations to size pumps, pipes, expansion tanks, etc.), and provides specifications for the procurement of equipment.

This chapter presents information and criteria that are helpful in performing the steps listed above. Critical decisions include selection of the heat transfer fluid and the system design elements. The engineer must decide whether to design the system for a single class of fluids, or to provide flexibility for changing to other fluids in the future.

Equipment vendors often provide the detailed thermal design of equipment (e.g., heat exchangers), typically using high-performance computer software, with the engineer's system design in hand.

Physical and thermal properties are provided here for a wide range of proprietary heat transfer fluids. The basic properties were obtained from information published by the manufacturers, converted to a consistent set of SI units, and fitted to formulae that relate properties to temperature. This gives a common basis for comparing different heat transfer fluids.

The information in this chapter is specific to closed loop systems that operate completely in the liquid state. Industry also uses condensing and evaporating systems. Some of the concepts in this chapter apply to those change-of-state systems, but the formulae and properties are limited to liquids.

Selecting the Heat Transfer Fluid

Consider the following factors when choosing a heat transfer fluid [2]:

1. Constraints. Is selection limited to a particular manufacturer or a specific family of fluids?
2. Compatibility. Is the fluid likely to get into the product or the environment? Are specific designations required, such as "food grade"? (Be sure to adequately define special designations.)
3. Temperature range. What temperature operating ranges are expected, both for the bulk fluid and at heat exchange surfaces (the "film temperature")? Are excursions possible?
4. Thermal duty. What overall heat transfer coefficient is required? See next section.
5. System volume. What overall quantity of heat transfer fluid is required to fill the system, including the expansion tank?

These properties of heat transfer fluids interact with the listed factors.

- Manufacturer. Each manufacturer has a limited portfolio of heat transfer fluids, and availability may be limited to certain geographic areas.
- Fluid type. Choices include water-based chemicals, ethylene or propylene glycol solutions, refined petroleum products, synthetic organic compounds, and silicones.
- Toxicity. Ethylene glycol is subject to strict controls, including containment to prevent spills or leaks from entering the environment. Some heat transfer fluids are considered safe for ingestion (e.g., food grade) or incidental human contact and may be more appropriate for use in food and pharmaceutical processing. Consider toxicity for ingestion, dermal, and inhalation contact.

- Temperature range. Manufacturers usually specify the range of temperatures over which their fluids may be used. However, also consider viscosity and vapor pressure.
- Degradation. Synthetic organic fluids degrade, especially at temperatures above 150 °C, and the degradation rate roughly doubles with each 10 °C increase. Silicone fluids do not undergo thermal degradation. Glycols oxidize, especially when used above 60 °C, to form organic acids; inhibitors are depleted eventually and the fluid will become corrosive.
- Viscosity. Viscosity affects the pumping rate and also heat transfer coefficients. Choose a fluid that has a viscosity of no greater than 10 mPa-s (cP) at the lowest temperature in your operating range. Higher viscosity fluids might not achieve turbulent flow. Also ensure the pumping system can handle the fluid throughout the operating range and also at ambient temperature (for when the system is started up, for example).
- Vapor pressure. By definition, closed liquid systems operate above the boiling point. Therefore, system pressure must be higher than the vapor pressure of the heat transfer fluid at the upper end of the operating temperature range. Although systems can be designed to contain nearly any pressure, standard practice is to limit pressures to about 1000 kPa (150 psig). Allowing for pressure drop and static pressure from elevated components (such as the expansion tank) leads to a rule of thumb to choose a fluid with a vapor pressure below 600 kPa (90 psi).
- Thermal properties. The temperature dependent properties of heat capacity, thermal conductivity, density, and viscosity, in conjunction with the system design, determine the heat transfer coefficient. These are discussed shortly.
- Flash point. Being the lowest temperature at which an ignitable concentration of the fluid's vapor (in air) exists above a liquid surface, the flash point determines whether the system requires hazardous area design features. The flash point may change as a fluid ages.

Types of Heat Transfer Fluids

Fluid Type	Examples	General Characteristics
Water		Ideal for unpressurized systems from 4 °C to 100 °C. Up to twice the thermal capacity and five times thermal conductivity, combined with low viscosity, compared to other fluids. Not flammable. Low ecological risk. Lowest cost. Disadvantages include corrosion and deposits of iron, chloride, and carbonates at higher temperatures. Mitigate with anti-corrosive additives (adds ecological risk) and use deionized water to eliminate deposits.
Water-based	Dowcal, Dowfrost, Dowtherm 4000, Dynalene Bioglycol, Dynalene HC	Combine good thermal characteristics of water with additional properties of glycol, for use range to about −25 °C. Ethylene glycol is considered toxic. Glycols degrade to form glycolic acid which is corrosive to steel; anti-corrosive additives are included with engineered glycol heat transfer fluids. Propylene glycol has a high viscosity. Lower price.
Paraffins, aliphatics, and mineral oils	Calflo, Chemtherm, Duratherm, Dynalene HF-LO, Dynalene LO-170, MultiTherm, Petro-Therm	Many fluids of this type meet FDA and USDA criteria for "incidental food contact." Do not form hazardous degradation byproducts. Most have nondiscernable odor. Non-toxic in case of skin contact or ingestion. High viscosity at low temperatures. Thermal stability not as good as aromatics.

(Continued)

—cont'd

Fluid Type	Examples	General Characteristics
Alkylated aromatics	Purity* FG Therminol XP Diethyl benzene Dowtherm J Dowtherm MX Dowtherm Q Dowtherm T Dynalene MV Dynalene HT Therminol 55	Excellent low temperature heat transfer properties and thermal stability. Less resistance to oxidation due to double bonds, especially above 80 °C. Cyclic compounds more resistant to thermal degradation compared with straight chain oils. Strong odor is irritating to some people.
Diphenyl and diphenyl oxide blends; terphenyls	Dowtherm ADowtherm GTherminol 59Therminol 66	
Silicones	Dimethyl polysiloxane Duratherm XLT Dynalene 600 Syltherm	Excellent service life in closed system in absence of oxygen. No odor. Very low toxicity. Low surface tension: have a tendency to leak through fittings. Higher price.

Heat Transfer Coefficient

The overall heat transfer coefficient for a specific unit operation is

$$\frac{1}{U} = \frac{1}{h_i} + \frac{1}{h_o} + R_w + R_f \qquad (12\text{-}1)$$

Where:

h_i = inside heat transfer coefficient
h_o = outside heat transfer coefficient

R_w = wall resistance
R_f = fouling resistance

The effect that h_i has on U is highly dependent on the value of h_o. In other words, if the biggest resistance to heat transfer is on the process side (h_o), then changes in h_i will have little effect on performance. On the other hand, if h_o is very high then h_i may become the rate-limiting resistance.

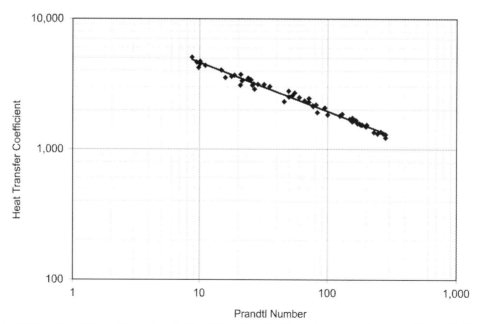

Figure 12-1. The relationship of Prandtl number to Heat Transfer Coefficient for organic and silicone fluids, using the Dittus-Boelter equation for cooling the fluid, at constant temperature and flow velocity.

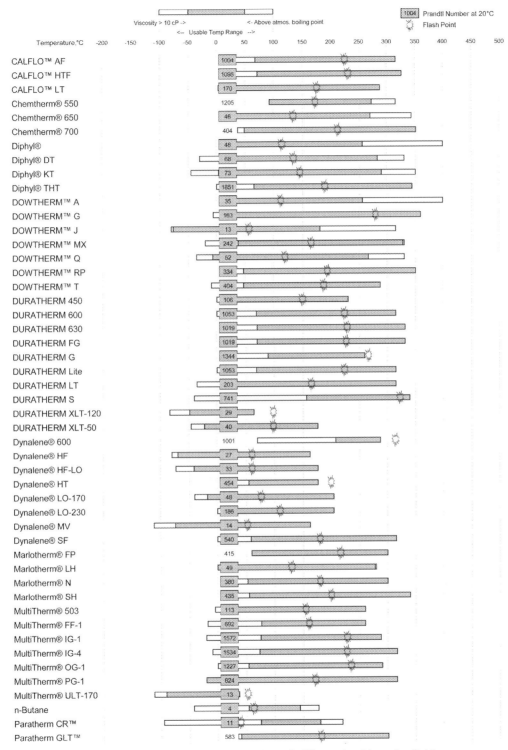

Figure 12-2. Properties for organic and silicone heat transfer fluids.

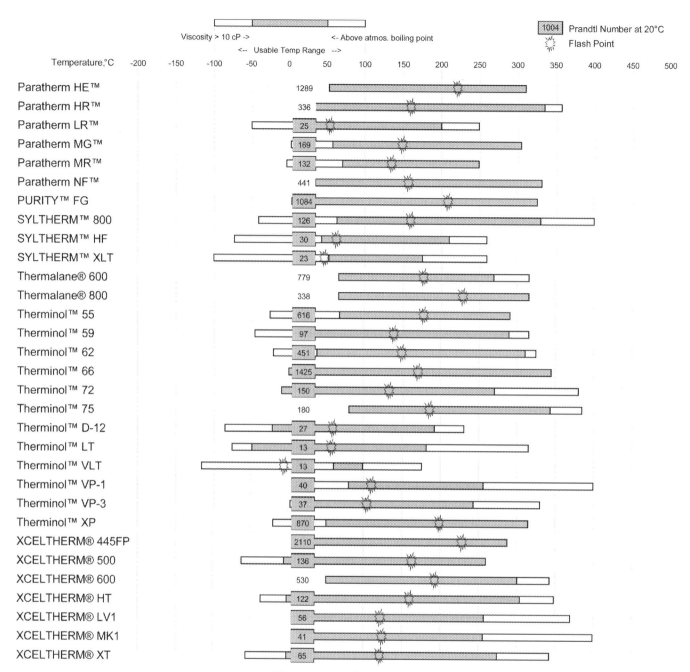

Figure 12-2. Continued.

h_i is calculated in 4 steps:

1. Calculate the Prandtl number (dimensionless);

$$N_{Pr} = \frac{c_p \, \mu}{k} \qquad (12\text{-}2)$$

Where:
c_p = heat capacity, J / g K
μ = dynamic viscosity, Pa s (= g/m s)

k = thermal conductivity, W/ m-K (= j m/s m^2 K)

2. Calculate the Reynolds N number (dimensionless);

$$N_{Re} = \frac{\rho \, u \, D}{\mu} \qquad (12\text{-}3)$$

Where:
ρ = density, kg/m^3
u = velocity, m/s

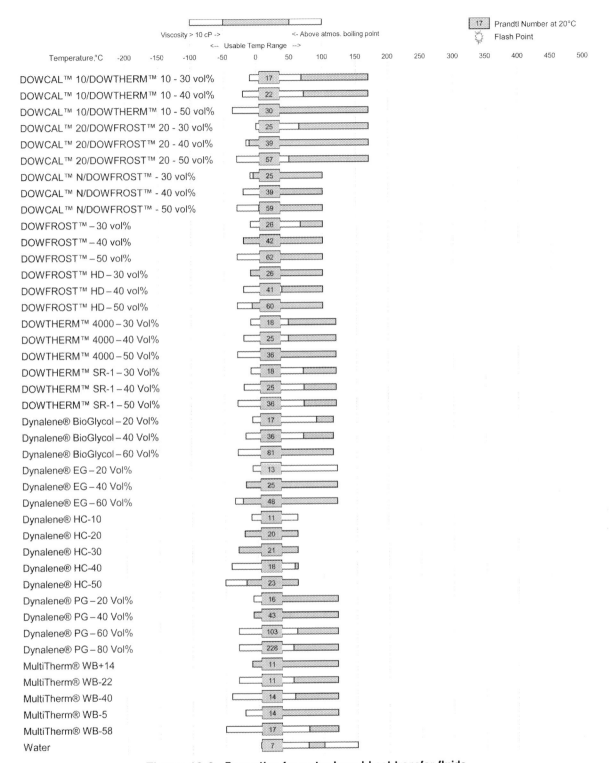

Figure 12-3. Properties for water-based heat transfer fluids.

D = diameter, m

μ = dynamic viscosity, Pa s (= g/m s)

3. Calculate the Nusselt number (dimensionless), using the Dittus-Boelter equation. This equation is suitable for turbulent flow ($N_{Re} > 10,000$). It is derived from the more rigorous Sieder-Tate correlation which requires that the wall temperature be determined (solution is iterative). This is good enough for approximate work where the goal is to decide on an appropriate heat transfer fluid. As stated, equipment vendors usually do the final rigorous calculations.

$$N_{Nu} = 0.023\, N_{Re}^{0.8} N_{Pr}^{n} \qquad (12\text{-}4)$$

Where:

$n = 0.4$ when the heat transfer fluid is being heated, and 0.3 when it is being cooled

4. Calculate h_i;

$$h_i = \frac{N_{Nu}\, k}{D} \qquad (12\text{-}5)$$

= inside heat transfer coefficient, W/m^2 K

It is evident that the heat transfer coefficient is exponentially related to the Prandtl number,

$h_i = Pr^x$. Figure 12-1 plots the Prandtl number for organic and silicone heat transfer fluids against h_i on a log-log chart. A similar plot for water-based fluids shows the same relationship with slightly different slope and intercept. For the purpose of comparing fluids, assuming turbulent flow, the exponent in the relationship $h_i = Pr^x$ may be taken as follows:

	Organic/Silicone	Water-Based
Heating the HTF	0.4	0.5
Cooling the HTF	0.5	0.6

A new chart compares many proprietary heat transfer fluids. Figure 12-2 gives data for organic and silicone fluids, and Figure 12-3 has data for water-based fluids including glycols. The charts include a bar representing the manufacturer's recommendation for usable temperature range. Where the viscosity exceeds 10 cP, or the temperature exceeds atmospheric boiling point, the bar is white. The Prandtl number at 20 °C and the flash point are also indicated.

System Design

Many of the suggestions in this section are taken from Bollard and Manning [1]. Also consult heat transfer fluid manufacturers' design and installation guides [3,4].

Heat transfer systems leak more than most other systems because flanges and seals experience severe temperature fluctuations, and the fluids are usually chosen for their thermal transfer properties which include low viscosity and low surface tension.

Locate the expansion tank at the high point in the system. Calculate the entire volume of the system and size the tank to allow at least 30% expansion. The tank should be sized to be ¼ full at ambient temperature, and ¾ full at operating temperature. The tank design pressure should be at least 100 kPa (15 psi) above the highest anticipated fluid vapor pressure. The best location for the piping connection from the expansion tank to the system is near the suction side of the circulating pump; this pipe should be generously sized to allow any vapors that get into the system to easily rise to the tank.

Heat transfer fluid should not normally circulate through the expansion tank, because it is desirable to allow static fluid in the tank to cool. This reduces vaporization of light ends from the fluid. However, during start-up, flowing the fluid through the tank helps eliminate entrained vapors from the system.

Design the entire system to permit full draining, by including drain plugs at low points, vent plugs at high points, and sloping long pipe runs.

Use inert gas (usually nitrogen) to blanket the expansion tank. This reduces oxidation of the heat transfer fluid, which occurs at temperatures higher than about 150 °C, and provides system pressure to reduce vaporization of light end components of the fluid. The rate of oxidation increases with temperature.

Carbon steel is generally avoided for service temperatures below − 45 °C (− 50 °F) because its body-centered crystalline structure becomes brittle. Since standard glass-lined carbon steel reactors utilize this material, the temperature cannot be safely reduced. When lower temperatures are needed, high-nickel steel can be specified, such as ASTM A645 5% Ni alloy. It is more common, however, to use a 300 series stainless steel in low temperature service.

Centrifugal pumps should be used to circulate the heat transfer fluid. Seal-less pumps that are magnetically coupled or "canned" are preferred *if* the fluid temperature is within the limits of the pump. Install a strainer at the suction to the pump. If using a pump with a double-mechanical seal, install a drip pan beneath the pump.

Use welded piping connections to the maximum practicable extent. Where demountable connections are needed, Class 300 ANSI Raised Face flanges are preferred, especially for hot systems using organic fluids. This is not for the pressure rating, but for the closer spacing of bolt holes and increased gasketing area compared with Class 150 flanges. Tighten the flanges at ambient temperature, and again after the system is hot. Use spiral-wound metallic gaskets.

If Class 150 flanges are used, consider using more compressible gaskets such as graphite or graphite-filled elastomers.

Insulation should be non-absorbent and non-wicking; closed-cell insulation is recommended. Small leaks can result in an absorbent insulation becoming saturated with fluid; a catastrophic fire may occur. Likely sources of leaks such as flanges, pump seals, instrument connections, and valve stems should not be insulated to facilitate the early detection of leaks. For long runs of welded pipe, with no fittings that could leak, open-cell insulation (such as fiberglass) is acceptable.

Use good quality water for dilution with glycol based fluids. Deionized or distilled water is best. Water containing excessive hardness or chlorides can result in premature inhibitor depletion and increased corrosion rates.

Temperature Control

The three system configurations presented in this section illustrate a range of design possibilities from which chemical engineers might choose. Consider these concepts if the application requires tight temperature control, multiple unit operations served by a central heat transfer system, or both heating and cooling duties.

A jacketed reactor operated in batch mode is illustrated. These are common in specialty chemical and pharmaceutical manufacturing facilities, often glass-lined steel for corrosion resistance.

Central systems can be comprised of one, two, or three different temperature loops (hot, cold, ambient), with local heaters or coolers if necessary. At the localized level, there are various ways to transfer heat to/from reactors and overhead condensers. Design details affect energy efficiency, safety, and effectiveness.

The best temperature control is achieved by adjusting the temperature of the fluid circulating through the reactor jacket. This is preferable to flow control because it results in more uniform temperature transfer throughout the jacket and can be finely controlled at the temperature set point.

Be sure the design provides turbulent flow to each demand point. Reactors with annular ("conventional") jackets usually need a local recirculation pump to give sufficient flow through the jacket. Dimple and half-pipe coil jackets *might* not need a pump depending on the capability of the central system to deliver heat transfer fluid with sufficient flow.

Initially assume a 70 kPa (10 psi) pressure drop through each control valve and unit operation (heat exchanger or reactor jacket). Similarly, assume that the heat transfer fluid temperature entering the process unit must be at least 6 °C (11 °F) warmer or cooler than the hottest or coldest process temperature. Fine tune the assumptions after the design is nearly complete.

The following three cases assume that both heating and cooling are required. If your system requires only heating or only cooling, then Case 1 without the local heater/chiller may be applicable.

Case 1: Single-Temperature Central System

It's usually easier to provide local heating with a centralized low temperature system, rather than local chilling. Local heating is effected in one of the following ways:

1. Shell-and-tube heat exchanger with steam. This is a good choice when the maximum required temperature of the fluid is less than 160 °C (320 °F).

If steam is used, then provision must be made to drain the heat exchanger shell or bypass the exchanger during periods of cooling to prevent freeze-up in the exchanger.

2. Immersion electric heater, packaged in a flow-through design or custom designed with a small tank. This is a good choice for heating loads less than 600 kW, and possible for loads of 2,000 kW or more. The advantage is simplicity and reliability.

3. Fired heater. Generally not used for local heating. Strongly consider using a centralized hot system if the heat duty is greater than 600 kW, possibly in conjunction with a central cold system (Cases 2 and 3).

Local cooling is done with a shell-and-tube heat exchanger using chilled water, or a mechanical refrigeration unit (for very small loads). The advantage of this is that a different heat transfer fluid can be used for cooling, and the hot and cold fluids are not comingled.

In either case the generalized control scheme shown in Figure 12-4 is used. A check valve is not needed in the recirculation loop. The desired reactor temperature is entered into the primary temperature controller. This

controller sends a set point to the secondary controller, which modulates the temperature of the recirculating heat transfer fluid either by adding cold fluid to the loop or by heating the recirculating fluid. Whenever cold fluid is added, displaced fluid is returned to the central system.

Case 2: Dual-Temperature Central System

A dual-temperature central system, using the same fluid, is advantageous for larger facilities because the heating and chilling equipment can be centralized in a mechanical room. Heating is typically done with a gas or oil-fired heater, capable of reaching temperatures above 350 °C (650 °F). Special environmental permits may be required due to the potential for emitting nitrogen oxides. Central chilling is usually accomplished with mechanical refrigeration machines, either air or water cooled. Central chilled HTF systems usually operate at about −30 °C (−22 °F).

For lower temperatures, a secondary heat exchanger can be installed at the reactor with liquid nitrogen being the cold source.

The "HTF Return" flow is handled in one of three ways. The simplest way is to return all fluid to a single tank, from which both the hot and cold systems draw. This is the least energy efficient method. The second method is to have two buffer tanks, with control valves selecting the return tank on the basis of temperature. The two tanks must be interconnected to equalize their volume. This is the costliest of the three. The third, and most complicated, method is to split the return flow to the hot and cold systems using modulating control valves that are actuated in concert with the supply valves. An equalization line between the hot and cold systems is still required because there may be slight differences in the actuation of the valves.

Case 3: Three-Temperature Central System

Some facilities use a three-temperature system. Hot and cold loops function almost exactly as described in Case 2. The middle temperature fluid is tempered with a closed cooling tower, exchanging its heat with the atmosphere. When the heating or cooling duty is satisfied using the ambient temperature loop, significant energy cost savings result since the refrigeration or fired heater systems are bypassed. The capital cost for the system is highest, however, due to the third centralized loop and added cooling tower.

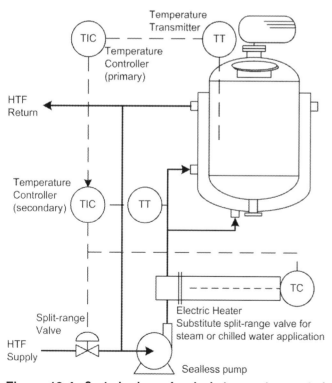

Figure 12-4. Control scheme for single temperature central system.

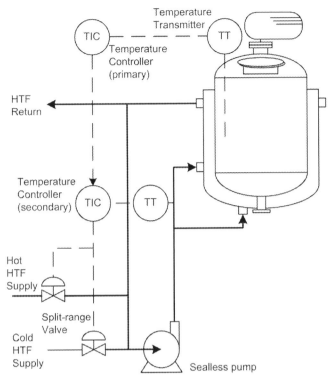

Figure 12-5. Control scheme for dual-temperature central system.

One way to pipe the system is to simply add a third "ambient" supply to the module depicted in Figure 12-5. That gives three modulating valves to control, which are split into three ranges as follows:

- An output in the range of 0 to 33% modulates the supply valve for the hot HTF from full open to full closed.
- At the other end of the scale, an output from 66 to 100% modulates the supply valve in the cold loop from closed to full open.
- The valve in the ambient line operates when the output is between 33% and 66%, but the action depends on whether this range was entered from the hot or the cold side. In either case, the valve will be closed until the controller output enters the 33% to 66% range. If the range is entered from below, increasing the output from 33% to 66% causes the control valve to stroke proportionally open; as the output increases above 66% the valve closes. Conversely, if the range is entered from above, decreasing the output from 66% to 33% causes the valve to open. As the output falls below 33% the valve is closed.
- In steady state, when the reactor calls for no heat transfer, the HTF will circulate through the pumped sub-loop at the reactor batch temperature. The temperature controller output will settle at either 33% or 66% and all valves will be closed.

The HTF Return line is designed as described under Case 2, with the complication of adding the third temperature fluid.

Physical Properties of Commercial Heat Transfer Fluids

Physical properties for commercial heat transfer fluids are readily available from the manufacturers' web sites. The important thermal properties are temperature dependent, so the data are fitted to the formulae listed below, providing a consistent way to use the information. The coefficients are listed in Tables 12-1 to 12-2. There is excellent agreement between the published data and the fitted curves, except for thermal conductivity of the aqueous solutions; however, the results are more than good enough for the calculations performed in this chapter. Manufacturers are listed in Table 12-3.

M, B, A, B, and C are coefficients from the tables.

Specific Heat

$$C_p = m\,t + b$$

Where:

C_p = specific heat of liquid, kJ / kg K

t = temperature, °C

divide Cp by 4.1868 for Btu / lb °F

Density

$$\rho = m\,t + b$$

Where:

ρ = density, kg/m³

t = temperature, °C

divide ρ by 16.02 for lb/ft³

Thermal Conductivity

$$k = m\,t + b$$

Where:

k = thermal conductivity, W / m K

t = temperature, °C

divide k by 1.731 for Btu / ft h

Table 12-1
Properties of organic and silicone heat transfer fluids

Fluid	Description	Specific Heat		Density		Thermal Conductivity		Viscosity		
		m	b	m	b	m	b	A	B	C
CALFLO™ AF	99.9% pure base oils, crystal-clear, free of aromatic compounds with additives	0.00330	1.837	-0.6189	876.7	-0.0000540	0.144	-2.810	862.7	-172.2
CALFLO™ HTF	99.9% pure base oils, crystal-clear, free of aromatic compounds with additives	0.00330	1.837	-0.6370	880.4	-0.0000524	0.144	-3.019	971.2	-162.4
CALFLO™ LT	Synthetic poly-alpha-olefins	0.00361	1.992	-0.6590	829.8	-0.0000858	0.143	-2.944	771.3	-150.5
Chemtherm® 550	Premium oil more thermally stable than mineral oils	0.00349	1.854	-0.6343	909.8	-0.0000753	0.133	-2.720	756.2	-187.1
Chemtherm® 650	Premium high temperature oil resists oxidation	0.00374	1.312	-0.7968	999.0	-0.0001308	0.129	-3.887	1148.4	-77.5
Chemtherm® 700	Premium high temperature single compound formulation	0.00353	1.482	-0.6994	1043.8	-0.0001036	0.133	-2.852	801.6	-167.5
Diphyl®	High-temperature HTF based on diphenyl oxide/diphenyl eutectic for liq and vap phase	0.00268	1.504	-1.0246	1094.6	-0.0001639	0.146	-2.956	695.5	-136.3
Diphyl® DT	Isomeric ditolyl ethers with broad spectrum of applications in liq phase	0.00276	1.530	-0.8712	1047.9	-0.0001449	0.137	-2.634	588.0	-159.4
Diphyl® KT	Synthetic mixture of isomeric bynzyl toluenes with high thermal stability	0.00295	1.530	-0.7879	1022.0	-0.0001323	0.132	-3.290	883.5	-119.0
Diphyl® THT	Partially hydrogenated terphenyls for pressureless high temperature applications	0.00338	1.535	-0.7633	1038.7	-0.0000993	0.124	-2.568	631.4	-209.0
DOWTHERM™ A	Biphenyl and diphenyl oxide eutectic mixture — liquid phase properties	0.00295	1.501	-0.9893	1,091.5	-0.0001600	0.142	-5.485	2,637.4	105.7
DOWTHERM™ G	Di- and tri-aryl compounds mixture	0.00350	1.476	-0.7747	1,062.3	-0.0001158	0.129	-4.209	1,446.7	-80.4
DOWTHERM™ J	Mixture of isomers of an alkylated aromatic	0.00363	1.804	-0.9149	873.7	-0.0002107	0.132	-3.071	638.5	-77.4
DOWTHERM™ MX	Mixture of alkylated aromatics	0.00325	1.545	-0.7759	979.0	-0.0001051	0.125	-4.547	1,505.8	-91.5
DOWTHERM™ Q	Mixture of diphenylethane and alkylated aromatics	0.00303	1.593	-0.7628	982.1	-0.0001487	0.124	-3.392	743.1	-135.9
DOWTHERM™ RP	Synthetic diaryl alkyl liquid	0.00297	1.561	-0.7434	1,046.5	-0.0001297	0.134	-3.199	983.2	-141.8
DOWTHERM™ T	Mixture of C14—C30 alkyl benzenes	0.00298	1.903	-0.6891	887.0	-0.0002207	0.139	-3.372	1,025.7	-139.9
DURATHERM 450	Refined oil, non-aromatic	0.00302	2.021	-0.6778	874.5	-0.0000753	0.145	-2.748	674.3	-151.0
DURATHERM 600	Refined and hydrotreated paraffinic oils, non-toxic and non-reportable	0.00330	1.838	-0.6743	866.4	-0.0000507	0.143	-2.906	903.6	-168.9
DURATHERM 630	Refined and hydrotreated paraffinic oils, non-toxic and non-reportable	0.00328	1.845	-0.6730	870.1	-0.0000529	0.144	-2.993	972.3	-160.4
DURATHERM FG	Refined paraffinic oils with additives, food grade	0.00328	1.845	-0.6730	870.1	-0.0000529	0.144	-2.993	972.3	-160.4
DURATHERM G	Polyalkylene glycol based fluid with additives	0.00180	1.906	-0.5457	938.2	-0.0001498	0.170	-2.133	859.9	-168.2
DURATHERM Lite		0.00330	1.838	-0.6743	866.4	-0.0000507	0.143	-2.906	903.6	-168.9

DURATHERM LT	Same as Duratherm 600, but with fewer additives									
DURATHERM S	Refined oil, non-aromatic	0.00343	2.019	−0.6842	827.0	−0.0000850	0.143	−3.015	794.6	−152.1
	Refined oil, non-aromatic especially resistant to oxidation	0.00170	1.629	−0.4700	976.7	−0.0002011	0.136	−3.829	3,727.5	178.5
DURATHERM XLT-120	Refined silocone oil, for cryogenic applications, food & pharma	0.00218	2.021	−0.4989	847.4	−0.0000653	0.136	−1.922	442.3	−120.8
DURATHERM XLT-50	Refined silocone oil, for cryogenic applications, food & pharma	0.00266	2.019	−0.9641	848.5	−0.0000979	0.136	−2.661	567.2	−137.0
Dynalene® 600	Silicone-based product with enhanced resistance to oxidation	0.00189	1.234	−1.0000	983.0	−0.0001773	0.157	−1.973	2,181.2	29.1
Dynalene® HF	Biodegradable aliphatic hydrocarbon blend, food grade	0.00455	1.924	−0.7024	816.8	−0.0001976	0.124	−11.004	7,574.1	365.5
Dynalene® HF-LO	Aliphatic hydrocarbon blend, non-toxic, odorless	0.00380	2.019	−0.7598	777.7	−0.0002000	0.113	−3.802	919.2	−81.4
Dynalene® HT	Synthetic organic hydrocarbon	0.00373	1.475	−0.7159	1,058.5	−0.0001319	0.133	−3.362	1,050.8	−143.2
Dynalene® LO-170	Aliphatic hydrocarbon blend	0.00380	1.804	−0.5017	784.0	−0.0002000	0.117	−4.009	975.0	−100.5
Dynalene® LO-230	Aliphatic hydrocarbon blend	0.00380	1.934	−0.6982	814.9	−0.0002000	0.151	−3.181	851.7	−146.0
Dynalene® MV	Biodegradable hydrocarbon blend	0.00354	1.724	−0.7726	860.8	−0.0002377	0.138	−4.495	1,230.9	−17.8
Dynalene® SF	Synthetic alkylated aromatics	0.00365	1.894	−0.6732	890.4	−0.0000798	0.136	−3.485	1,205.8	−123.2
Marlotherm® FP	Isoparaffinic chemical structure, clear liquid, bland odor	0.00376	1.837	−0.6568	861.6	−0.0000609	0.134	−4.463	1,669.1	−79.9
Marlotherm® LH	Synthetic, organic, heat transfer medium	0.00335	1.547	−0.8073	1,013.8	−0.0001152	0.134	−2.765	699.2	−125.2
Marlotherm® N	Synthetic, organic, heat transfer medium, ideal 150 to 300 deg C range	0.00367	1.892	−0.6717	890.1	−0.0000813	0.137	−4.370	1,719.2	−67.9
Marlotherm® SH	Synthetic, organic, heat transfer medium	0.00371	1.477	−0.7144	1,058.3	−0.0001309	0.133	−3.144	1,006.8	−143.9
MultiTherm® 503	Paraffinic hydrocarbon	0.00381	2.024	−0.6545	806.9	−0.0000618	0.143	−3.715	1,037.4	−112.6
MultiTherm® IG-1	White mineral oil	0.00364	1.813	−0.5253	871.7	−0.0000759	0.140	−2.890	907.4	−174.4
MultiTherm® IG-4	White mineral oil	0.00455	1.864	−0.7102	877.7	−0.0000732	0.136	−3.238	1,030.8	−162.6
MultiTherm® OG-1	Hydrocracked mineral oil with oxidation inhibitor/stabilizer	0.00356	1.794	−0.5360	871.7	−0.0000365	0.140	−2.668	863.5	−172.9
MultiTherm® PG-1	White mineral oil, food grade	0.00363	1.805	−0.4203	881.7	−0.0000723	0.134	−3.159	876.1	−167.0
MultiTherm® ULT-170	Hydrocarbon blend	0.00350	1.725	−0.7704	860.5	−0.0002000	0.138	−1.696	284.0	−121.5
Paratherm CR™	Synthetic hydrocarbon blend	0.00372	1.846	−1.0082	848.3	−0.0000765	0.142	−2.785	568.9	−70.8
Paratherm GLT™	Alkylated aromatic for closed loop liquid phase heating	0.00321	1.874	−0.6956	892.8	−0.0001413	0.133	−3.774	1,137.4	−140.3
Paratherm HE™	Hydrotreated heavy paraffinic distillate – mineral oil	0.00365	1.811	−0.6413	875.9	−0.0000754	0.134	−3.196	1,055.9	−156.1
Paratherm HR™	Alkylated aromatic, for closed loop heating	0.00222	1.899	−0.7665	971.1	−0.0000995	0.122	−3.733	1,159.4	−121.8
Paratherm LR™	Paraffinic hydrocarbon, for closed loop heating and cooling	0.00387	1.925	−0.7377	777.6	−0.0000774	0.153	−4.051	992.6	−80.6
Paratherm MG™	Linear alkene, food grade	0.00203	2.334	−0.5631	809.5	−0.0000854	0.135	−2.995	757.8	−148.6
Paratherm MR™	Linear alkene, fully saturated	0.00192	2.339	−0.7393	816.4	−0.0000551	0.134	−2.798	620.4	−163.7
Paratherm NF™	Hydrotreated mineral oil, food grade	0.00529	1.718	−0.6627	897.7	−0.0000509	0.106	−11.043	7,675.3	244.6
PURITY™ FG	Pure base oil, food grade, with additives	0.00339	1.806	−0.6773	864.7	−0.0000564	0.139	−3.024	966.4	−162.6
SYLTHERM™ 800	Silicone fluid – dimethyl polysiloxane	0.00171	1.574	−0.9867	957.6	−0.0001882	0.139	−6.161	4,125.9	190.9

(Continued)

Table 12-1
Properties of organic and silicone heat transfer fluids—cont'd

Fluid	Description	Specific Heat		Density		Thermal Conductivity		Viscosity		
		m	b	m	b	m	b	A	B	C
SYLTHERM™ HF	Silicone fluid – dimethyl polysiloxane	0.00246	1.633	−1.0110	892.0	−0.0002480	0.112	−4.135	1,402.5	2.1
SYLTHERM™ XLT	Silicone fluid – dimethyl polysiloxane	0.00210	1.730	−1.0272	875.8	−0.0002343	0.114	−3.694	958.1	−54.6
Therminol™ 55	Alkyl aromatic derivatives	0.00354	1.835	−0.6951	885.8	−0.0001166	0.131	−3.319	932.6	−160.7
Therminol™ 59	Mixture of diphenyl alkanes	0.00333	1.614	−0.7729	989.9	−0.0001014	0.123	−3.741	1,121.9	−96.0
Therminol™ 62	Mixture of di- and tri-isopropyl biphenyl	0.00222	1.902	−0.7695	970.7	−0.0000987	0.125	−4.480	1,412.2	−112.8
Therminol™ 66	Hydrogenated terphenyl and polyphenyls	0.00358	1.486	−0.7010	1,022.8	−0.0000791	0.120	−2.961	798.1	−188.8
Therminol™ 72	Mixture of diphenyl ether, terphenyl, biphenyl, and phenanthrene	0.00271	1.498	−0.9098	1,097.9	−0.0001214	0.142	−4.539	1,495.1	−83.8
Therminol™ 75	Mixture of terphenyl, quaterphenyl, and phenanthrene	0.00241	1.543	−0.8010	1,109.2	−0.0000892	0.139	−4.974	2,294.8	4.2
Therminol™ D-12	Hydrotreated heavy naphtha (petroleum)	0.00405	2.022	−0.7783	774.6	−0.0001687	0.112	−4.242	1,070.6	−58.8
Therminol™ LT	Diethyl benzene	0.00371	1.719	−0.9065	880.3	−0.0001993	0.128	−4.457	1,355.6	16.4
Therminol™ VLT	Mixture of methyl cyclohexane and trimethyl pentane	0.00423	1.866	−0.8789	763.1	−0.0002125	0.107	−4.859	1,392.5	19.1
Therminol™ VP-1	Mixture of diphenyl ether and biphenyl	0.00276	1.494	−0.9271	1,089.5	−0.0001533	0.142	−4.084	1,380.2	−35.3
Therminol™ VP-3	Mixture of cyclohexylbenzene and bicyclohexyl	0.00405	1.531	−0.8524	959.9	−0.0001338	0.120	−5.231	2,426.4	97.6
Therminol™ XP	White mineral oil	0.00423	1.737	−0.6589	891.8	−0.0000947	0.118	−3.269	948.1	−163.0
XCELTHERM® 445FP	Hydroprocessed naphthenic mineral oil; high flash point	0.00376	2.043	−0.4894	875.6	−0.0000622	0.134	−2.993	961.2	−171.1
XCELTHERM® 500	Hydrogenated polyalpha olefin	0.00419	2.076	−0.6104	802.3	−0.0000559	0.138	−3.259	832.2	−139.4
XCELTHERM® 600	Hydroprocessed paraffinic white mineral oil	0.00343	1.923	−0.6078	864.0	−0.0000809	0.138	−3.851	1,092.5	−146.3
XCELTHERM® HT	Synthetic alkyl aromatic	0.00331	1.337	−0.7195	1,017.0	−0.0000860	0.119	−3.108	665.0	−170.7
XCELTHERM® LV1	Diphenyl ethane and diphenyl oxide mixture	0.00280	1.517	−0.8346	1,080.2	−0.0001521	0.141	−4.003	1,273.7	−65.5
XCELTHERM® MK1	Diphenyl oxide and diphenyl	0.00281	1.489	−0.9565	1,092.6	−0.0001579	0.142	−4.119	1,398.0	−35.0
XCELTHERM® XT	Synthetic alkyl aromatic	0.00309	1.641	−0.7291	1,011.0	−0.0001542	0.137	−3.783	1,086.9	−92.4

Table 12-2
Properties of water-based heat transfer fluids

Fluid	Description	Specific Heat		Density		Thermal Conductivity		Viscosity		
		m	b	m	b	m	b	A	B	C
DOWCAL™ 10/DOWTHERM™ 10 – 30 vol%	Ethylene glycol with inhibitor package	0.00276	3.618	−0.6949	1,059.3	0.0003237	0.458	−4.181	909.7	−107.6
DOWCAL™ 10/DOWTHERM™ 10 – 40 vol%	Ethylene glycol with inhibitor package	0.00326	3.440	−0.7212	1,073.7	0.0002825	0.422	−4.727	1,202.6	−83.2
DOWCAL™ 10/DOWTHERM™ 10 – 50 vol%	Ethylene glycol with inhibitor package	0.00377	3.253	−0.7483	1,087.1	0.0002502	0.389	−5.291	1,555.1	−56.1
DOWCAL™ 20/DOWFROST™ 20 – 30 vol%	Propylene glycol with inhibitor package	0.00269	3.810	−0.7431	1,043.9	0.0003067	0.450	−2.868	390.4	−193.6
DOWCAL™ 20/DOWFROST™ 20 – 40 vol%	Propylene glycol with inhibitor package	0.00326	3.659	−0.7986	1,053.1	0.0002204	0.410	−2.201	260.6	−221.9
DOWCAL™ 20/DOWFROST™ 20 – 50 vol%	Propylene glycol with inhibitor package	0.00381	3.488	−0.8417	1,060.7	0.0001461	0.374	−2.665	403.3	−202.6
DOWCAL™ N/DOWFROST™ – 30 vol%	Propylene glycol with inhibitor package, food grade	0.00266	3.813	−0.6437	1,040.9	0.0005276	0.440	−3.083	464.4	−181.2
DOWCAL™ N/DOWFROST™ – 40 vol%	Propylene glycol with inhibitor package, food grade	0.00318	3.671	−0.6931	1,049.9	0.0003961	0.401	−2.870	429.3	−194.1
DOWCAL™ N/DOWFROST™ – 50 vol%	Propylene glycol with inhibitor package, food grade	0.00374	3.499	−0.7355	1,056.9	0.0002731	0.365	−3.181	558.9	−181.1
DOWFROST™ – 30 vol%	Propylene glycol with inhibitor package	0.00275	3.793	−0.6573	1,042.8	0.0004924	0.435	−3.045	463.8	−181.2
DOWFROST™ – 40 vol%	Propylene glycol with inhibitor package	0.00331	3.635	−0.6489	1,047.0	0.0004741	0.386	−2.958	466.7	−188.6
DOWFROST™ – 50 vol%	Propylene glycol with inhibitor package	0.00386	3.455	−0.6577	1,050.4	0.0003857	0.344	−3.424	657.1	−168.0
DOWFROST™ HD – 30 vol%	Propylene glycol with inhibitor package, dyed bright yellow	0.00285	3.727	−0.6422	1,052.6	0.0004924	0.435	−3.045	463.8	−181.2
DOWFROST™ HD – 40 vol%	Propylene glycol with inhibitor package, dyed bright yellow	0.00369	3.527	−0.6358	1,060.2	0.0004741	0.386	−2.958	466.7	−188.6
DOWFROST™ HD – 50 vol%	Propylene glycol with inhibitor package, dyed bright yellow	0.00452	3.301	−0.6496	1,067.1	0.0003857	0.344	−3.424	657.1	−168.0
DOWTHERM™ 4000 – 30 Vol%	Ethylene glycol with inhibitors, dyed fluorescent orange	0.00289	3.571	−0.5842	1,064.7	0.0006947	0.435	−3.993	832.6	−119.1
DOWTHERM™ 4000 – 40 Vol%	Ethylene glycol with inhibitors, dyed fluorescent orange	0.00343	3.375	−0.6082	1,081.0	0.0006278	0.396	−4.412	1,071.2	−98.8
DOWTHERM™ 4000 – 50 Vol%	Ethylene glycol with inhibitors, dyed fluorescent orange	0.00395	3.170	−0.6182	1,094.5	0.0005784	0.358	−4.250	991.9	−118.1
DOWTHERM™ SR-1 – 30 Vol%	Ethylene glycol with inhibitors, dyed fluorescent pink	0.00284	3.590	−0.5415	1,055.2	0.0006947	0.435	−3.993	832.6	−119.1
DOWTHERM™ SR-1 – 40 Vol%	Ethylene glycol with inhibitors, dyed fluorescent pink	0.00335	3.403	−0.5496	1,68.6	0.0006278	0.396	−4.412	1,071.2	−98.8
DOWTHERM™ SR-1 – 50 Vol%	Ethylene glycol with inhibitors, dyed fluorescent pink	0.00386	3.206	−0.5440	1,079.0	0.0005784	0.358	−4.250	991.9	−118.1
Dynalene® EG – 20 Vol%	Inhibited ethylene glycol	0.00233	3.759	−0.5246	1,045.9	0.0008909	0.474	−3.519	637.4	−135.0
Dynalene® EG – 40 Vol%	Inhibited ethylene glycol	0.00343	3.376	−0.5886	1,080.9	0.0006945	0.395	−3.928	860.3	−122.0

(Continued)

Table 12-2
Properties of water-based heat transfer fluids—cont'd

Fluid	Description	Specific Heat		Density		Thermal Conductivity		Viscosity		
		m	b	m	b	m	b	A	B	C
Dynalene® EG — 60 Vol%	Inhibited ethylene glycol	0.00447	2.953	−0.6235	1,110.0	0.0005182	0.333	−4.399	1,040.1	−122.1
Dynalene® HC-10	Aqueous-based, engineered for low temperature applications	0.00300	3.280	−0.6913	1,363.2	0.0010000	0.505	−4.276	1,294.2	−23.9
Dynalene® HC-20	Aqueous-based, engineered for low temperature applications	0.00228	3.177	−0.4980	1,248.8	0.0009540	0.502	−5.291	2,948.8	163.7
Dynalene® HC-30	Aqueous-based, engineered for low temperature applications	0.00269	3.019	−0.5956	1,282.9	0.0011000	0.498	−0.512	242.6	−157.8
Dynalene® HC-40	Aqueous-based, engineered for low temperature applications	0.00230	2.890	−0.7000	1,330.0	0.0009875	0.490	−2.987	974.8	−58.3
Dynalene® HC-50	Aqueous-based, engineered for low temperature applications	0.00200	2.670	−0.5500	1,362.5	0.0009800	0.484	−8.684	4,625.6	162.9
Dynalene® PG — 20 Vol%	Propylene glycol with corrosion inhibitors	0.00220	3.931	−0.5305	1,028.4	0.0008282	0.468	−3.744	656.3	−145.5
Dynalene® PG — 40 Vol%	Propylene glycol with corrosion inhibitors	0.00329	3.638	−0.6083	1,045.8	0.0005526	0.383	−4.058	747.1	−159.3
Dynalene® PG — 60 Vol%	Propylene glycol with corrosion inhibitors	0.00441	3.253	−0.6522	1,056.4	0.0002990	0.312	−4.271	936.2	−150.4
Dynalene® PG — 80 Vol%	Propylene glycol with corrosion inhibitors	0.00505	2.763	−0.7901	1,069.0	0.0000026	0.258	−4.610	1,143.8	−143.2
MultiTherm® WB+14	Water based, no flash point, for low temperature performance	0.00253	3.280	−0.4795	1,200.7	0.0010000	0.506	−27.535	49,896.9	1,485.8
MultiTherm® WB-22	Water based, no flash point, for low temperature performance	0.00223	3.041	−0.4983	1,285.4	0.0010000	0.499	−1.550	217.1	−191.2
MultiTherm® WB-40	Water based, no flash point, for low temperature performance	0.00220	2.891	−0.5738	1,325.5	0.0010794	0.486	−1.197	266.8	−166.7
MultiTherm® WB-5	Water based, no flash point, for low temperature performance	0.00227	3.181	−0.5281	1,249.6	0.0010000	0.505	−1.067	193.6	−191.6
MultiTherm® WB-58	Water based, no flash point, for low temperature performance	0.00208	2.662	−0.6000	1,365.0	0.0010326	0.484	−2.020	517.6	−128.4
Water	City water	0.00017	4.193	−0.4580	1,006.3	0.0011752	0.569	−3.724	582.2	−137.0

Table 12-3
Manufacturers of proprietary heat transfer fluids

Trade Name	Manufacturer	Website
CALFLO™	Petro-Canada	http://lubricants.petro-canada.ca
Chemtherm®	Coastal Chemical, a Brenntag company	http://www.coastalchem.com
Diphyl®	Lanxess Deutschland GmbH	http://www.lanxess.com
DOWCAL™	Dow Chemical	http://www.dow.com/heattrans/
DOWFROST™	Dow Chemical	http://www.dow.com/heattrans/
DOWTHERM™	Dow Chemical	http://www.dow.com/heattrans/
DURATHERM	Duratherm, division of Frontier Resource & Recovery Services	http://www.heat-transfer-fluid.com
Dynalene®	Dynalene	http://www.dynalene.com
Marlotherm®	Sasol	http://www.marlotherm.com
Mobiltherm	Exxon Mobil	http://www.mobil.com
MultiTherm®	MultiTherm	http://www.multitherm.com
Paratherm	Paratherm	http://www.paratherm.com
Petro-Therm™	Petro-Canada	http://lubricants.petro-canada.ca
PURITY™	Petro-Canada	http://lubricants.petro-canada.ca
SYLTHERM™	Dow Chemical	http://www.dow.com/heattrans/
Thermalane®	Coastal Chemical, a Brenntag company	http://www.coastalchem.com
Therminol	Solutia	http://www.therminol.com
XCELTHERM®	Radco Industries	http://radcoind.com

Viscosity

$$\ln(\mu) = \frac{A + B}{(t + 273.15 + C)}$$

$$\mu = \exp\left(\frac{A + B}{(t + 273.15 + C)}\right)$$

Where:

μ = dynamic viscosity, mPa s (= cP)

t = temperature, °C

multiply by $\left(\dfrac{1000}{\rho}\right)$ for centistokes

Nomenclature

C_p = specific heat of liquid, kJ / kg K
D = diameter, m
h_i = inside heat transfer coefficient
h_o = outside heat transfer coefficient
k = thermal conductivity, W / m K
R_w = wall resistance
R_f = fouling resistance
t = temperature, °C

U = overall heat transfer coefficient
u = velocity, m/s
ρ = density, kg/m^3
μ = dynamic viscosity, mPa s (= cP)
N_{Nu} = Nusselt number, dimensionless
N_{Pr} = Prandtl number, dimensionless
N_{Re} = Reynolds number, dimensionless

References

[1] Bollard, Don and Manning, William P. Boost Heat-Transfer System Performance, *Chemical Engineering Progress*, November 1990.

[2] Cuthbert, John. Choose the Right Heat-Transfer Fluid, *Chemical Engineering*, July 1994.

[3] Dow Chemical. Equipment for Systems Using Dowtherm Heat Transfer Fluids, www.dow.com

[4] Solutia. Liquid Phase Systems Design Guide, www.therminol.com.

13

Biopharmaceutical Systems

Introduction

Systems and facilities intended for processing of pharmaceutical materials must be designed to prevent non-conforming product from reaching the market. Non-conformance means that the product fails to meet specifications and standards for identity, strength, quality, potency, and purity. For biopharmaceuticals, which are typically manufactured using fermentation or bioreaction processes, pure cultures are required for successful and productive outcomes.

This chapter is primarily concerned with features that contribute to pure cultures. Although focused on biopharmaceuticals, very similar principles are applied to sterile or aseptic processes (e.g., water systems, injectable drugs, etc.), traditional drug products (e.g., tablets and capsules), and even, to some degree, food production.

The terms "pure culture" and "pure culture capability" are used instead of "sterile" or "sterility assurance" to acknowledge that bioreactions are, by nature, the process of growing large populations of helpful microorganisms or cells as opposed to unwanted varieties. Contamination by adventitious agents costs time, money, and lost productivity. Moreover, contaminants and their sources can be very difficult to locate and eliminate.

The first portion of this chapter is divided into four basic sections:

1. Design aspects for sterile operations
2. Common contamination root causes
3. Troubleshooting foreign growth (FG) events
4. Application of rigorous microbiology to better understand and improve pure culture

The International Society for Pharmaceutical Engineering has granted permission to draw heavily from Reference [5] for the first portion; Dale Seiberling supported preparation of the CIP section [13]. The Disposable Systems section is original material.

The second portion of the chapter touches on design challenges that are unique to drug processing:

1. Containment, to protect people from unwanted contact with potent drugs and intermediates
2. Risk assessment
3. Validation

Facility Design and Sterilization Best Practices

Prevention of FG (Foreign Growth) is the most important factor in long-term successful pure culture performance. The design and installation of equipment and facilities are key to this prevention. Consider the process from the inside out; each layer – equipment boundary, local protection with hood or glovebox, room envelope, and air handling systems – is more complex with greater surface area and volume. This means that it is easier to prevent problems with good design close to the product than to rely on more distant systems.

General Design Considerations

Consult the latest revision to the Bioprocessing Equipment Standard, BPE [1]. The surface smoothness is usually specified for pipes, fittings, valves, and equipment that come into contact with product directly or indirectly. This includes purified water and clean steam systems. BPE provides a designation system for surface roughness, Ra, defined as the log of the arithmetic mean of the surface profile. (Use double-mechanical seals for agitator shafts. Ensure that seals in pure culture operations are lubricated with a sterile fluid, such as steam or clean condensate; see Table 13-1.) Ra is usually measured with a profilometer.

Dead legs in hygienic piping systems are defined as places where a branch pipe contains a quiescent pocket

Table 13-1
***Ra* readings for product contact surfaces [1]**

Surface Designation	Mechanically Polished	
	Ra Maximum	
	μ-inch	μ-meter
SF0	No requirement	No requirement
SF1	20	0.51
SF2	25	0.64
SF3	30	0.76

Surface Designation	Mechanically Polished and Electropolished	
	Ra Maximum	
	μ-inch	μ-meter
SF4	15	0.38
SF5	20	0.51
SF6	25	0.64

All *Ra* readings are taken across the lay, wherever possible.
No single *Ra* reading shall exceed the maximum *Ra* value in this table.
Other *Ra* readings are available if agreed upon between the owner/user and manufacturer, not to exceed values in this table

between the main pipe and the end of the branch (typically at a closed valve, cap, or instrument). Measure the dead leg from the main pipe wall to the point of blockage in the branch. For example, if a diaphragm valve blocks the branch, measure to the centerline of the valve (which is where the seat and diaphragm meet). The acceptable length of a dead leg has decreased over the years. BPE now requires that dead legs be no longer than twice the diameter of the branch (L/D $<= 2$) [1].

Process piping to be sterilized should be configured to completely drain back into the equipment if possible, minimizing the number of separate drain points. Sterilized piping should be sloped to eliminate holdup points. Slopes need to be much greater if against the direction of steam flow. Never branch a line from the bottom because it could promote condensate buildup. A key quality of pipe insulation is the ability to wick moisture and freely drain so that it won't retain leaks (wet insulation is less effective and provides potential cold spots in steam seals).

Elastomers, including O-rings, gaskets, and valve diaphragms, are often critical elements of the sterile boundary, simply because they have no backup in case of failure. Thus, they need to be designed with the optimum material of construction for conditions of the bioreactions (which usually means the temperature exposure from SIP) and replaced on a set frequency rather than be allowed to run to failure.

Use double-mechanical seals for agitator shafts. Ensure that seals in pure culture operations are lubricated with a sterile fluid, such as steam or clean condensate.

System Design Considerations

BPE provides many specific recommendations for process systems used in the biopharmaceutical, pharmaceutical, and personal care product industries. The ones listed here are counter to practices that are common in the general chemical processing industry [1].

- Use one of these methods to physically prevent cross-contamination of product streams:
  a) removable spool piece,
  b) U-bend transfer panel,
  c) double block-and-bleed valve system, or
  d) mix-proof valving.
- Fluid bypass piping (around traps, control valves, etc.) is not recommended.
- Redundant in-line equipment is not recommended due to the potential creation of dead legs.

- The use of check valves in hygienic process piping systems requires caution and is not recommended.
- Install orifice plates in a drainable position.
- Use eccentric reducers in horizontal piping to eliminate pockets in the system.
- Ball valves are not recommended in fluid hygienic piping systems. However, if used, specify full-port, three-piece ball valves.
- Plate-and-frame type heat exchangers are difficult to clean-in-place (CIP) and sterilize-in-place (SIP).

Clean-in-Place (CIP) System Design Considerations

An automated CIP system can successfully clean pipelines, tanks, and other process equipment. It is important that the CIP system be designed simultaneously with the process system to ensure they are tightly integrated.

Dale Seiberling wrote that these factors influence the design, selection, and application of CIP equipment [12]. Also see Appendix B.

- Required Delivery Flowrate: The required pumping capacity will be determined by the size of the transfer lines and tanks to be cleaned. If a single CIP recirculating unit is applied to clean lines and tanks, the tank CIP requirement for the largest tanks will generally establish the maximum delivery rate.
- Delivery Pressure: The CIP supply pump discharge head must exceed the head loss through the longest piping circuit, and supply the sprays in the largest and most distant tank at the required pressure. Pressures at supply pump discharges are normally in the range of 350 to 550 kPa (50 to 80 psig), well within the capability of a sanitary centrifugal pump.
- Required Sequence of Treatment: Nearly all CIP cleaning is accomplished with water-based solutions by a program consisting of;
  (a) A pre-flush with the lowest grade water available, or recovered solution.
  (b) An alkaline solution wash at a variety of time and temperature combinations.
  (c) A post-rinse with water.
  (d) A recirculated acid rinse, generally at ambient temperature, to neutralize final traces of the alkaline wash.

Subsequent pure water rinses may be required to achieve the desired removal of all traces of chemical from the equipment surface.

- Number of CIP Tanks Required: All of the above sequences can be accomplished from a single tank, of as little as $2.5\,m^3$ (60 gallons) capacity, if the water supplies are adequate to meet the above defined delivery requirement for prolonged periods of time. However, if a DI water, WFI water, RO water, or other type of purified water supply is substantially lower than the CIP pump delivery rate, then the CIP unit must contain a solution recirculation tank and, in addition, one or more tanks for the required forms of high quality water, these tanks being sized to fill at low rates and empty at high rates for the required duration to complete all flushes and rinses.
- Delivery Temperature: Flushing (pre-rinsing) of most organic fat, carbohydrate or proteinaceous nature soil is generally accomplished with water at ambient temperature, or below $46\,°C$ ($115\,°F$), to avoid "cooking" or setting the soil on the equipment surface. Heating the solution to final cleaning temperature can be easily accomplished during recirculation for chemical feed purposes via a shell-and-tube heat exchanger. Most CIP cleaning will be accomplished at temperatures between $57\,°C$ and $80\,°C$ ($135\,°F$ and $175\,°F$), though hot water sanitizing may require the delivery of water at $88\,°C$ to $91\,°C$ ($190°$ to $195\,°F$). Shell-and-tube heat exchangers and steam lines must be insulated for safety reasons. It is desirable to avoid insulation of CIP tanks, solution lines, pumps and valves, and assure employee safety by controlling personnel access to the facilities during CIP and design of the facilities.
- Physical Space: The equipment required to meet the above criteria must fit within the available space and provide adequate accessibility for inspection and maintenance.

See Figure 13-1 for an example installation.

Use fixed spray balls to clean vessels, using the chemical action of detergents rather than physical action to effect the cleaning. The rule of thumb is to spray at a rate of 0.25 to $0.75\,m^3/h/m^2$ (0.1 to $0.3\,gpm/ft^2$) of internal surface area with patterns arranged to spray the upper one-third of the tank. Vertical tanks that are free of mixers, baffles, and other internal components are cleaned satisfactorily at a rate of 1.8 to $2.2\,m^3/h/$(m of tank circumference), [2.5 to $3.0\,gpm/$(ft of tank circumference)]. Most spray balls are designed to operate at 170 to $210\,kPa$ (25 to 30 psi) differential pressure [12]. The fixed

Figure 13-1. Two-tank CIP skid. The circulating pump is visible below and behind the electrical panel, and the heat exchanger is located behind the tanks (not seen). The drums contain concentrated cleaning chemicals. Curbs protect the facility in the event of a spill; in this case the curbed area has an open drain to the plant's wastewater collection system.

spray ball has the following advantages compared to rotating spray devices:

- No moving parts
- Constructed completely of stainless steel or other alloys
- Performance not greatly affected by minor variations in supply pressure
- Once installed and validated, will continue to function satisfactorily for a long period of time
- Sprays all surfaces all the time thus reducing the time to clean a vessel

Select sprays, design and install supply/return piping and pumps, and size tank outlets so that all tanks in a given system are cleaned at approximately the same flow-rate and spray supply pressure, or a minimal range of pressure and flow. The volume of water required to rinse a piping circuit is normally found to be 1–½ to 2 times the volume contained in that piping. This may be increased to 4 to 5 times the volume of the circuit to meet pharmaceutical and biotech final rinse test criteria [12].

Seiberling's recommendations are summarized in Table 13-2.

Table 13-2
Seiberling's ten commandments for CIP design [12]

Always remember that water runs downhill...
... and that it is easier to pump water into a tank, than to pump it out
Pitch tank CIP-return connections continuously to pump inlet
Design to close all valves against flow
Avoid 3-port divert valves like the common plague
Keep tank head nozzles few in number, short in length, and large in
    diameter, for they are not easy to clean
Locate CIP systems in (near) and (when possible) beneath the center of
    CIP loads
Eliminate all "dead ends" (branches of more than 1–½ pipe diameters)
    for they will trouble you forever
Pitch all lines to easily opened drain points
Design and install supports to eliminate "friendly" piping that waves when
    starting pumps and opening and closing valves, for friendly systems
    are short-lived

Sterilize-in-Place (SIP) System Design Considerations

Optimal sterilization systems:

- Using steam, quick heat-up of all points in the sterile boundary to 121.1 °C (250 °F), with a minimum of 15 minutes moist heat sterilization time.
- Free drainage of condensate
- Easy displacement of air. Steam is lighter than air; both low and high point vents may be needed.
- Replace collapsing steam with sterile air. Collapsing steam creates vacuum which must be avoided to protect equipment from damage and to prevent leakage through the sterile boundary into the system.

Process piping connected to sterilized equipment must be sterilized up to and through the closest valve that isolates the sterile from the non-sterile system. Another way to isolate the sterile system is through an appropriate 0.2-micron rated, sterilizing grade, filter. Other sterile boundaries include vessel walls themselves, mechanical seals (subject to pressure gradient), feed nozzles, internal cooling coils, rupture discs, steam traps, exhaust lines, and O-rings (elastomers) on instrument ports.

Use pure saturated steam that is not diluted with air or other gases. One reason superheated steam is undesirable for sterilization is because it has further to cool in order for it to transfer its heat of vaporization, making it less efficient than saturated steam.

Systems must be designed for quick and complete drainage to a low point where a steam trap is installed. Air must be completely displaced by saturated steam for sterilization to be effective. Air must be either pulled by vacuum or displaced effectively by the steam itself. Typically, air is discharged through a sterilizing filter. Recognize that air is denser than steam when both are at the same temperature – air vents should be provided at *low points* (the same locations as condensate drains).

Temperature measurement must include the coldest spot in the system to ensure that all points are held above sterilization temperatures. Redundant temperature measurement is essential to verify that sterilization temperatures are maintained.

Disposable Systems

Disposable bioreactors, buffer make-up and hold tanks, and attachments are gaining in popularity because they can reduce risk of cross-contamination between batches while providing flexibility, minimizing turn-around time, reducing cleaning costs, and easing validation restrictions. Additionally, disposables typically have fewer connections (sterile boundary points) than fixed vessels which provide incrementally better FG protection.

Generally, a disposable vessel is compared to a fixed stainless steel tank using the parameters listed in Table 13-3. While it's feasible to create a side-by-side cost comparison, the other parameters may take precedence. In particular, you must confirm that the materials used in the disposable system are compatible with the process. See Reference [4] for details.

Disposable System Economic Evaluation

If the qualitative parameters in Table 13-3 favor single-use technology, perform an economic comparison with fixed systems. Use a life-cycle approach that accounts for first costs and operating expenses. Table 13-4 outlines a typical economic comparison; the values provided are illustrative but are not intended for use in a real evaluation which requires analysis and input from equipment vendors.

Open or Closed?

A sterile boundary is clearly defined when the system is contained within vessels and pipes. Where a process can be conducted entirely within a closed system, no contamination should be possible. The CIP and SIP

Table 13-3
Comparison of disposable vessel with fixed stainless steel tank system.
Details vary by installation

Parameter	Disposable	Fixed Stainless Steel
Scope	Disposable plastic bag in a fixed frame. May include disposable internal elements such as an agitator impeller or sparger. Disposable plastic tubing for filling and emptying. Disposable filters.	Polished stainless steel tank with hygienic agitator and stainless steel filter housings, piping, pumps, etc. May be jacketed or plain.
Size	Readily available to 2,500 liters	Unlimited
Use / Function	Extremely flexible to adapt to changing process demands. Single use usually means disposing the plastic bag between batches, or at intervals corresponding to cleaning intervals for the fixed tank alternative. Limited heat transfer and very little pressure capability (partially depends on the design of the fixed frame). Excellent agitation achievable. Limited number of nozzles or instrument connections; limited variety of compatible instrument sensors. Bags are pre-sterilized (often by irradiation); if process sterility is required then appropriate environmental controls are needed when the bag is installed (see "Open or Closed" section). Bag replacement time might be shorter than CIP/SIP time for fixed tank.	Extremely flexible when initially designed to meet process demands, but may be difficult and expensive to change later if the process requirements change. Heat transfer jackets and internal coils provide aggressive heating and/or cooling capability. Pressure and full vacuum capability. Excellent agitation achievable. Wide variety of nozzle configurations and instrument selections available. Can be designed and operated as a completely "closed" system that requires little or no external environmental control.
Cleaning	Not required because the bags and appurtenances are clean and sterile when purchased	Recirculating CIP system usually used, requiring large quantity of water and 30 to 60 minutes of time. SIP uses clean steam and may require several hours including drying the tank after sanitization is complete.
Quality Control	The end user relies on the manufacturer of the disposable items for QC	QC is the end user's responsibility
Availability	Supply of disposable components is subject to interruption	Fixed tank systems are on site; risk of interruptions from maintenance issues
Validation	Plastic components must be tested for leachable compounds, and their effect on drug product evaluated. Standard equipment qualification protocols (Installation Qualification and Operational Qualification) required. No cleaning validation.	Standard equipment qualification protocols (Installation Qualification and Operational Qualification) required. Cleaning validation and clean "hold time" studies required.
Waste	Disposal of used plastic bags and components, either by incineration or landfill.	Wastewater, some with chemical composition requiring treatment, may be copious. Sinclair estimates that disposable systems consume 87% less water than fixed stainless steel systems [14].

concepts discussed above assume a closed system. PDA defines a "closed system" as follows [11]:

- Is constructed, installed and qualified in a manner which demonstrates integrity is maintained throughout the full range of operating conditions, and over a time period inclusive of the longest expected usage (i.e., manufacturing campaign). The qualification is done according to a formal protocol, following generally accepted engineering principles, and is documented.

- Is sterilized-in-place or sterilized while closed prior to use using a validated procedure.
- Can be utilized for its intended purpose without compromising the integrity of the system.
- Can be adapted for fluid transfers in and/or out while maintaining asepsis.
- Is connectable to other closed systems while maintaining integrity of all closed systems (e.g., Rapid Transfer Port, steamed connection, etc.).
- Is safeguarded from any loss of integrity by scheduled preventive maintenance.

Table 13-4
Example economic comparison between disposable (single use) and fixed stainless steel 750 liter buffer hold/feed tank

Cost Element	Disposable	Fixed Stainless Steel
Capital costs	Jacketed frame to hold disposable bag. Piping to facility heating/cooling system and compressed air. Piping to vent and drain. Minimal instrumentation (in addition to instruments that are provided with the frame). Power supply. Network connection to Process Automation System. Assumes portable or disposable vessels upstream and downstream with no fixed piping. No CIP system required. $150,000	Stainless steel jacketed vessel, hygienic design. Sanitary agitator, bottom mounted, magnetic drive. Ancillary equipment including vent filter. CIP system (packaged skid) Piping to facility heating/cooling system and compressed air. Piping to vent and drain. Piping to CIP system and upstream/downstream process vessels. Instrumented and tied to Process Automation System. $550,000
Engineering, construction management, start-up and qualification	$35,000	$180,000
Operating costs	Depreciation and maintenance. Company policy determines if engineering and other indirect costs are depreciated. At straight-line over 7 years: $26,400 /yr	Depreciation and maintenance. Company policy determines if engineering and other indirect costs are depreciated. At straight-line over 7 years: $104,000 /yr
	Consumables – purchase of disposable components $2,500 /batch	Consumables – purchase of disposable filter cartridges $300 /batch
	Water for cleaning $0	Water for cleaning $1,000 /batch See Note 1
	Disposal of single-use components see Note 2	Wastewater treatment for CIP discharge $2 /batch

Note 1: The cost for Purified Water or Water for Injection (WFI) varies widely among pharmaceutical manufacturers, depending on factors such as energy cost, size of the purification system, and whether total life cycle cost is considered or just the incremental cost for producing a liter of water. Cost per liter of WFI can range from $0.05 to $1.50.
Note 2: Disposal costs are usually based on weight and depend on the disposal method (landfill or incineration) and hazard classification.
Note 3: Operating labor may also be considered in an evaluation but should normally be limited to differences in head count rather than calculated from operating hours.
Note that the costs listed are illustrative; each installation must be evaluated based on its individual details and assumptions (US$ in 2012).

- Utilizes sterilizing filters that are integrity tested and traceable to each product lot for sterilization of process streams.

An "open" system is one that lacks one or more of the features listed above. If the product requires sterility, it's necessary to

1) return a system to a "closed" state and conduct CIP/SIP prior to introducing product,
2) treat the surrounding environment as part of the system and provide measures to ensure appropriate cleanliness, or

3) post-sterilize the product after processing, typically with filtration or heat (e.g., pasteurization).

This is the critical concept that drives biopharmaceutical facilities to classify (or not classify) the cleanliness of the surrounding environment. The environment has a boundary, analogous to the sterile boundary of a specific system, that may be as large as the room housing the process or as small as a glove box attached to a vessel opening. A good starting place for further information on this topic is The International Society for Pharmaceutical Engineering (ISPE) at www.ispe.org.

Contamination Root Causes

A FG is essentially a failure:

- Of the process to either kill all adventitious organisms at the start through the sterilization process, or

- To successfully keep the process isolated from outside invaders

The system fault can be grouped into one (or several) of the following: design and sterile boundary, equipment, or human error.

Design and Sterile Boundary Faults

Any breach of, or migration across, a system's sterile boundary has the potential to bring FG into the system. Therefore, the first goal in protecting a system from FG is to know the boundary. Every element of the sterile boundary must be understood, including how it might change over time. An imperfect sterile boundary condition – for example, a small crack in a weld – might maintain sterility if the pressure differential is always favorable, but if the pressure equalizes, or if vacuum were to develop in the process at some point, even for an instant, then sterility will no longer be maintained.

Leaks and defects are a natural consequence of inevitable system decay. Identifying and eliminating leaks is a continuous challenge; a proactive leak detection program where systems are periodically inspected and leaks repaired prior to operation is essential for successful pure culture capability. System pressure checks and light gas (hydrogen, helium) checks are standard tools to check for leaks in the sterile boundary.

Equipment Faults

Equipment faults as a source of foreign organisms may simply be the mechanism by which a sterile boundary lead develops, such as a sterilizing filter flaw, a weld defect, or an imperfect O-ring. In addition, as equipment ages, faults and defects begin to arise that could compromise sterility. Also, over time the system might decay in subtle ways, creating equipment defects that could alter the dynamics of the system to create sterilization issues. See Table 13-5 for examples of equipment faults.

Human Error (Procedures, Execution)

PM (Preventative Maintenance) is clearly essential to maintaining reliability, safety, and pure culture capability. However, PM work itself is only half of the strength; the other half is designating an effective PM *schedule*: understanding the system enough to determine both what needs to be done and also the correct frequency with which it needs to be completed. See Table 13-6 for some examples.

Table 13-5
Examples of equipment faults

Fault	Consequences
Debris in spray ball	Cleaning patterns altered, allowing media hold-up
Weld defect	Gap, hole, or crack can harbor pocket of unsterilized FG
Incomplete draining	Small patches of media accumulates in vessel. Over time it can become insulating and prevent heat penetration during SIP. Eventually, it can harbor FG.
Bolted or screwed connections in vessel	Bolts, screws, and washers will occasionally and unpredictably loosen from repeated heating and cooling cycles, creating pockets and crevices for environmental bacterial contaminants to fester

Table 13-6
Examples of preventative maintenance practices to maintain sterile boundaries

Boundary Element	PM Practices
Steam traps	For every batch, inspect critical sterile boundary traps to ensure they are set up properly, then temperature check (via adjacent temperature sensors or a temperature stick) the trap to ensure it is functioning.
	For traps that are unused for an extended period of time, pay special attention upon restart. Consider running a sterile hold test after long periods of facility idle time.
	PM plan should specify replacing critical traps at regular intervals due to the potential cost (in lost product) of "running to failure."
Valves	Diaphragms in diaphragm valves will wear out over time so it is essential to inspect and replace them on a periodic basis. Timing should be based on frequency of use and exposure to high temperatures.
	Ball and socket surfaces in ball valves can develop defects that could harbor FG. Inspect periodically and replace when wear is observed.
Vessel	Inspect the internals of vessels by experienced sterility experts (in addition to vessel experts) to ensure they remain free of corrosion and defects that could cause media hold-up and eventually lead to a FG.
Elastomers	Replace regularly to ensure they don't wear or crack in service. A "run to failure" strategy is not recommended.

In an ideal world, procedures and batch records would be completely objective and able to be followed in a standard, repeatable way, every time. However, in the real world, some operational steps require a certain manual technique gained through experience or coaching in order to be performed optimally.

GMP operations require a formal change control process to ensure that product safety, identity, strength, purity, and quality, and process safety are not negatively impacted by process or equipment alterations. It is equally important to carefully scrutinize the impact if bioprocess operations, such as specific steps in a batch procedure, are changed.

Contamination Investigation and Recovery

The first step in troubleshooting a FG event is to determine if there are any abnormalities observed in process operation. Adverse trends can often suggest where the FG event originated. Conversely, FG or sterile boundary flaws also can be non-detectable by continuous process monitoring measurements. The worst luck of all is to have a system that fails intermittently or in some non-repeatable pattern.

Table 13-7 provides a sample checklist for systematically attacking a FG investigation. The key actions contained in the checklist are discussed below.

Table 13-7
Foreign growth investigation checklist [5]

If possible, as soon as foreign growth is detected, examine on-run condition of process (vessels, feed tanks, headers, etc.)
   ✔ Valves set-up properly
   ✔ Steam traps set-up properly and sufficiently hot
   ✔ Leaks
Isolate and identify foreign organisms
Gather relevant process data, including tank/process history
Check automation/computer profiles of feed tanks, inoculum, and fermentor
   ✔ Batch plots
   ✔ SIP temperatures, including temperature control valve positions
   ✔ Backpressures
   ✔ Feeds and timing, including feed control valve position if continuously feeding
   ✔ Other process interventions
Review manufacturing batch record and procedures for observations/remarks
Track recent history of facility (environmental monitoring, cleaning, etc.)
Note any equipment or process changes
Identify recent maintenance that has been performed on the system; check work notes for observations
Check for recent process upsets or deviations
Interview operations personnel who set up and monitored process
Integrity check and inspect any process air filters
Leak check tanks, valves, flanges, and piping
Check calibration on temperature probes
Inspect pH, DO, etc. probes (install new probes if applicable)
Inspect rupture discs
Internal vessel inspection
   ✔ Obvious visual defects − initial inspection
   ✔ Other tank defects − conduct a very thorough examination of the tank walls and interior hardware
   ✔ Agitator shaft and seal areas
   ✔ In a tank with older welds: visual inspection, dye penetrant check, flame check, X-ray examination
   ✔ If the tank has internal coils, pressure or leak test
   ✔ If the tank has internal bolted connections, inspect them for hold-up
   ✔ Swab suspect areas and test for organism of interest
Perform SIP cycle, check all areas within the sterile boundaries to ensure areas are heating up to target temperatures, utilizing probes, temperature sticks, IR technology, etc.
Carefully consider changes or shifts to processes and facilities outside of your immediate control (air, water, utilities, media, etc.)
Brainstorm other less likely scenarios with investigation team; follow-up and check off items
Formulate "return to service" strategy

Time is of the Essence

If the FG event is detected while the process is operating (rather than being discovered during a post-production analytical contamination test), it is critical to inspect the "on-run" condition of the process, including feed tanks, seed vessels, bioreactors, headers, valves, etc. Look for any set-up faults, unusual observations, leaks or other upsets, process alarms, cold spots, or visual faults.

Go for Data

The next step is generally to capture as much data as possible about the process and FG, such as:

- Age of FG
- Pattern of FG
- Any recent changes or unusual observations
- Identity of FG
- History of vessels
- Recent audits, inspections, and environmental monitoring data
- Utility upsets
- Foaming issues

Widen the Search

Investigate automation and process profiles, including batch plots, sterilization temperatures, control valve positions, back pressures, feed flows and timing, any process interventions, or unusual previous metabolic trends.

Performing post pressure checks or more sensitive checks with light gases (hydrogen or helium) will help to locate leaks that may have appeared at SIP or on-run.

A standard industrial grade mix of five % hydrogen in nitrogen is used for modern leak detecting. This mix is inexpensive, non-flammable, easily available, and still holds the important features needed for using hydrogen as a trace gas.

Get Hands-On

Walk the system; look for anything out of the ordinary, including: leaks, cold spots, steam traps set up incorrectly, incorrect connections, etc. Inspect and integrity-check air filters. Remove (and check calibration on) temperature probes, pH, DO, etc.

Time to Open Up

If efforts to locate the root cause from the above actions are not successful, a more thorough internal inspection process may be warranted. Internal inspections should include any of the following relevant checks: obvious visual defects, subtle defects detected by a very thorough examination of all wall and internals surfaces, agitator and shaft vulnerabilities, weld defects or hold-ups, coil leaks, and bolted connection defects.

Experimental

This is hypothesis and scenario testing. For instance, if the sterilization process is suspected, conduct sterilization runs in which as much of the area within the sterile boundary as possible can be checked to ensure it is meeting the minimum temperature.

Microbiology Investigation

By definition, FG testing is screening for the presence of a small population of unknown organisms within a high background of known organisms. It is essential to have a method for isolating the foreign organism so it can be identified and evaluated. Selective broths and agars are indispensable in this activity.

Once the foreign organism is isolated, appropriate identification testing should be completed. Biochemical and/or genetic identification methods are useful in comparing one isolated organism to another to confirm a potential common source. Confirmation of the contaminating organism as genetically identical within the limits of the method may be helpful in focusing the investigation on a common system or, alternatively, focusing the investigation on independent root causes.

Risk Assessment

Risk assessments – formal or informal – are used to help engineers identify and understand design features that could affect product quality. *Risk* is defined as the combination of the probability of occurrence of *harm* and the *severity* of that harm [6]. By systematically evaluating the elements comprising a project, an interdisciplinary team can characterize the risks and assign numerical rankings. With a pre-defined threshold, the numerical rankings inform the team which risks are acceptable and which ones require reduction.

An internationally accepted risk assessment model is presented by ICH [6], pictured in Figure 13-2. There are several key aspects to the model:

- Potential risks to patient safety are identified, analyzed, and evaluated by an interdisciplinary team

that includes experts in appropriate areas, such as the quality unit, engineering, manufacturing, regulatory affairs, legal, and sales. There are many possible sources for the potential *harm* to patients: raw materials not meeting specification, failures in engineered systems, manufacturing defects, computerized data corruption, labeling errors, product adulteration while it is in the supply chain, etc. Therefore, comprehensive assessments may require multiple teams who evaluate specific aspects of the manufacturing process. *Harm* includes potential damage that can occur from loss of product quality or availability.

- Either the probability of occurrence or the severity of the potential harm may be reduced through the use of *risk control mechanisms,* such as automated rejection of product that fails an on-line test, elimination of a dangerous raw material, or enhanced operating procedures that ensure a more robust production environment.
- The assessment should be an open process that is shared with stakeholders throughout the organization. Good two-way communication is important, so that potential risks identified by people outside the assessment teams can be captured and evaluated.
- Risk assessments are ongoing activities, which are initiated with a new project and updated when information is developed or changes. If a plant experiences a problem that results in a product recall or customer complaint, part of the resolution to the problem is to evaluate the risk assessment and update it if warranted.
- A variety of tools are available to support the risk assessment process. Due to the use of interdisciplinary teams who may not have established working relationships, facilitated review sessions are usually advised. Formal qualitative tools such as FMEA and HazOp are effective for risk identification. Quantitative statistical approaches can be used if historical data is available for the process being evaluated.

The degree of rigor and formality of the quality risk management should reflect available knowledge and be commensurate with the complexity and/or criticality of the issue being addressed. Just as with safety assessments, risk

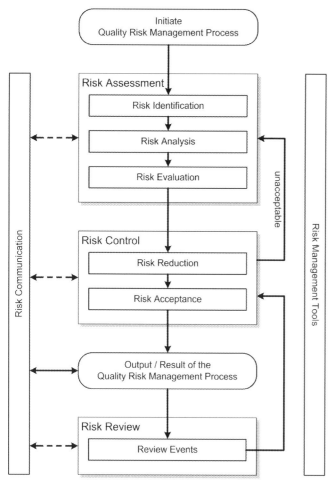

Figure 13-2. Overview of a typical quality risk management process [6].

assessments are often layered with simple overview reviews conducted early in the project using checklists or what-if tools, with additional reviews performed when further information is available. The additional reviews may be prioritized based on the results from the early work.

Validation

Biopharmaceutical facilities must be "validated" to ensure the equipment and processes reproducibly manufacture drug products that meet established criteria for identity, purity, strength, activity, and quality. The pharmaceutical industry, with cooperation from regulatory agencies, has a highly developed – but changing – paradigm for validation.

FDA's new definition for process validation is [3]:

The collection and evaluation of data, from the process design stage throughout production, which establishes scientific evidence that a process is capable of consistently delivering quality products.

They define a 3-stage product lifecycle that emphasizes a statistical, risk-based approach.

- Stage 1 – Process Design: The commercial process is defined during this stage based on knowledge gained through development and scale-up activities.
- Stage 2 – Process Qualification (PQ): During this stage, the process design is confirmed as being capable of reproducible commercial manufacturing.
- Stage 3 – Continued Process Verification: Ongoing assurance is gained during routine production that the process remains in a state of control.

Stage 1 constitutes the science behind the drug's manufacturing process. Using DOE (Design of Experiments) and statistical analyses, data is collected and evaluated to demonstrate the factors that affect the product's attributes. Stage 1 includes establishment of a strategy for process control for each unit operation and the overall process.

In Stage 2, the facility is designed, constructed, and qualified. A verification activity precedes manufacturing confirming that systems and equipment are built and installed in compliance with design specifications and that they operate in accordance with process requirements throughout their anticipated operating ranges. ISPE's *Baseline Guide* provides recommendations for successful qualification [7]. Other important references are ICH Q9 [6], ASTM E2500-07 [2]), GAMP®5 [9], and ISPE's guide to a science and risk based approach for delivery of systems and equipment [8].

Stage 2 also covers the initial manufacturing of product with the qualified facility. PQ (process qualification):

combines the actual facility, utilities, equipment (each now qualified), and the trained personnel with the commercial manufacturing process, control procedures, and components to produce commercial batches [3].

Bringing a new drug to market culminates with PQ, and FDA expects that manufacturers use the cumulative data from all relevant studies to establish manufacturing conditions such as operating ranges, set points, and process controls.

Stage 3 is an ongoing effort continually assuring that the process remains in a state of control during commercial manufacture. A program to collect and analyze data is required, and FDA expects that manufacturers use statistical methods to detect deviations from normal operation.

References

[1] American Society of Mechanical Engineers (ASME), Bioprocess Equipment, ASME BPE-2009.

[2] ASTM Standard E2500-07. Standard Guide for Specification, Design, and Verification of Pharmaceutical and Biopharmaceutical Manufacturing Systems and Equipment. West Conshohocken, PA: ASTM International, www.astm.org; 2007. doi: 10.1520/E2500-07.

[3] Food and Drug Administration (FDA). Process Validation: General Principles and Practices. Guidance for Industry, Draft, November, 2008.

[4] Goldstein A, Schieche D, Harter J, Samavedam R, Wilkinson L, Manocchi A. Disposable Systems in

Biopharm Production – Design, Compliance, and Validation Methods. *American Pharmaceutical Review*, September/October, 2005.

[5] Hines M, Holmes C, Schad R. Simple Strategies to Improve Bioprocess Pure Culture Processing. *Pharmaceutical Engineering*, May/June, 2010;30(3).

[6] International Conference on Harmonisation of Technical Requirements for Registration of Pharmaceuticals for Human Use (ICH), Guidance for Industry: Q9 Quality Risk Management, November, 2005.

[7] International Society for Pharmaceutical Engineering (ISPE). *Baseline Guide Volume 5: Commissioning and Qualification*, www.ispe.org; March, 2001.

[8] International Society for Pharmaceutical Engineering (ISPE). *Baseline Guide Volume 7: Risk-Based Manufacture of Pharmaceutical Products (Risk-MaPP)*, www.ispe.org; September, 2010.

[9] International Society for Pharmaceutical Engineering (ISPE). *GAMP® 5: A Risk Based Approach to Compliant GxP Computerized Systems*, www.ispe.org; February, 2008.

[10] Personal communication with Dale Seiberling.

[11] Parenteral Drug Association (PDA). Process Simulation Testing for Sterile Bulk Pharmaceutical Chemicals. Technical Report No. 28, www.pda.org; 2006.

[12] Seiberling D, CIP Evangelist, www.seiberling4cip.com.

[13] Seiberling D. *Clean-in-Place for Biopharmaceutical Processes*. Informa Healthcare; 2007.

[14] Sinclair A, Leveen L, Monge M, Lim J, Cox S. The Environmental Impact of Disposable Technologies. BioPharm International.com; November 2, 2008.

14

Vacuum Systems

Introduction

Vacuum systems are important in all segments of the chemical process industry. The major types of vacuum devices are ejectors, liquid-ring pumps, and dry pumps. Multiple devices are often combined to achieve vacuum goals.

Chemical engineers have three primary challenges when designing vacuum systems:

1) specification and selection of the vacuum source,

2) piping and installation details, and

3) instrumentation and controls.

Many standard practices apply to the systems, but each application should be evaluated on its merits to avoid potential problems with performance, reliability, safety, and emissions.

Vacuum Units

Vacuum performance is specified in terms of pressure and flow.

Pressure is either absolute or relative to atmospheric pressure. It's an important distinction, especially when the discharge from the vacuum equipment is to a pressure other than atmospheric pressure at sea level. Vacuum pumps provide a pressure differential, and if left unstated it is assumed that the discharge is to an absolute pressure of 760 mm Hg, or 29.92 inches of Hg. However, if the pump is located at high altitude, or if there is back pressure on the discharge, then this assumption becomes invalid.

Absolute pressure is usually given as kPa abs, psia, Torr, or mm Hg. This is a fixed pressure that is appropriate for chemical processes involving vapor-liquid equilibria.

Vacuum is usually given in feet of water or inches of mercury and is relative to the actual barometric pressure at the location of the plant. The relationship between absolute pressure and vacuum is:

Absolute Pressure = Actual Barometric Pressure
$$- \text{Vacuum}$$

This equation relates altitude to atmospheric pressure:

$$P = 101.3 \left(1 - 2.256 \, (10)^{-5} \, h\right)^{5.256} \text{ SI Units, or}$$

(14-1)

$$P = 14.696 \left(1 - 6.876 \, (10)^{-6} \, h\right)^{5.256} \text{ US Units}$$

(14-2)

Where:

P = atmospheric pressure, kPa or psia
h = altitude, m or ft

Conversion between Mass Flow, Actual Volume Flow, and Standard Volume Flow

In most of the world, standard conditions are defined at 0 °C and 1.013 bar (101.3 kPa) pressure. Standard volumes are 22.1 liters per g-mol (22.1 m^3 per kg-mol) or 359 ft^3 per lb-mol.

In the US, standard conditions are usually defined as 60 °F and 14.696 psia. The standard volume is 379 ft^3 per lb-mol.

Conversion between standard volume flow, actual volume flow, and mass flow requires the molecular weight of the gas, its temperature, and its pressure.

Example

A vacuum pump is rated for 200 acfm at 68 °F and 100 mg Hg. Calculate the scfm where standard temperature is 0 °C, and calculate the pump's capacity in lb/h of dry air. The molecular weight of air is 29 lb per lb-mol.

$$scfm = 200 \, acfm \left(\frac{100 \, mmHg}{760 \, mmHg}\right) \left(\frac{460 + 32}{460 + 68}\right)$$

$$= 24.5 \, scfm$$

$$mass\ flow = 24.9\ scfm \left(\frac{1\ lbmol}{359\ scf} \right) \left(\frac{29\ lb}{lbmol} \right)$$

$$= 1.98\ lb\!/\!{}_{min}$$

Comparison of Vacuum Equipment

Table 14-1 compares each general type of vacuum equipment. Each of the major types of device is available in several alternative formats. Some of the alternatives are discussed in the corresponding sections in this chapter.

Table 14-1
General comparison of vacuum source equipment

Characteristic	Ejector	Liquid-Ring	Dry
Ultimate (shut-off) pressure, Pa (mm Hg) absolute	0.13 to 6,700 (0.001 to 50) depending on number of stages	130 to 6,700 (1 (oil-sealed) to 50) (single-stage water sealed)	6.7 to 67 (0.05 to 0.5)
Typical practical limit for process applications, Pa (mm Hg) absolute	0.40 to 10,000 (0.003 to 75)	1,300 to 6,700 (10 to 50)	13 to 200 (0.1 to 1.5)
Single-unit capacity range, m³/h (ft³/min)	17 to 1,700,000 (10 to 1,000,000)	5 to 30,000 (3 to 18,000)	85 to 2,400 (50 to 1,400)
Process application	Suitable for any process gases that are compatible with water, and do not pose an environmental discharge concern.	Suitable for most process gases because the seal liquid can be chosen for its process compatibility	Especially advantageous for clean processes that cannot withstand contamination from plant steam or seal liquids. Not suited if the process gas has an autoignition temperature above 200 °C (390 °F)
Major advantages	Relatively low capital cost Simple to maintain Extremely reliable	Robust, proven technology Little increase in temperature of the discharge gas No damage from liquids or small particulates Fabricated from any castable metal Easy maintenance Quiet	Energy efficient No liquid seal fluid Rotor does not contact the casing No contamination of process gases, so the vapors can be recovered using downstream condensation Corrosive gases handled — if they don't condense in the pump Quiet
Major disadvantages	Designed to operate at a specific optimum point; deviation results in significant performance deterioration Water disposal Installation cost might be significant, when considering need for barometric seals, steam piping, and possibly boiler capacity Load specific; very sensitive to variations in process conditions and pressure Often run continuously, even if the process requires intermittent vacuum	Relatively high energy consumption Pumps are often oversized to compensate for possible seal liquid temperature variation Waste disposal Process gas mixes with seal liquid (can also be an advantage) Vacuum is limited by the vapor pressure of the seal liquid	Higher operating temperature Cannot handle particulates or liquid slugs Most difficult to repair Relatively high capital cost Noisy; may require a silencer Risk of explosion if flammable vapors are present; can be mitigated with flame arrestors and oxygen control

Ryans and Bays compared the thermal efficiency of ejectors, liquid ring pumps, and three types of dry pumps [5]. When comparing them on the basis of power consumed for operating pressures from 5 mm Hg to 600 mm Hg, they found that dry pumps are 3 to 10 times more efficient than ejectors. However, electric energy might cost 3 to 6 times more than steam energy, so they recommend that if a project is proposed to replace steam jet ejectors with dry pumps on the basis of energy savings, a careful analysis must be done. Dry pumps are significantly more efficient than liquid ring pumps across the range from 1 to 50 mm Hg.

Vacuum Jets

Steam jet ejectors are simple, low cost, and reliable devices that can draw vacuums to 0.003 mm Hg absolute, or less. (See Table 14-2) They should be considered whenever a steady source of low pressure steam is available, especially when the steam is not needed for other uses. Jet ejector vendors will design the system using their proprietary computer modeling tools. It is up to the process engineer to evaluate whether this is appropriate technology for the purpose, and to provide process data to the vendor.

Here is a list of the typical data needed to select and size a steam ejector system:

1. General process description. This will orient the vendor and help define general boundaries. Provide the location and altitude as well as service conditions such as weather protection or environmental controls.
2. Fluid composition and properties. Specify the expected range of fluid composition at the inlet to the ejector as well as the steady state composition. For each component, provide physical properties unless they are compounds found in standard databases. Identify expected physical state (gas, liquid, solid).
3. Fluid flow rate, temperature, and pressure. Specify the range and normal expected values. Think about the driving force that provides the flow. If the ejector is intended to reduce the pressure in a vessel, for example, flow will be generated by the ejector's vacuum and it will come initially from the headspace of the vessel but later from leaks into the system. However, supplementary material may be sourced from evaporation in the vessel or a small purge flow of nitrogen directed into the headspace.
4. Evacuation requirements. If the system must be evacuated within a certain time, provide the system volume, its initial conditions, and the maximum time allowed for evacuation. In some cases a second ejector may be required, piped in parallel, for the initial evacuation.
5. Available steam pressure and quality. State the minimum steam pressure and whether it is dry, saturated, or superheated (with the degree of superheat). Although motive steam pressure as low as 35 kPag (5 psig) can pull a vacuum, higher pressures of 100 kPag to 350 kPag (15 psig to 50 psig) are more practical [6].
6. Materials of construction for the three primary parts: steam nozzle, body, and venture tail (or diffuser).
7. Condensing requirements. Multi-stage ejectors are more efficient with interstage condensers. Surface condensers or direct contact condensers may be specified if there is a preference, but shell-and-tube surface condensers are most common with cooling water on the tube side. Condensers are preferably installed with a barometric seal leg (i.e., located more than 10 m, 34 ft, above the condensate receiver for 100% water condensate). After-condensers are optional, and are used to condense the operating steam and any other condensable vapors before discharging non-condensables to

Table 14-2
Typical application of single and multi-stage steam jet ejectors [1]

Ejector Stages	Suction Pressure	
	Pa, Absolute	mm Hg, Absolute
1	100,000 to 3,300	735 to 25
2	16,600 to 400	125 to 3
3	3,500 to 100	26 to 0.8
4	530 to 10	4 to 0.075
5	50 to 1.3	0.4 to 0.010
6	13 to 0.4	0.1 to 0.003

atmosphere or a vent collection system. Specify the temperature of cooling water available for condensing (see Table 14.4).

8. Electrical power (volts, phases) available and electrical area classification.

9. Instrumentation requirements.

10. Backpressure conditions. If the final stage will operate against a backpressure, rather than discharging to atmosphere, specify the pressure.

Birgenheier, et al. provided guidance for design and installation of steam jet ejectors and accessories [2]. See Table 14-3.

Table 14-3
Design and installation of steam jet ejectors [2]

Design Aspect	Recommendations
Ejector installation angle	Any angle is acceptable, but low points in the vacuum piping must be avoided to keep condensate and any entrained solids from collecting
Drainage	Ejector bodies must have provisions for draining condensed steam or process vapors. Manual or automatic drains may be used. Install at low points. Drain batch systems between batches; continuously operating systems may be drained periodically during operation.
Supports	Steam jet ejectors are not intended to support associated piping systems. External loads on an ejector could cause misalignment which can adversely affect performance. Consider both mechanical and thermal stresses.
Cleaning	Provide space and access for cleaning the inside of the ejectors, especially if there is a potential for solids accumulation
Condenser mounting	Install to allow for complete condensate drainage. Mounting should account for the weight of the condenser if fully flooded. Shell-and-tube condensers may be installed horizontally or vertically and the vapors may be condensed on the tube or shell side. Tailpipes connected to inter- and after-condensers should be run separately to the hotwell to prevent recirculation of non-condensable vapors.
Condenser tailpipes	The tailpipe arrangement is crucial. There should be no horizontal runs of pipe. A vertical tailpipe is ideal, but if necessary sections of 45-deg run are acceptable as long as the first bend occurs more than 5 pipe diameters (or 1.2 m, 4 ft) below the condenser outlet flange.
Hotwell	Tailpipes extend into the hotwell. The hotwell should be sized such that the volume between the hotwell overflow pipe and the bottom of the tailpipe contains at least 1.5 times the volume of the tailpipe, but in no case less than 0.3 m (1 ft). The distance from the bottom of the tailpipe to the bottom of the hotwell should be no less than half the diameter of the tailpipe or 0.3 m (1 ft), whichever is greater.
Vacuum piping	In general, the diameter of the piping from the process unit to the suction of the first stage ejector should be at least as large as the suction connection. If multiple ejectors are piped in parallel to increase capacity, the vacuum pipe cross-sectional area should be at least equal to the sum of the cross-sectional areas of the suction connections on the ejectors. To minimize pressure drop, all piping between the process and the ejectors, and piping between multiple ejector stages, should have as few valves and fittings as possible, and all connections should be as short as possible. Wherever possible, use long-radius elbows; provide drains at low points.
Precondenser	If a precondenser is used, the potential pressure drop across it must be calculated and considered in the overall process design.
Steam	Use dry, saturated steam unless the system specification calls for superheated steam. For dry steam the inlet line should be taken off the top of the steam header. If moisture is present in the steam, a separator and trap should be used to improve the steam quality to better than 99.5%.
Cooling Water	Install a temperature indicator at the cooling water outlet from each condenser; it is used to determine the adequacy of the cooling water flow.

Table 14-4
Comparison of direct contact and surface condensers for steam jet ejector systems

Direct Contact Condensation	Surface Condensation
Lower capital cost	Condensing water and process vapors are kept separate, making recovery of the process condensate more feasible
Seldom or never need cleaning	
Corrosion is usually minimized due to dilution with the condensing water	If a barometric leg is not practicable, the condensate pump for a surface condenser is smaller since it only pumps process condensate and not the condensing water
Condensable vapors with relatively high vapor pressure which are partially soluble in water, such as ammonia, can be more effectively condensed due to diluting effect of condensing water	If the condensate contains corrosive, poisonous, radioactive, or biologically active components, disposal is easier since the condensate does not contain the condensing water

Liquid Ring Vacuum Pumps

Liquid ring pumps are commonly used for steps requiring roughing vacuum (down to about 25 mm Hg absolute pressure). In combination with a blower or ejector, they are also used in systems requiring deeper vacuum, to about 1 mm Hg. The pumps can be purchased "bare" or mounted on a frame with appurtenances for a full featured vacuum source.

Performance is tied to the seal liquid composition and temperature. While water is the most common seal liquid, almost any liquid with low volatility at the service temperature can be used. This flexibility is an important feature of liquid ring pumps, providing an opportunity to remove noxious components from the process stream as it passes through the pump.

Vacuum pump manufacturers will size the pump based on information provided by the user. Published curves may be used for preliminary sizing. However, read the curves carefully, because some manufacturers state performance for "dry air" and others use a "saturated air" basis. The difference is significant. Seal water evaporates to saturate "dry air" which then takes up space in the pump impeller bucket. Less water evaporates into "saturated air." Therefore, for two pumps with identical ratings (x acfm with dry or saturated air), the one with the saturated air rating has a higher capacity.

Here are the process parameters needed to properly size a liquid ring vacuum pump:

1. Pressure at the pump suction. Account for pressure drop through piping from the process unit.
2. Temperature at the pump suction
3. Flow rate at design conditions, usually at steady state after the system pressure is pumped down, expressed in mass or standard volumetric units. The composition of the gas must also be given as mass or volume fractions and the molecular weight of each.

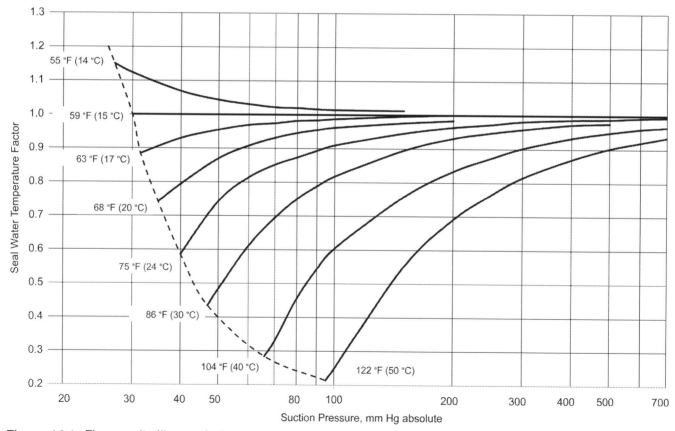

Figure 14-1. The capacity (flow rate) of a single-stage liquid ring vacuum pump, and the ultimate vacuum capability (dotted line), are highly dependent on the temperature of the seal water. (Source: Sterling Fluid Systems).

Include information about the moisture load of the gas (dry or saturated).

4. Vapor pressure data for each component
5. Discharge pressure at the pump
6. Seal fluid data including composition, temperature, and physical properties (specific gravity, specific heat, viscosity, thermal conductivity, molecular weight, and vapor pressure data).

Recommended seal water parameters:

Temperature: 15 °C (60 °F) or colder; pump is derated from published curves if the seal water temperature is higher or the vapor pressure of an alternative seal liquid is higher than that of water at 15 °C
pH: 7 or higher
Chlorides: <10 ppm
Total dissolved solids: <200 ppm
Total hardness: <200 ppm

Correcting for Seal Liquid Temperature and Composition

If the seal liquid temperature differs from 15 °C (60 °F), or if it is not water, use Figure 14-1 to estimate the capacity correction factor for a single-stage liquid ring vacuum pump; for two-stage pumps use Figure 14-2. For seal liquids other than water, the same chart can be used by first determining the temperature at which water has the same vapor pressure as the actual seal liquid at its circulation temperature.

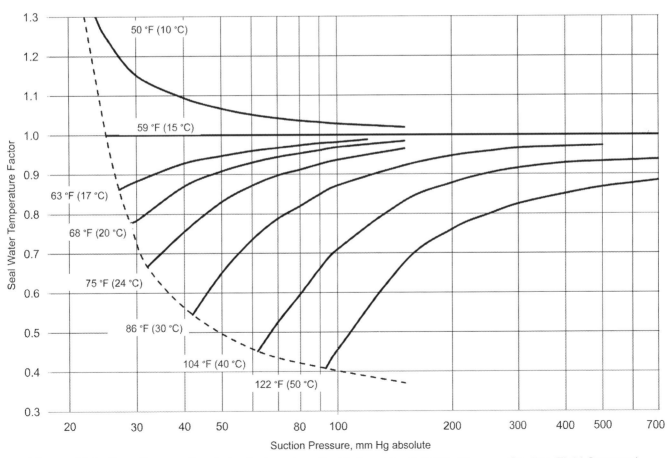

Figure 14-2. Capacity correction factor for two-stage liquid ring vacuum pump. (Source: Sterling Fluid Systems).

Dry Vacuum Pumps

Ryans and Bays have published an excellent review of dry vacuum pumps, with comparisons with liquid ring pumps, steam ejectors, and combination systems [5]. Dry pumps use the operating principles of rotary lobe blowers, claw compressors, or screw compressors. They are usually cast iron or ductile iron due to the very close tolerances that are required. Temperature control, with a water jacket, injection of cooled process gas or nitrogen into the working volume of the pump, or with interstage coolers, is required. They achieve an ultimate vacuum as low as 0.05 mm Hg and capacities in the range of 85 m^3/h to 2,400 m^3/h (50 acfm to 1,400 acfm).

Reasons for choosing a dry vacuum pump include [5]:

- Eliminate process contamination (from liquid-sealed pumps or ejectors)
- Environmental constraints
- Solvent and product recovery applications

Because dry vacuum pumps run hot – the discharge temperature from screw pumps sometimes reaches 350 °C (660 °F) – care must be taken with flammable process vapors. Ryans and Bays recommend that significant caution be employed if vapors have an autoignition temperature that is less than 200 °C, even though pumps are available with internal temperature controlled to less than 135 °C.

If indicated by process conditions, dry vacuum pumps can be protected by installing catch pots in the suction line to capture slugs of liquid. Catch pots are often designed with tangential inlets and demisters that minimize liquid carryover to the pump. Filters, screens, and wet scrubbers can be used to remove solid particles from the suction line.

Typical Jet Systems

Figure 14-3 provides a convenient comparison of capacity and suction pressure ranges for typical commercial steam jet systems handling any non-condensable gas, such as air. Figure 14-3 is based on each of the designs using the same amount of motive steam at 100 psig. Each point on each curve represents a point of maximum efficiency. Therefore, the curves represent a continuum of designs rather than a single one. A single design normally wouldn't operate over the wide ranges shown. For example, a reasonable range for good efficiency would be 50–115% of the design capacity [2 and 4].

Steam Supply

Steam jets require a constant pressure steam supply for best performance. Pressure below design will usually lower performance and pressure above design usually doesn't increase capacity and can even reduce it. Lieberman has recommended an intelligent way to handle the problem of variable steam pressure, which is shown in Figure 14-4. The steam to the jet is controlled at a pressure slightly below the lowest supply pressure so the jet has a constant steam pressure. Lieberman points out that there is hardly any steam penalty in providing this better operating system. For example, a jet designed for 8.5 barg steam will use only about 15% more steam than one designed for 10 barg steam [4].

Measuring Air Leakage

The most reliable way to establish the amount of air leakage to specify for a new vacuum system is to draw on experience with an existing similar system.

To determine the amount of air leakage in an existing system, estimate the total volume of the system. Pull the vacuum to a pressure of approximately 20 kPa absolute

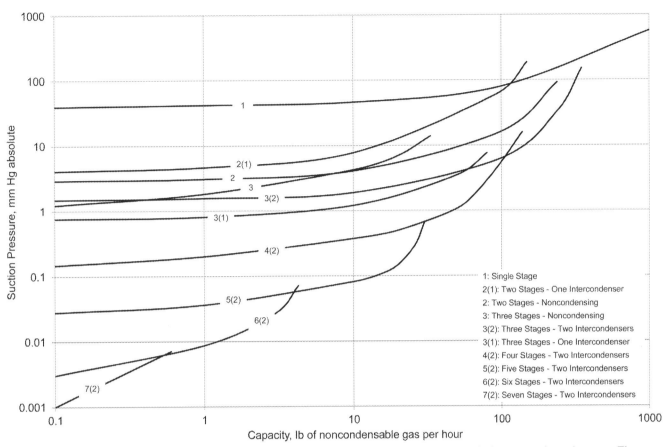

Figure 14-3. A wide range of pressures can be achieved by using various combinations of ejectors and condensers. The same steam consumption is used for each design here. Note: Curves are based on 85 °F condensing water. If warmer water is used, curves shift to the left – cooler water, shift right.

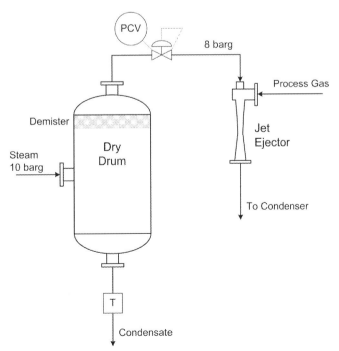

Figure 14-4. Wet steam of variable pressure will ruin a vacuum steam jet's performance.

(150 mm Hg). Then isolate the system at the vacuum source. To ensure that the isolation valve isn't leaking, run the test twice with the vacuum source running and then with the vacuum source turned off. The results should be identical.

Measure the time required for a rise in pressure in the vessel (say 15 kPa or 110 mm Hg). It is essential that the absolute pressure does not rise above 50 kPa absolute (380 mm Hg) during this time. The following formula will then give the leakage:

$$W_L = C\,V(\Delta P)/t \qquad (14\text{-}3)$$

Where:

W_L = leakage, kg/h or lb/h
C = conversion factor, 0.71 for SI or 0.006 for US units
V = system volume, m^3 or ft^3
ΔP = pressure rise, kPa or mm Hg
t = time, min

If the volume of the system is not known, the leakage can still be determined, but two tests are required.

First, run the test described above. Then introduce a known air leak into the system. This can be done by means of a calibrated air orifice. Run a second test with the known air leak, obtaining a new pressure rise and time. The unknown leak is then given by

$$W_l = \frac{W'}{(\Delta P'\,t/\Delta P\,t')} - 1 \qquad (14\text{-}4)$$

Where:

W' = known leak, kg/h or lb/h
$\Delta P'$ = second pressure rise, kPa or mm Hg
t' = second time, min

Rules of thumb for air leakage are as follows. For moderately tight, small chemical processing systems (say 15 m^3 or 500 ft^3), an air capacity of 5 kg/h (10 lb/h) is adequate. For large systems, use 10 kg/h (20 lb/h). For very tight, small systems, an air capacity of 1 to 2.5 kg/h (2 to 5 lb/h) is reasonable.

A reverse philosophy in sizing vacuum equipment is occasionally applicable to a system in which even a small quantity of air leakage will upset the operation or contaminate the product. In such a system it may be desirable to size the vacuum with a deliberately limited air handling capacity, so that the system cannot be operated until an injurious rate of air leakage is corrected.

Time to Evacuate

For a positive displacement vacuum pump, such as liquid ring, evacuation time is estimated with:

$$t = \frac{V}{q\,\ln\!\left(\dfrac{P_0}{P_1}\right)} \qquad (14\text{-}5)$$

Where:

t = time to evacuate the system, seconds
V = volume of the system, m^3 or ft^3
q = volumetric flow rate of the vacuum pump, m^3/s or ft^3/s
P_0 = initial absolute pressure of the system in consistent units
P_1 = final absolute pressure of the system in consistent units

To estimate the time required for an ejector to evacuate a system from atmospheric pressure down to the design pressure, assume that the average air handling capacity during the evacuation period is twice the design air handling capacity. Assume also

that the actual air leakage into the system during the evacuation is negligible. The approximate evacuation time is [3]:

$$T_e = \frac{2.3\,V}{C_a} \qquad (14\text{-}6)$$

Where:

T_e = time to evacuate the system, minutes
V = volume of the system, ft^3
C_a = ejector design air capacity, lb/h

If this approximate evacuation period is too long, it may be shortened by adding a larger last stage to the ejector or by adding a non-condensing ejector in parallel with the primary ejector. One non-condensing ejector may be used as an evacuation and spare ejector serving several adjoining systems. The evacuation performance is specified by indicating the system volume, the desired evacuation time, and the absolute pressure to which the system must be evacuated.

Instrumentation and Controls

Several alternative vacuum control schemes are listed in Table 14-5.

Table 14-5
Alternative vacuum control schemes may be used in combination

Control Scheme	Ejectors	Mechanical Pumps	Comments
Recycle from discharge to suction	√	√	Sizing and installation are critical. Reduces efficiency.
Bleed external fluid (air, steam or other) into suction	√	√	Less reliable. For ejectors, condensable vapor bleed is preferred.
Control valve on suction line to introduce artificial pressure drop	√	Not Recommended	Works best if actual mass flow is greater than 50% of design flow
On-off control of select stages within a multistage ejector	√	Not Applicable	
Variable-speed drive	Not Applicable	√	Most efficient method; for liquid-ring pumps must maintain speed above a minimum, otherwise the liquid ring collapses causing instability

Nomenclature

C_a = ejector design air capacity, lb/h

h = altitude, m or ft

P = atmospheric pressure, kPa or psia

P_0 = initial absolute pressure of the system in consistent units

P_1 = final absolute pressure of the system in consistent units

ΔP = pressure rise, kPa or mm Hg

$\Delta P'$ = second pressure rise, kPa or mm Hg

q = volumetric flow rate of the vacuum pump, m^3/s or ft^3/s

T_e = time to evacuate the system, minutes

t = time to evacuate the system, seconds

V = volume of the system, m^3 or ft^3

W' = known leak, kg/h or lb/h

W_L = leakage, kg/h or lb/h

References

[1] Berkley F. Ejectors Have Wide Range of Uses. *Petroleum Refiner*, December, 1958:95.

[2] Birgenheier D, Butzbach T, Bolt D, Bhatnagar R, Ojala R, Aglitz J. Designing Steam Jet Vacuum Systems. *Chemical Engineering*, July, 1993.

[3] Evans F. *Equipment Design Handbook for Refineries and Chemical Plants*. vol. 1, 2nd ed. Gulf Publishing Co; 1979.

[4] Lieberman N. *Process Design for Reliable Operations*. 2nd ed. Gulf Publishing Co; 1988.

[5] Ryans J, Bays J. Run Clean with Dry Vacuum Pumps. *Chemical Engineering Progress*, October, 2001:32–41.

[6] Schutte, Koerting. Steam Jet Ejectors. Bulletin 5E-H. downloaded from, www.s-k.com March, 2011.

15
Pneumatic Conveying

Introduction

Granular and powdered solids are conveyed in gas streams throughout the chemical industry. This important unit operation, once an art, is now grounded in engineering, with the selection and design of its various components well understood. Simulation tools are available to aid the designer, and the technology has been successfully extended to encompass cohesive, abrasive, and friable materials.

A major difference between pneumatic and hydraulic or slurry conveying of solids is that the carrier gas expands continuously along the pipe length. The flow regime, dense phase or dilute phase, in the pipe depends on the ratio of solids to gas and the particle characteristics.

Pneumatic conveying systems are comprised of four parts:

1. **Prime mover.** Compressors, fans, blowers, and vacuum pumps are used to move the gas stream. The engineer must determine the gas flow rate and pressure (positive or negative) required to convey the process stream.
2. **Feed.** Solids are introduced into the gas stream by vacuum or through specially designed valves, pumps, or blow vessels. The solids must be accelerated rapidly to the conveying velocity, which causes a large pressure drop.
3. **Conveying piping.** The pipe sizing and design (e.g., slopes, bends) must be suitable for the characteristics of the solids and the intended conveying rate(s).

4. **Disengagement.** Solids are removed from the gas stream by cyclone separators or fabric filters. This part of the system includes capturing the solid (into a bin, for example), and disposition of the gas stream (as an emission or recycled for reuse).

Some pros and cons of pneumatic conveying are shown in Table 15.1 below.

Some rules of thumb (from [2] and [7]):

- Pneumatic conveyors are generally more suited to the conveyance of fine particles over shorter distances (up to a few hundred meters, or 1,000 feet).
- The majority of existing systems have capacities within the range of 1,000 kg to 100,000 kg (2,000 lb to 200,000 lb) per hour over distances less than 1,000 m (3,000 ft) with average particle size less than 10 mm (0.4 in).
- Typical conveying velocities, expressed as free air at atmospheric pressure, for low positive pressure systems are: 10 m/s (2000 ft/min) for light materials, 15 to 20 m/s (3000 to 4000 ft/min) for medium density materials, and 25 m/s (5000 ft/min) for high density solids.
- It is often stated that the maximum practicable particle size is 15 mm (0.5 in.), but rocks up to 70 mm (3 in.) size, live chickens, and manufactured parts of unusual geometry have been conveyed.

Table 15-1
Advantages and disadvantages of transporting solids using pneumatic conveying ([2] and [7])

Advantages	Disadvantages
Material can be transported with minimal degradation (with appropriate choice of system)	Product degradation as a result of incorrectly designed system
Little or no exposure of product to the environment	Pipe and component wear
Can transport several thousand feet	Not suitable for long distance (beyond a few thousand feet) due to gas expansion
Excellent for multiple sources and multiple destinations	High power consumption
Ability to transport material which might be air, moisture, etc., sensitive	High skill level required to design, operate, and maintain systems
Compared to mechanical conveyors, relative ease in system routing, especially elevation changes	
Interfaces well with a variety of transportation modes — trucks, railcars, ships	
Highly reliable with few moving parts	
Multiple use: one pipeline can be used for a variety of products	
Security: pipelines can be used to convey high-value products	
Easily automated and controlled	

- The inside diameter of the conveying pipe should be at least three times larger than the largest particle to avoid blockage inside the pipe.
- The ideal candidates for pneumatic conveying are free-flowing, non-abrasive, and non-fibrous materials. However, with the development of new types of conveyors operating at low gas velocities, cohesive, abrasive, and friable materials can be handled.
- It is best to have the most direct route from the feeder to the delivery point, with as few bends as possible.

- Inclined vertical piping can be used as long as the flow is dilute. For denser phase systems large angles, especially between 60° and 80°, can be troublesome, because refluxing can occur.
- The use of more than two bends in close sequence is catastrophic to the smooth operation of the system.
- Conveying capacity is most strongly influenced by the pipe diameter, distance to be conveyed, and the available pressure drop; air flow rate is a secondary function (although it is very important for achieving optimum performance).

Flow Regimes

Pneumatic conveyors are classified as *dilute phase* or *dense phase*. Their general characteristics are compared in Table 15-2.

In dilute phase systems, the conveyed solid particles are uniformly suspended in the carrier gas. They operate at relatively high velocity to prevent the particles from settling onto the bottom of a horizontal run, or choking flow in a vertical section. The majority of pneumatic conveyors operate in the dilute phase.

Dense phase systems convey the particles at lower velocity, which is insufficient to carry the solids uniformly. Instead, they flow as full-bore slugs or plugs, or as moving-bed waves like sand dunes blown by the wind. Free-flowing fine particles and granules, with high gas permeability (such as plastic pellets) or retention (such as cement), are candidates for dense phase conveying; tests are typically conducted (by vendors) to ensure the technology is applicable to the specific system being considered.

Table 15-2
Characteristics of pneumatic conveying flow regimes [2]

Dilute Phase	Dense Phase
High velocity (700 m/min to 1,000 m/min, or 2,400 ft/min to 3,200 ft/min for fine powders or granules)	Low velocity (200 m/min, 600 ft/min typical)
Can have very high attrition	Low attrition
Pressure typically less than 100 kPa (15 psig)	High pressure (typically 100 kPa to 600 kPa)
Low cost	Comparatively high cost
Larger pipe size	Small pipe size
Low loadings (mass ratio solid to gas up to 15)	High loadings (mass ratio solid to gas 100 or more)

Types of Systems

There are three major types of system configuration: pressure (or push), vacuum (or pull), and combination pressure and vacuum (push/pull). Relevant details are supplied in Fig. 15.1–15.3 and Table 15.3.

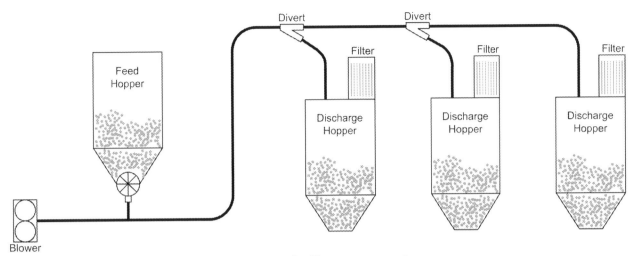

Figure 15-1. Positive pressure system.

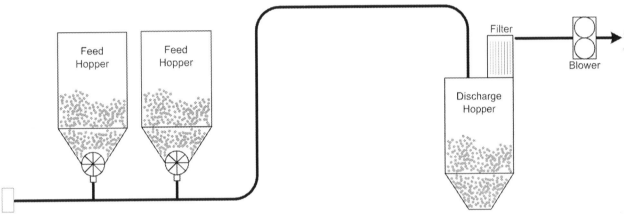

Figure 15-2. Negative pressure system.

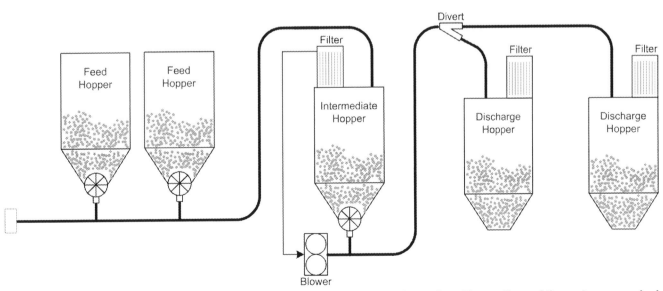

Figure 15-3. Push-pull system. Separate blowers can be used for the negative and positive portions of the system, or a single blower as shown here.

Table 15-3
Features of the major types of pneumatic conveying systems [5]

System Type	Features and Applications
Positive pressure	• The most common type of pneumatic conveying system • Feeding material into the positively pressured pipe can be challenging • Delivery to a number of alternative receivers is easily accomplished using diverter valves (one receiver at a time) • Sourcing from multiple feed hoppers is possible, but complicated by propensity for feed valves to bypass air and not usually recommended • "Low pressure" refers to systems operating below about 100 kPag (15 psig) • "High pressure" refers to systems operating above about 100 kPag (15 psig). The practical upper limit is about 660 kPag (100 psig) when discharging to a vessel at atmospheric pressure.
Negative pressure	• Used when multiple feed hoppers are required, and when introducing material from an open bin or pile through a wand with suction nozzle (like a vacuum cleaner) • Rotary valve or screw feeders are cheaper for negative pressure systems compared with those for positive pressure • Filter on the discharge hopper is larger compared with a positive pressure system due to the lower gas density • Leakage is nearly all inward so material and gas is not lost to the environment (particularly advantageous when conveying toxic or combustible materials) • Backup filters usually required to protect the blower in the event the primary filter fails • Multiple discharge hoppers are possible, but not usually recommended due to complexity of piping and isolation valves • Typically, the minimum pressure is 400 mm Hg to 500 mm Hg absolute
Push-pull	• Used when conveying from multiple sources to multiple destinations • Capable of transporting over long distances • Separate vacuum and pressure blowers can be used if process requirements render a single mover impractical

System Design Procedure

Data for the solids must be obtained or estimated to complete the design. Data are available for common materials in the literature. Any proprietary or unusual materials should be tested to obtain the design data and to assess flowability. The calculations are best applied to free flowing particles with uniform size and density.

The basic steps needed to design a dilute phase pneumatic conveying system are [2]:

1. Select a configuration: pressure, vacuum, or push-pull
2. Create a preliminary layout, including elevation changes
3. Guess a pipe size using an estimated conveying gas velocity
4. Calculate the saltation velocity
5. Estimate the system pressure drop
6. Calculate the volume of the prime mover
7. Select the prime mover
8. Calculate the pressure drop; check to see if pressure drop, velocity, and pipe size agree with estimates, iterate if necessary

If the pressure drop is more than about 20% of the absolute inlet pressure, step changes in pipe size should be considered. As pressure reduces along the pipeline the gas expands thus increasing velocity. Increasing the pipe size reduces the velocity and provides a way to manage both the velocity and pressure drop through the conveying system. The design calculations can be done in sections to determine the best point to increase the size and to determine the total pressure drop through the line. An example is shown in Figure 15.4.

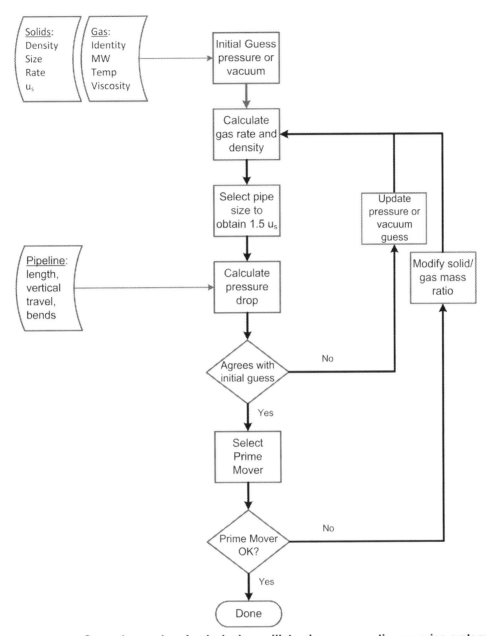

Figure 15-4. General procedure for designing a dilute phase pneumatic conveying system.

Saltation Velocity

Saltation velocity is defined as the minimum superficial gas velocity required to keep particles in suspension for a horizontal gas-solid flow. On a phase state diagram (Figure 15-5), the saltation velocity is defined as the point where the lowest pressure drop occurs in the transport pipeline. At lower velocities, conveying is in the dense phase and the pressure drop rises dramatically. At higher velocities, dilute phase conveying occurs but pressure drop rises due to the increased velocity.

The saltation velocity depends on both the solids loading (usually expressed as the mass ratio of solids to conveying gas) and the pipe diameter. The example phase state diagram, which charts pressure drop for solids conveyance from 4,000 kg/h to 80,000 kg/h, spans a solids loading ratio of about 0.5 to 9 (at 150 kPa).

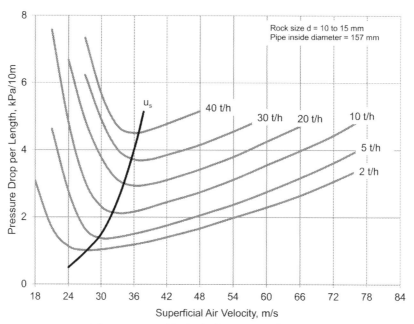

Figure 15-5. Typical phase state diagram.

Experimental data should be obtained before designing a system. However, for preliminary work, the Rizk correlation may be used as a first approximation. Its agreement with published measurements is reported to have a root mean square relative deviation of less than 60% [7]. Jacob said that the correlation "works well" for conveyance of plastic pellets [2]. However, these comparisons are for systems operating at near atmospheric pressure. Vacuum systems require a higher velocity due to the lower gas density.

The Rizk correlation estimates saltation velocity; easily solve it for velocity, given the mass loading, by using Goal Seek in Excel [7]:

$$\mu = \frac{1}{10^a}\left(\frac{u_{salt}}{\sqrt{g\,D}}\right)^b \qquad (15\text{-}1)$$

Where

u_{salt} = saltation velocity, m/s or ft/s

$\mu = \dfrac{M_s}{M_f}$

M_s = solids mass flow rate
M_f = conveying gas mass flow rate
$g = 9.81$ m/s^2 or 32.17 ft/s^2
D = inside diameter of conveying pipe, m or ft
$a = 1440\,x + 1.96$ (SI Units) or $a = 439\,x + 1.96$ (US Units)
$b = 1100\,x + 2.5$ (SI Units) or $a = 335\,x + 2.5$ (US Units)
x = mean particle diameter, m or ft

Pressure Drop

Compute the pressure drop by summing calculations for contributions from the carrier gas pressure drop, acceleration of the solids, friction of the solids against the pipe wall (separate factors for straight pipe and elbows), lifting of solids through vertical sections, and miscellaneous factors. See [3].

$$\Delta P = \Delta P_{ff} + (\Delta P_{sa} + \Delta P_{sf} + \Delta P_{sb} + \Delta P_{sv}) + \Delta P_{misc}$$

$$(15\text{-}2)$$

Pressure Drop of Carrier Gas

The mass flow of carrier gas is known from the solids loading factor. Using an assumption for the inlet pressure and temperature, calculate the density of the carrier gas. Then, choose the largest standard pipe size that achieves the desired gas velocity.

Recheck the saltation velocity based on the chosen pipe size and solids loading factor (Equation 15-1) to be sure there is a cushion between it and the calculated velocity in the pipe. The actual velocity should be at least 50% greater than the saltation velocity. If it isn't, consider adjusting the input parameters, for example by decreasing the solids loading factor.

Because the pressure drop due to gas flow is a fraction of the total pressure drop, it is not necessary to use compressible flow equations. Instead, treat the gas as incompressible for the purpose of the calculation.

Calculate the Reynolds number with:

$$N_{Re} = \frac{\rho \, D \, u_f}{\mu} \tag{15-3}$$

D = pipe diameter, m or ft
u_f = gas velocity, m/s or ft/s
μ = gas dynamic viscosity, kg/m-s or lb/ft-h
ρ = density of gas, kg/m^3 or lb/ft^3, or gas $= \dfrac{P \, M}{R \, T}$

Then, the friction factor:

$$f = 8\left[\left(\frac{8}{N_{Re}}\right)^{12} + \frac{1}{(a+b)^{3/2}}\right]^{1/12} \tag{15-4}$$

Where:

$$a = \left[2.457 \ln \frac{1}{(7/N_{Re})^{0.9} + (0.27 \, \varepsilon/D)}\right]^{16}$$

$$b = \left(\frac{37530}{N_{Re}}\right)^{16}$$

ε = surface roughness, m or ft

Finally, the pressure drop due to flow of carrier gas, using the pipeline equivalent length as defined in Chapter 1:

$$\Delta P_{ff} = \frac{f \, L \, u_f^2 \, \rho}{2 \, g_c \, D} \tag{15-5}$$

Acceleration of Solids

As the solids are metered into the pipe they must be accelerated to the flowing speed. However, the particles achieve something less than the gas velocity, a difference that is called "slip." Slip typically has a value of between 0.5 and 1.0; it is the ratio of particle velocity to gas velocity.

If the slip value is unknown, estimate it with:

$$\frac{u_s}{u_f} = \left(1 - S \, x^{0.3} \, \rho_s^{0.5}\right) \tag{15-6}$$

u_s = particle velocity, m/s or ft/s
S = constant, 0.0638 for SI units or 0.1794 for US units
x = particle diameter, m or ft
ρ_s = particle density, kg/m^3 or lb/ft^3

Pressure drop due to acceleration of the solids is:

$$\Delta P_{sa} = \frac{M_s}{M_f}\left(\frac{\rho_s}{g_c}\right)\left(\frac{u_f}{u_s}\right) \tag{15-7}$$

Friction of Solids in Straight Pipe

The friction factor applied to solids flow in the pipe is determined experimentally. If unknown, use a friction factor = 0.2. For this equation, the actual pipe length is used, not the equivalent length from gas flow, since pressure drop through bends is calculated separately.

$$\Delta P_{sf} = \lambda_s\left(\frac{M_s}{M_f}\right)\left(\frac{\rho_f}{g_c}\right) u_f^2 \left(\frac{L_{actual}}{2 \, D}\right) \tag{15-8}$$

Friction of Solids in Elbows

Estimate the equivalent length of the bends, diverter valves, and hoses. See Table 15-4.

The pressure drop due to solids flow though the bends is:

$$\Delta P_{sb} = L_{eqb}\left(\frac{\Delta P_{sf}}{L_{actual}}\right) \tag{15-9}$$

Table 15-4
Equivalent length of bends and components [3]

Bend or Component	Equivalent Length
Short radius elbow, R < 4D	30 D to 60 D
Long radius elbow, R > 8D	15 D to 30 D
Divert valve	15 D to 30 D
Flexible hose	Actual length times 4
Specialty bend (Vortice ell, etc.)	40 D to 60 D

Lifting Solids through Vertical Sections

Add up the length of vertical pipe where the flow is upwards and call it Z. The added pressure drop due to lifting the solids is then:

$$\Delta P_{sv} = \frac{M_s}{M_f} g \, Z \left(\frac{\rho_f}{g_c}\right) \left(\frac{u_f}{u_s}\right) \qquad (15\text{-}10)$$

Miscellaneous Factors

Additional pressure drop is added for the separator (cyclone or fabric filters), piping upstream and downstream of the solids flow path, and any other special condition.

Prime Movers

After determining the flow and pressure requirements, a prime mover can be selected. The most common prime mover is a roots blower. Fans and liquid ring vacuum pumps are also used.

A major consideration is that the mover must be able to perform with and without solids being present in the flowing stream. The pressure drop through the system is much higher when conveying solids than when the solids feed is discontinued. Therefore, the blower will operate along its performance curve, with the implications being much more severe with a centrifugal machine compared with a positive displacement one.

Blowers increase the gas temperature unless after-coolers are incorporated into the design. However, if the solids addition rate is significant (and it usually is), the gas temperature will quickly resolve to a value close to the solids temperature. This effect needs to be accounted for if the gas velocity is close to the saltation velocity, since cooling will decrease the gas velocity.

As described in Chapter 5, a system curve should be constructed for flow with no solids present. This can be superimposed on the selected prime mover performance curve; the flow rate with no solids is predicted by the point where the two curves cross. During solids transport the prime mover will operate at the flow and pressure determined in the preceding procedure; this point should also lie on the performance curve. The engineer should ensure that the prime mover will operate successfully – that is, without surging or overloading the motor – anywhere along the performance curve between these two points.

For pressures above about 100 kPa (15 psig), compressors are usually used. Compressed gas that is piped through a receiver with pressure and flow control can quickly respond to demand variations, so the potential performance issue with blowers – and, especially, fans – is of little concern. However, compressed gas is very expensive and it should only be used for pneumatic conveying when required by process requirements.

Feed Systems

Feed systems are comprised of the vessel that holds the solids and a feeding device that meters solids into the conveying gas stream while sealing that stream from the vessel. In vacuum systems, the vessel may be as simple as an open top drum or bag and the feeding device a wand with no moving parts. Pressure systems require a complex assembly with considerations ranging from flowability through the vessel to power required to drive the feeding device.

In most pneumatic conveying applications the single component that contributes to the largest pressure loss is

the feeding device [7]. Therefore, the selection and design of the feeder should be approached with caution.

Feeding devices include: rotary valves, venturis, vacuum nozzles, screw conveyors, powder pumps, and blow tanks. There are many factors to consider before making a final selection, including [7]:

- System pressure: low (up to 100 kPa), medium (100 kPa to 300 kPa), or high (up to 1,000 kPa)

- Product characteristics: particle size; cohesive; friable; free-flowing; degradation concerns; temperature limitations
- Physical layout of the system, especially whether the feed system must fit within a restricted vertical space
- Cost
- Continuous or batch process
- Feeding control accuracy requirements

Rotary valves are widely used, providing both a pressure seal and metered feeding. Rotary valves are suitable for low pressure systems that convey free-flowing solids. They can be fitted into tight spaces, offer fine feed control, and can be operated with batch or continuous systems. All rotary valves leak air; the leakage rate is dependent on the pressure differential across the valve and the valve speed, in addition to the design and construction features of the valve.

Valves are sized based on the size of the cavities and the rotor speed. A rule of thumb is that the rotor tip speed should not exceed 40 m/min. For a 150-mm (6-inch) valve this translates to about 85 rpm; a 1000-mm (40-inch) valve should be limited to 12.7 rpm. However, standard practice is to limit the speed to 25 rpm for valves up to 300 mm (12 inches). The speed should be less for materials with a low bulk density and for systems with high differential pressure.

For pressure systems fed from a rotary valve, the pressure differential across the valve should be less than about 80 kPa (12 psi) to minimize leakage of conveying gas through the valve. For vacuum systems, the differential should be kept below about 40 kPa (12 in. Hg or 6 psi).

A procedure for estimating leakage through a rotary valve is given in [7]. The results from the calculations are comparable with experimental results, as charted in Figure 15-6. Cast iron or steel valves typically have a 0.13-mm clearance; stainless steel valves require increased clearance, to about 0.18 mm, to minimize the risk of galling. If the operating temperature is greater than about 50 °C to 60 °C then greater clearance may be needed to accommodate differential expansion of the rotor and housing.

Silos

Silos and bins hold material which, for a positive pressure system, flows by gravity into the feed device. The shape of the bottom of the silo can be conical, pyramidal,

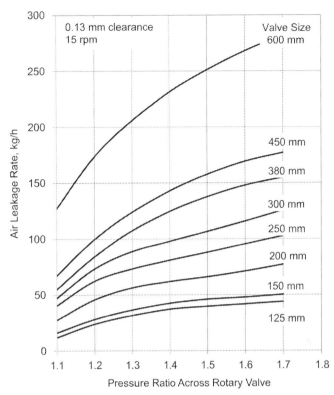

Figure 15-6. Typical air leakage through a rotary valve. (After [7]).

or wedge, each with a sloped surface to the silo or bin vertical walls. Vendors should be consulted to ensure that the silo and feed device are properly matched and that they have characteristics consistent with the properties of the solids that are being conveyed.

The important properties of the solids are: angle of internal friction, wall friction angle, and flow function which measures cohesive strength of the bulk material as a function of consolidation pressure. The first two properties are used to determine the maximum hopper angle that will provide mass flow; the last properties determine the minimum size of the opening in the bottom of the silo.

The silo should normally be designed for mass flow, where all of the solids are in motion as the material travels down toward the feed device. This requires that the angle of the conical (or pyramidal, or wedge) portion of the silo be steep enough so that friction does not cause the solids near the wall to remain in place while material in the center flows to the discharge.

Jenike's correlation for the angle of the silo hopper needed to ensure mass flow, expressed in degrees as half of the internal angle, is charted in many references (for example, [4]). Oko, et al. regressed the curves to obtain equations for use in computer programs, such as Excel [6]. Any hopper angle that is steeper than indicated by the equation will give mass flow; shallower angles result in funnel flow (i.e., ratholing).

$$\theta = A + B\,\delta + C\,\delta^2 \qquad (15\text{-}11)$$

Where:

θ = Hopper angle, degrees (see note below table)
δ = Wall friction angle, degrees
A, B, C = parameters from Table 15-5

Table 15-5
Parameters for use in equation 15-11

Angle of Internal Friction, degrees	Conical Hoppers			
	A	B	C	Valid Wall Angles
30	52.7	− 0.678	− 0.331	0 to 30
40	47.8	− 0.902	− 0.0122	0 to 35
50	46.1	− 1.1	− 0.0027	0 to 38
60	43.3	− 0.97	− 0.0033	0 to 40
	Wedge Shaped Hoppers			
30	64.7	− 1.56	0.0023	0 to 27
40	63.4	− 1.48	0.0005	0 to 36
50	59.6	− 1.34	0.0004	0 to 45
60	56.8	− 1.22	− 0.00003	0 to 48

Note: For conical hoppers, subtract 4 degrees from the result to account for an unstable region between mass flow and funnel flow.

Conveyor Piping

Consider stepping the conveyor pipe size to maintain velocity within a desired range. As the carrier gas expands, velocity increases. This practice is most common on long systems; short runs (<100 m, 300 ft) rarely step up the pipe size.

At the feed point allow at least 30 pipe diameters of straight horizontal pipe as the acceleration zone where the solids achieve the conveying velocity. Limit the number of bends and avoid placing them close together. Allow at least 20 pipe diameters between bends to facilitate reacceleration of the solids. Use only horizontal and vertical lines; avoid inclined lines [3].

Use a consistent inside diameter for pipes, fittings, diverters, and gates to eliminate lips or ledges where product could hang up. However, include provisions for cleaning out the pipeline in the event of plugging.

For non-abrasive materials that do not degrade during conveying, the type of bend makes very little difference and should be based on cost [3]. Choose elbows with the lowest pressure drop, such as very long radius sweep bends, if the maximum pressure is limited.

Virtually all pipeline wear occurs at the bends; abrasive materials can quickly wear through an elbow especially at high velocity. Velocity increases toward the discharge end of the system, so this is where wear is first apparent. Wear is proportional to the gas velocity raised to a power of two to four [3].

Dhodapkar, Solt, and Klinzing published a comprehensive comparison of fourteen types of bends used in pneumatic conveying systems [1]. They tabulated advantages and disadvantages of each type, and discussed the major factors associated with erosion in bends. A summary is given in Table 15-6. Wagner discussed standard and specialty elbows [8]. Standard elbows, both short and long radius, are recommended for dense phase conveying systems, and dilute phase when the product is non-abrasive and not temperature sensitive, and also for cohesive or highly elastic powders. Specialty elbows such as a blind tee, Vortice Ell, Hammertek Smart Elbow®, Gamma bend, or Pellbow® are recommended for abrasive, temperature sensitive, and fragile powders conveyed in dilute phase.

Table 15-6
Bend suitability based on material characteristics [1]

Bend Type	Cohesive or Sticky or Moist	Fragile or Friable Solids	Hard and Abrasive Solids	Soft and Rubbery Solids	Product Purity Required / No Cross Contamination
Blind tee	NS	S*	S	NS	NS
Blind radius bend / blind lateral	NS	S*	S	NS	NS
Mitered bend (90-deg turn)	NS	NS	NS	S	S
Elbow (R/D < 3)	NS	NS	NS	S	S
Radius bend: short radius (R/D = 3 to 7)	S	S*	NS	S	S
Radius bend: long sweep (R/D = 15 to 24)	S	S*	NS	S	S
Radius bend with liners	S*	NR	S	NR	S
Radius bend with wearable backing	S	NR	S	NR	S*
Radius bend with internal baffles	NS	NR	S	NS	NS
Short-radius bends with pocket for material (Vortice Ell www.rotaval.co.uk, Hammertek Smart Elbow® www.hammertek.com)	NS	S	S	NR	S*
Transition designs (Gamma Bends www.coperion.com)	NS	S	NS	S	S
Transition designs (Pellbow® www.pelletroncorp.com)	NS	S	S	S	S
Rubberized or flexible bend	S	S*	S*	NR	S*

R = radius of curvature of the bend; D = diameter of the pipe;
S = Suitable; S^* = Suitable under certain conditions; NS = Not Suitable;
NR = Not Required.

Disengagement

At the discharge end of the system, the solids must be disengaged from the conveying gas. This is usually accomplished using a combination of impingement or centrifugal force and a fabric filter. Sometimes the operations are combined in a single vessel; other times the fabric dust filter is separated from the cyclone separator. Solids are collected in a hopper with design parameters as described in the Feed Systems section.

The design of the disengagement system directly affects the overall conveying system, especially with respect to pressure drop, possible loss of fines, particle segregation, particle breakdown, and cost. However, the selection of the system is based primarily on the degree of separation required and the potential for environmental consequences [7].

Rizk [7] argues that the conventional way of specifying dust collectors, with a removal efficiency value such as "99.98%" (typical for performance of gas cleaning efficiency), does not match with current environmental legislation. Although laws do allow for "Maximum Achievable Control Technology (MACT)", they also stipulate the total amount of material that may be discharged during a given time period. These factors must be carefully considered in conjunction with the potential environmental and health hazards posed by the materials being conveyed.

Performance of the separators depends on the particle size. Larger particles are captured with a cyclone and smaller ones by a fabric filter. See Table 15-7.

Fabric filters are sized on the basis of the gas-to-cloth ratio, defined as

$$\gamma^* = \frac{\dot{V}}{A_F} \tag{15-12}$$

Table 15-7
Percentage efficiencies for various collectors [7]

Equipment	Percentage Efficiency At		
	50 μm	5 μm	1 μm
Inertial collector	95	16	3
Medium efficiency cyclone	94	27	8
High efficiency cyclone	98	42	13
Shaker type fabric filter	> 99	> 99	99
Reverse jet fabric filter	100	> 99	99

Where:

$\dot{V}$ = volumetric flow rate of gas at ambient conditions, m^3/min or ft^3/min

A_F = fabric filter area, m^2 or ft^2

Since the dimensions of the gas-to-cloth ratio are velocity, it is also referred to as the face velocity or approach velocity. The range of values, in m/min, for γ^* is $0.5 \leq \gamma^* \leq 4$. Lower values are used for fine particulates and cohesive materials. As a rule of thumb, a value between 0.8 and 2 is used for industrial applications.

Nomenclature

A_F	=	fabric filter area, m^2 or ft^2
D	=	inside diameter of conveying pipe, mm
g	=	9.81 m/s^2 or 32.17 ft/s^2
M_f	=	conveying gas mass flow rate
M_s	=	solids mass flow rate
u_f	=	gas velocity, m/s or ft/s
u_s	=	particle velocity, m/s or ft/s
u_{salt}	=	saltation velocity, m/s or ft/s
x	=	mean particle diameter, m or ft
S	=	constant, 0.0638 for SI units or 0.1794 for US units
$\dot{V}$	=	volumetric flow rate of gas at ambient conditions, m^3/min or ft^3/min
γ^*	=	gas-to-cloth ratio, m/min or ft/min
δ	=	Wall friction angle, degrees
ρ	=	density of gas, kg/m^3 or lb/ft^3, or gas $= \dfrac{P\,M}{R\,T}$
ρ_s	=	particle density, kg/m^3 or lb/ft^3
θ	=	Hopper angle, degrees
μ	=	gas dynamic viscosity, kg/m-s or lb/ft-h
μ	=	$\dfrac{M_s}{M_f}$

References

[1] Dhodapkar S, Solt P, Klinzing G. Understanding Bends in Pneumatic Conveying Systems. *Chemical Engineering*, April, 2009:53–60.

[2] Jacob K. Introduction to the Pneumatic Conveying of Solids, webinar for AIChE, September 8, 2010.

[3] Maynard E. Designing Pneumatic Conveying Systems. *Chemical Engineering Progress*, May, 2006:23–33.

[4] Mehos G. Designing Dust Collectors. *Chemical Engineering Progress*, September, 2011:32–8.

[5] Mills D. Pneumatic Conveying: Know Your Options. *Chemical Engineering*, May, 2005:58–63.

[6] Oko C, Diemuodeke E, Akilande I. Design of Hoppers Using a Spreadsheet. *Research in Agricultural Engineering*, Czech Academy of Agricultural Sciences 2010;56:53–8.

[7] Rizk F, Marcus R, Leung L, Klinzing G. Pneumatic *Conveying of Solids: A theoretical and practical approach*. 3rd ed. Springer; 2010.

[8] Wagner P. Selecting Elbows for Pneumatic Conveying Systems. *Chemical Engineering Progress*, September, 2007:28–32.

16

Blending and Agitation

Introduction

Agitators are used for blending, emulsifying, promoting mass transfer, promoting chemical reactions, and mixing particulates. Agitated systems are most often specified as stirred tanks or batch blenders. This chapter addresses some of the more common applications, and gives advice for establishing process response criteria, choosing impellers, and sizing the equipment.

Vendors traditionally provide the calculations and designs for agitation systems to meet the customer's requirements. It is the customer's responsibility to establish quantifiable requirements. Although vendors use proprietary methods to design their equipment, engineers can apply published equations, such as those presented here, for preliminary work and to check the vendor's work.

The rise of Computational Fluid Dynamics (CFD) has enabled sophisticated analysis of mixing systems. However, the empirical approaches developed in the mid-20th century remain useful for approximate work, especially if standard impellers (e.g., propellers, pitched blade turbines) are used. Agitator manufacturers have invested heavily in their proprietary designs; coefficients needed for the empirical equations are not usually available in the open literature.

An Excel workbook with worked examples accompanies this chapter.

Basics

The geometric nomenclature used in this chapter is illustrated in Figure 16-1. See Table 16-1 for a rough guide to mixing parameters. Table 16-2 indicates where decisions are made in the design process.

For vertical vessels with center-mounted agitators, baffles are normally required to prevent the fluid from swirling in the tank. Off center or angled agitators produce asymmetrical flow, so baffles are not required. Baffle width

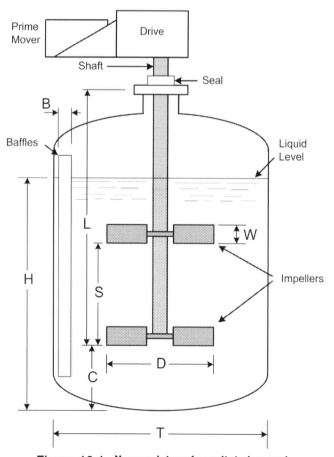

Figure 16-1. Nomenclature for agitated vessel.

Table 16-1
Rough guide to mixing parameters in stirred tanks [14]

Operation	Power hp/1000 gal	Tip Speed ft/s
Blending	0.2 to 0.5	
Homogeneous reaction	0.5 to 1.5	7.5 to 10
Reaction with heat transfer	1.5 to 5.0	10 to 15
Liquid-liquid mixtures	5	15 to 20
Liquid-gas mixtures	5 to 10	15 to 20
Slurries	10	

is typically 8% to 10% of the tank diameter (B = T/12 or B = T/10) or less. Spacing between the baffle and tank wall is typically about 1.5% of the tank diameter (T/72). A distance of ¼ to 1 full baffle width is left between the bottom of the baffles and the bottom of the tank. Generally, four baffles are installed in vertical vessels, at 90° increments.

Prime movers are usually electric motors. Other prime movers are: air-driven motors, hydraulic motors, steam turbines, and diesel or gas engines. Gear boxes are used to drive the agitator, or variable speed prime movers can drive the shaft directly. Fixed speed drives usually consist of a combination of helical and spiral-bevel gears, or right-angle worm gears. In the US, the gear boxes are built with specific gear combinations in accordance with recommendations from the American Gear Manufacturers Association (AGMA). The drives operate at the speeds listed in Table 16-3.

Table 16-2
Responsibilities for specification and design of agitators are generally assigned to the customer or the manufacturer

Agitator Specification or Design Element	Customer	Agitator Mfg
General description of mixing problem	Defines	Provides advice
Process fluids (type: liquid/solid; interaction: miscible/immiscible; properties: temperature, density, viscosity, particle size). Viscosity characteristic (Newtonian, shear thinning, etc.). Hazards (flammability, toxicity, etc.)	Defines	Uses for calculations
Type of agitation (blending, dissolving, dispersing, suspending, emulsifying, etc.)	Specifies	Complies
Degree of agitation (blend time, ChemScale®, scale of solids suspension, gas dispersion intensity, high shear, etc.)	Specifies	Provides advice
Mixing environment (tank and its size and dimensions; batch/continuous; in-line)	Defines	Provides advice
Mixer type (portable, top entry, side entry, bottom entry, static)	Defines	Provides advice
Instruments (tachometer, torque meter, power meter, etc.)	Specifies	Complies
Site conditions (available area for agitator removal/maintenance; indoor/outdoor; heated/unheated; dusty; sun; etc.)	Specifies	Complies
Tank dimensions, access for impeller	Depends on project	Depends on project
Materials of construction, finishes	Defines	Complies
Impeller selection (type, number, size, location)	Approves (sometimes specifies)	Defines (confirms and complies if specified)
Baffles (number, size, location)	Specifies if existing tank with baffles	Defines
Required operating speed(s)	Approves	Defines
Shaft diameter and type (solid, hollow) and features (couplings, steady-bearings)	Approves	Defines
Calculates critical speeds		Defines
Steady bearing and in-tank couplings; bolted impeller blades	Allows/disallows	Determines requirement
Reaction forces (shaft deflection, mounting flange, seal, bearings)	Provides to tank vendor, if applicable	Defines
Prime mover (electric motor, air motor, hydraulic motor, turbine)	Defines type, electrical requirements, and fixed-speed or VFD	Size, enclosure, mounting
Drive system (gearbox, bearings)	Approves	Defines
Seal (packing, single-mechanical, double-mechanical, cartridge, lip, hydraulic, sealless)	Specifies	Defines

Table 16-3
Agitator speeds using AGMA gear reducers (+/− 5%)

Gearbox Type	Fixed Speeds (revolutions per minute)
Helical or combination helical and spiral-bevel gears	350, 230, 190, 155, 125, 100, 84, 68, 56, 45, 37, 30, 25, or 20
Worm (right-angle) gears	350, 233, 175, 146, 117, 88, 70, 58, 44, 35, 29, or 25

Impeller Types

Table 16-4 lists several classes of impeller that are used in stirred tanks. For each class, typical applications are given. The major impeller manufacturers have developed proprietary designs for their products (most are hydrofoils or high shear impellers), and promote the efficiency and effectiveness they provide. Generic designs, such as Rushton impellers, remain in widespread use, however, and design data for them are more readily available.

Table 16-4
Impeller classes and specific types used in transitional and turbulent flow conditions [9]

Impeller Class	Specific Types	Application
Axial flow	Propeller, pitched blade turbine (PBT), hydrofoil	Liquid blending, solids suspension, heat transfer. Axial impellers are sometimes installed within draft tubes in the vessel.
Radial flow	Flat-blade turbine, disk turbine (Rushton), hollow-blade turbine (Smith), concave disc	Gas dispersion and liquid-liquid dispersion. Higher shear and turbulence with lower pumping compared with axial flow. Above a D/T ratio of 0.55, pitched blade turbines become radial flow impellers.
High shear	Cowles, disk, bar, pointed blade impeller, rotor-stator	Operate at high speed for grinding, dissolving, dispersion, homogenization, emulsification, and cell disruption. Low pumping, so often used along with axial flow impellers.
Specialty	Retreat curve impeller, sweptback impeller, spring impeller, glass-lined turbine	Glass-lined vessels. Spring impellers used for solids that frequently settle to the tank bottom.
Up/down	Disk, plate, circle	Low shear, typically used in extraction columns. Also used in disposable bioreactors.
High viscosity	Helical ribbon, anchor, screw (auger), double helix (two shafts)	Blending of miscible liquids with high viscosity, typically used for creams and pastes with viscosity > 10,000 cP

Dimensional Analysis

The relationships in this section are used to compare the performance of systems, especially when evaluating alternative impellers or speeds for an application.

Reynolds Number and Turbulence

It is essential to know the flow regime in the mixer. In fully turbulent conditions, with Reynolds number greater than 10,000, the agitator's power is proportional to fluid density, rotational speed, and impeller diameter. At small values of Reynolds number, below about 10, laminar conditions exist and power is proportional to fluid viscosity, rotational speed, and impeller diameter. Transitional conditions are present when the Reynolds number is between 10 and 10,000.

Calculate Reynolds number with:

$$N_{Re} = \frac{\rho N D^2}{\mu} \qquad (16\text{-}1)$$

Where:

ρ = fluid density, kg/m^3 or lb/ft^3
N = agitator speed, revolutions/s
D = impeller diameter, m or ft
μ = fluid dynamic viscosity, kg/m-s or lb/ft-s

Pumping Number

The amount of fluid pumped by an impeller is dependent on several factors, the most important being the specific design of the impeller and its size relative to the tank. A dimensionless group, called the pumping number, is defined as:

$$N_Q = \frac{Q}{N D^3} \qquad (16\text{-}2)$$

Where:

Q = Pumping capacity, m^3/s or ft^3/s

The relationship between the pumping number and Reynolds number is determined experimentally and should be available from the impeller manufacturer. However, at Reynolds numbers greater than 10,000, the pumping number is usually constant and should fall into the ranges given in Table 16-5.

Table 16-5
Pumping number under turbulent conditions ($N_{Re} > 10{,}000$) [9]

Impeller Type	Pumping Number, N_Q
Propeller	0.4 to 0.6
Pitched blade turbine (PBT)	0.79
Hydrofoil impellers	0.55 to 0.73
Flat-blade turbine	0.7
Disk flat-blade turbine (Rushton)	0.72
Hollow-blade turbine (Smith)	0.76
Retreat curve impeller	0.3

These impellers should not be used for laminar conditions because mixing will be very poor.

Power and Power Number

The power number is a function of Reynolds number and is constant for Reynolds number greater than 10,000. From the power number, calculate shaft power with:

$$P = \frac{N_P \rho N^3 D^5}{g_c} \quad (16\text{-}3)$$

Where:

P = shaft power, W or ft-lb$_f$/s (divide by 550 for hp)
g_c = conversion factor, 1 m/s^2 or 32.17 ft/s^2

Relationships between the tabulated power numbers (Table 16-6) and other conditions include [9]:

- NP changes slightly in the transitional region ($100 < N_{Re} < 10{,}000$)
- For six-bladed Rushton (straight-bladed turbine) impellers,

$$N_P \propto \left(\frac{W}{D}\right)^{1.45}$$

- For four-bladed turbine impellers with 45° pitch,

$$N_P \propto \left(\frac{W}{D}\right)^{0.65}$$

- If n = number of blades, for three to six blades:

$$N_P \propto \left(\frac{n}{D}\right)^{0.8}$$

Table 16-6
Power number at fully turbulent flow for various impellers

Impeller	Power Number, N_P	Ref
Propeller, T/D = 3; C/D = 0.75 to 1.3; 4 standard baffles; H/D = 2.7 to 3.9; 3 blades; pitch = D	0.35	12
Marine propeller; T/D = 3; T/C = 3; D/W = 5; 1.0 pitch; 4 standard baffles	0.34	9
Propeller, T/D = 3; C/D = 0.75 to 1.3; 4 standard baffles; H/D = 2.7 to 3.9; 3 blades; pitch = 2D	0.9	12
Marine propeller; T/D = 3; T/C = 3; D/W = 5; 1.5 pitch; 4 standard baffles	0.62	9
Pitched-blade turbine; D/W = 8; 4 standard baffles; 6 blades; 45° pitch	1.5	12
Pitched-blade turbine; T/D = 3; T/C = 3; D/W = 5; 4 blades, 45° pitch; 4 standard baffles	1.27	9
Pitched-blade turbine; T/D = 3; T/C = 3; D/W = 5; 6 blades, 45° pitch; 4 standard baffles	1.64	9
Flat-blade turbine; D/W = 8; 4 standard baffles; 6 blades	3	12
Curved full-blade turbine; 4 standard baffles	2.7	12
Flat-blade turbine; L/D = 0.25; C/D = 0.75 to 1.3; T/D = 3; D/W = 5; H/D = 2.7 to 3.9; 6 blades; D/W = 5; 0° pitch; 4 baffles with T/B = 6	7	12
Flat-blade turbine; L/D = 0.25; C/D = 0.75 to 1.3; T/D = 3; D/W = 5; H/D = 2.7 to 3.9; 6 blades; D/W = 5; 0° pitch; 4 baffles with T/B = 10	6	12
Flat-blade turbine; L/D = 0.25; C/D = 0.75 to 1.3; T/D = 3; D/W = 5; H/D = 2.7 to 3.9; 6 blades; D/W = 5; 0° pitch; 4 baffles with T/B = 25	4	12
Flat paddle; T/D = 3; C/D = 0.75 to 1.3; L/D = 1; D/W = 4; 4 standard baffles; 2 blades	1.8	12
Hollow-blade turbine (Smith); 4 standard baffles	4.1	9
High-shear disk at N_{Re} = 10,000; 4 standard baffles	0.2	9
Hydrofoil – Lightnin A310; 4 standard baffles	0.3	9
Hydrofoil – Chemineer HE3; 4 standard baffles	0.3	9

- And for six to twelve blades:

$$N_P \propto \left(\frac{n}{D}\right)^{0.7}$$

- For pitched-blade turbines, changing the blade angle, θ, changes the power number by:

$$N_P \propto (\sin \theta)^{2.6}$$

- For pitched-blade turbines, the elevation of the impeller above the tank bottom has a small effect on power number:

$$N_P \propto \left(\frac{C}{D}\right)^{-0.25}$$

- If there are multiple impellers on a shaft, the power numbers may or may not be additive, depending on impeller type and spacing. Typical spacing between impellers is one impeller diameter. Axial flow impellers will typically draw less power than the sum of individual impellers. However, flat-blade radial flow impellers may draw more than the sum of the individual impellers.

Blending

A common objective is to mix two or more miscible liquids in a vessel. This operation is characterized by the "blend time", defined as the time required to achieve a specific degree of homogeneity. The mixing equipment (i.e., impellers) and calculation procedure are dependent on the fluid viscosities and Reynolds number. Glenville and Nienow summarized the research behind the equations given in this section [7].

Another way to specify the degree of mixing is with a qualitative scale. Table 16-7 is a standardized scale that, together with physical property data and tank dimensions, a mixer manufacturer uses to quote appropriate equipment based on calculations and experience.

There are several experimental methods for measuring the degree of homogeneity. By introducing a tracer into the vessel, the uniformity of the solution can be determined by analyzing the mix at various locations. Some methods, such as conductivity measurement, can be done nearly continuously at specific locations in the tank. Grab samples are required for other methods; these are less precise. See Table 16-8, from [4].

Newtonian Fluids in Turbulent or Transitional Flow

For a given system, with tank and impeller dimensions and power number specified, use these relationships between the blend time and impeller speed. The most common condition with low viscosity fluids is turbulent flow; use Equation 16-4 or 16-5 to determine whether the agitation is turbulent. For laminar flow ($N_{Re} < 100$), blend time using a turbine impeller is lengthy and unpredictable. Instead, use a helical ribbon impeller; see Laminar

Table 16-7
Standard qualitative scale for mixer specification

ChemScale®	Process Response
1 to 2	Mild/minimum blending and motion. Produces a flat, but moving surface
3 to 5	Intermediate/moderate blending of miscible liquids when specific gravity differences are less than 0.6. Produces surface rippling at water-like viscosities.
6 to 8	Moderate to vigorous agitation for uniform blending of miscible liquids when specific gravity differences are less than 0.6. Produces surface rippling at lower viscosities.
9 to 10	Very vigorous agitation for uniform blending of miscible liquids when specific gravity differences are less than 1.0. Produces violent surface motion at lower viscosities.

Source: Chemineer.

Table 16-8
Methods for determining blend time [4]

Technique	Tracer	Blend Time Reached When
Grab sample	Any material that can be analyzed	Samples do not vary more than +/− X% from final concentration
Dye introduction	Dyed fluid	Uniform color is attained
Conductivity cell	Concentrated salt solution	Measured conductivity that represents concentration is within +/− X% of final concentration
Acid-base indicator	Acid (or base)	Neutralization is complete as determined by color change of indicator

+/−X% is the spcified degree of homogeneity

Flow Regime on page 264. The transition region, between laminar and turbulent, is modeled with Equation 16-7.

After calculating the Reynolds number (Equation 16-1), use the turbulent flow correlation (Equation 16-6) when:

$$N_P^{(1/3)} N_{Re} \geq 6370 \tag{16-4}$$

If the Reynolds number is unknown, but the desired blend time to achieve 95% homogeneity is specified, calculate the Fourier number and use the turbulent flow correlation when:

$$\frac{1}{N_{Fo}} \geq 1225 \tag{16-5}$$

Where:

$$N_{Fo} = \frac{\mu\, \theta_{95}}{\rho\, T^2}$$

θ_{95} = time to achieve 95% homogeneity, s

The turbulent flow correlation is:

$$\theta_{95} = \frac{C_1\, T^{1.5}\, H^{0.6}}{N_P^{(1/3)}\, N\, D^2} \tag{16-6}$$

$C_1 = 5.20$ for T, H, and D in meters; 4.62 for T, H, and D in feet

In the transition region, with $N_{Re} > 100$,

$$\theta_{95} = \frac{183^2}{N\, N_P^{(2/3)}\, N_{Re}} \left(\frac{T}{D}\right)^2 \tag{16-7}$$

For different degrees of homogeneity, adjust to the result with:

$$\theta_Z = \theta_{95}\, \frac{\ln[(100 - z)/100]}{\ln(0.05)} \tag{16-8}$$

Where z is the desired degree of homogeneity (e.g., $z = 99$ for 99% homogeneity).

Shear-Thinning Fluids

Shear thinning, or pseudoplastic, fluids are characterized by consistency and flow behavior indices, K and n. These values are determined experimentally by measuring the fluid viscosity at different shear rates, satisfying the power law relationship:

$$\mu_A = K\, \dot{\gamma}^{n-1} \tag{16-9}$$

Where:

μ_A = apparent viscosity
K = consistency index
$\dot{\gamma}$ = shear rate, s^{-1}
n = flow behavior index

Shear rate in the vessel varies with distance from the impeller; it depends on velocity gradients and is lowest near the vessel walls and at the fluid surface. Thus, the apparent viscosity is lowest near the impeller and highest at the walls.

For a system where the impeller and speed are known, the blend time is calculated in two steps: 1) calculate the apparent viscosity at the tank wall, and 2) use the equations in the previous sub-section for Newtonian fluids to compute the blend time.

1. Calculate the apparent viscosity at the tank wall

This is done by first estimating the shear stress at the wall, using the agitator's torque. The symbol τ (Greek *tau*) is usually used for both torque and shear stress. Therefore, to avoid confusion shear stress is denoted by SS. Subscript w refers to the conditions at the tank wall.

$$SS_w = \frac{1}{1.622} \left(\frac{\tau}{T^3}\right) \tag{16-10}$$

$$\dot{\gamma}_w = \left(\frac{SS_w}{K}\right)^{1/n} \tag{16-11}$$

$$\mu_w = K\, \dot{\gamma}_w^{\,n-1} \tag{16-12}$$

2. Calculate the Reynolds number with Equation 16-1 then follow the procedure in the previous sub-section to compute the blend time.

The calculation procedure for a new system, where the impeller speed is the unknown, is complicated because torque, shear stress, shear rate, and apparent viscosity at the wall change with the speed. In turn, the blend time equations require speed. Therefore, this is an iterative calculation, where the apparent viscosity at the wall is calculated using an assumed speed. Then, with a specified blend time, the equations in the previous sub-section are used to compute the required speed; this is compared with the assumed speed and the procedure is iterated until the two agree. For fixed-speed agitators there are specific speeds possible based on available gear boxes (see Table 16-3). In this case only the discrete available speeds are used.

For highly viscous, shear-thinning fluids, where the flow behavior index is less than about 0.3, a cavern forms around the impeller. Outside the cavern, little or no mixing occurs and the fluid remains stagnant. With turbine impellers, the height to diameter ratio of the cavern ranges from about 0.4 to 0.6. Models for estimating the size of the cavern are discussed in [7].

Laminar Flow Regime

Determine the Reynolds number at the transition point between laminar and transitional regimes, for a standard turbine impeller, with:

$$N_{Re,LT} = \left(\frac{4.8\, T}{N_P^{1/3}\, D} \right)^2 \text{ (Wichterle and Wein) and,}$$

$$N_{Re,LT} = \frac{183}{N_P^{1/3}} \text{ (Hoogendoorn and den Hartog)}$$

The lower of the two results will indicate the transition point where an impeller specifically designed for laminar flow, such as an anchor or helical ribbon, must be used.

The higher number would indicate the point where the special design should be strongly considered.

Calculate the shaft power for a helical ribbon impeller with [7]:

$$P = \frac{K_P\, \mu\, N^2\, D^3}{g_c} \tag{16-13}$$

Where:

$$K_P = 82.8\, \frac{h}{D} \left(\frac{c}{D} \right)^{-0.38} \left(\frac{p}{D} \right)^{-0.35} \left(\frac{w}{D} \right)^{0.20} n_b^{0.78}$$

D = impeller diameter
h = overall height of the helical ribbon impeller
c = wall clearance (typically 2.5% to 5% of the vessel diameter)
p = pitch (ribbon height in one 360° turn)
w = width of the blade (typically 10% of the impeller diameter)
n_b = number of blades (typically two)

For Newtonian and shear-thinning fluids, the blend time when in the laminar region is independent of Reynolds number and viscosity [7].

$$\theta = \frac{896000\, K_P^{-1.69}}{N} \tag{16-14}$$

Estimate the apparent viscosity of shear-thinning fluids operating in the laminar region with the Metzner and Otto approach for calculating shear rate, with a constant, $k_s = 30$ for helical ribbons [7]. Use the apparent viscosity to compute the Reynolds number and verify that the mixer is operating in the laminar regime.

$$\dot{\gamma} = k_s\, N \tag{16-15}$$

$$\mu_A = K\, \dot{\gamma}^{n-1} \tag{16-16}$$

Solids Suspension

This section gives an approach to sizing an agitator for a batch liquid-solids system in which no dissolution or reaction occurs. It is applicable when the settling velocity of the solids is greater than about 0.15 m/min (0.5 ft/min). The desired distribution in the vessel of particles of uniform size and heavier than the liquid, such as in a crystallizer, is specified by the engineer.

The procedure indicates the impeller size, location, speed, and power.

Most solids-liquids processes have objectives that go beyond desired distribution. See Table 16-9. Surface control, mass transfer, and shear rate are other factors that must be considered. The engineer should inform the agitator manufacturer of all process requirements.

Table 16-9
Process objectives must be understood before specifying an agitator [2]

Process	Objectives
Dispersion	Desired distribution in the vessel, such as partial suspension (some solids rest on bottom of tank), complete suspension (all solids are off the bottom), and uniform suspension (solids suspended throughout the tank).
Dissolution	Rate of dissolution, or partial dissolution (called leaching), of a soluble solid in the liquid. The density and viscosity of the liquid may change considerably during the process.
Crystallization and precipitation	Control of the rate of nucleation and growth of the particles, and minimization of particle breakage or attrition. Usually important to assure liquid phase mixing to achieve uniform concentration and avoid local high concentration regions.
Solid-catalyzed reaction	Uniform suspension of catalyst particles. Agitation reduces the diffusional mass transfer boundary layer, enhancing the solid-liquid mass transfer.
Suspension polymerization	Produce and maintain a dispersion of uniform size monomer droplets and suspension of monomer drops and, eventually, polymer particles.

Terminal Settling Velocity

Calculate the settling velocity of the solids, adjusted for solids concentration. Use these equations to obtain the basic velocity [10]:

$$w_s = \left(A + \frac{B}{S_*} \right)^{-1} \sqrt{(s-1)\, g\, d_N} \qquad (16\text{-}17)$$

Where:

$$S_* = \frac{d_N}{4\, v} \sqrt{(s-1)\, g\, d_N}$$

w_s = settling velocity, cm/s or ft/s

A, B = coefficients dependent on particle shape factor and roundness (see Table 16-10)

s = specific gravity of solids $\left(s = \dfrac{\rho_s}{\rho_l} \right)$

g = gravitational acceleration, 980 cm/s^2 or 32.17 ft/s^2

d_N = nominal particle diameter, cm or ft

v = kinematic viscosity of liquid phase, cm^2/s or ft^2/s

For solids concentrations greater than 15%, correct the calculated settling velocity using Table 16-11 to account for the observation that it is more difficult to agitate slurries with higher concentrations [6].

Table 16-10
Coefficients for terminal settling velocity equation [10]

Solids Characterization	A	B
Crushed sediment	0.995	5.211
Rounded sediment	0.954	5.121
Well-rounded sediment	0.890	4.974
Spherical particles	0.794	4.606

Table 16-11
Correction factor for settling velocity of solids in slurries [6]

Solids %	Correction Factor
2	0.8
5	0.84
10	0.91
15	1.0
20	1.10
25	1.20
30	1.30
35	1.42
40	1.55
45	1.70
50	1.85

Scale of Agitation

Using the qualitative descriptions in Table 16-12, characterize the scale of agitation from 1 (lowest) to 10.

The size of the solids suspension problem is defined by the equivalent volume:

$$V_{eq} = s_{sl}\, V \qquad (16\text{-}18)$$

Where:

s_{sl} = specific gravity of the slurry (water = 1)

V = volume of slurry to be agitated, gal

Number and Location of Impellers

The number of impellers required for solids suspension is found from the aspect ratio of liquid height divided by

Table 16-12
Process requirements set degree of agitation for solids suspension [6]

Scale of Agitation	Description
1 to 2	Agitation levels 1–2 characterize applications requiring minimal solids-suspension levels to achieve the process result. Agitators capable of scale levels of 1 will: • Produce motion of all of the solids of the design-settling velocity in the vessel • Permit moving fillets of solids on the tank bottom, which are periodically suspended
3 to 5	Agitation levels 3–5 characterize most chemical process industries solids suspension applications. This scale range is typically used for dissolving solids. Agitators capable of scale levels of 3 will: • Suspend all of the solids of design settling velocity completely off the vessel bottom • Provide slurry uniformity to at least one-third of fluid batch height • Be suitable for slurry drawoff at low exit nozzle elevations
6 to 8	Agitation levels of 6–8 characterize applications where the solids suspension level approaches uniformity. Agitators capable of scale level 6 will: • Provide concentration uniformity of solids to 95% of the fluid batch height • Be suitable for slurry drawoff up to 80% of fluid batch height
9 to 10	Agitation levels 9–10 characterize applications where the solids suspension uniformity is the maximum practical. Agitators capable of scale level 9 will: • Provide slurry uniformity of solids to 98% of the fluid batch height • Be suitable for slurry drawoff by means of overflow

tank diameter, H/T. Ideally, $H/T = 1$ for a single impeller. Additional impellers are needed for each increment of approximately $H/2T$. In other words, if $H/T = 1.5$ then two impellers are required. See Table 16-13.

Table 16-13
Number of impellers for solids suspension [6]

Number of Impellers	Clearance Beneath Impeller		Maximum Ratio H/T
	Lower Impeller	Upper Impeller	
1	$H/4$	–	1.2
2	$T/4$	$(2/3)\,H$	1.8

Impeller Speed and Agitator Power

Assuming that the liquid depth is equal to the tank diameter ($H = T$), the rotational speed is calculated using the following procedure. Assign an impeller diameter. For solids suspension, an axial flow impeller is usually used. This data is for a pitched-blade turbine impeller. Taking the bottom head as an extension of the cylindrical shell (ignoring the volume reduction from the head's elliptical shape), the tank diameter is:

$$T = \left(\frac{4\,V}{\pi}\right)^{(1/3)} \quad (H = T) \tag{16-19}$$

The impeller diameter should be between about 0.25T and 0.6T. A graph in [6] relates the Scale of Agitation and D/T ratio to an expression called phi (Note: use indicated units):

$$\phi = \frac{N^{3.75}\,D^{2.81}}{w_s} \tag{16-20}$$

Where:

N = speed, rpm
D = impeller diameter, inches
w_s = settling velocity, ft/min

Rather than using the graph, compute phi with:

$$\phi = \exp(m\,\ln(SoA) + b)\,10^{10} \tag{16-21}$$

Where:
SoA = Scale of Agitation (1 to 10)

$$m = A + B\left(\frac{D}{T}\right) + C\left(\frac{D}{T}\right)^2$$

with A, B, and C selected from Table 16-14.

Table 16-14
Coefficients for the phi relationship

Scale of Agitation	A	B	C	b
<= 6	8.46	−28.9	17.0	1.23
> 6	3.79	−29.5	17.7	3.89

b is also selected from Table 16-14

Use Equation 16-21 to calculate the required impeller speed.

If the tank diameter and diameter are given, and the aspect ratio (H/T) > 1.5, perform the calculations as above for $H/T = 1$ but assume a second impeller is required for the next calculation, power.

The agitator power is correlated by the following expression, which includes an allowance for losses through the drive:

$$H_P = \frac{n\,N^3\,D^5\,s_{sl}}{(394)^5} \qquad (16\text{-}22)$$

Where:

H_P = minimum horsepower of the prime mover
n = number of impellers

If a variable speed drive is planned, the size of an electric motor is the first standard size at or above the calculated horsepower. However, fixed speed drives are limited to certain power/rpm combinations. Some of these are listed in Table 16-15 and Table 16-16.

To use the tables, select a drive power and speed from those listed for the desired scale of agitation and calculated

Table 16-15
Prime-mover power and shaft speed (hp/rpm) for solids suspension at terminal velocity = 10 ft/min, pitched-blade turbine [6]

Scale of Agitation	Equivalent Volume, gal							
	500	1,000	2,000	5,000	15,000	30,000	75,000	100,000
1	1/350	1/190	2/190	5/125	10/84	20/100	50/68	60/84
			1/100	3/84	7.5/68	15/68	40/84	50/68
				3/68	5/45	10/45	40/56	40/56
				2/45	3/37	7.5/37	20/37	30/37
2	1/230	1/100	2/125	7.5/125	20/100	40/84	100/100	125/68
			1.5/84	5/100	15/68	30/68	75/68	100/56
				5/84	10/45	25/56	60/56	75/45
				3/56	7.5/37	20/37	50/45	75/37
3	1/190	2/190	2/84	3/37	25/100	60/125	100/68	75/30
			1.5/56		20/68	50/100	100/56	60/20
					15/56	50/84	75/45	
					10/37	30/45	60/30	
4	1/155	2/155	5/155	7.5/84	30/100	60/84	150/84	200/68
		1.5/100		5/56	25/84	50/68	125/68	150/56
					15/45	40/56	75/37	125/45
						30/37		100/30
5	1/125	1.5/84	3/84	15/155	40/100	75/100	75/30	300/100
		2/125		10/100		60/68	60/20	250/84
				7.5/68		50/56		150/45
				5/45		30/30		125/37
6	1/100	2/100	5/125	10/84	40/84	75/68	250/84	300/68
		1.5/68	3/68		30/68	60/56	200/68	250/56
			3/56		25/56	50/45	150/45	200/45
			2/45		20/37	40/37	125/37	150/37
7	2/190	2/84	7.5/155	15/84	60/125	100/68	350/84	200/30
		1.5/56	7.5/125	10/56	50/100		200/45	150/30
			5/84	7.5/45	40/56		150/37	150/25
				7.5/37	30/45		100/20	
8	1.5/84	3/84	7.5/84	25/125	75/100	125/68	300/68	400/56
	2/125		5/56	20/100	60/84	100/56	250/56	350/45
				15/68	50/68	75/45	150/30	300/45
				10/45	30/37	75/37	125/25	250/37
9	2/84	7.5/155	15/155	40/155	75/68	75/30	400/56	
		5/125	10/100	30/100	60/56		300/45	
		5/100	7.5/68	25/84	50/45		250/37	
		3/68		20/68	40/37		200/30	
10	5/125	7.5/125	20/100	50/100	150/84	250/84	600/84	
		5/84	15/84	40/84	125/68	200/68	500/68	
			10/84	30/68	100/56	150/45	350/45	
				25/56	75/45	125/37		

Table 16-16
Prime-mover power and shaft speed (hp/rpm) for solids suspension at terminal velocity = 25 ft/min, pitched-blade turbine [6]

Scale of Agitation	Equivalent Volume, gal							
	500	1,000	2,000	5,000	15,000	30,000	75,000	100,000
1	1/230	2/190 1/190 1/100	2/125 2/84 1.5/84 1.5/56	5/125 3/84 3/68 2/45	20/100 15/68 10/45 7.5/37	30/100 25/84 20/68 15/45	75/100 60/56 50/45 40/37	125/68 100/56 75/68 75/37
2	1/190	2/125	3/84	15/155 10/100 7.5/68 5/45	30/100 25/84 20/68 15/45	60/84 50/68 40/56 30/37	150/84 125/68 100/56 75/37	250/84 200/68 150/45 125/37
3	1/100	1.5/84	5/125 3/68 2/45	10/84	40/84 30/68 25/56 20/37	75/84 60/56 50/45 40/37	250/84 200/68 150/56 125/45	400/100 200/45 150/37 100/20
4	2/190	2/84 1.5/56	7.5/155 5/100 3/56	7.5/45	60/125 50/100	75/68	300/100 150/45 125/37	300/68 250/56 150/30 125/25
5	2/155	2/68 2/56	7.5/125 5/84	15/84 10/56 7.5/37	75/125 40/84 30/45	100/68	400/100 200/45 150/37 100/20	150/25
6	2/125 1.5/84	3/84	5/56	25/125 20/100 15/68 10/45	60/84 50/68 40/56 30/37	125/68 100/56 75/45 75/37	300/68 250/56 150/30 125/25	400/56 350/45 250/37 200/30
7	2/84	7.5/155 5/125 5/100 3/68	15/155 10/100 7.5/84 7.5/68	30/100 25/84 20/68 15/56	75/68 60/56 50/45 40/37	75/30	400/56 300/45 250/37 200/30	
8	3/100	7.5/125 5/84	10/84	60/155 40/100 30/68 25/56	100/68 75/56	250/84 200/68 150/45 125/37	600/84 500/68 350/45	
9	5/155	10/125 7.5/100	15/84	75/190 60/125 50/100 40/84	150/84 125/68 100/56 75/45	400/100 350/84 200/45 150/37		
10	7.5/155 5/125	15/155 10/100	30/155 30/100 25/125 20/100	75/125 75/100 60/84 50/84	300/100 250/84 200/68 150/56	500/68 400/56 300/68 250/56		

equivalent volume. Then, use Equation 16-23, rearranged as:

$$D = 394 \left(\frac{H_P}{n\, N^3\, s_{sl}} \right)^{0.2} \qquad (16\text{-}23)$$

Gas Dispersion

The problem of injecting gas near the bottom of a tank of low viscosity liquid, and forming a large interfacial area so that the gas may be absorbed or reacted, is called gas dispersion. Agitators are used to break up gas bubbles and create the interfacial area between gas and liquid where mass transfer occurs. Gas dispersion in agitated vessels should be considered when moderate intensity is required, for fast reactions. High intensity situations, for very fast reactions and short residence times, may require different equipment, such as static mixers or thin-film contactors. Simple bubble columns, possibly with packing, may be appropriate for low intensity contacting, for slow reactions needing high residence time.

Radial flow impellers are usually used for gas dispersion. These include disk turbines and proprietary concave blade turbines. Gas is preferably introduced through a sparge ring located beneath the impeller. The sparge ring diameter should be less than the impeller diameter, ideally about 75% of the impeller diameter. A rule of thumb is to size the sparger holes such that the gas velocity through the holes is at least three times that through the pipe that is used for the sparge ring; this ensures even flow through all of the holes.

Recommendations for agitated vessels [11]:

- Locate sparger ring beneath a radial flow impeller. The ring diameter should be about 75% of the impeller diameter.
- Clearance beneath the impeller should be about one-fourth the tank diameter ($C = T/4$).
- Use four standard baffles.
- Aspect ratio for a single impeller should be about 1 ($H/T = 1$).
- A larger aspect ratio may be needed for obtaining more heat transfer surface area, providing a longer contact time for the gas, giving a staged counter-current system, or circumventing a mechanical limitation on available tank diameter. In this case, more than one impeller is required.
- Recommended impellers include disk turbines, hollow-blade radial flow designs, and upward pumping hydrofoils. Downward pumping hydrofoils or pitched-blade turbines may be unstable.
- Spacing between multiple impellers should be greater than their diameter. Vendors often recommend a combination of a radial flow impeller with one or more upward pumping hydrofoils.

Design calculations require physical property and process data including:

- Gas: flow rate (Q_G) (volumetric, computed at the pressure and temperature at point where it is introduced into the vessel).
- Liquid: density (ρ_L), viscosity (μ_L).

- Vessel: diameter (T), filled height (H).
- Impeller: type, diameter (D), distance from the tank bottom (C), Power number (N_P).

Superficial gas velocity is usually below 0.1 m/s (0.3 ft/s). The homogeneous regime in the vessel, where small, uniformly sized bubbles predominate in the mixing zone, exists when the velocity is below 0.02 m/s to 0.03 m/s (0.06 ft/s to 0.09 ft/s). At higher velocities, the heterogeneous regime occurs, where there are large bubbles in addition to the small uniformly sized ones. Flooding occurs when the gas flow is high enough to overwhelm the impeller; mixing is very poor in this regime. Calculate the superficial gas velocity with:

$$U_S = \frac{4\,Q_G}{\pi\,T^2} \tag{16-24}$$

The following two formulae depend on the agitator speed. Calculate the dimensionless flow number with:

$$N_{Fl,G} = \frac{Q_G}{N\,D^3} \tag{16-25}$$

Calculate the impeller Froude number with:

$$N_{Fr} = \frac{N^2\,D}{g} \tag{16-26}$$

Flooding occurs when the following relationship is true [11]. This condition is avoided by decreasing the gas flow rate, decreasing the impeller diameter, increasing the speed, or increasing the tank diameter.

$$N_{Fl,G} > 30\,N_{Fr}\left(\frac{D}{T}\right)^{3.5} \tag{16-27}$$

The transition between the homogeneous and heterogeneous regimes, with a Rushton impeller, occurs when [11]:

$$N_{Fl,G} > \sim 0.025\left(\frac{D}{T}\right)^{-0.5} \tag{16-28}$$

The agitator power is calculated from:

$$P = \frac{N_P\,(RPD)\,N^3\,D^5\,\rho_L}{g_c} \tag{16-29}$$

Where:
  $RPD = 0.18\,N_{Fr}^{-0.25}\,N_{Fl,G}^{-0.20}$ if in the heterogeneous regime, or
  RPD is linearly interpolated from the value computed at the flow number at the heterogeneous transition point and 0, with $RPD = 1$ at $N_{Fl,G} = 0$

In-Line Mixers

In-line (or static) mixers may be the best equipment choice for systems that require short residence time (typically less than 0.2 seconds) where only two or three components must be blended. The mixers come in various forms, but usually resemble a piece of pipe with internal elements that force the fluids together as they flow through the device. Most static mixers are housed in the same size or one-size-larger pipe than the adjacent runs of piping and are the same material and wall thickness [3]. By definition, there are no moving parts; this results in a highly reliable and very inexpensive piece of equipment.

Static mixers can be used for liquid-liquid, gas-liquid, gas-gas, liquid-solid, and solid-solid applications. The mixers work best with steady-state processes and miscible streams that aren't prone to fouling the mixer internals, but dispersions of immiscible liquids are possible.

Before calling a mixer manufacturer, gather data and answer questions that define the process requirements. See Table 16-17.

Table 16-17
Specification requirements for static mixers

Characteristic	Specification Requirements
Stream properties	For each stream to be mixed, specify the design physical state, flow rate, pressure, temperature, viscosity, density, miscibility, interfacial surface tension, vapor pressure (if applicable).
	Specify flow variations including magnitude and time duration.
	Specify the maximum allowable pressure drop (or minimum discharge pressure) and desired discharge temperature.
	Specify if any chemical reaction will occur and, if so, the heat of reaction and required residence time.
Hazards	Specify hazardous properties of the individual and mixed streams such as flash point, corrosivity, and toxicity.
	Specify any potential side reactions, especially those that could occur in the event of a leak or spill.
	Specify Code requirements.
Process goal	Specify the degree of mixing required, with quantitative and measurable criteria. The Variation Coefficient (CoV) is a standard statistical test used by the static mixer manufacturers; it is defined as the standard deviation divided by the mean. CoV values from 0.01 to 0.05 are often cited as reasonable targets. For CoV = 0.01, 95% of the concentration measured from all samples will be within $+/-2\%$ of the mean concentration. For CoV = 0.05, 95% of the concentration measurements will be within $+/-10\%$ of the mean concentration.
	Specify criteria for sampling: number of samples, sample size, and location(s)
	If applicable, specify the time limit for achieving the mixing.
Mechanical design	Specify desired materials of construction, whether the internal elements should be removable, and end connections.
	Specify internal surface finishes (e.g., BPE SF — see Chapter 13).
	Specify any space restraints and desired orientation (vertical or horizontal).
	Specify if the mixer must be jacketed and, if so, the properties of the heating or cooling medium.
	Specify other requirements such as insulation, insulation support rings, painting, nameplates, and instrument connections.

High Shear Mixers

The science behind high shear mixers is either poorly understood or shrouded by corporate secrecy. Unlike mixers with turbine impellers there are few published correlations or data from which to establish design criteria. Instead, small scale testing is generally used to determine parameters for commercial units. The manufacturers of high shear mixers maintain proprietary databases with performance information gathered over the years; they can assess measurements made in their test equipment and make empirical choices for the commercial units. Research is underway, however, and a more science-based approach may emerge [1].

Rotor-stator devices constitute a large proportion of high shear mixers used in agitated vessel applications.

Rotor-stators consist of a high speed impeller that forces the process fluid through slots or holes in a surrounding stationary housing. This imparts large shear stress on the fluid, used to facilitate the making of emulsions, dispersions, homogenizations, and products requiring grinding or cell disruption. The equipment manufacturers have their own proprietary designs, and they are best able to select an appropriate unit for a specific application.

Typical high-shear mixers have small impellers, 10% to 20% of the tank diameter, and operate at high speeds, 1,000 rpm to 3,600 rpm [3].

Typical applications are given in Table 16-18.

Table 16-18
Application of rotor-stator mixers [1]

Characteristic	Rotor-Stator Application
Shear stress	$20,000 \text{ s}^{-1}$ to $100,000 \text{ s}^{-1}$
Tip speed	10 m/s to 50 m/s (30 ft/s to 160 ft/s)
Major uses	Production of latex, adhesives, personal care products, cleaning products
	Dispersion and microdispersion
	Agricultural pesticides
Process fluid viscosity	Less than 150 Pa-s (150,000 cP)
Pumping in batch reactor	Low. Vessels larger than 40 liters (10 gal) are often equipped with an auxiliary axial impeller to provide flow, in addition to a rotor-stator device.

Mixing Solids

The mixing of particulate systems differs from liquid systems in three important respects [8]:

1. The rate of mixing is entirely dependent on the flow characteristics of the particulates and the handling pattern imposed by the mixer. Liquids diffuse into each other without external forces.
2. Powders and granules differ widely in physical characteristics such as size, density, and shape. Mixing motions that depend on identical particulate properties are unlikely to achieve their objective; this would more likely produce a "grading" or segregation of the particles. Miscible liquids will ultimately achieve a random distribution within a system regardless of differences in their molecular structures.
3. The discrete nature of particles means that the ultimate element of a particulate mixture is orders of magnitude larger than the ultimate molecular element of a liquid system. Samples withdrawn from a randomized particulate mixture will have a much coarser texture, or poorer mixture quality, than equivalent samples taken from gaseous or liquid systems.

Some powders segregate when mixed. Segregation can occur immediately, or the constituents may mix initially then segregate with further mechanical input. Well mixed powders in a bin may separate when conveyed through a pipe or discharged directly into another vessel. A major influence on the mechanism of mixing and segregation is the flow characteristics of the powder. Free flowing powders tend to separate, while cohesive solids do not. Different mixing mechanisms are used, with the flow characteristics being the primary driver for choosing one over another. See Figure 16-2 for a decision flow chart.

Some rules of thumb for particulate mixing are [8]:

- Materials with a size greater than 75 μm will readily separate.
- From 10 μm to 75 μm some segregation is likely, being just detectable at the lower limit.
- Below about 10 μm no appreciable segregation will occur.
- A process that suffers from serious flow problems will not suffer from segregation. A process with segregation problems will flow easily.
- The addition of very small quantities of moisture can transform a strongly separating mixture into a cohesive and non-segregating mixture.

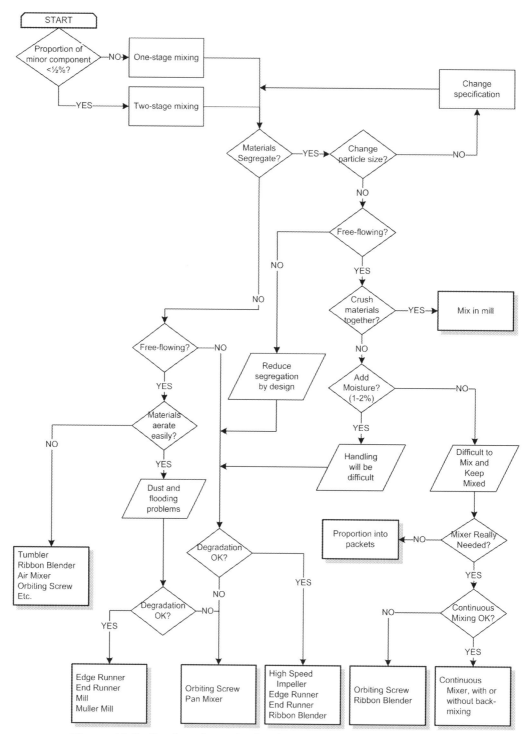

Figure 16-2. Powder mixer selection decision chart (adapted from [8]).

Mechanical Design

Torque

Torque is calculated from fluid properties, rotational speed, and the power number with:

$$\tau = \frac{N_P \, \rho \, N^2 \, D^5}{2 \, \pi} \qquad (16\text{-}30)$$

For multiple impellers, torque is cumulative.

Or, in US units, use [13]:

$$\tau_{Q(\max)} = \sum \frac{63025 \, P}{N} \qquad (16\text{-}31)$$

Where:

$\tau_{Q(\max)}$ = maximum total torque, in.-lb

Shaft Diameter

Using the distance from the individual impellers to the first bearing (see Figure 16-3), calculate the maximum bending moment (subscript i refers to each impeller on a common shaft) with:

$$M_{\max} = \sum \frac{C \, P_i \, L_i' \, f_H}{N \, D_i} \qquad (16\text{-}32)$$

Where:

C = conversion factor, 0.048 (SI units) or 19,000 (US units)

f_H = the hydraulic service factor, specific to each impeller, dimensionless [3]. See Table 16-19.

The minimum shaft diameter is the greater of [13]:

$$d_s = \left[\frac{16 \, \sqrt{\left(\tau_{Q(\max)}\right)^2 + \left(M_{\max}\right)^2}}{\pi \, \sigma_s} \right]^{1/3} \qquad (16\text{-}33)$$

$$d_t = \left\{ \frac{16 \left[M_{\max} + \sqrt{\left(\tau_{Q(\max)}\right)^2 + \left(M_{\max}\right)^2} \right]}{\pi \, \sigma_t} \right\}^{1/3} \qquad (16\text{-}34)$$

d_s = shaft diameter based on shear, m or in
d_t = shaft diameter based on torque, m or in
σ_s = allowable shear stress, N-m or in.-lb$_f$
σ_t = allowable tensile stress, N-m or in.-lb$_f$

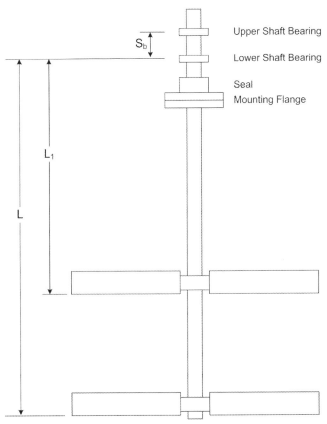

Figure 16-3. Nomenclature for impeller weight and critical speed calculations.

Table 16-19
Hydraulic service factors, f$_H$ [3]

Condition	High Efficiency Impeller	45° Pitched Four-Blade Impeller
Standard	1.5	1.0
Significant time at the liquid level	2.5 to 3.5	2.0 to 3.0
Operation in boiling systems	2.0 to 3.0	1.5 to 2.5
Operation in gas sparged systems	2.5 to 3.5	2.0 to 3.0
Large volume solid additions	3.0 to 5.0	3.0 to 5.0
Impacting of large solids	5.0 to 7.0	5.0 to 7.0
Startup in settled solids	5.0 to 7.0	5.0 to 7.0
Operation in a flow stream	1.5 to 7.0	1.0 to 7.0

For carbon steel and the common austenitic stainless steels, use stress limits of 6,000 psi for shear and 10,000 psi for tensile stress. These values account for dynamic loads, stress risers due to keys, set screws, and manufacturing tolerances. Allowable stress values may be extended to other materials by the ratio of yield strengths [13].

Round the calculated minimum diameter to the next higher standard shaft diameter. In the US, shafts are usually made of bar stock and come in ½-inch increments.

Impeller Weight

The equivalent weight of the impellers is given by [3]:

$$W_e = \sum W_i \left(\frac{L_i}{L}\right)^3 \qquad (16\text{-}35)$$

Where the subscript i refers to each impeller on a common shaft. W_i is the weight of impeller i, N (1 $kg_f = 9.81$ N) or lb_f. (1 $lb_f = 32.17$ lb_m).

Obtain impeller weights from the manufacturer if possible. For small forged or welded impellers, use Figure 16-4. For larger impellers, typically bolted to a hub, use the following equations to estimate the weight of the blades and Figure 16-5 for the weight of the hubs.

For 3-bladed hydrofoils, 380 mm to 2,300 mm (15 in. to 90 in.) diameter, this equation is about $+/-25\%$ accurate if the blade width to impeller diameter ratio is about one-sixth. Wide-blade hydrofoils can weigh two to three times the estimate given by the equation [3].

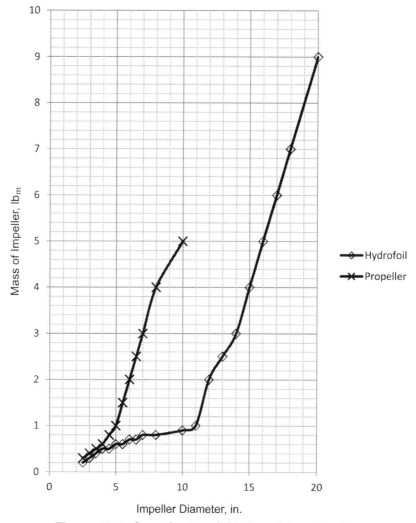

Figure 16-4. Approximate weight of small impellers [3].

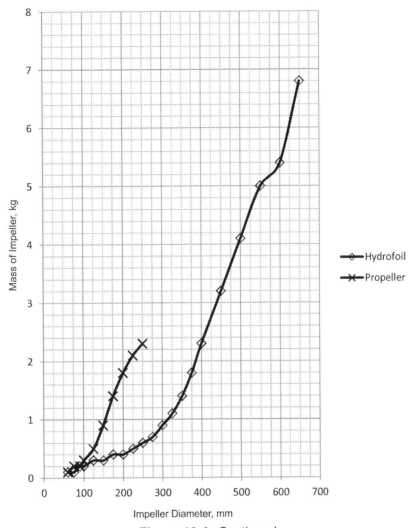

Figure 16-4. Continued.

For the total weight for four blades of a 45° pitched-blade turbine, 380 mm to 2,300 mm (15 in. to 90 in.) diameter, this equation has an accuracy of about +/−15%. The equation may be adjusted (by ratio) for two-blade or six-blade PBTs [3].

$$W_b = \sqrt{\frac{C\,D^3\,P_i}{N}} \qquad (16\text{-}36)$$

Where:

W_b = weight of impeller blades, kg or lb$_{\mathrm{m}}$
C = conversion factor,
hydrofoil: 0.14 (SI units), or 0.50 (US units)
four-blade PBT: 0.084 (SI units), or 0.30 (US units)
D = impeller diameter, m or in.

P_i = power calculated for the individual impeller, W or hp
N = speed, rps (SI units) or rpm (US units)

Critical Speed

Mixers must not be operated continuously at a speed that is near the natural frequency, or the frequency of free vibration, of the shaft and impeller. Lateral natural frequencies affect all mixer shafts (Table 16-20). The general rule is to keep agitators at least 20% away from the critical speed and its harmonics. Small portable mixers operating above 250 rpm are usually quickly accelerated through the first critical speed. Large mixers operating below 150 rpm are usually below the first critical speed [3].

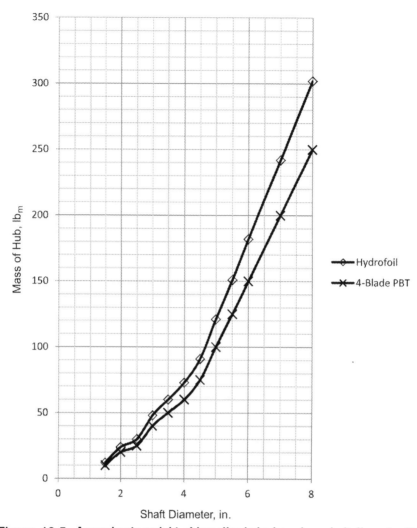

Figure 16-5. Approximate weight of impeller hubs based on shaft diameter [3].

See Figure 16-6 for a flow chart of the shaft design procedure.

The calculation given here is for a stiff agitator (i.e., well supported) under static conditions. The manufacturer should provide detailed calculations, including dynamic analysis, especially if this preliminary calculation is near the desired operating point. The manufacturer can also calculate the natural frequency for systems with hollow shafts and/or steady bearings.

Calculate the first critical speed with [3]:

$$N_C = \frac{C\, d^2 \sqrt{\dfrac{E_m}{\rho_m}}}{L\,\sqrt{L + S_b}\,\sqrt{W_e + \dfrac{w\,L}{4}}} \qquad (16\text{-}37)$$

Where:

N_C = critical speed, rps or rpm
C = conversion factor, 5.33 (SI units) or 37.8 (US units)
d = shaft diameter, m or in.
E_m = modulus of elasticity (tensile), N/m^2 or psi
ρ_m = shaft density, kg/m^3 or lb/in.^{3}
L = shaft length, m or in.
S_b = bearing span, m or in.
W_e = equivalent weight of impellers at point of calculation, kg or lb$_m$
w = weight of shaft per unit length, kg/m or lb/in.

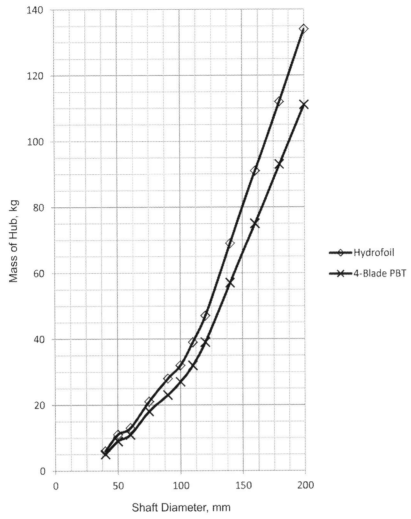

Figure 16-5. Continued.

Table 16-20
Metal properties for natural frequency calculation [3]

Metal Type	Modulus of Elasticity, E_m		Density, ρ_m	
	psi x 10^6	N/m^2 x 10^{12}	lb/in.^{3}	kg/m^3
Carbon steel	29.8	0.205	0.283	7833
Stainless steel 304/316	28.6	0.197	0.290	8027
Hastelloy C	30.9	0.213	0.323	8941
Hastelloy B	30.8	0.212	0.334	9245
Monel 400	26.0	0.179	0.319	8830
Inconel 600	31.0	0.214	0.304	8415
Nickel 200	29.7	0.205	0.322	8913
Alloy 20	28.0	0.193	0.289	7999

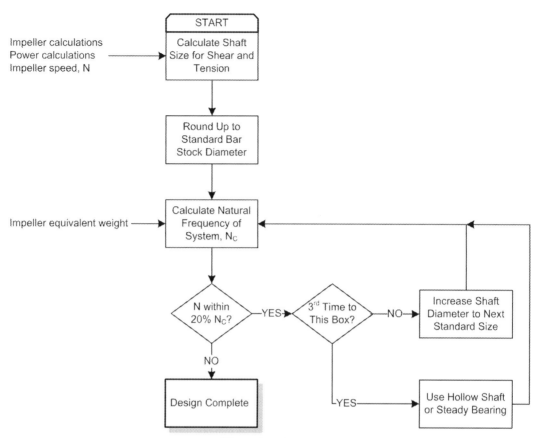

Figure 16-6. Mixer shaft design procedure [3].

Nomenclature

A, B = coefficients dependent on particle shape factor and roundness

B = baffle width, m or ft

C = conversion factor, 5.33 (SI units) or 37.8 (US units)

D = impeller diameter, m or ft

d = shaft diameter, m or in.

d_N = nominal particle diameter, cm or ft

d_s = shaft diameter based on shear

d_t = shaft diameter based on torque

E_m = modulus of elasticity (tensile), N/m^2 or psi

f_H = the hydraulic service factor, specific to each impeller, dimensionless

H = height of liquid in the tank, m or ft

h = overall height of helical coil impeller or projected blade height of a turbine impeller

H_P = minimum horsepower of the prime-mover

g = gravitational acceleration, 9.81 mm/s^2 or 32.17 ft/s^2

g_c = conversion factor, 1 m/s^2 or 32.17 ft/s^2

L = shaft length to mounting surface, m or in.

L = shaft length, m or in.

L' = shaft length from hydraulic force to first bearing, m or in.

N = agitator speed, revolutions/s, rpm, or revolutions/h

N_C = critical speed, rps or rpm

n = number of impellers

N_{Fo} = Fourier number, dimensionless

N_{Fr} = Froude number, dimensionless

$N_{Fl,G}$ = Flow number for gas, dimensionless

N_{Re} = Reynolds number, dimensionless

N_P = Power number, dimensionless

N_Q = Pumping number, dimensionless

M_{max} = maximum bending moment, N-m or in.-lb$_f$

P = shaft power, W or ft-lb$_f$/s (divide by 550 for hp)

s = specific gravity of solids $\left(s = \dfrac{\rho_s}{\rho_l}\right)$

S_b = bearing span, m or in.

s_{sl} = specific gravity of the slurry (water = 1)

U_S = superficial velocity, m/s or ft/s

V = volume of slurry to be agitated, gal

W = projected width of impeller blade

W_e = equivalent weight of impellers at point of calculation, kg or lb$_m$

w = weight of shaft per unit length, kg/m or lb/in.

w_s = settling velocity, cm/s or ft/s

ρ = fluid density, kg/m^3 or lb/ft^3

ρ_m = shaft density, kg/m^3 or lb/in^3

σ_s = allowable shear stress, N-m or in.-lb

σ_t = allowable tensile stress, N-m or in.-lb

τ = torque, N-m or in.-lb

μ = fluid dynamic viscosity, kg/m-s or lb/ft-s

v = kinematic viscosity of liquid phase, cm^2/s or ft^2/s

References

[1] Atiemo-Obeng V, Calabrese R. Rotor-Stator Mixing Devices. Chapter 8. In: Paul E, Atiemo-Obeng V, Kresta S, editors. *Handbook of Industrial Mixing: Science and Practice*. Hoboken, NJ: North American Mixing Forum, John Wiley & Sons, Inc.; 2004.

[2] Atiemo-Obeng V, Penny WR, Armenante P. Solid-Liquid Mixing. Chapter 10. In: Paul E, Atiemo-Obeng V, Kresta S, editors. *Handbook of Industrial Mixing: Science and Practice*. Hoboken, NJ: North American Mixing Forum, John Wiley & Sons, Inc.; 2004.

[3] Dickey D, Fasano J. Mechanical Design of Mixing Equipment. Chapter 21. In: Paul E, Atiemo-Obeng V, Kresta S, editors. *Handbook of Industrial Mixing: Science and Practice*. Hoboken, NJ: North American Mixing Forum, John Wiley & Sons, Inc.; 2004.

[4] Dickey D, Fenic J. Dimensional Analysis for Fluid Agitation Systems. *Chemical Engineering*, January 5, 1976:139–45.

[5] Dickey D, Hicks R. Fundamentals of Agitation. *Chemical Engineering*, February 2, 1976:93–100.

[6] Gates L, Morton J, Fondy P. Selecting Agitator Systems to Suspend Solids in Liquids. *Chemical Engineering*, May 24, 1976:144–50.

[7] Grenville R, Nienow A. Blending of Miscible Liquids. Chapter 9. In: Paul E, Atiemo-Obeng V, Kresta S, editors. *Handbook of Industrial Mixing: Science and Practice*. Hoboken, NJ: North American Mixing Forum, John Wiley & Sons, Inc.; 2004.

[8] Hamby N, Edwards M, Nienow A. *Mixing in the Process Industries*. Oxford: Butterworth–Heinmann; 2001 (digital edition).

[9] Hermajani R, Tatterson G. Mechanically Stirred Vessels. Chapter 6. In: Paul E, Atiemo-Obeng V, Kresta S, editors. *Handbook of Industrial Mixing: Science and Practice*. Hoboken, NJ: North American Mixing Forum, John Wiley & Sons, Inc.; 2004.

[10] Jiménez J, Madsen O. A Simple Formula to Estimate Settling Velocity of Natural Sediments. *Journal of Waterway, Port, Coastal and Ocean Engineering*, American Society of Civil Engineers March/April 2003:70–8.

[11] Middleton J, Smith J. Gas-Liquid Mixing in Turbulent Systems. Chapter 11. In: Paul E, Atiemo-Obeng V, Kresta S, editors. *Handbook of Industrial Mixing: Science and Practice*. Hoboken, NJ: North American Mixing Forum, John Wiley & Sons, Inc.; 2004.

[12] Pietranski J. Mechanical Agitator Power Requirements for Liquid Batches. K-103 Course Notes, PDH Online, www.pdhonline.org.

[13] Ramsey W, Zoller G. How the Design of Shafts, Seals and Impellers Affects Agitator Performance. *Chemical Engineering*, August 30, 1976:101–8.

[14] Walas S. *Chemical Process Equipment: Selection and Design*. Butterworth-Heinemann; 1990.

17
Process Evaluation

Introduction

This chapter discusses deliverables that process engineers typically produce for a significant capital project. *Deliverables* are specific work products, such as drawings and data sheets. The work plan presented here has a critical underlying requirement:

Engineers must communicate with management, stakeholders, colleagues, and suppliers to ensure that requirements and expectations are fully aligned every step of the way. Communication is a two-way exchange of facts, opinions, and feelings. The primary purpose of the deliverables listed in this chapter is to facilitate communication by collecting and organizing the engineer's work into a structured set of progressively more detailed documents.

The deliverables are usually grouped into project phases such as Initiation, Preliminary Design, and Detailed Design (see Table 17-1). An alternative group of phases is Concept, Design Development, and Contract Documents. There may be intermediate milestones that are marked by funding requests or management approvals.

The process engineer's deliverables fit into a larger project context, with accompanying documents from other disciplines (e.g., architectural, mechanical, electrical), cost estimates, and project schedules. Each deliverable will normally undergo multiple iterations as input is received from the engineer's work and comments from management, stakeholders, colleagues, and suppliers. Difficult aspects, where deeper analysis and discussion are required

Table 17-1
Typical deliverables from process engineers by project phase

Project Phase	Typical Deliverables
Initiation or Concept	Outline of user requirements (products, capacities, technologies, etc.) Block Diagrams Orientation-grade cost estimate ($+/-35\%$ to 50%) and schedule
Preliminary Design	User Requirement Specifications Process Flow Diagrams (PFDs) Material and Energy Balances Basis of Design (BOD) Equipment Data Sheets (preliminary) General Arrangement Drawings (preliminary) Site/Plot Plans (conceptual) Preliminary cost estimate ($+/-$ 20% to 30%) and schedule
Detailed Design	Block-Sequential Diagrams (batch plants only) Piping & Instrumentation Diagrams (P&IDs) Equipment Specifications Instrument Specifications General Arrangement Drawings (final) Equipment Lists Instrument Lists Line Lists Appropriation-grade cost estimate ($+/-$ 10% to 15%) and schedule

to reach consensus, are often documented in separate reports; these are part of the body of supporting documentation that also include calculations, meeting minutes, and management directives.

Accuracy of Cost Estimates

Estimates are usually prepared at project milestones such as initiation, conceptual design completion, detailed design completion, and construction initiation. Decision makers expect a certain level of accuracy in the estimates; they often treat estimates as "not to exceed" values that include all contingencies and uncertainties. Engineers typically assign an uncertainty range to an estimated cost, such as "$+/-30\%$", meaning that the realized cost for the facility should fall within the range and it could be higher or lower than the reported amount. Estimates also include contingencies for items that are "in scope" but not estimated due to lack of sufficient granularity in the source documents.

Two of the biggest pitfalls in estimating project costs are the definition of scope and the treatment of uncertainty and contingency. Engineers should work hard to ensure alignment of all team members, stakeholders, and approvers.

Definition of Scope

What is included in the scope of a cost estimate? This can be difficult to answer, especially when trying to explain why an estimate increased (or decreased) after the project was funded. In an ideal project, the "as-is" state is fully known, everyone associated with the project shares expectations of what the "completed" project looks like and how it functions, and the pathway to achieving completion is agreed and understood. This information is written down and everyone signs an affidavit attesting to their agreement. Unfortunately, most projects are far from ideal, and difficulties arise when individuals identify deviations from their understandings or expectations.

The kinds of questions to address when a project is initiated include:

- What are the project goals with respect to throughput, product specifications (e.g., purity), operating schedule, flexibility (to accommodate additional products), reliability and redundancy, operating costs (automation), etc.?
- Where are the project's battery limits? For example, does a tank farm project include environmental controls for the vents? Are the controls local or centralized? Do centralized controls need to accommodate additional existing or future tanks?
- What technology is required? For example, must the latest inventions from the company's R&D department be utilized in the project, or does the older (and well established) method suffice?
- How will unexpected (hidden) conditions be funded if found, such as asbestos during demolition or rock during excavation? Some projects include an allowance for hidden conditions while others explicitly recognize that if such conditions are found additional funding will be provided.

Treatment of Uncertainty and Contingency

Uncertainty and contingency are different concepts, related by the fact that they modify the cost obtained by simply summing the estimated value of identified line items. *Uncertainty* is like a confidence interval; it addresses the range of possible costs for a line item (for example, the purchase price of a centrifugal pump could range from $10,000 to $30,000 depending on the size, materials of construction, features, manufacturer, reseller, and other factors). It is sometimes possible, after knowing this range, to fix the price (or greatly reduce the range) by allowing flexibility to the specifications in order to achieve the desired cost. Uncertainty is greatest when very little is known about the item. It should decrease as the project is developed.

Uncertainty also addresses the potential for scope variation, in the sense that line items may be added (or removed) to the list of items estimated as more is known about the project. For example, the number of portable bins included in the estimate for a pharmaceutical oral solid dosage facility may change as the layout and operating philosophy are developed; the accuracy for the cost of each bin could be high, but the overall estimate for portable bins might be poor if the quantity of bins changes significantly.

Contingency is added to a cost estimate in one or more line items that specifically state that the item is an allowance for contingency. There are four major definitions for contingency, with variations and combinations. This is why it is critical to discuss and agree with the project team the use of contingency. The major definitions are:

1. Recommended: Contingency is an allowance for unknowns that are within the agreed project scope. This includes anything that is not specifically itemised in the cost estimate, but is needed to complete the project. It is similar to the scope variation (above), but not the same as accounting for estimate confidence. All projects experience this kind of cost because estimates are made before the design is complete. The value of contingency for this allowance is usually decreased as the project scope solidifies – never decreasing to zero – and project stakeholders should expect the money to be spent.
2. Contingency is an allowance for unforeseen events, such as the discovery of hidden conditions (asbestos or rock, in the example above). This type of contingency is normally expected to *not* be spent, and many companies allocate it across a range of projects. Through experience or a risk analysis, the company might estimate that 20% of projects require some expenditure for unforeseen events, and hold an appropriate amount of money in a separate "reserve" fund. In this sense it is much like insurance.
3. Contingency is an allowance to account for the estimate confidence. If used in this way, the reported cost estimate would have a range of $+x\%/-y\%$, with $x < y$. For instance, say the estimate range is judged to be $+/-20\%$ and includes an allowance for undefined scope (item 1 above). If another 10% contingency is added, the resulting total would have a range of $+10\%/-30\%$.
4. Contingency is an allowance for scope creep. This is poor practice, but some people like to use contingency in this way to reduce the need to negotiate for more funding if scope creep occurs, and to provide extra assurance for achieving the cost estimate target.

Cost estimates are never 100% accurate. Even those projects that are fully defined receive a range of proposals from vendors and contractors. It is common to receive bids that vary more than $+/-10\%$ from the median or mean. However, it is often feasible to make adjustments to the project scope or contract approach in order to achieve a certain value. As a project nears completion and there is money left over, it could be spent on additional scope items. Conversely, if the project is trending above the budget, scope could be eliminated or deferred. In each case the end result is a project falsely realized for its budget; these practices may provide immediate gratification, but negative consequences in the long term. The company is misinformed about the true cost of the project. Future budgets that use this project as a benchmark will be hurt. If scope was eliminated to make the number then inevitably the items will be funded and added later, usually at higher cost than if they had simply been done initially (with a budget overrun).

Table 17-2
Matrix of deliverables typically needed to support cost estimates of various uncertainty ranges

Process Engineering Deliverable	Orientation > $\pm$ 35%	Conceptual $\pm$25% to $\pm$35%	Preliminary $\pm$20% to $\pm$30%	Detailed $\pm$10% to $\pm$20%	Definitive $\pm$5% to $\pm$10%
Product data and process description	X	X	X	X	X
Plant capacity	X	X	X	X	X
Location − general	X	X			
Location − specific			X	X	X
Basic design criteria			X	X	X
Block flow diagrams	X				
PFDs		X	X		
P&IDs				X	X
Equipment list		X	X	X	X
Site conditions and data				X	X
Preliminary equipment sizes	X	X	X		
Finalized equipment sizes; data sheets, specifications				X	X
Site plot plan and elevations			X	X	X
General Arrangements			X	X	X
Scope − testing and verification		X	X	X	X

User Requirements

Usually called "User Requirements Specification" (URS), this document is often misused or misconstrued. The purpose of a URS is to establish the minimum acceptable attributes that the end user of the equipment, system, or facility wants to accept. The attributes listed in a URS should be specific and measurable; they can cover any aspect of the design, construction, or performance of the system. Best Practice is to list each requirement in a numbered statement. This allows the creation of a trace matrix during the project execution that ties together the URS and design, construction, and testing documents. Misuse occurs when the URS is substituted for an equipment data sheet or specification.

The engineer will use the URS along with many other source documents such as Codes, calculations, maintenance records from the plant, and other project specifications to develop the body of material required to procure, install, and test the equipment and systems. It is likely that aspects of the equipment specifications will exceed requirements listed in the URS (see Table 17-3). There are several reasons for this, including: assurance that performance perturbations will not cause the equipment's output to dip below the URS value, incorporation of wear factors to keep the performance above URS requirements as the equipment ages, standardization to off-the-shelf vendor designs that exceed actual need,

Table 17-3
Checklist for creation of a URS

Style Point	Description
Precision, ambiguity, and clarity	Each listed item is exact, not vague or open to interpretation Each item has only one interpretation The meaning of each item is understood; jargon is eliminated or defined Statements are easy to read
Consistency	No item is in conflict with another item within the URS or with an item in another URS for the project
Relevancy	Each item pertains to the equipment, system, or facility being described
Testability	It is possible to verify that each item is delivered using objective and measurable criteria
Feasibility	Each item can be implemented with the technologies, tools, resources, and people that are assignable to the project, and implementation can be achieved within the project's cost and schedule constraints
What, not How	Each item presents a requirement (the What) that must be satisfied, and is free of the problem solution (the How) unless a specific solution is a requirement
Understood and agreed	Each item is both understood and agreed to by the stakeholders involved in the project

adherence to Codes or standards that are more stringent that the user's requirement, etc.

A typical Table of Contents for a URS will include the following. See http://www.ispe.org/jett/jett-sample-documents for templates and examples.

Cover Page with:
- Title
- Issue Date
- Document Number and Project Number if applicable
- Revision Level
- Author
- Approvals

Revision History

Table of Contents

Introduction
- Provide a general overview of the equipment or system that is being specified, including the scope of the equipment or system (answering the question, "What are the system boundaries?")
- Describe who produced the document, under what authority, and for what purpose.
- Describe the contractual status of the document. Include or reference as an attachment the "legal" paragraph that communicates purchasing terms and conditions in this section.
- Describe the relationship of this document to other documents. This is important for equipment that is part of an integrated process or line and will help the Supplier to understand and ask questions that may otherwise be overlooked. [Note: It may not be appropriate to send the URS to a supplier, especially if it contains sensitive or superfluous information.]

Overview
- This section should provide a high level description of the system explaining why it is required and what is required of it. Include the background, key objectives, and the main functions and interfaces.
- Use: Describe what applications the equipment or system will be used to support. Include a brief explanation of the general functions.
- Capacity: Provide a brief description of the equipment/system capacity requirements.

Operational Requirements
- Capacity: Provide as much detail as is known about the range of products, commodities to be used (e.g., rolls of certain width, diameter and weight), process materials to be used or involved (e.g., liquid additives), and production rates.
- Process Requirements: Discuss quality of products and concentrations. Define any product limitations such as temperature, humidity, etc.
- Process Control: What are the desired measurement ranges, setpoints, parameter control tolerances, etc.? At a minimum list the critical process parameters. Consider documenting how each of the parameters was determined – this helps the Supplier know if he must, for example, provide a custom-engineered design rather than use an off-the-shelf model that doesn't quite meet the stated goal.
- Functions: This section is especially applicable to automated systems and includes:
  - Operation – specific requirements such as number of cycles, number of recipes, operator

interaction (e.g., through a control panel, touchscreen, workstation, etc.).

  ◦ Power failure and recovery – what is expected to happen when power fails and how should the system behave when power is restored.

  ◦ Emergency stop – specific E-Stop strategy.

  ◦ Alarms and warnings – how should the system behave when critical alarms are engaged; what informational messages should be sent when non-critical alarms (e.g., alert levels) are reached.

• Data and security: If the automated system will store electronic records required for regulatory (i.e., FDA) compliance, then the requirements of 21CFR Part 11 must be satisfied.

  ◦ User interface – access levels, frequency of data point collection, hardcopy/electronic data collection requirements, data retention time on the system, data storage media.

  ◦ User interface strategy – language requirements, displayed units, interface with other equipment (e.g., Ethernet), security levels, data collection (record, print, graph, color, etc.).

• Environment: provide details of the physical environment including layout, cleaning conditions, and operating environment (e.g., cGMP).

Constraints

• Milestones and timeline.

• Equipment constraints: Describe any operating constraints such as available utilities.

• Compatibility and support: Include any requirements for compatibility with other systems; list any preferred vendors.

• Availability: Is the equipment expected to operate 24×7?

• Procedural constraints: This section includes product contact materials of construction, noise level, EMI/RFI levels, containment and labeling requirements.

• Maintenance: Describe expected hardware and software maintenance support including ease of maintenance, expansion capability (software), likely enhancements (software), expected lifetime, and long term support requirements.

Life Cycle

• Development: Specify any standards or methodologies that the Supplier is expected to follow, including applicable company procedures.

• Testing: Describe Supplier testing requirements such as Factory Acceptance Tests.

• Delivery: Include listing of required documents and how they are expected to be formatted (e.g., Microsoft Word).

• Support: Describe what support activities are required after acceptance. Consider start-up support, training, technical support (telephone/modem, replacement parts), and on-site support (preventative maintenance, calibration, system improvements/upgrades).

Block Diagrams

Block diagrams are the simplest and easiest way to depict a process. They are particularly useful for communicating the major unit operations to non-technical audiences, and for illustrating documents such as project plans and process descriptions.

Keep the uniformly sized blocks arranged neatly on a grid to enhance readability and comprehension. The text in the blocks can be limited to equipment names, or supplemented with information that clarifies the function and intent. Standard software packages that include drawing tools, such as Microsoft Office, or flowcharting software, like Microsoft Visio, are useful tools for creating block diagrams.

The first example (Figure 17-1) illustrates the bare minimum content. Notice that two block sizes are used; the larger size is for major functions and the smaller blocks are minor functions. The blocks don't all correspond to single equipment units although this process uses piped transfer between the steps.

The second example (Figure 17-2) is more complex. In this case the primary process units are depicted in a large shaded box that represents a process room. Solid connectors represent pipes while dotted lines indicate movement of portable tanks (IBC and Bin).

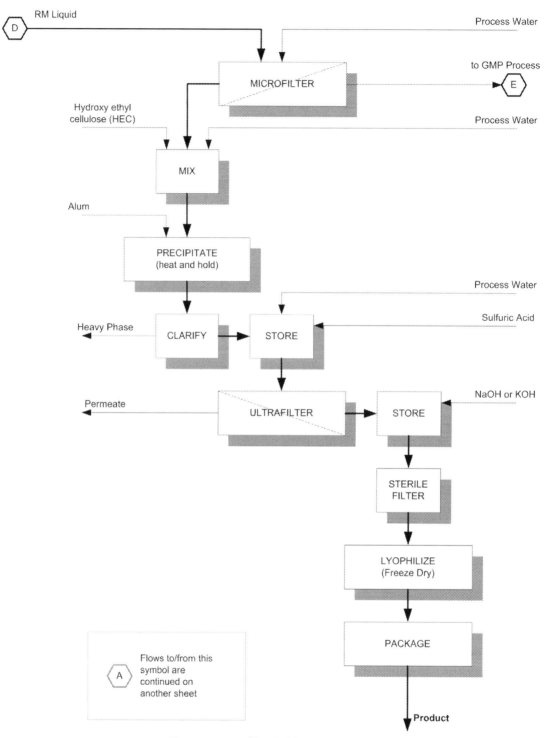

Figure 17-1. Simple block diagram.

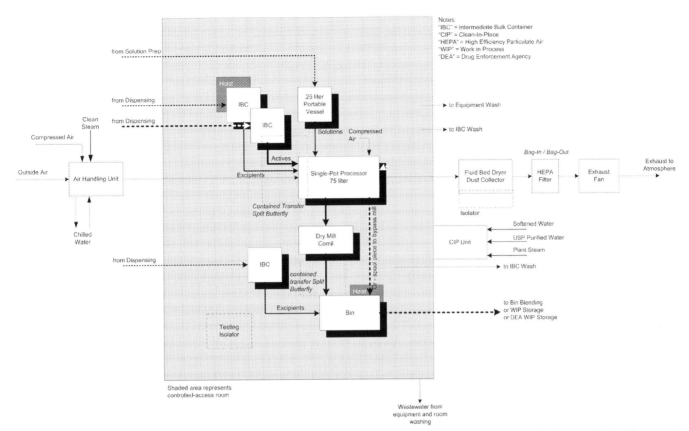

Figure 17-2. Block diagram for a single-pot processor system, used to prepare the ingredients for drug tablets. The environment and operators must be protected from exposure to highly active ingredients.

Process Flow Diagrams (PFDs)

Process Flow Diagrams (PFDs) depict major equipment and controls for the process. Major equipment is usually that which would appear in a process simulation (e.g., Aspen) such as columns, reactors, vessels, heat exchangers, pumps, compressors, agitators, filters, etc. In fact, the output from a simulator is often used as the project's PFDs with no additional enhancements.

Information included on PFDs typically includes: equipment numbers and brief (a few words) descriptions, material and energy balances, equipment sizes (often rough, or preliminary), materials of construction, flow direction arrows on pipelines, and simplified primary process control loops. Critical in-line components like

control valves may be shown. Physical installation concepts including indoor/outdoor, room walls, and platforms are sometimes shown. Notes are added to explain any important features that are not obvious from the drawing.

People like to visualize an entire process on a single drawing, so PFDs will contain much more equipment on a drawing than the subsequently created P&IDs. This leaves less space for ancillary equipment, features, and pipelines which would only serve to dilute the main message of the drawing. Therefore, PFDs usually omit utility lines, secondary instrumentation, and specialty items like expansion joints. See the example in Figure 17-3.

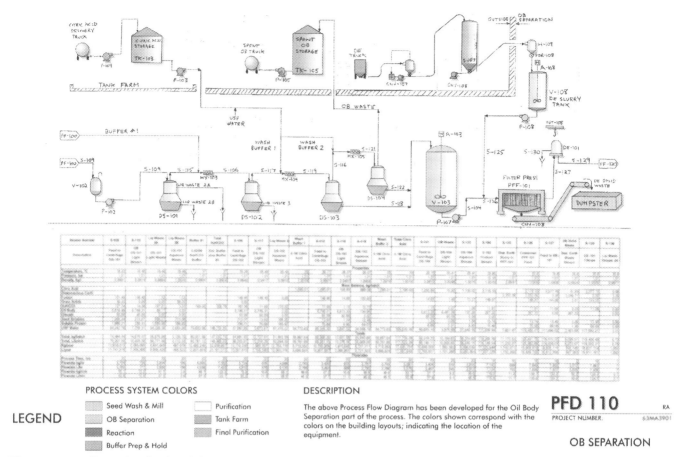

LEGEND

PROCESS SYSTEM COLORS

- Seed Wash & Mill
- OB Separation
- Reaction
- Buffer Prep & Hold
- Purification
- Tank Farm
- Final Purification

DESCRIPTION

The above Process Flow Diagram has been developed for the Oil Body Separation part of the process. The colors shown correspond with the colors on the building layouts; indicating the location of the equipment.

PFD 110   RA

PROJECT NUMBER.   6.3MA.3903

OB SEPARATION

Figure 17-3. Example of a hand-drawn Process Flow Diagram with material and energy balance. Several sheets were needed to illustrate the entire process; the Legend defined colors used to shade equipment in different parts of the plant; a corresponding conceptual layout drawing used the same colors to make it easy for reviewers to follow the process.

Basis of Design

The Basis of Design (BOD) is a technical report that brings together the project objectives, assumptions, constraints, data, alternatives, and design solutions. It is usually prepared by the entire technical project team, with process engineers contributing significant portions, including PFDs, material and energy balances, process descriptions, and preliminary equipment data sheets. The primary purpose of the BOD is to provide the information needed to fully execute the engineering for the project. Therefore, it is a critical and essential document that has the power to facilitate – or disrupt – the detailed design phase of the project.

Typical sections in the BOD include:

- Introduction and background
- Overall process and design parameters, including capacity and throughput analysis, raw material requirements, and effluent identification and constraints
- Overall process design criteria used to prepare material and energy balances, and key data and assumptions required to prepare major categories of process calculations (e.g., reaction stoichiometry, and vapor-liquid equilibria)

- Process description that explains the function of each unit operation
- Process Flow Diagrams and material and energy balances
- Safety
- Process control criteria and strategy
- Environmental protection design basis

- Materials of construction selection
- Overall facility description, including building Code compliance
- Design criteria for civil, structural, architectural, mechanical, electrical, and fire protection disciplines
- Support facilities
- List of applicable Codes (all disciplines)

Block-Sequential Diagrams

PFDs can be difficult to explain for a batch process. Instead of a steady-state flow of materials in and out of the equipment, a batch process consists of consecutive phases with each phase involving different equipment, flows, and heat transfer. The Block-Sequential Diagram is designed to communicate the process phases, using a time-anchored block diagram.

Batch processes utilize a train of equipment. Each piece in the train cycles through a recipe consisting of the phases, beginning and ending with an idle state. For instance, a mix tank may start in a clean state then over the course of a few hours:

- Operator sets up the tank with filters and fittings needed for a batch
- Pressure test
- Fill with water
- Add solid ingredients from bulk bins
- Heat and mix
- Add oil phase from another tank
- Mix at higher shear rate to emulsify mixture
- Sample and check water content
- Add water or evaporate water as needed to meet specification
- Cool
- Adjust pH
- Transfer to hold tank
- Remove filters and components that are cleaned in a washer
- Clean vessel using automated clean-in-place system
- Dry and return to idle state, ready for next batch

This recipe, and the recipes for all of the equipment in the train, is described on the block-sequential diagram. Estimate the time required for each step. Then create a stacked bar chart with the elements of the bar representing the steps, one bar for each piece of equipment. The bars are plotted with a timeline so the interactions between equipment pieces line up. A partial example is given in Figure 17-4; notice that the steps labeled "diafiltration" and "concentration" are interactions between the two pieces of equipment.

The entire block-sequential diagram may span days in the vertical axis, with as many columns as necessary to cover the entire process. Typically, more than one batch is shown so it's clear how equipment transitions from one batch to the next and how many batches may be in process simultaneously (at Hour 10, Batch 1 may be located in Column 6 while Batch 2 is being processed in Column 1). The diagram can be visually inspected to see where operators are needed, peak utility loads occur, and when idle time is expected. Blocks can be moved up or down (dragging adjacent interactions along) to force operations to begin at certain times (such as the beginning of a shift) or to alleviate bottlenecks (such as limited resource availability).

A great way to use the block-sequential diagram is to review with the team of engineers, operators, laboratory support technicians, and other stakeholders who will run and support the plant. Step through the plant operation with the diagram; other project documents like equipment arrangements and PFDs should be available for reference. This exercise, which may take several days to complete, will provide insights that will strengthen the design concept and help each of the stakeholders plan their own work. Operators will be in a better position to write batch instructions. Lab technicians will discover who will obtain in-process samples and what the expected analytical turnaround time is. Engineers will learn that additional instrumentation or interlocks are required. Everyone will feel they are aligned and invested in the plant design.

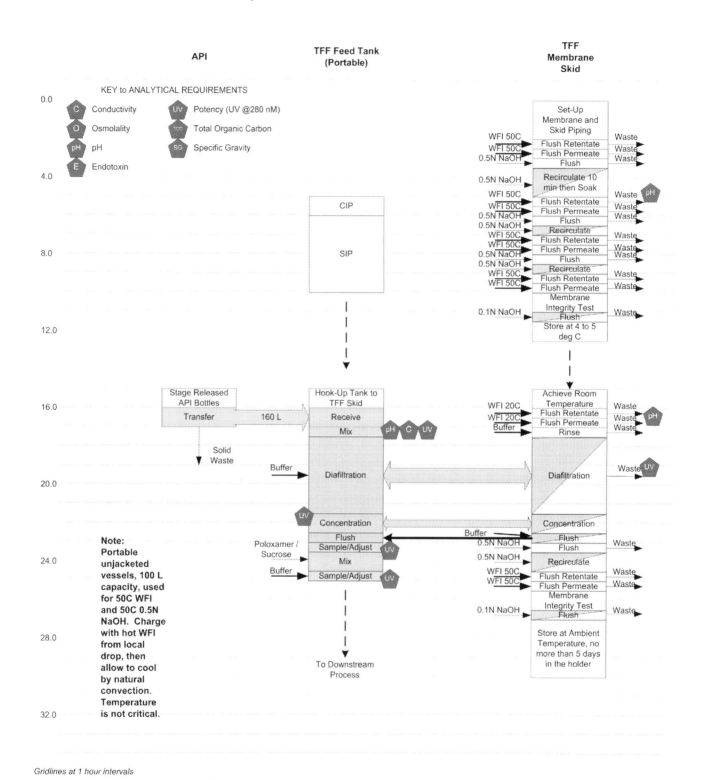

Gridlines at 1 hour intervals

Figure 17-4. Excerpt from a Block-Sequential Diagram. Labels from a traditional block diagram are arrayed across the top ("API," "TFF Feed Tank," "TFF Membrane Skid"). The blocks are expanded down the timeline to show how the equipment is used over time. Material volumes and flow rates could also be shown. "TFF" means Tangential Flow Filter. "WFI" means Water for Injection, a highly purified water used in pharmaceutical processes; "API" means Active Pharmaceutical Ingredient.

Piping & Instrumentation Diagrams (P&IDs)

Piping & Instrumentation Diagrams (P&IDs) are the most fundamental and definitive depiction of the plant, covering all systems and components. They span the process, from raw material receipt to product delivery and from utility generation to vapor, liquid, and solids effluent. They are capable of communicating a range of information to management, stakeholders, users, engineers, and contractors; the effectiveness of the communication is partly governed by the care used in creating the drawings.

Process engineers are usually responsible for preparing P&IDs, with significant input from automation and mechanical engineering groups. Other engineering disciplines have similar schematic diagrams that complement P&IDs, including air flow diagrams (for HVAC systems), riser diagrams (for building plumbing systems), single-line diagrams (for electrical power), and loop diagrams (for instrument wiring).

P&IDs are critical drawings that must be produced to meet certain regulatory requirements (e.g., OSHA

Table 17-4
P&IDs can convey information more effectively if the designer thinks beyond the piping and instruments

Consideration	Effective P&ID Practices
Readability. Users more readily understand the content of a P&ID when the drawing is artistically arranged	Limit the number of major pieces of equipment to maintain focus. For example, use one P&ID for each tank in a tank farm especially if the tanks are independent. Preserve white space on the drawing so it doesn't look too busy.
	Avoid crossing pipelines as much as practicable, especially major lines. Use at least three different line weights for piping: major process lines, minor process and utility lines, supporting lines (e.g., air connections, drains).
	Draw equipment with actual size in mind. P&IDs are never drawn to scale, but if three vessels of different volumes are depicted the graphical depictions should be different sizes.
	Draw equipment with elevation in mind. Gravity flow pipelines should be represented such that the origin is above the destination on the drawing.
	Provide adequate space for all the instruments, including bubbles for transmitters, switches, alarms, etc. See Chapter 21, "Controls," for more on this.
	Locate valves, connections, and instruments in relative location to equipment. Example: a check valve at a pump discharge should be drawn right next to the pump, not several inches away. Although it would be technically correct to show the check valve anywhere as long as the pipe connectivity, branches, and instruments are in the correct sequence, it is much friendlier to the user to convey a sense of location.
	Use an imaginary grid and keep the elements aligned. Equipment on grade should be lined up on a grade line. Pipelines should be straight with as few jogs as possible. When connecting to another drawing, try to position the continuation flags on each drawing so they match up when the two sheets are placed next to each other.
Operations	Indicate movement of portable vessels and equipment using dashed lines.
	Show manual dumping and pouring operations.
	Illustrate platforms, stairs, ladders, and other building features that operators interface with, to provide context for the equipment and help user groups understand how they will interact with the process.
	Use notes to describe operator functions such as making temporary hose connections. Notes can also indicate requirements for locating operating points (e.g., sample and shutoff valves, sight glasses), providing space for in-process materials (e.g., drums or pallets), and leaving room for maintenance (e.g., heat exchanger tube pulls).
	Instrument details are contained in other documents such as loop diagrams. P&IDs show the control loops including sensors, transmitters, controllers, and final control elements. Everything the operator will see or touch — every indicator, recorder, controller, alarm, and push button that is connected with the process — must be shown.
Safety	Use notes to list requirements such as location of the discharge from pressure safety valves.
	Interlocks are automated permissives (e.g., valve may open only if the pressure in the tank is below a certain value); indicate them within the instrument and control loops usually using a numbered diamond symbol and accompanying list.
	Show emergency venting for equipment (e.g., dust collectors) and rooms.
Environment	How can the environment impact the process? Use notes for special requirements including building air classification (e.g., clean rooms), wind screens, rain caps, seismic bracing, etc.
	How can the process impact the environment?
	Illustrate physical boundaries such as interior and exterior building walls to show relative location of equipment. Show drainage locations and features like trenches and sumps.

Process Safety Management, 29CFR119). Regulatory agencies expect P&IDs to accurately depict the "as-is" condition of the facility, updated whenever the plant is modified.

The exact content of a P&ID is somewhat dependent on its final use, the type of facility, and the scope of related drawings. For example, details of the instrument loops may be omitted if individual instrument loop diagrams are created.

Some of the art of P&ID creation has been lost with the growth of computer aided design (CAD). When pencil – or ink! – was used to draw P&IDs on mylar, vellum, or linen sheets, the designer carefully planned the placement of the major elements before beginning. This planning provided time to reflect on some of the subtle aspects that make a good P&ID (Table 17-4). Now, CAD lets the designer pull the equipment elements from a template and plop them on the screen. Pipeline endpoints are entered and the program "auto-routes" the lines between equipment. Using layers for different services (process, water, drains, etc.) gives different colors for the various lines – which is fine for on-screen viewing but lost when printed in black-and-white; unless suitable line weights are assigned to the layers all of the printed pipelines look the same.

Factory Assembled Skids and Modules

Systems ranging from simple vacuum pump assemblies to complete refinery units are often procured as pre-assembled modules. Small modules are called "skids". Modular assembly may be attractive for several reasons:

- The manufacturer's expertise with the technologies, components, design, and construction can be more effectively harnessed with a factory-assembled approach.
- The manufacturer will provide certain warranties, sometimes including process guarantees, when he is in control of the module construction.
- Cost certainty is more easily assured when modular systems are purchased on a fixed price and turnkey basis.
- Schedule compression is possible when modules are fabricated in parallel with building construction; a conventional build-in-the-field alternative may require that the building be constructed first.

P&IDs are generally prepared by the module manufacturer. While not always practicable, best practice is for the manufacturer to use the same symbols, nomenclature, and

drawing conventions that the engineer is using for the remainder of the project. It is also good practice to provide project specifications to the manufacturer and ensure that he uses the same piping, component, and instrumentation specifications as the overall project.

The method for incorporation of the module P&IDs into the project documentation set is dependent on the scope of the modules, the number and type of tie-ins to other systems, and the style of the project engineer. Large modular units that span many P&IDs are often documented solely by the manufacturer.

Smaller units, such as reactor modules and vacuum skids, are usually incorporated into the project P&IDs. There are three general approaches for incorporating such systems. Consider the reactor module in Figure 17-5. This is the manufacturer's P&ID for a stirred tank reactor, associated tanks and heat exchangers, interconnecting piping, and controls that are within his scope. Tie-ins to the customer's systems are shown; there are more than twenty-five tie-in points. The three approaches for incorporating into the customer's data set are:

1. Indicate the module on a P&ID that shows all of the tie-in points but no information about the module as seen in Figure 17-6. This approach is attractive because of its simplicity, lack of redundancy, and focus on field-erected piping and systems. It is the least expensive to engineer and changes to the module during the design phase don't affect the customer's P&ID unless the tie-in points are affected.
2. Use the same approach, but include basic information about the module in the box that references the manufacturer's P&ID. See Figure 17-7. In this case, the primary equipment in the module is shown, similar to a PFD, in order to convey a sense of the overall function of the process on the customer's P&ID.
3. Bring all of the detail from the manufacturer's P&ID onto the customer's P&ID. This can be achieved by replicating the information, or by using CAD referencing techniques (e.g., X-REF) to exactly copy the source document. This approach is useful for very small modules that have few pieces of equipment and simple instrumentation. It is cumbersome for larger systems and would be unworkable for the example reactor module.

The example P&IDs in this section illustrate another practice that can simplify a drawing that would otherwise

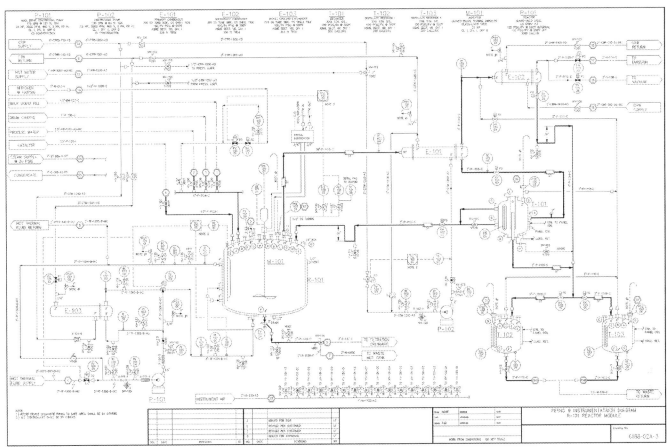

Figure 17-5. The manufacturer of a factory-assembled batch reactor system created this P&ID. It depicts the manufacturer's scope of supply. Hexagon symbols show the interfaces ("tie-in" points) with the customer's piping systems.

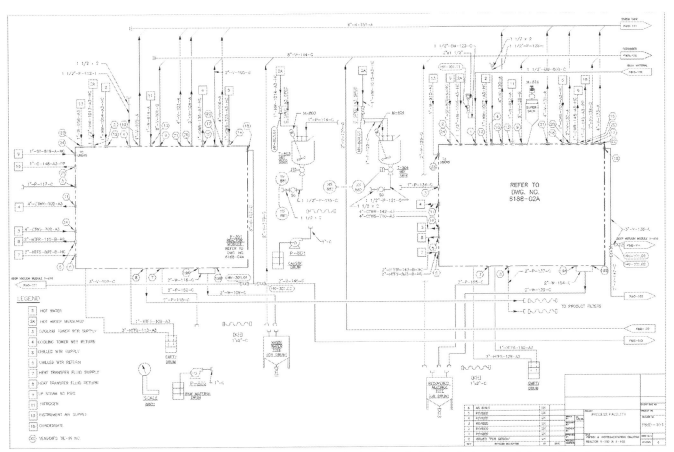

Figure 17-6. The customer's P&ID references the manufacturer's drawings but doesn't show any information about the factory-assembled reactor systems other than the tie-in points.

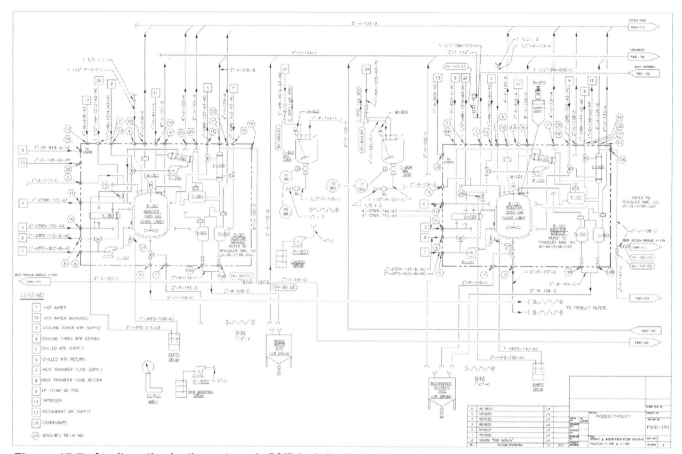

Figure 17-7. An alternative for the customer's P&ID includes basic information about the reactor system, similar to a PFD.

be very cluttered: utility tie-ins are indicated with coded boxes. Other P&IDs on the project are used to document the distribution of the utilities, and details such as location of branches, valves, instruments, and pipe sizes are shown on the utility diagrams.

The example P&IDs fail to follow some of the conventions recommended in Table 17-4, including limiting a drawing to a single unit operation, varying line weights based on importance of the pipeline, and showing manual operations. Figure 17-8 shows the finished module, created from these specifications.

Figure 17-8. The factory-assembled reactor system delivered to the site. One advantage to using this modular approach is that significant piping and instrumentation work is completed in parallel with construction of the building, saving time.

Data Sheets and Specifications

Data sheets are used to capture process, mechanical, electrical, and control requirements for equipment and instruments. The sheets incorporate information about the selected components, usually with input from vendors after purchase. Preliminary data sheets are often issued to vendors to obtain budgetary prices; the

LEVEL INSTRUMENTS (CAPACITANCE TYPE)

				CLIENT	EQUIP. NO	PAGE
REV	PREPARED BY	DATE	APPROVAL	W.O.	REQUISITION NO.	SPECIFICATION NO.
0						
1				UNIT AREA	PROCURED BY	INSTALLED BY
2						

1	GENERL	Tag No.				
2		Service				
3		Line No./Vessel No.				
4		Applicaton				
5		Function				
6		Fail-Safe				
7	PROBE	Model No.				
8		Orientation				
9		Style				
10		Material				
11		Sheath				
12		Insertion Length				
13		Inactive Length				
14		Gland Size & Material				
15						
16		Conduit Connection				
17	AMP	Location				
18		Enclosure				
19		Conduit Connection				
20		Power Supply				
21	SWITCH	Type				
22		Quantity & Form				
23		Rating: Volts/Hz or DC				
24		Amps Watts HP				
26		Load Type				
27		Contacts:				
28		Open On Incr.				
29		Close Level Decr.				
30	TRANS	Output				
31		Range				
32		Enclosure class				
33	OPTIONS	Compensation Cable				
34		Local Indicator				
35		I/P Transducer				
36		Signal Lights				
37						
38	SERVICE	Upper Fluid				
39		Dielectric Constant				
40		Lower Fluid				
41		Dielectric Constant				
42		Press Max. Normal				
43		Temp. Max Normal				
44		Moisture				
45		Material Buildup				
46		Vibration				
47						
48		Manufacturer				
49		Model No.				
50						
51						
52						

Process engineer's scope

Figure 17-9. Instrument data sheets are prepared by process, instrument, and other engineering disciplines.

prices can be used when preparing detailed cost estimates. Examples of process data sheets are in other chapters and listed in the index (Heat Exchangers, Vessels, etc.).

Process engineers also complete a portion of instrument data sheets, providing performance requirements.

Figure 17-9 is an example. While process engineers may be tasked with completing the entire instrument data sheet, at a minimum they give requirements related to the process performance.

Capital Cost Estimates

As indicated in Table 17-2, cost estimates iterate as information becomes available. In theory, the final cost of the project will be within the stated range of the first orientation estimate. Each new estimate should fit within the range of the previous iteration. Some managers insist that the upper limit of a new estimate is at or below the target cost of the previous estimate; this attitude leads to the use of excessive or hidden contingency factors.

The two most critical factors to ensure a project is realized for its estimated value are 1) completion of an appropriate amount of engineering for the estimate, and 2) understanding and control of the project scope.

Adequate engineering is the topic of Table 17-2, although this is just a guide. Use judgment to decide where deeper engineering development is warranted. For example, if there are competing technologies such as packed tower or trayed tower absorption, it may be prudent to perform enough design to select a single technology to reduce the cost estimate risk.

Scope control is the other critical factor. It is essential to track scope changes to avoid surprises. The corollary is that estimates must reflect the full project scope. Be sure that all elements of the project are identified during the project initiation phase; if significant aspects such as site preparation, commissioning, spare parts procurement, etc. are inadvertently omitted then a nasty cost surprise could ensue. Small scope additions can add up; any change that is clearly outside the approved project scope should be formally authorized. In order to identify such changes, prepare and maintain a detailed scope definition document and attach it to all authorization requests.

Cost estimates should be prepared by professional estimators with significant input from process engineers. Software can be used (Aspen Capital Cost Estimator from Aspentech, www.aspentech.com, is a leading example), but care is still required to ensure that the best assumptions are used and that they are correctly processed by the program.

Organization of a Cost Estimate

Orientation estimates, also called screening estimates, are typically reported as a single value. See the example below.

Estimates based on design deliverables have increasing detail commensurate with the supporting documents. They are typically organized into three broad categories: Direct Costs, Indirect Costs, and Owner Costs – see Table 17-5. The Direct and Indirect Costs sum to the Capitalized Total Project Cost.

There are many rules of thumb used in the preparation of capital cost estimates – for an example see Table 17-6. All, however, should be used cautiously since the underlying assumptions for the "rule" may not apply (and the underlying assumptions are not known!).

Example of a Screening Estimate for a Refinery (From [1], with Permission)

The total construction cost of a refinery with a production capacity of 200,000 bbl/day in Gary, Indiana, completed in 2001 was $100 million. It is proposed that a similar refinery with a production capacity of 300,000 bbl/day be built in Los Angeles, California, for completion in 2003. For the additional information given below, make an order of magnitude estimate of the cost of the proposed plant.

1. In the total construction cost for the Gary, Indiana, plant, there was an item of $5 million for site preparation which is not typical for other plants.

Table 17-5
Organization of capital cost estimates for design through turnover

Cost Estimate Category	Content
Direct Costs	Major equipment
	Buildings, with civil, structural, and architectural elements
	Furnishings, such as laboratory casework (1)
	Mechanical systems, including HVAC, plumbing, and process piping
	Electrical systems, including substations, generators, conduit, cable, and building controls
	Process automation, including instruments and controls
	Initial charge of utility and operating materials such as heat transfer fluids and chromatography resins
	Initial inventory of spare parts
Indirect Costs	Legal services (for preparation of contracts and other project documents)
	Engineering and procurement services (outsourced)
	Construction management services and construction overhead costs such as management fees, offices, and temporary power
	Inspection and testing services
	Commissioning and validation services
	Permits
	Insurance and taxes during construction
Owner Costs (separated for clarity from Indirect Costs; some may be capitalized)	Initial feasibility studies
	Owner's employees for engineering, project management, procurement, maintenance, operations, etc.
	Travel and living costs for Owner's employees for visiting engineering contractors and equipment suppliers
	Construction financing
	Land and site preparation costs that are outside the scope of the Direct Costs category (may be separately authorized), including site search and selection services
	Demolition if outside the scope of the Direct Costs category
	Repairs and maintenance for existing systems and facilities that support the new project (include major system upgrades, such as capacity increases, required for the project in Direct Costs)
	Furnishings, such as office furniture (1)
	Testing and start-up materials such as process raw materials and packaging components
	One-time process licensing fees

(1) Rule of thumb is that furnishings permanently affixed to the building are included with Direct Costs, and removable furnishings like desks and chairs are with Owner's Costs. Portable process equipment is included with Direct Costs.

2. The variation of sizes of the refineries can be approximated by the exponential rule, with m = 0.6.
3. The inflation rate is expected to be 8% per year from 1999 to 2003.
4. The location index was 0.92 for Gary, Indiana and 1.14 for Los Angeles in 1999. These indices are deemed to be appropriate for adjusting the costs between these two cities.
5. New air pollution equipment for the LA plant costs $7 million in 2003 dollars (not required in the Gary plant).
6. The contingency cost due to inclement weather delay will be reduced by the amount of 1% of total construction cost because of the favorable climate in LA (compared to Gary).

On the basis of the above conditions, the estimate for the new project may be obtained as follows:

1. Typical cost excluding special item at Gary, IN is $100 million – $5 million = $ 95 million
2. Adjustment for capacity based on the exponential law yields
$(\$95)(300,000/200,000)^{0.6} = (95)(1.5)^{0.6} = \$121.2$ million
3. Adjustment for inflation leads to the cost in 2003 dollars as
$(\$121.2)(1.08)^4 = \$164.6$ million
4. Adjustment for location index gives
$(\$164.6)(1.14/0.92) = \$204.6$ million

Table 17-6
Examples of rules of thumb for preparation of capital cost estimates

Rule of Thumb	Application Notes
Facility estimates extend 1 meter (3 feet) outside the building footprint(s)	Sitework such as underground utilities, fire protection piping loops, and electrical substations are typically excluded from estimates when the estimate is prepared on a "square-foot" or other factored basis. Whenever piperacks, or "yard piping" is required to connect buildings and outdoor tanks, they should be estimated separately.
Estimates for individual pieces of equipment, unit operations, and complete facilities can be scaled using the exponential rule $$y = y_n \left(\frac{Q}{Q_n}\right)^m$$ y = cost Q = capacity m = exponent Subscript n = known value	The exponent is typically taken as 0.6 for chemical process facilities, but may range from about 0.5 to 0.9 depending on the type of facility or component. Some equipment is available in discrete sizes which should be considered when determining Q. It's always better to discuss with vendors, if possible, before blindly assigning an exponent (especially when scaling a component cost).
The Total Direct Cost is equal to four times the cost of major process equipment	This factor, the "Lang factor," ranges from about 3 to 6 depending on the scope of the process equipment estimate and the complexity of the facility. If the process equipment is estimated as modules rather than discrete components the factor is toward the low end of the range. Plants that are predominately solids based (using conveyors instead of piping) have a factor around three. For facilities with a relatively simple process but significant building infrastructure the factor is toward the high end (and less reliable). The Lang factor method was published over fifty years ago, but remains a decent indicator of total plant cost if a good equipment list, with costs, is available.
The ratio of labor and material costs is 50:50	If all components and materials are listed and costed from a material take-off from the contract documents, the total direct installed cost of those items is roughly twice the purchase cost. Expensive materials (e.g., titanium piping) that are not significantly more difficult to install than normal materials will have a lower ratio. If labor is significantly expensive (due to geography or overtime conditions, for instance), the ratio will be higher. This factor is best used as a check rather than a primary estimating method.
Engineering costs 8% to 12% of the total Direct Cost	Engineering costs vary widely for many reasons. Project complexity and engineering difficulty are important along with other factors discussed separately.

5. Adjustment for new pollution equipment at the LA plant gives
$204.6 + $7 = $211.6 million
6. Reduction in contingency cost yields
($211.6)(1 − 0.01) = $209.5 million

Since there is no adjustment for the cost of construction financing, the order of magnitude estimate for the new project is $209.5 million.

Engineering

Engineering effort is required from project inception through close-out. Due to budget concerns and the rise in capabilities of computerized design solutions there is continual pressure to reduce the cost of engineering. There is widespread belief among managers and stakeholders that the need for engineering is diminishing as CAD and BIM (Building Information Management) software become more mainstream. Engineering is seen as a repetitive task that computers can replicate. Process calculations are performed quickly and accurately by process simulation software.

The opposing view is that the computerized tools require as much, or more, time from professional engineers to complete a project design, but significant advantages result including reduction of problems when constructing the facility and greater user satisfaction with the final result. Some of the engineering hours are being transferred to the construction phase of the project, with construction managers hiring engineers to build 3-dimensional "coordination" models. Similarly, equipment vendors are sometimes providing 3D models of their equipment that can be easily referenced into the project database. Other engineering hours that are reduced by using tools are replaced with new requirements such as training, increased checking of vendor submittals, and greater emphasis on commissioning and qualification.

Everyone can agree that the computer tools are transforming the industry by improving communication, reducing rework, and supporting more efficient project execution strategies. But sound professional engineering judgment is still required, and sometimes the tools create more confidence in a design that is warranted (like when calculators replaced slide rules and engineers suddenly started submitting calculations with five or more significant digits even though the assumptions carried only two). Engineering is as important as ever, and sufficient project funding must be provided.

Hendrickson gives five characteristics that are unique to the planning of constructed facilities. They should be kept in mind even at the very early stage of the project life cycle [1]:

- Nearly every facility is custom designed and constructed, and often requires a long time to complete.
- Both the design and construction of a facility must satisfy the conditions peculiar to a specific site.
- Because each project is site specific, its execution is influenced by natural, social and other locational conditions such as weather, labor supply, local building codes, etc.
- Since the service life of a facility is long, the anticipation of future requirements is inherently difficult.

- Because of technological complexity and market demands, changes of design plans during construction are not uncommon.

Engineering typically costs 8% to 12% of the Total Direct Cost of a project. The breakdown of total engineering is (very approximately):

Concept	10%
Preliminary	20%
Detailed	50%
Construction Support	10%
Close-Out	10%

Startup and Commissioning costs are excluded from the above factors. Chapter 23, "Startup", gives guidelines for estimating startup costs.

Professional services are often purchased that are related to engineering but do not contribute directly to the creation of construction documents. These services are excluded from the above factors, but their costs can be significant: site search, project management, cost and schedule creation, cost and schedule control, procurement, factory acceptance testing for equipment, and testing of construction work (e.g., concrete testing, weld inspections).

Reference

[1] Hendrickson C. *Project Management for Construction*. Version 2.2, ebook at, http://pmbook.ce.cmu.edu; 2008.

18
Reliability

Introduction

There are many dimensions to the topic of process reliability. From the design perspective, reliability means identifying, prioritizing, and mitigating the process details that contribute to process interruptions. Maintenance engineers are driven to know the weak points in the plant, and to prepare for problems with training, spare parts, preventative maintenance, and work order procedures. Those responsible for manufacturing want to minimize unforeseen downtime and maximize productivity. Everyone has an interest in plant safety and the health and wellbeing of those who work there; process reliability directly affects plant safety.

Methods for addressing reliability concerns often require prediction of where problems might occur. Statistical models give the probability that a particular unit operation, piece of equipment, or specific component will fail. But statistics require quantitative data germane to the issue being considered, with enough observations to be statistically significant. It is quite difficult to obtain appropriate data.

Structured methodologies are used, including Failure Modes and Effects Analysis (FMEA), Fault Tree Analysis (FTA), and Cause-and-Effect Diagrams (also known as Fishbone Charts or Ishikawa Diagrams). These are the same methods used in Process Hazards Analysis (see Chapter 20).

Qualitative recommendations are also available. The US Department of Defense (DoD) views RAM (reliability, availability, and maintainability) as essential elements of mission capability. DoD outlines the following basic techniques for system designers [12]:

- Simplify the design
- Improve the design by eliminating failure modes
- Implement redundancy judiciously
- Design for fault-tolerance
- Design the items to be fail-safe

- Derate components or elements (i.e., practice of limited electrical, thermal, and mechanical stresses on electronics to levels below their specified ratings)
- Provide early warnings of failure through fault diagnosis/condition monitoring
- Use standard parts and reduce variation in parts and components
- Adopt a modular design approach
- Use robust design techniques
- Use improved technology and better materials
- Make suitable performance trade-offs (e.g., less stress – some decrease in performance traded for longer life that still satisfies the system's promised capabilities)
- Use proven Testability guidelines for minimizing false alarms (thresholds, timing, n-of-n faults before reporting, etc.)

The techniques listed above are largely self-explanatory. However, the case against redundancy is interesting, and helps reinforce some of the other recommendations. Again from DoD:

Although redundancy increases mission or functional reliability, it decreases what is referred to as basic or logistics reliability. That is, the total number of failures, mission and non-mission, will increase because more items have been added to the design. In the case of standby redundancy, where switching and detection is involved, additional failure modes are introduced. Similarly, redundancy can increase the complexity of the manufacturing process and thereby increase the risk of introducing quality problems during production. Thus, it is essential that redundancy be used judiciously and only when no other approach can ensure an adequate level of the reliability for critical functions.

Initial Design

The best and most effective time to provide for reliability is in the initial design. The importance of initial design is illustrated by a study [4] undertaken by Sohio at their Toledo refinery. Their first listed major finding from the study was:

A disproportionately high percentage (39%) of all failure-producing problems encountered during the 11 year life of the processes were revealed during the first year of operation. This high failure rate during the early life of a new unit is felt to be characteristic of most complex processes. From the standpoint of reliability control, these early failures could potentially be prevented by improvements in design, equipment selection, construction, and early operator training.

Here are things the process design engineer can do to help assure reliability:

1. Understand existing operations of similar process units before embarking on a new design [6].
   - Inspect the equipment in the field.
   - Talk with operators.
   - Crawl through towers during turnarounds.
2. Include tech service (design) engineers in turnaround inspections [7].
3. Reflect the unique nature of every process application [6], such as:
   - Local climate
   - Associated processes
   - Feedstock variabilities
   - Environmental constraints
   - Heat and material balances
   - Plant culture (e.g., how operators are trained and accustomed to operate the plant)
4. Carefully define the process and project requirements, including roles and responsibilities, to ensure that the design team (owner and engineering contractor) have the same expectations for the outcome.
5. Perform hazards and operability analyses at an early stage in the project and after the design is complete.
6. Ensure that design elements are checked against requirements. This includes independent confirmation that calculations are correct, comparison of specifications and drawings with the process and project requirements, and evaluation of vendor submittals to be sure the procured instruments and equipment meet specifications.

7. Institute a quality control procedure during construction that inspects equipment and materials as they are received to confirm they conform to approved vendor submittals, and observes installation to confirm it conforms with drawings and specifications.

Process and project requirements include definitions of performance standards for equipment and systems. Often called "User Requirement Specifications", performance standards must be clearly stated and measurable if possible. Guidelines for establishing performance standards include [8]:

- Write functional performance statements using a verb, an object, and a desired standard of performance. Example: "To pump water from Tank X to Tank Y at not less than 800 liters per minute".
- Be quantitative where possible. Avoid qualitative statements like; "To produce as many widgets as required by production".
- If qualitative standards are unavoidable, take special care to ensure that everyone shares a common understanding of what is meant by words like "acceptable".
- Be explicit if performance standards are *absolute* (does an enclosed system absolutely contain the content, or is some amount of leakage tolerable), *variable*, or imply *upper and lower limits*.

Design engineers should provide an allowance for deterioration or variable performance. If the performance requirements are taken as measures of successful operation, equipment and systems should be designed so that they fulfill those requirements after an initial run-in period, or at the low points of a fluctuating output. For example, heat exchangers are normally designed to function with a certain amount of fouling, and centrifugal pumps must deliver the required flow whether the feed tank level is high or low.

Risk Analysis

Risk analysis attempts to identify and quantify the outcomes from a certain set of events. It boils down to answering these questions: What can go wrong? How likely is this? What are the outcomes or consequences if this occurs?

Assess risk using an appropriate methodology, such as one of those listed in Table 18-1.

Table 18-1
Risk assessment methods [1]

Method	Scope
Safety/Review Audit	Identifies equipment conditions or operating procedures that could lead to a casualty, or could result in property damage or environmental impacts.
Checklist	Ensures that organizations are complying with standard practices.
What-If	Identifies hazards, hazardous situations, or specific accident events that could result in undesirable consequences.
Hazard and Operability Study (HazOp)	Identifies system deviations and their causes that can lead to undesirable consequences, and determines recommended actions to reduce the frequency and/or consequences of the deviations.
Preliminary Hazard Analysis (PrHA)	Identifies and prioritizes hazards leading to undesirable consequences early in the life of a system. It determines recommended actions to reduce the frequency and/or consequences of the prioritized hazards. This is an inductive modeling approach.
Probabilistic Risk Analysis (PRA)	Quantifies risk; developed by the nuclear engineering community for risk assessment. This comprehensive process may use a combination of risk assessment methods.
Failure Modes and Effects Analysis (FMEA)	Identifies the equipment failure modes and the impacts on the surrounding components and the system. This is an inductive modeling approach.
Fault Tree Analysis (FTA)	Identifies combinations of equipment failures and human errors that could result in an accident. This is a deductive modeling approach.
Event Tree Analysis (ETA)	Identifies various sequences of events, both failures and successes, that could lead to an accident. This is an inductive modeling approach.
The Delphi Technique	Assists experts to reach consensus on a subject such as project risk while maintaining anonymity by soliciting ideas about the important project risks that are collected and circulated to the experts for further comment. Consensus on the main project risks may be reached in a few rounds of this process.
Interviewing	Identifies risk events through interviews with experienced project managers or subject-matter experts. The interviewees identify risk events based on experience and project information.
Experience-Based Identification	Identifies risk events based on experience including implicit assumptions.
Brainstorming	Identifies risk events using facilitated sessions with stakeholders, project team members, and infrastructure support staff.

Sustainable Practices

While certain aspects of reliability can be engineered into the plant design, sustainable practices must be followed to maximize productivity throughout the plant's life cycle. There are many approaches to this, such as Six Sigma, TPM (total productive maintenance), and RCM (reliability-centered maintenance). These practices require plant-wide participation and top-down management support to be fully effective. The goals of the programs are similar: to be cost effective; to provide greater safety and environmental integrity; to maximize plant availability; to improve product quality; and to increase the useful life of high-value equipment.

Basic policies, procedures, and tools underpin the broad programs. These include:

- Measuring and categorizing downtime. A popular method is OEE (Overall Equipment Effectiveness).

- Monitoring the health of operating equipment (see Appendix 4 in [8] for a discussion of about 100 condition-monitoring techniques).
- Hiring qualified personnel.
- Providing appropriate training.
- Defining clear roles and responsibilities.
- Establishing equipment history files.
- Identifying, sourcing, and stocking (if not available locally) critical spare parts.
- Implementing preventative maintenance (PM) procedures, including instrument calibration. PM can be expensive; RCM provides a method for selecting which proactive actions are most cost effective [8].
- Providing effective work order systems.
- Assessing the potential impact when process or equipment modifications are proposed (change control procedure).

These policies and procedures are not limited to certain groups (such as the maintenance department), but apply to all personnel who can influence plant reliability. Organizations require participation, teamwork, and flexibility from all employees.

References

[1] Ayyub B. *Risk Analysis in Engineering and Economics*. Boca Raton, FL: Chapman and Hall/CRC Press; 2003.

[2] Center for Chemical Process Safety (CCPS). *Guidelines for Chemical Process Quantitative Risk Analysis*. 2nd ed. Wiley; 1999.

[3] Center for Chemical Process Safety (CCPS). *Guidelines for Process Equipment Reliability Data, with Data Tables*. Wiley; 1989.

[4] Cornett C, Jones J. Reliability Revisited. *Chemical Engineering Progress*, December, 1970.

[5] Ebeling CE. *An Introduction to Reliability and Maintainability Engineering*. Waveland Press; 2009.

[6] Lieberman N. *Process Design for Reliable Operation*. 3rd ed. Lieberman Books; 2008.

[7] Miller J. Include Tech Service Engineers in Turnaround Inspections. *Hydrocarbon Processing*, May, 1987.

[8] Moubray J. *Reliability-centered Maintenance*. 2nd ed. New York: Industrial Press; 1997.

[9] OREDA (Offshore Reliability Data). *Offshore Reliability Data Handbook*. 5th ed. 2009.

[10] Reliability Information Analysis Center. Nonelectronic Parts Reliability Data (NPRD). 1995.

[11] Robinson C, Ginder A. *Implementing TPM: The North American Experience*. New York: Productivity Press; 1995.

[12] U.S. Department of Defense. DoD Guide for Achieving Reliability, Availability, and Maintainability. 2005.1

19
Metallurgy

Introduction

Engineers usually think of failure modes (e.g., corrosion) when they specify materials of construction for process equipment. This chapter discusses failure categories in the context of specific applications.

In addition to chemical compatibility, consider these factors before making a material selection:

- Uniformity. The plant may have standardized on certain materials.
- Maintainability. Exotic materials (meaning, materials that are new and unfamiliar to the plant) may be difficult to maintain due to the need for additional training, parts, or equipment.
- Flexibility. A higher grade material may be justified on the basis that the unit could be used for more chemistries than initially planned.
- Cost.
- Environmental impact. Mining, processing, transportation, and disposal affect the environment.
- Counterfeit materials are in the market. For example, Type 304 SS is marked and sold as Type 316 SS. Equipment specifications should require that the seller must provide Material Certifications for all critical materials in the fabrication. This usually means wetted parts, but could include external structures as well. Metallic materials include plate, sheet, pipe, flanges, forgings, castings, bolts, welding rods, etc. Material Certifications can also be counterfeited; physical testing should be performed to confirm that critical materials meet specification, on a spot-check basis, especially when working with new or unknown suppliers. Be especially wary if the price seems too low in comparison with other bids.

A very rough ranking of alloys by increasing resistance to general corrosion would be 304L, LDX 2101, 316L, 317L, 2205, 20, AL-6XN, 625 and C-276 [11]. Table 19-1 gives an application orientation for various metals in wet processes.

Visit Matweb (www.matweb.com) for an extremely useful searchable database of over 80,000 metals, plastics, ceramics, and composites. The compilation is comprised primarily of data sheets supplied by manufacturers and distributors.

Table 19-1
Wet corrosion performance guide [11]

Environment	Not Suggested	Good	Better	Best
Chlorides (pitting, crevice corrosion)	304L	Alloy 20, 316L, LDX 2101, 600	400, 2205, 317L	AL-6XN, 625, C-276, Titanium, C22, 686, ZERON 100
Chloride Stress Corrosion Cracking	304L, 316L	LDX 2101, 904L, 2205, 317L	AL-6XN, Alloy 20, ZERON 100	400, 600, 625, 686, C-276, C22
Hydrochloric Acid	Titanium (b), 600, Alloy 20, 2205, LDX 2101, 317L	200 (a), 400 (a), 625, ZERON 100	C22, C-276, 686	Zirconium (a), Hastelloy B-2 (a), Tantalum, Titanium (b)
Hydrofluoric Acid	200, 600, 2205, etc.	C-276, C22, 686, 400 (N2 purged)	400 (a), Silver (a)	Gold, Platinum
Sulfuric Acid	Titanium, 600	316L, 317L, LDX 2101, 2205	AL-6XN, 625	Alloy 20, C-276, Tantalum, ZERON 100
Phosphoric Acid (commercial)	200, 400, 316L, 317L	904L, 2205	AL-6XN, Alloy 20, ZERON 100	G-30, 625
Nitric Acid	904L, AL-6XN, 200, 400, 600	304L, Alloy 20, 2205, ZERON 100	625	Zirconium, Tantalum
Caustic	304L, 316L, 317L, Tantalum	Alloy 20, 2205, LDX 2101, ZERON 100	600, 625, 400, 686, C22, C-276	200 (a)

(a) Presence of oxygen or oxidizing salts may greatly increase corrosion. (b) Titanium has excellent resistance to hydrochloric acid containing oxidizers such as FeCl3, HNO3, etc. However, titanium has very poor resistance to pure, reducing, HCl. This chart is intended as guidance for what alloys might be tested in a given environment. It must NOT be used as the major basis for alloy selection, or as a substitute for competent corrosion engineering work.

Cost Comparison

The prices of major components of stainless steels and alloys – nickel, chromium, molybdenum, titanium, aluminum, and tantalum – fluctuate wildly. There are large differences in price (expressed as $/unit mass) between different forms of the same material. In addition, published price lists don't reflect surcharges or discounts.

Cost comparisons are greatly dependent on the quantity of compared materials required to make a particular piece of equipment (e.g., vessels, pipes, fittings, valves). The material's strength and density help determine the quantity needed. Another factor is the labor to fabricate the equipment; this may differ depending on how easy it is to cut, shape, cast, mill, grind, and weld the materials. For life cycle comparisons, the durability and corrosion resistance of the material should be considered to include a time component to the analysis.

Table 19-2 is provided as a starting point, with the caveat that any price comparison is only directional. If accurate cost comparisons are needed, solicit firm quotations from multiple vendors for supply against written specifications.

Table 19-2
Selected alloys with composition and relative cost data

Material	UNS	Iron	Nickel	Chromium	Molybdenum	Manganese	Silicone	Carbon	Aluminum	Other	Relative Cost by Weight
Austenitic Stainless Steels											
304	S30400	70	9	18.3	—	—	—	0.02	—		0.6
304L SS	S30403										
316	S31600	69	10.2	16.4	2.1	—	—	0.02	—	—	1.0
316L SS	S31603										
317L SS	S31703	65	11.6	18	3.1	1.5	0.4	0.02	—	N 0.05	1.5
Super Austenitic Stainless Steels											
AL6XN	N08367	48	24	20.5	6.3	—	—	0.02	—	N 0.22	3.0
Alloy 20	N08020	40	33	19.5	2.2	—	—	0.02	—	Cu 3.3 CB+To 0.5	3.3
Martensitic Stainless Steels											
410 SS	S41000	85	—	12.5	—	1.0	1.0	0.15			1.0
Nickel Alloys											
Nickel Alloy 200	N02200	0.2	99.2	—	—	0.3	—	0.1		Cu 0.2	6.2
A400	N04400	2.2	63	—	—	2.0	0.5	0.3	—	Cu 32	5.0
A600	N06600	8	76	15.5	—	0.3	0.2	0.05	—	—	5.0
A625	N06625	61	21.5	9	—	—	—	—	—	Cb 3.6	6.8
C-276	N10276	7	55	17	16.5	—	—	—	—	W 4.5	6.0
Other Metals											
Ti6Al4V Grade 5	R56401	<.25	—	—	—	—	—	<.08	6	V 4 Ti 90	6.0
Aluminum									100		0.5
Copper	C12200									Cu 100	1.2
Zinc										Zn 100	0.5
Magnesium										Mg 100	1.0

Embodied Energy

Metal use impacts the environment throughout the life cycle of the material. The supply side includes acquisition of the ores or minerals, through processing into metals or alloys with specific composition, and manufacturing of products. During usage, the manufactured products might cause environmental harm because of maintenance practices, cleaning, and remanufacturing to restore utility. At the end of its useful life, a metallic component may enter the solid waste stream or require energy for recycling into new material.

The most accurately determined metric for environmental impact is embodied energy. The Life Cycle Assessment (LCA) approach is commonly used to estimate energy consumption from mining, processing, etc. from "cradle to gate". This encompasses the material from when it is first mined to when it is ready to be shipped to customers in the form of bulk metal (e.g., ingots).

The recycle material method accounts for a certain percentage of metal being recycled. Recycling generally uses much less energy than primary production. Hammond and the Sustainable Energy Research Team have a database of construction materials that compiles published reports of embodied energy estimates, among other metrics, for a wide range of materials [5]. See Table 19-3.

The United Nations offers an assessment of the best available science from a global perspective to identify priorities among industry sectors, consumption categories, and materials. It is recommended for decision makers trying to determine how they can make a meaningful contribution to sustainable consumption and production [13].

Table 19-3
Gross energy requirement to mine, process, smelt, and refine the listed metals, or manufacture resin and form into products the listed plastics

| Metal | Embodied Energy using Recycling Method [5] | |
	MJ/kg	Notes
Aluminum	155	Assumes 25.6% extrusions, 55.7% rolled, and 18.7% castings with 33% recycled content
Copper	42	Assumes 37% recycled content (virgin copper tube & sheet 57 MJ/kg; recycled copper 16.5 MJ/kg).
Lead	25	Assumes 61.5% recycled content
Nickel	164	Cited source no longer on line
Polyethylene, HDPE pipe	84.4	Feedstock energy (55.1 MJ/kg) included
Polyvinyl Chloride (PVC) pipe	67.5	Feedstock energy (24.4 MJ/kg) included
Stainless Steel	56.7	Highly conflicting data. Based on Type 304 SS
Steel	24.4	Worldwide recycled content 42.7%
Steel Pipe	19.8	Assumes recycled content 59% (virgin steel pipe 45.4 MJ/kg)
Titanium	361 to 745	Large range of data, small sample size
Zinc	53.1	Assumes 30% recycled content (virgin zinc 72 MJ/kg; recycled zinc 9 MJ/kg)

Embrittlement

Embrittlement is the loss of a material's ductility, due to a chemical or physical change, leading to crack propagation without appreciable plastic deformation. Commonly cited embrittlements include:

- Hydrogen embrittlement. Susceptible high strength steels (tensile strength over 1,100 MPa or 150 kpi, [1], nickel, and titanium alloys crack when hydrogen molecules assemble within the metal structure. See page 309.

- Sulfide stress cracking. This is a form of hydrogen embrittlement where steels react with hydrogen sulfide to form metal sulfides and hydrogen.

- Cryogenic embrittlement. Certain metals transform to a body centered cubic structure when chilled. Carbon steel (ASTM A442, A516, A517), in particular, loses ductility below about −45 °C. Austenitic stainless steels are often specified for cryogenic service, but suitable carbon steel grades are also available (ASTM A537, A203, A533, A543).

Stress-Corrosion Cracking

When an alloy fails by a distinct crack, you might suspect stress-corrosion cracking as the cause. Cracking will occur when there is a combination of corrosion and tensile stress (either externally applied or internally applied by residual stress). It may be either intergranular or transgranular, depending on the alloy and the type of corrosion.

Austenitic stainless steels (the 300 series) are particularly susceptible to stress-corrosion cracking. Frequently, chlorides in the process stream are the cause of this type of attack. Remove the chlorides and you will probably eliminate stress-corrosion cracking where it has been a problem.

Beware of insulation. Specify it to be chloride-free, or paint stainless steel pipe with a special coating made for this purpose prior to insulating it. Another potential cause of stress cracking is concrete. Stress cracks in stainless steel vessels that were resting on concrete pads have been reported; the amount of chloride in the concrete was in the ppm range.

Many cleaning solutions and solvents contain chlorinated hydrocarbons. Be careful when using them on or near stainless steel. Sodium hypochlorite, chlorethene. methylene chloride and trichlorethane are just a few in common use. The most common cleaner used with dye checking material is trichloroethane, explaining the reason that cracks sometimes appear after dye checking welds to inspect their quality.

Austenitic stainless steels have failed during downtime because the piping or tubes were not protected from chlorides. A good precaution is to blanket austenitic stainless steel piping and tubing during downtime with an inert gas (nitrogen).

If furnace tubes become sensitized and fail by stress corrosion cracking, the remaining tubes can be stabilized by a heat treatment of 24 hours at 1,600 °F. In other words, if you can't remove the corrosive condition, remove the stress.

Chemically stabilized steels, such as Type 304L, have been successfully used in a sulfidic corrodent environment but actual installation tests have not been consistent.

Straight chromium ferritic stainless steels are less sensitive to stress corrosion cracking than austenitic steels (18 Cr-8 Ni) but are noted for poor resistance to acidic condensates.

In water solutions containing hydrogen sulfide, austenitic steels fail by stress corrosion cracking when they are quenched and tempered to high strength and hardness (above about Rockwell C24).

Hydrogen Attack

Cecil M. Cooper, Los Angeles

Atomic hydrogen will diffuse into and pass through the crystal lattices of metals causing damage, called hydrogen attack. Metals subject to hydrogen attack are the carbon, low-alloy, ferritic, and martensitic stainless steels even at subambient temperatures and atmospheric pressure. Sources of hydrogen in the processing of hydrocarbons include the following:

- Nascent atomic hydrogen released at metal surfaces by chemical reactions between the process environment and the metal (corrosion or cathodic protection reactions)
- Nascent atomic hydrogen released by a process reaction such as catalytic desulfurization
- Dissociation of molecular hydrogen gas under pressure at container metal surfaces

Types of Damage

Hydrogen attack is characterized by three types of damage, as follows:

Internal stresses with accompanying embrittlement may be only temporary, during operation. Embrittlement is caused by the presence of atomic hydrogen within the metal crystal lattices, with ductility returning once the source of diffusing hydrogen is removed. Such embrittlement can become permanent when hydrogen atoms combine in submicroscopic and larger voids to form trapped molecules which, under extreme conditions, may build up sufficient stresses to cause subsurface fissuring.

Blistering and other forms of local yielding. Diffusing atomic hydrogen combines to form molecular hydrogen gas in all voids until local yielding or cracking results [12]. Such voids include laminations, slag pockets, shrinkage cavities in castings, unvented annular spaces in duplex tubes, unvented metal, plastic or ceramic linings, unvented voids in partial-penetration welds, and subsurface cracks in welds.

Decarburization and fissuring. Diffusing hydrogen combines chemically with carbon of the iron carbides in steels to form methane at hydrogen partial pressures above 7 bar (100 psia) and temperatures above 220 °C to 360 °C (430 °F to 675 °F), depending upon hydrogen pressures. This reaction begins at the metal surface and progresses inward, causing both surface and subsurface decarburization and the buildup of the gas in voids and grain boundaries, and resulting in eventual subsurface fissuring.

Conditions for hydrogen damage. Only the first two types of damage described above occur under aqueous phase (wet) conditions [3]. When the source of diffusing hydrogen is corrosion of the containing equipment, high strength, highly-stressed parts, such as carbon steel or low-alloy gas compressor rotor blades, exchanger floating-head bolts, valve internals, and safety valve springs are particularly susceptible to embrittlement [9]. Experience has confirmed, however, that if yield strengths are limited to 6,000 bar (90,000 psig), maximum, and working stresses are limited to 80% of yield, cracking failures are minimized. Welds and their heat affected zones should be stress relieved, hardnesses should be limited to Rockwell C22, or less, and they should be free from hard spots, slag pockets, subsurface as well as surface cracks, and other voids.

Tests of such steels as ASTM-A302, A212, and T-1 at elevated pressures have revealed no appreciable effects of hydrogen on the ultimate tensile strength of unnotched specimens, but have shown losses as high as 59% in the ultimate tensile strength of notched specimens [14]. Special quality control measures are necessary in the fabrication of such containers to minimize notch effects, and design pressures should be lowered 30 to 50% below the pressures considered safe for the storage of inert gases. The safest procedure would seem to be to use vented, fully austenitic stainless steel liners in such containers, and austenitic stainless steel auxiliary parts for which linings are not practical.

The use of fine-grained, fully killed carbon steels (such as ASTM-A516 plate) is not justifiable as a first choice for aqueous-phase hydrogen services. This is because these premium grades are equally as subject to embrittlement and subsurface fissuring as the lower grades. In the absence of blistering, such embrittlement is not easily discernible. Although the lower grades may blister, this

blistering serves as its own warning that steps should be taken to protect the equipment from hydrogen damage.

When corrosion is the source of diffusing hydrogen, the elimination of this reactive hydrogen is necessary to stop its diffusion into the metal. This can be done either by eliminating the corrosion, or by eliminating chemicals in the process stream that "poison" the corroding metal surface: that is, by eliminating chemical agents which retard or inhibit the combining of nascent hydrogen atoms at the corroding surface into molecular hydrogen which cannot enter the metal.

Corrosion can be eliminated by the use of corrosion resistant barriers and/or alloys. This becomes necessary in situations in which measures to eliminate "poisoning agents" are not permissible or effective. Ganister linings (one part lumnite cement and three parts fine fire clay) over expanded metal mesh, and, to a limited extent, plastic coatings have been used successfully where the normal operating pH is maintained above 7.0. Austenitic strip or sheet linings and metallurgically bonded claddings are used extensively, and are usually adequate for the most severe operating conditions. Surface aluminizing of small parts, such as relief valve springs, has been reported to be effective in protecting against hydrogen embrittlement.

Compounds such as hydrogen sulfide and cyanides are the most common metal surface poisoners, occurring in process units which are subject to aqueous-phase hydrogen attack. In many process units, these compounds can be effectively eliminated, and hydrogen diffusion stopped, by adding ammonium polysulfides and oxygen to the process streams. This converts the compounds to polysulfides and thiocyanates, provided the pH is kept on the alkaline side.

Non-aqueous (dry), elevated-temperature hydrogen attack can produce all three of the types of damage previously described. Since the iron carbides in steels give up their carbon to form methane at lower temperatures and pressures than do the carbides of molybdenum, chromium, and certain other alloying elements, progressively higher operating temperatures and hydrogen partial pressures are permissible by tying up, or stabilizing, the carbon content of the steel with correspondingly higher percentages of these alloying elements [2]. The fully austenitic stainless steels are satisfactory at all operating temperatures.

To avoid decarburization and fissuring of the carbon and low-alloy steels, which is cumulative with time and, for all practical purposes, irreversible, the limitations of the Nelson Curves should be followed religiously, as a minimum [2]. Suitable low-alloy plate materials include

ASTM-A204-A, B, and C and A387-A, B, C, D, and E, and similarly alloyed materials for pipe, tubes, and castings, depending upon stream temperatures and hydrogen partial pressures, as indicated by the Nelson Curves.

Stainless steel cladding over low-alloy or carbon steel base metal is usually specified for corrosion resistance where necessary in elevated-temperature hydrogen services. This poses the additional hazards of voids created by unbonded spots in roll-bonded cladding, sensitization of the cladding to stress corrosion cracking by the final stress relief heat treatment of the vessel, and cracking of the cladding welds by differential thermal expansion. Rigid quality control in steelmaking and fabrication is therefore extremely important, as is frequent and thorough maintenance inspection during operation, for pressure containing equipment in elevated temperature hydrogen services.

Pitting Corrosion

Pits occur as small areas of localized corrosion and vary in size, frequency of occurrence, and depth. Rapid penetration of the metal may occur, leading to metal perforation. Pits are often initiated because of inhomogeneity of the metal surface, deposits on the surface, or breaks in a passive film. The intensity of attack is related to the ratio of cathode area to anode area (pit site), as well as environmental variables. Halide ions such as chlorides often stimulate pitting corrosion. Once a pit starts, a concentration-cell is developed, since the base of the pit is less accessible to oxygen.

Pits often occur beneath adhering substances where the oxidizing capacity is not replenished sufficiently within the pores or cavities to maintain passivity there. Once the pit is activated, the surface surrounding the point becomes cathodic and penetration within the pore is rapid.

Pitting environments. In ordinary sea water, the oxygen dissolved in the water is sufficient to maintain passivity, whereas beneath a barnacle or other adhering substance, metal becomes active since the rate of oxygen replenishment is too slow to maintain passivity, hence activation and pitting result.

Exposure in a solution that is passivating and yet not far removed from the passive-activity boundary may lead to corrosion if the exposure involves a strong abrasive condition as well. Pitting of pump shafts under packing handling sea water is an example of this abrasive effect.

The 18-8 stainless steels pit severely in fatty acids, salt brines, and salt solutions. Often the solution for such chronic behavior is to switch to plastics or glass fibers that do not pit because they are made of more inert material.

Copper additions seem to be helpful to avoid pitting. Submerged specimens in New York Harbor produced excessive pitting on a 28% Chromium stainless but no pitting on a 20 Cr-1 Cu alloy for the period of time tested.

Higher alloys. Some higher alloys pit to a greater degree than lower alloyed materials. Inconel® 600 pits more than type 304 in some salt solutions. Just adding alloy is not necessarily the answer or a sure preventative.

In a few solutions such as distilled, tap, or other fresh waters, the stainless steels pit but it is of a superficial nature. In these same solutions carbon steels suffer severe attack.

Pitting in amine service. In the early 1950s, the increased use of natural gas from sour gas areas multiplied the number of plants using amine systems for gas sweetening. Here, severe pitting occurred in carbon steel tubed reboilers; in fact, failures occurred in two or three months. The problem was solved by lowering temperatures or by going to more alloyed materials.

In this same amine service it was found that carbon steel pits much more readily if it has not been stress relieved.

Aluminum was tried in the reboiler of an aqueous amine plant but it was found to pit very quickly.

Type 304 stainless steel does not seem to be a satisfactory material in all instances either. Monel® (Alloy 400) and Type 316 appear most suitable in this service where pitting must be avoided.

Salt water cooling tower. A somewhat unusual occurrence where the proper choice of materials was certainly made but pitting still occurred involved a salt water cooling tower. The makeup water for this cooling tower was brackish and high in sulfur, from a highly industrialized area. By allowing the sulfur to concentrate in this closed circuit cooling system, 90-10 Cu-Ni tubes were pitted, with the pits penetrating through the tubes in three months.

A comparison unit with identical design used once-through water cooling. This is the same high-sulfur water that was used as makeup for the above cooling tower.

Here the sulfur was not allowed to concentrate and the unit is still working satisfactorily after several years.

Tests for pitting. A test which is in popular use to determine pitting characteristics is the 10% Ferric Chloride test, which can be conducted at room temperature. This test is usually used for stainless or alloy steels.

Salt Spray (Fog) Testing, such as set forth in ASTM B-117, B-287, and B-368, is a useful method of determining pitting characteristics for any given alloy. This is also useful for testing inorganic and organic coatings, etc., especially where such tests are the basis for material or product specifications.

Avoiding pitting. Often, this is a matter of design tied in with proper material selection. For example, a heat exchanger that uses cooling water on the shell side will cause tube pitting, regardless of the alloy used. If the cooling water is put through the tubes, no pitting will occur if proper alloy selection is made.

Creep and Creep-Rupture Life

Walter J. Lochmann, The Ralph M. Parsons Company, Los Angeles

Creep is that phenomenon associated with a material in which the material elongates with time under constant applied stress, usually at elevated temperatures. A material such as tar will creep on a hot day under its own weight. For steels, creep becomes evident at temperatures above 350 °C (650 °F). The term creep was derived because, at the time it was first recognized, the deformation which occurred under design conditions occurred at a relatively slow rate.

Depending upon the stress load, time, and temperature, the extension of a metal associated with creep finally ends in failure. Creep-rupture and stress-rupture are the terms used to indicate the stress level needed to produce failure in a material at a given temperature for a particular period of time. For example, the stress required to produce rupture for carbon steel in 10,000 hours (1.14 years) at a temperature of 480 °C (900 °F) is substantially less than the ultimate tensile strength of the steel at the corresponding temperature. The tensile strength of carbon steel at 480 °C (900 °F) is 3,700 bar (54,000 psi), whereas the stress to cause rupture in 10,000 hours is only 800 bar (11,500 psi).

To better understand creep, it is helpful to know something of the fracture characteristics of metals as a function of temperature for a given rate of testing. At room temperature, mechanical failures generally occur through the grains (transcrystalline) of a metal, while at more elevated temperatures, failure occurs around the grains (intercrystalline). Such fracture characteristics indicate that at room temperature there is greater strength in the grain boundaries than in the grain itself. As temperature is increased, a point is reached called the "equicohesive temperature," at which the grain and the grain boundaries have the same strength. Below the equicohesive temperature, initial deformation is elastic, whereas above the equicohesive temperature deformation becomes more of a plastic rather than an elastic property.

The factors that influence creep are:

- For any given alloy, a coarse grain size possesses the greatest creep strength at the more elevated temperatures, while at the lower temperatures a fine grain size is superior.
- Creep becomes an important factor for different metals and alloys at different temperatures. For example, lead at room temperature behaves similarly to carbon steel at 540 °C (1,000 °F) and to certain of the stainless steels and super alloys at 650 °C (1,200 °F) and higher.
- Relatively slight changes in composition often alter the creep strength appreciably, with the carbide forming elements being the most effective in improving the strength.

Creep-rupture life. Because failure will ultimately occur if creep is allowed to continue, the engineering designer must not only consider design stress values such that the creep deformation will not exceed a limiting amount for the contemplated service life, but also that fracture does not occur. In refinery practice, the design stress is usually based on the average stress required to produce rupture in 100,000 hours, or the average stress needed to produce a creep rate of 0.01% per 1,000 hours (considered to be equivalent to 1% per 100,000 hours) with appropriate factors of safety. A criterion based on rupture strength is preferable, because rupture life is easier to determine than low creep rates.

Hydrogen reformers. The refinery unit in which the creep problem is most prevalent is the steam-methane, hydrogen producing reformer. In reforming furnace design, creep and creep-rupture life of the catalyst tubes, outlet pigtails, and collection headers usually sets the upper limit for the possible operating temperatures and pressures of the reforming process.

In a reforming furnace, all the economically available materials for catalyst tubes will creep under stress. However, from creep-rupture curves, or stress-rupture curves as they are usually called, the designer can predict the expected service life of the catalyst tubes. The relative magnitudes of the effects of stress and temperature on tube life are such that at operating temperatures of 900 °C (1,650 °F), a small increase in the tube-metal temperature drastically reduces service life. For example, a prediction that is 40 °C (100 °F) too low will result in changing the expected life from 10 years into an actual life of 1½ years. Conversely, a prediction that is 40 °C (100 °F) too high will increase the tube thickness unnecessarily by 40% with a corresponding increase in cost. The tube material cost for a hydrogen reformer heater is almost 30% of the total; therefore, judicious use of the optimum tube design and selection of materials is essential for adequate service life and for significant cost savings.

ASTM A297 Gr. HK or A 351 HK-40, a 26 Cr-20 Ni alloy with a carbon range of 0.35 to 0.45%, is the material almost always specified for catalyst tubes. A recent API Survey indicated that, for most plants, the tube wall was designed on the basis of stress to produce rupture in 100,000 hours. Other design bases were 50% of the stress to produce rupture in 10,000 hours or 40 to 50% of the stress to produce one % creep in 10,000 hours.

Failures of catalyst tubes have occurred principally because of overheating and consequent creep-rupture cracking. Overheating may have been caused by local hot spots in the furnace as a result of faulty burners, inadequate control of furnace temperature, ineffective catalyst, or plugging of the catalyst tube inlet pigtail or catalyst tube. Header and transfer line failures have resulted from the inability of the materials to withstand strains from thermal gradients and internal loads during cyclic operation or from poor design and/or material selection.

Solution annealed Incoloy® 800 is the material almost universally selected for the outlet pigtails. Pigtail failures have been particularly troublesome and have frequently been the result of creep failure associated with the stresses resulting from thermal expansion and bending moments transmitted from the catalyst tubes or collection headers. HK-40 material, although it has a higher creep strength, is not used for pigtails because it has insufficient high temperature ductility for this sensitive application.

Special consideration must also be given to selecting welding electrode materials with adequate creep strength and ductility for joining HK-40 to Incoloy® 800. INCO-WELD® A is the material most commonly used, with some refineries using INCO 82, 182, or 112. For joining HK-40, a high-carbon, Type 310 stainless steel filler material is generally used. For joining Inconel® 800, INCO-WELD® A, INCO 82, or INCO 182 are the filler materials used.

The present trend of material selection for collection headers is leaning toward Incoloy® 800. The cast alloys used, HK and HT, have failed in most instances because of their inherently low ductility – especially after exposure to elevated temperature. It now appears that wrought alloys should be used in preference to cast alloys unless the higher creep strength of the cast alloy is required and the inherently low ductility of the aged cast alloy is considered in the design.

For hydrogen reformer transfer lines, materials used are Incoloy® 800, HK, and HT cast stainless steels, Wrought 300 series stainless steels and internally insulated carbon, carbon-½ Mo, and 1¼ Cr-½ Mo steels. Reported failures of transfer lines indicate that failures are associated with design that did not account for all the stresses from thermal expansion, or failure or cracking of the insulation in internally insulated lines. The general trend is toward headers of low alloy steel internally lined with refractory.

Piping, exchanger tubing, and furnace hardware. Above temperatures of 480 °C (900 °F), the austenitic stainless steel and other high alloy materials demonstrate increasingly superior creep and stress-rupture properties over the chromium-molybdenum steels. For furnace hangers, tube supports, and other hardware exposed to firebox temperatures, cast alloys of 25 Cr-20 Ni and 25 Cr-12 Ni are frequently used. These materials are also generally needed because of their resistance to oxidation and other high temperature corrodents.

Furnace tubes, piping, and exchanger tubing with metal temperatures above 425 °C (800 °F) now tend to be an austenitic stainless steel, e.g., Type 304, 321, and 347, although the chromium-molybdenum steels are still used extensively. The stainless steels are favored not only because their creep and stress-rupture properties are because superior at temperatures over 480 °C (900 °F), but more importantly because of their vastly superior resistance to high-temperature sulfide corrosion and oxidation. Where corrosion is not a significant factor, e.g., steam generation, the low alloys, and, in some applications, carbon steel may be used.

Pressure vessels. Refineries have many pressure vessels, e.g., hydrocracker reactors, cokers, and catalytic cracking regenerators, that operate within the creep range, i.e., above 340 °C (650 °F). However, the phenomenon of creep does not become an important factor until temperatures exceed 425 °C (800 °F). Below this temperature, the design stresses are usually based on the short-time, elevated temperature, tensile test.

For desulfurizers, coking, and catalytic reforming units, many of the pressure vessels operating with metal temperatures of 370 °C to 480 °C (700 °F to 900 °F) are constructed from carbon-½ Mo or 1¼ Cr-½ Mo alloy steels. These steels have markedly better creep-rupture strength properties than carbon steel. For example, compare the design stress value of 1,000 bar (15,000 psi) for 1¼ Cr-½ Mo steel at 480 °C (900 °F) with the 450 bar (6,500 psi) value for carbon steel at the corresponding temperature.

For hot wall vessels, the increased strength may be such that the use of chromium and molybdenum alloy steels will be cheaper. Also, these steels may be required to prevent hydrogen attack and to reduce oxidation and sulfidation.

For refinery units such as hydrocrackers in which the hydrogen partial pressure is much higher, e.g., above 93 bar (1,350 psi) and the operating temperatures are above 425 °C (800 °F), the 2¼ Cr-1 Mo steel is commonly used. The higher alloy content is necessary to prevent hydrogen attack in these applications. The 2¼ Cr-1 Mo steel also has better creep and stress-rupture properties than the 1¼ Cr-½ Mo and carbon-½ molybdenum steels, but is about 20% higher in cost. It is therefore selected in most of its refinery applications over the over alloyed steel for its better resistance to corrosion and hydrogen attack. Finally, this alloy can be obtained in plate thicknesses of 25 to 30 cm (10 to 12 inches), whereas there is now about a 15-cm (6-inch) thickness limitation for the 1¼ Cr-½ Mo alloy.

Carbon steel, where not limited by sulfur corrosion or hydrogen attack, can be the most economical material for elevated temperature service. Where the creep and stress-rupture temperature conditions are high enough to severely limit the service life or require too low design stress values, it is often advantageous to refractory lined carbon steel and thus reduce the metal temperature rather than use materials with higher creep strength. Furnace stacks; sulfur plant combustion chamber, reaction furnaces, converters and incinerators; catalytic cracking regenerators; and some cold-shell catalytic reformer reactors are all examples of refinery equipment operating in the creep range that use refractory lined carbon steel.

Naphthenic Acid Corrosion

Cecil M. Cooper, Los Angeles

Naphthenic acid is a collective name for the organic acids present in some but not all crude oils [4]. In addition to true naphthenic acids (naphthenic carboxylic acids represented by the formula X-COOH in which X is a cycloparaffin radical), the total acidity of a crude may include various amounts of other organic acids and sometimes mineral acids. Thus the total neutralization number of a stock, which is a measure of its total acidity, includes (but does not necessarily represent) the level of naphthenic acids present. The neutralization number is the number of milligrams of potassium hydroxide required to neutralize one gram of stock as determined by titration using phenolphthalein as an indicator, or as determined by potentiometric titration. It may be as high as 10mg KOH per gram stock for some crudes. The neutralization number does not usually become important as a corrosion factor, however, unless it is at least 0.5mg KOH/g.

Theoretically, corrosion rates from naphthenic acids are proportional to the level of the neutralization number of feed stocks; but investigators have been unable to find a precise correlation between these factors. Predicting corrosion rates based on the neutralization number remains uncertain. Published data, however, indicate a scattered trend toward increasing corrosion with increasing neutralization number [6].

Temperatures required for corrosion by naphthenic acids range from 230 °C to 400 °C (450 °F to 750 °F), with maximum rates often occurring between 270 °C and 280 °C (520 °F and 535 °F) [4]. Whenever rates again show an increase with a rise in temperature above 340 °C (650 °F), this is believed to be caused by the influence of sulfur compounds which become corrosive to carbon and low alloy steels at that temperature.

Where is naphthenic acid corrosion found? Naphthenic acid corrosion occurs primarily in crude and vacuum distillation units, and less frequently in thermal and catalytic cracking operations. It usually occurs in furnace coils, transfer lines, vacuum columns and their overhead condensers, sidestream coolers, and pumps. This corrosion is most pronounced in locations of high velocity, turbulence, and impingement, such as at elbows, weld reinforcements, pump impellers, steam injection nozzles, and locations where freshly condensed fractions drip upon or run down metal surfaces.

What does the corrosion look like? Metal surfaces corroded by naphthenic acids are characterized by sharp edged, streamlined grooves or ripples resembling erosion effects, in which all corrosion products have been swept away, leaving very clean, rough surfaces.

Which alloy to use. Unalloyed mild steel parts have been known to corrode at rates as high as 800 mils per year. The low-chrome steels, through 9-Cr, are sometimes much more resistant than mild steel. No corrosion has been reported with either 2¼-Cr or 5-Cr furnace tubes, whereas carbon steel tubes in the same service suffered severe corrosion. The 12-Cr stainless steels are scarcely, if at all, better than the low-chromes. But the 18-8 Cr-Ni steels, without molybdenum, are often quite resistant under conditions of low velocity although they are sometimes subject to severe pitting.

Type 316 (18-8-3 Cr-Ni-Mo) has, by far, the highest resistance to naphthenic acids of any of the 18-8 Cr-Ni alloys and provides adequate protection under most circumstances. It provides excellent protection against both high-temperature sulfur corrosion and naphthenic acids whereas the 18-8 Cr-Ni alloys without molybdenum are not adequate for both. The high-nickel alloys, except those containing copper (such as Monel®) are also highly resistant but have little advantage in this respect over Type 316. Copper and all copper alloys, including aluminum-copper alloys such as Duralumin (5-Cu), are unsuitable.

Aluminum and aluminum-clad steels are highly resistant to naphthenic acids under most conditions. Aluminum-coated steels give good service until coatings fail at coating imperfections, cracks, welds, or other voids. In general, the use of aluminum and aluminized steels for the control of corrosion by naphthenic acids, as well as by other elevated temperature corrodents, such as hydrogen sulfide, is somewhat unpredictable and less reliable than Type 316 stainless steel. Several large refiners have discontinued the use of aluminum materials for such services after thorough field trials.

Some of the restrictions on the use of aluminum are caused by manufacturing and fabrication problems and by its low mechanical strength. However, aluminum is widely used and is competitive with Type 316 stainless steel in many instances. The explosion-bonding process has made the aluminum cladding of steel practical, and improvements in the diffusion coating process are producing more reliable aluminum coatings for such parts as furnace tubes and piping.

Thermal Fatigue

Howard S. Avery, Abex Corp., Mahwah, N.J.

Thermal fatigue characteristically results from temperature cycles in service. Even if an alloy is correctly selected and operated within normal design limits for creep strength and hot-gas corrosion resistance, it can fail from thermal fatigue.

Thermal fatigue damage is not confined to complex structures or assemblies. It can occur at the surface of quite simple shapes and appear as a network of cracks.

Thermal shock. As a simplified assumption, thermal fatigue can be expected whenever the stresses from the expansion and contraction of temperature changes exceed the elastic limit or yield strength of a material that is not quite brittle. If the material is as brittle as glass, and the elastic limit is exceeded, prompt failure by cracking can be expected (as when boiling water is poured into a glass tumbler). The term thermal shock is logically applied to the treatment that induces prompt failure of a brittle material.

A metal tumbler would not crack, because its elastic limit must be exceeded considerably before it fails. However, repetitive thermal stress, when some plastic flow occurs on both heating and cooling cycles, can result in either cracking or so much deformation that a part becomes unserviceable.

Thermal stress. The formula for thermal stress can be rearranged to calculate the tolerable temperature gradient for keeping deformation within arbitrary limits.

$$S = a\,M\,T\,K(1 - v) \tag{19-1}$$

Where:

S = Stress in psi, or elastic limit in psi, or yield strength in psi.

a = Coefficient of thermal expansion in microinches per inch per °F.

M = Modulus of elasticity (Young's Modulus) in psi, or modulus of plasticity.

Note: When the elastic modulus is used, the elastic limit or proportional limit should be used with it in the formula. When the plastic modulus or Secant Modulus is used, it should be used with the corresponding yield strength.

T = Temperature difference in °F.

K = Restraint coefficient.

v = Poisson's ratio.

The usefulness of this formula is restricted by the difficulty of obtaining good values to substitute into it. They must apply to the alloy selected, and be derived from carefully controlled tests on it. The stress value, S, reflects an engineer's judgment in the selection of elastic limit or some arbitrary yield strength. The modulus value must match this. The restraint coefficient, K, is seldom known with any precision.

Fatigue life. The formulae for estimating fatigue life are more complex and usually require several assumptions. Without attempting to evaluate the merit of such formulae, it is suggested here that alloy selection or metallurgical variables operating within an alloy may invalidate the assumptions. One of the critical factors that enters such calculations is the amount of plastic strain (usually first in compression and then in tension as the temperature reverses) that occurs. Another is the tolerable flow before cracking develops.

Thermal gradients may be measured or calculated by means of heat flow formulae, etc. After they are established it is likely to be found from the formula that for most cyclic heating conditions the tolerable temperature gradient is exceeded. This means that some plastic flow will result (for a ductile alloy) or that fracture will occur. Fortunately, most engineering alloys have some ductility. However, if the cycles are repeated and flow occurs on each cycle, the ductility can become exhausted and cracking will then result. At this point it should be recognized that conventional room temperature tensile properties may have little or no relation to the properties that control behavior at higher temperatures.

Ductility. As a warning against the quick conclusion that high ductility is the premium quality to be desired, it should be pointed out that the amount of deformation is closely related to the hot strength with which deformation is resisted. A very strong alloy may avoid yielding, or yield so little that there is no need for much ductility.

Design approach. The engineering solution to thermal fatigue problems becomes chiefly a matter of design to minimize the adverse factors in the stress formula and alloy selection aimed carefully at the operating conditions. Much effort should be devoted to reducing the thermal gradients or the temperature differential to the practical minimum. This, and avoiding restraint, is a rewarding area where the designer can demonstrate his skill.

The restraint coefficient, K, in the thermal stress formula is very potent. It can be varied over a wider range than any of the other parameters. If a designer can build in flexibility, and thus substitute elastic deflection for plastic flow, he can achieve a major safety factor.

Thermal expansion. Alloys differ in their thermal expansion, but the differences are modest. Coefficients for the ferritic grades of steel are perhaps 30% below those of the austenitic steels at best, while expansion of the nickel-base austenitic types may be no more than 12 to 15% less than those of the less expensive, iron-base, austenitic, heat-resistant alloys. Unfortunately, the most economical grades for use in the 1,200 °F to 2,000 °F range tend to have the highest expansion.

Poisson's ratio can vary somewhat but at present does not provide much choice. There is evidence that it is affected by crystal orientation but specifying and providing control are hardly practical. The value of 0.3 is frequently used for convenience in calculations.

Alloy selection. The designer must make two choices before using the stress formula. The first is a tentative selection of the alloy, because this determines the strength modulus and coefficient values. The second is the decision about strength and modulus limits, as typified by the term elastic limit vs. yield strength for some defined permanent set (e.g., 0.1% or 0.2%). After preliminary calculations for several candidate alloys, the range of choice may be narrowed. It is then advisable to seek out and examine results from comparative thermal fatigue tests using temperature conditions close to those of the intended operation. If such data can be found, or if arrangements for the tests can be made, the resultant ranking is more likely to integrate the various factors (particularly the role of strength and ductility) than any other procedure.

If some other criterion such as creep-rupture strength is of primary importance, the alloy choice may be restricted. Here it would be necessary to have thermal fatigue comparisons only for the alloys that pass the primary screening. When alloy selection reaches this stage some further cautions are in order.

For maximum temperatures below 800 °F, suitable ferritic steels are usually good selections. Above 800 °F their loss of strength must be considered carefully and balanced against their lower thermal expansion. It should be recognized that if they are heated through the ferrite to austenite transformation temperature their behavior will become more complex and the results will probably be adverse.

Environments. Among the environmental factors that can shorten life under thermal fatigue conditions are surface decarburization, oxidation, and carburization. The last can be detrimental because it is likely to reduce both hot strength and ductility at the same time. The usual failure mechanism of heat-resistant alloy fixtures in carburizing furnaces is by thermal fatigue damage, evidenced by a prominent network of deep cracks.

Aging (extended exposure to an elevated temperature) of an austenitic alloy can strengthen the material and increase its resistance to damage. Thus such alloys can be shown to improve after a heat treatment. However, there is little point in imposing such a treatment before use. The alloy will age anyway during its first few days of service. The sophisticated use of this information leads to aging of the thermal fatigue specimens before they are evaluated in a laboratory test rather than heat treating parts before service. (This remark does not apply to the "quench anneal" or solution heat treatment given to some wrought stainless steels to alter their metallurgical condition before service.)

The range in properties of a high-temperature alloy, within the compass of a nominal designation or a simplified specification, is much greater than most engineers recognize. Some of this is due to the influence of minor elements within the specification, some is due to production variables, and some may be due to subsequent environmental factors. Among the last, the cycling temperatures that lead to thermal fatigue can also have an important effect on creep behavior, usually operating to shorten life expectancy. Variations in room temperature properties may not be especially serious, but scatter can be extremely significant at high temperature. Unfortunately, the statistical probability limits have not been worked out for many of the available alloys. Moreover, the hot strength values offered to an engineer sometimes have a very sketchy basis.

Abrasive Wear

Howard S. Avery, Abex Corp., Mahwah, N.J.

Abrasive wear can be classified into three types. *Gouging abrasion* is a high stress phenomenon that is likely to be accomplished by high comprehensive stress and impact. *Grinding abrasion* is a high stress abrasion that pulverizes fragments of the abrasive that become sandwiched between metal faces, and *erosion* is a low stress scratching abrasion.

Gouging abrasion. Most of the wearing parts for gouging abrasion service are made of some grade of austenitic manganese steel because of its outstanding toughness coupled with good wear resistance.

Grinding abrasion. The suitable alloys range from austenitic manganese steel (which once dominated the field) through hardenable carbon and medium alloy steels to the abrasion-resistant cast irons.

Erosion. The abrasive is likely to be gas borne (as in catalytic cracking units), liquid borne (as in abrasive slurries), or gravity pulled (as in catalyst transfer lines). Because of the association of velocity and kinetic energy, the severity of the erosion may increase as some power (usually up to the third) of the velocity. The angle of impingement also influences the severity. At supersonic speeds, even water droplets can be seriously erosive.

There is some evidence that the response of resisting metals is influenced by whether they are ductile or brittle. Probably most abrasion involved with hydrocarbon processing is of the erosive type. There is a large assortment of alloys available for abrasive service in the forms of wrought alloys, sintered metal compacts, castings, and hard-surfacing alloys. They can be roughly classified as follows, with the most abrasion resistant materials toward the top and the toughest at the bottom:

- Tungsten carbide and sintered carbide compacts
- High-chromium cast irons and hardfacing alloys
- Martensitic irons and hardfacing alloys
- Austenitic cast irons and hardfacing alloys
- Martensitic steels
- Pearlitic steels
- Ferritic steels
- Austenitic steels, especially the 13% manganese type

Since toughness and abrasion resistance are likely to be opposing properties, considerable judgment is required in deciding where, in this series, the best prospect lies, especially if economic considerations are important. The choice is easiest at the extremes. Cobalt-base and nickel-base alloys may also be considered. Their chief merits are heat-resistance and corrosion-resistance, respectively. For simple abrasion resistance, the iron-base alloys are generally more economical. The basis of classification – total alloy content – sometimes put forward in the hard-facing field, is misleading, especially as it implies that merit increases with alloy content. This can be disproved for abrasion. Carbon has a dominant role in determining abrasion resistance: moreover, it adds almost nothing to the cost of an alloy in the form of a casting or a surfacing electrode for welding. Unfortunately, many proprietary alloys do not reveal the amount of carbon used or its role.

Hardness. Though carbon is important, it must be used with proper insight to be most effective. In the form of graphite it usually is detrimental. As hard carbides, the form, distribution, and crystallographic character are important. Even hardness must be used with discretion for evaluating wear resistance; it should be considered as an unvalidated wear test until its relation to a given service has been proven. Simple and widely used tests (e.g., for Brinell or Rockwell hardness) tell almost nothing about the hardness of microscopic constituents.

For erosive wear, Rockwell or Brinell hardness is likely to show an inverse relation with carbon and low alloy steels. If they contain over about 0.55% carbon, they can be hardened to a high level. However, at the same or even at lower hardness, certain martensitic cast irons (HC 250 and Ni-Hard) can outperform carbon and low alloy steel considerably. For simplicity, each of these alloys can be considered a mixture of hard carbide and hardened steel. The usual hardness tests tend to reflect chiefly the steel portion, indicating perhaps from 500 to 650 BHN. Even the Rockwell diamond cone indenter is too large to measure the hardness of the carbides; a sharp diamond point with a light load must be used. The Vickers diamond pyramid indenter provides this, giving values around 1,100 for the iron carbide in Ni-Hard and 1,700 for the chromium carbide in HC 250. (These numbers have the same mathematical basis as the more common Brinell hardness numbers.) The microscopically revealed differences in carbide hardness account for the superior erosion resistance of these cast irons compared with the hardened steels.

There is another interesting difference between the two irons; Ni-Hard (nominally 1½ Cr, 4½ Ni, 3C) has a matrix of iron carbide that surrounds the areas of the steel constituent. This brittle matrix provides a continuous path if a crack should start; thus the alloy is vulnerable to impact and is weak in tension. In contrast, HC 250 (nominally 25 Cr, 2½ C) has the steel portion as the matrix which contains island crystals of chromium carbide. As the matrix is tougher, HC 250 has more resistance to impact and its tensile strength is about twice that of Ni-Hard. Moreover, by a suitable annealing treatment the matrix can be softened enough to permit certain machining operations and then rehardened for service.

Overlay. Design should also consider whether a tough substrate covered with a hard overlay is suitable or desirable. The process is quite versatile, and contributes the advantages of protection to a depth of from $1/32$ to about $3/8$ inch, incorporation of a tough base with the best of the abrasion resistant alloys, economical use of the expensive materials like tungsten carbide, ready application in the field as well as in manufacturing plants, and, usually, only simple welding equipment is needed. Sometimes high labor costs are a disadvantage.

There are size limitations. If large areas are surfaced by automatic welding, only tough alloys can be applied without cracking. The cracks tend to stop at the tougher base, but there is no simple answer to the question about the erosion resistance of a surface containing fine cracks.

Erosion and corrosion combined require special consideration. Most of the stainless steels and related corrosion-resistant alloys owe their surface stability and low rate of corrosion to passive films that develop on the surface either prior to or during exposure to reactive fluids. If conditions change from passive to active, or if the passive film is removed and not promptly reinstated, much higher rates of corrosion may be expected.

Where erosion by a liquid-borne abrasive is involved, the behavior of a corrosion-resistant alloy will depend largely on the rate at which erosion removes the passive film and the rate at which it reforms.

If the amount of metal removal by erosion is significant the surface will probably be continually active. Metal loss will be the additive effect of erosion and active corrosion. Sometimes the erosion rate is higher than that of active corrosion. The material selection judgment can then disregard corrosion and proceed on the basis of erosion resistance, provided the corrosion rates of active surfaces of the alloys considered are not much different. As an example of magnitudes, a good high-chromium iron may lose metal by erosion only a tenth as fast as do the usual stainless steels.

Pipeline Toughness

There has been renewed interest in pipeline toughness in recent years. Pipeline flaws have caused failures, and increased toughness makes pipelines more tolerant of flaws, thus helping to prevent or mitigate ruptures.

Predicting the appropriate level of ductile fracture resistance involves an analysis of fluid properties, operating conditions, and material properties. Kiefner describes a simple equation for natural gas pipelines containing mostly methane with very few heavy hydrocarbons [7]:

$$R_{max} = 0.0072 \, S_o^2 (Rt)^{1/3} \qquad (19\text{-}2)$$

Where:

R_{max} = Required level of absorbed energy for a two-thirds-size Charpy V-notch specimen (i.e., 10 mm × 6.7 mm × 55 mm) to assure arrest of a ductile fracture, ft-lb
S_o = Maximum operating hoop stress level as determined by the Barlow formula using the maximum working pressure (P_w), ksi

$$S_o = \frac{P_w D}{2t}$$

D = outside pipe diameter, in.

$$R = \frac{D}{2}, \text{ in.}$$

t = Pipe wall thickness, in.

The source article discusses use of API Specification 5L-Supplementary Requirement 5 to obtain adequate fracture toughness. Certain small diameter, thin-wall pipes cannot use API 5L SR5, so the article discusses alternatives for obtaining adequate toughness. If the purchaser is limited to only chemical analyses, the following guideline is given:

Materials with very high carbon contents (C > 0.25), very high sulfur (S > 0.015), and very high phosphorus (P > 0.01) should be avoided. [7]

Common Corrosion Mistakes

Kirby gives sound advice for designers who are not experts in metallurgy. He gives seven pitfalls to avoid [8]:

1. Not understanding details of the corrosion service. Stating only the predominant acid without the other details (such as presence of chloride ion) is an example. The author explains how to use the acronym SPORTSFAN:
   S = Solvent
   P = pH
   O = Oxidizing potential
   R = Reducing potential
   T = Temperature
   S = Salts in solution
   F = Fluid flow conditions
   A = Agitation
   N = New aspects or changes to a chemical process
2. Concentrating on overall or general corrosion, and ignoring pitting, crevice corrosion, stress corrosion, cracking, etc.
3. Ignoring alkaline service. Just because strong alkalis do not cause severe overall corrosion in carbon steel or stainless steel, don't overlook stress corrosion, cracking, or effects on other materials.
4. Not considering water or dilute aqueous solutions. This can be overlooked if the other side of the tube or coil has strong chemicals, such as sulfuric acid.
5. Confusion about L-grade of stainless steels. The L-grades, such as 304L, have lower carbon (0.03% vs. 0.08%) than the standard grades (e.g., 304). The L-grades are used to prevent "sensitization" from carbide precipitation during welding. This minimizes strong acid attack of the chromium-depleted areas along the welds. Don't forget to specify the L-grade for the filler metal as well as the base plate. Some confusion exists that the purpose of the L-grade is to handle chloride stress corrosion cracking at percent or multiple ppm levels.
6. Not accounting for the oxidizing or reducing potential of acidic solutions. For non-chromium-containing alloys that are capable of withstanding reducing acids, a small amount (ppm level) of oxidizing chemical can have devastating effects.
7. Neglecting trace chemicals. Watch for ppm levels of chloride with stainless steels or ppm levels of ammonia with copper-base alloys, for example.

References

[1] American Galvanizers Association (AGA). Different forms of embrittlement. Technical FAQ at, www.galvanizeit.org; 2004.

[2] API Div. of Refining Publication 1941. Steels for Hydrogen Service at Elevated Temperatures and Pressures in Petroleum Refineries and Petrochemical Plants. July, 1970.

[3] Bonner WA, et al. Prevention of Hydrogen Attack on Steel in Refinery Equipment. Paper presented at API Div. of Refining 18th Mid-year Meeting. May, 1953.

[4] Derungs WA. Naphthenic Acid Corrosion – an Old Enemy of the Petroleum Industry. *Corrosion*, Dec, 1956:41.

[5] Hammond G, Jones C. *Inventory of Carbon & Energy (ICE)*, Version 2, Department of Mechanical Engineering. UK: University of Bath; January, 2011.

[6] Heller JJ. Corrosion of Refinery Equipment by Naphthenic Acid. Report of Technical Committee T-8, NACE, Materials Protection, Sept, 1963:90.

[7] Kiefner JF, Maxey WA. Specifying Fracture Toughness Ranks High In Line Pipe Selection. *Oil and Gas Journal*, Oct. 9, 1995.

[8] Kirby GN. Avoid Common Corrosion Mistakes for Better Performance. *Chemical Engineering Progress*, April, 1997.

[9] NACE Publication 1F166. Sulfide Cracking-Resistant Metallic Materials for Valves for Production and Pipeline Service. *Materials Protection*, Sept, 1966.

[10] Norgate TE, Jahanshahi S, Rankin WJ. Assessing the environmental impact of metal production processes. *Journal of Cleaner Production*, 2007;15:8–9.

[11] Alloys Rolled. Alloy Performance Guide. Bulletin No. 128Use 08/10, www.rolledalloys.com; 2009.

[12] Skei and Wachttr. Hydrogen Blistering of Steel in Hydrogen Sulfide Solutions. *Corrosion*, May, 1953.

[13] Hertwich E, van der Voet E, Suh S, Tukker A, Huijbregts M, Kazmierczyk P, Lenzen M, McNeely J, Moriguchi Y. UNEP. Assessing the Environmental Impacts of Consumption and Production: Priority Products and Materials. A Report of the Working Group on the Environmental Impacts of Products and Materials to the International Panel for Sustainable Resource Management; 2010.

[14] Walter RJ, Chandler WT. Effects of High Pressure Hydrogen on Storage Vessel Materials. Los Angeles: Paper presented at meeting of American Society for Metals, March, 1968.

20
Safety

Introduction

A prerequisite for every process design is that the safety of people, the environment, and equipment is assured. While it may not be possible or practicable to absolutely prevent every incident, by using appropriate assessment and design tools engineers do strive to strike a consistent balance between safety, technology, and cost.

This chapter gives an overview of the most comprehensive of the US safety standards, OSHA's Process Safety Management (PSM). The chapter then gives specifics that pertain to a range of safety related issues and calculations.

Process Safety Management (PSM)

Compliance with OSHA's Process Safety Management (PSM) [15] regulation is mandatory for every process facility that meets one of these tests:

- Quantities of specific listed highly hazardous chemicals are greater than the quantities listed in the regulation. Examples: 250 lb methyl isocyanate, 500 lb nitric acid (94.5% by weight or greater), 1,500 lb chlorine gas, or 15,000 lb methyl chloride.
- Flammable liquids or gases are present in one location in excess of 10,000 lb, except for hydrocarbon fuels used solely for onsite consumption as a fuel, and flammable liquids stored in atmospheric tanks or

transferred at a temperature below their normal boiling point (without benefit of chilling or refrigeration).

Certain facilities are exempt, including retail facilities, oil or gas drilling and servicing facilities, and normally unoccupied remote facilities.

It's a 14-point performance-oriented system, summarized in Table 20-1. While clearly emphasizing safety, PSM also makes good *business sense*, because compliance reduces the frequency and severity of accidents. The elements comprising PSM range widely and may have far reaching implications.

Table 20-1
The 14 elements of OSHA's process safety management regulation [15]

Element	Description
Employee participation	Employers must consult with their employees and their representatives regarding the employers' efforts in the development and implementation of the process safety management program elements and hazard assessments. Employers must also train and educate their employees and inform affected employees of the findings from incident investigations required by the process safety management program.
Process safety information	Complete and accurate written information concerning process chemicals, process technology, and process equipment is essential to an effective process safety management program. The information to be compiled about the chemicals, including process intermediates, needs to be comprehensive enough for an accurate assessment of the fire and explosion characteristics, reactivity hazards, the safety and health hazards to workers, and the corrosion and erosion effects on the process equipment and monitoring tools.
Process Hazard Analysis	A process hazard analysis (PHA), sometimes called a process hazard evaluation, is one of the most important elements of the process safety management program. A PHA is an organized and systematic effort to identify and analyze the significance of potential hazards associated with the processing or handling of highly hazardous chemicals.
Operating procedures	Operating procedures describe tasks to be performed, data to be recorded, operating conditions to be maintained, samples to be collected, and safety and health precautions to be taken. The procedures need to be technically accurate, understandable to employees, and revised periodically to ensure that they reflect current operations.
Training	All employees, including maintenance and contractor employees, involved with highly hazardous chemicals need to fully understand the safety and health hazards of the chemicals and processes they work with for the protection of themselves, their fellow employees and the citizens of nearby communities.
Contractors	For contractors, whose safety performance on the job is not known to the hiring employer, the employer will need to obtain information on injury and illness rates and experience and should obtain contractor references. Additionally, the employer must assure that the contractor has the appropriate job skills, knowledge and certifications (such as for pressure vessel welders). Contractor work methods and experiences should be evaluated.

Table 20-1
The 14 elements of OSHA's process safety management regulation [15]—cont'd

Element	Description
Pre-startup safety review	P&IDs are to be completed along with having the operating procedures in place and the operating staff trained to run the process before startup. The initial startup procedures and normal operating procedures need to be fully evaluated as part of the pre-startup review to assure a safe transfer into the normal operating mode for meeting the process parameters.
Mechanical integrity	Elements of a mechanical integrity program include the identification and categorization of equipment and instrumentation, inspections and tests, testing and inspection frequencies, development of maintenance procedures, training of maintenance personnel, the establishment of criteria for acceptable test results, documentation of test and inspection results, and documentation of manufacturer recommendations as to meantime to failure for equipment and instrumentation.
Non-routine work and hot work permit	Non-routine work which is conducted in process areas needs to be controlled by the employer in a consistent manner. The hazards identified involving the work that is to be accomplished must be communicated to those doing the work, but also to those operating personnel whose work could affect the safety of the process.
Management of change	Change includes all modifications to equipment, procedures, raw materials and processing conditions other than "replacement in kind." These changes need to be properly managed by identifying and reviewing them prior to implementation of the change. A typical change form may include a description and the purpose of the change, the technical basis for the change, safety and health considerations, documentation of changes for the operating procedures, maintenance procedures, inspection and testing, P&IDs, electrical classification, training and communications, pre-startup inspection, duration if a temporary change, approvals and authorization.
Incident investigation	The incidents for which OSHA expects employers to become aware and to investigate are the types of events which result in or could reasonably have resulted in a catastrophic release. Employers need to develop in-house capability to investigate incidents that occur in their facilities.
Emergency planning and response	Each employer must address what actions employees are to take when there is an unwanted release of highly hazardous chemicals. Employers at a minimum must have an emergency action plan which will facilitate the prompt evacuation of employees when there is an unwanted release of highly hazardous chemical.
Compliance audits	Employers need to select a trained individual or assemble a trained team of people to audit the process safety management system and program. The audit is to include an evaluation of the design and effectiveness of the process safety management system and a field inspection of the safety and health conditions and practices to verify that the employer's systems are effectively implemented.
Trade secrets	Employers must make all necessary information available to persons responsible for compiling the process safety information, development of PHAs, development of operating procedures, and those involved with incident investigations.

Process Hazard Analysis (PHA)

The overarching requirements of a PHA are to identify, evaluate, and control the hazards of a process. When appropriately conducted, a PHA brings together the viewpoints of subject matter experts who are familiar with all relevant aspects of the process being evaluated. The PHA team should include knowledgeable process engineers, equipment engineers, instrumentation and control engineers, safety professionals, plant operators, environmental specialists, project engineers, and others who were involved in the design and execution of the project.

The team can choose from a variety of methodologies and formats for performing the analyses and documenting the results. See Table 20-2. Factors that influence the choice include:

- Project complexity, size, and status (e.g., whether it is early in the design phase or already under construction).
- The company's desire to base decisions on quantitative cost/benefit data rather than relying on qualitative information.
- Company and personal experience with similar plants and technologies, and their degree of similarity.
- Team experience and/or preference for using a particular tool.
- Recommendations from others, including proposed facilitators.

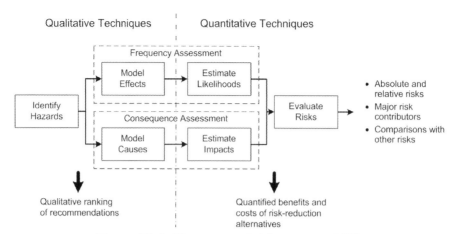

Figure 20-1. The process of risk assessment [3].

The minimum expected outcome for a PHA is that the team:

a) identify pathways that could conceivably lead to hazards, and the controls (both design and administrative) associated with those pathways,

b) assess the likelihood that the controls will fully or partially fail, and

c) qualitatively evaluate and document the range of safety and health effects due to those failures.

The team should consider additional measures for hazards that are found to be inadequately controlled.

It's not unusual to utilize a high level qualitative method, such as a "Checklist" approach, to perform a preliminary PHA during the preliminary design phase of a project, graduate to a more structured qualitative method such as HazOp when the P&IDs are finished, and then later employ a detailed quantitative tool, such as Fault Tree Analysis when the design is completed (see Figure 20-1). However, the quantitative methods should usually be targeted to selected cases, when decision making justifies the associated high effort and cost. Risk assessments should also be done after the facility is constructed, but prior to introducing hazardous substances into the plant.

Table 20-2
Popular methodologies for process hazards analysis

Method	Type	Description
Checklist	Qualitative	Compares a system design against a list of known potential hazards and failures to identify which of the known potentials are feasible in the system under study. Depends on the quality of the checklist and experience of the study team.
What-If / Checklist	Qualitative	Brainstorming activity that poses "What if…" questions to the study team with the goal being to identify what could go wrong and the hazardous consequences if that occurs. Checklists are sometimes used to help structure the brainstorming and ensure that all relevant what if questions are asked.
Hazard and Operability Study (HazOp)	Qualitative	Structured and systematic technique for identifying potential hazards and operability problems. Facilitated by using "guide words" (e.g., "too high") to step through a system and find risk events that would result if there were deviations from design or operating intentions. Simple and intuitive method that is helpful when hazards are difficult to quantify. Used to identify hazards as the first step in performing a quantitative assessment.
Failure Mode and Effects Analysis (FMEA)	Semi-Quantitative	Structured method for identifying potential failures for each function performed by a system and the potential consequences of those failures. A qualitative scoring (e.g., low, medium, high) system results in a Risk Priority Number (RPN) for each failure mode; RPNs are sorted to rank the potential failures for use in design or operation of the facility. Used to determine which failures should be analyzed further using FTA or ETA.

Table 20-2
Popular methodologies for process hazards analysis—cont'd

Method	Type	Description
Incident Frequencies from Historical Record	Quantitative	Where accumulated experience is relevant and statistically meaningful, the number of recorded incidents can be divided by the exposure period to give a failure estimate. Straightforward technique without the need for detailed frequency modeling. Example: use of historical information to estimate the probability that an unconfined vapor cloud explosion will follow a release (pipe rupture). Technological change, either in the scale of plant or the design may make some historical data inapplicable [6].
Fault Tree Analysis (FTA)	Quantitative	Based on the combinations of failures of basic system components, safety systems, and human reliability. Used to estimate incident frequencies (e.g., major leakage of a flammable material). Requires historical reliability information and/or mechanistic models of plant component data and operator response. Relies on selection of reliability parameters and treatment of events as repairable or non-repairable. Provides complementary information from the initial tree generation (qualitative) and final analysis (quantitative) aspects of the method [6].
Event Tree Analysis (ETA)	Quantitative	Identifies and quantifies possible outcomes following an initiating event. Consequences can be direct (e.g., fire) or indirect (e.g., domino incidents). Used to quantitatively estimate the distribution of incident outcomes (e.g., frequencies of explosions, pool fires, flash fires). Pre-incident event trees highlight the value and potential weaknesses of protective systems, especially indicating outcomes that lead directly to failures with no intervening protective methods. Post-incident event trees highlight the range of outcomes that is possible from a given incident, including domino incidents [6].
Human Reliability Analysis	Quantitative	Provides quantitative values of human error for inclusion in FTA and ETA, or to identify potential recommendations for error reduction. Requires specialists to conduct the study and is best utilized for critical systems when the human error component is important [6].

Qualitative methods are limited to identification of hazards and pathways; quantitative methods are required to determine frequencies or likelihoods. Severities, or consequences, are usually determined using mathematical modeling or statistical analysis (if data are available).

Safe Design Practices

Ulrich and Vasudevan cite the high cost of performing a HazOp and the lack of sufficient information early in a project as barriers to conducting PHAs [16]. They offer instead some useful rules of thumb to help engineers design safer facilities.

By recognizing where hazardous conditions usually exist, engineers can pay particular attention to the plant characteristics that may lead to them. Ulrich and Vasudevan point to latent energy as the key indicator, including:

- Kinetic energy, especially where there is rotating equipment.
- Potential energy, notably associated with structural failures.
- Work, as may be stored in springs or capacitors.
- Heat, and rapid heat build-up or release from, for example, runaway reactions.

- Enthalpy, or internal energy, which fuels runaway reactions, deflagrations, and explosions.

They go on to summarize four basic steps in achieving "Inherently Safer Predesign (ISPD)".

1. *Identify* the potential hazards in the design, using checklists and data such as chemical reactivity charts, toxicity information, and flammability diagrams. Fire triangles illustrate that fuel, oxidizer, and ignition source are required to support combustion and cause a chemical explosion.
2. *Eradicate* hazards by providing design features such as inert gas blanketing and failsafe design. For example, see the discussion on Safety Instrumented Systems in Chapter 21.
3. *Minimize – Simplify, Moderate, Attenuate* by keeping volumes small, minimizing ignition

sources, and providing safety equipment and instruments. Ignition sources include heat (e.g., failed pump bearings), repair activities (e.g., grinding, welding), open flames (e.g., furnaces), faulty electrical devices, static electricity, and natural causes (e.g., lightning).

4. *Isolate* by segregating hazardous operations.

Emergency Vent Systems

Emergency venting, through relief valves or rupture discs, is generally a response to:

- Fire outside the vessel; heat from the fire boils material in the tank causing a pressure increase
- Runaway reaction
- Blocked-in line; thermal expansion causes a pressure rise

Deflagrations or explosions inside equipment also result in a pressure rise, but the increase is too rapid for a relief valve or rupture disc to respond to. Protect against them by:

a) designing the system to eliminate conditions that support combustion (e.g., nitrogen purging, page 331),

b) using deflagration pressure containment (page 329), and

c) utilizing specialized, fast-acting, explosion suppression systems.

Fire Outside the Vessel

Use API 520/521/2000 for a refinery storage tank, or NFPA 30 or OSHA 1910.106 to size the vent of a vessel used in other industries. The calculation estimates the amount of heat from the fire that will transfer into the tank.

The heat of vaporization of the vessel contents is used to compute its boiling rate, which is the basis for sizing the vent. Tables in the standards give an answer in terms of venting rate (standard m^3/h or ft^3/h as air); these tables were created with the assumption that the tank stores n-hexane. Chemicals that are less volatile will vent at a lesser rate.

This calculation is valid only for vents that are 100% vapor. Foaming systems carry liquid along with the vapor and require a larger vent.

Here is the procedure [14] applicable to "atmospheric" tanks with a design pressure below 103.4 kPa gage (15 psig):

1. Calculate the effective wetted surface area of the tank [2].
   For a spherical tank: 55% of the total surface area or the surface area to 9.14 m (30 ft) above grade, whichever is greater.
   For a horizontal tank: 75% of the total surface area or the surface area to 9.14 m (30 ft) above grade, whichever is more.
   For a vertical tank: the first 9.14 m (30 ft) above grade of the vertical shell surface area; if the tank is on legs use engineering judgment to evaluate the portion of the bottom head to include. The wetted area, A, is expressed in m^2 (ft^2).

2. Calculate the heat input from a pool fire using Table 20-3:

Table 20-3
Heat input from pool fire to above-ground storage tank

SI Units		US Units	
Wetted Surface, m^2	Heat Input, W	Wetted Surface, ft^2	Heat Input, Btu/h
$< 18.6\,m^2$	$Q = 63150\,A$	$< 200\,ft^2$	$Q = 20000\,A$
$18.6\,m^2$ to $< 93\,m^2$	$Q = 224200\,A^{0.566}$	$200\,ft^2$ to $< 1000\,ft^2$	$Q = 199300\,A^{0.566}$
$93\,m^2$ to $< 260\,m^2$	$Q = 630400\,A^{0.338}$	$1000\,ft^2$ to $< 2800\,ft^2$	$Q = 963400\,A^{0.338}$
$>= 260\,m^2$, DP > 7 kPag	$Q = 43200\,A^{0.82}$	$>= 2800\,ft^2$, DP > 1 psig	$Q = 21000\,A^{0.82}$
$>= 260\,m^2$, DP $<= 7$ kPag	$Q = 4129700$	$>= 2800\,ft^2$, DP $<= 1$ psig	$Q = 14090000$

Table 20-4
Correction factors for tank venting calculation

Feature	API 2000	OSHA 1910.106
Area around tank is drained to a safe location	0.5	0.5
Approved insulation with thermal conductivity of 9 W/m²-K/cm (4 Btu/h-ft²-°F/in)	0.3 for 2.5 cm (1 in)	0.3
	0.15 for 5.1 cm (2 in)	
	0.075 for 10.2 cm (4 in)	
	0.05 for 15.2 cm (6 in)	
	0.0375 for 20.3 cm (8 in)	
	0.03 for 25.4 cm (10 in)	
	0.025 for 30.5 cm (12 in)	
Approved water spray directed onto the tank	NA	0.3
Approved insulation and approved water spray	NA	0.15
Concrete tank or fireproofing	Use factor for an equivalent conductance value of insulation	NA
Earth-covered above ground storage tank	0.03	NA

Choose one value; features are not multiplicative.

3. Apply a correction factor that reduces the rate due to one of the features listed in Table 20-4.
4. Calculate the venting rate, expressed as standard m³/h (ft³/h) of air, using (the leading factor includes units conversions):

$$VentRate = C\frac{Q\,F}{L}\left(\frac{T}{M}\right)^{0.5}$$

Where:

C = units conversion factor, 906.6 for SI units or 3.091 for US units
Q = heat input from Step 2, W or Btu/h
F = environmental correction factor from Step 3
L = latent heat of vaporization, J/kg or Btu/lb
T = absolute temperature at relieving conditions, usually the boiling point at the relieving pressure, K or R
M = molecular weight

Runaway Reaction

The rate of an irreversible reaction between liquid components typically doubles for each 10 °C temperature rise. If the reaction is exothermic, and there are sufficient quantities of starting materials, there will be an exponential temperature rise. This can lead to rapid boiling, deflagration, and catastrophic equipment failure due to overpressurization.

Consider the possibility that a runaway reaction could occur when sizing pressure relief devices on reactors. Every exothermic reaction has the potential to run away; initiators include impurities that catalyze the reaction, loss of cooling, and raw material addition errors.

Engineers should consult experts who use the latest tools to evaluate the venting requirements. In particular, methods documented by the Design Institute for Emergency Relief Systems (DIERS) should be utilized. OSHA (CFR 1910.119) recognized technical reports published by the AIChE on topics such as two-phase flow for venting devices as constituting "generally recognized and accepted good engineering practice". OSHA also recommended "re-evaluation of the size and capacity of the emergency relief system using the methodology of the AIChE's Design Institute for Emergency Relief Systems" for a device involved in a runaway reaction venting incident.

Blocked-In Line

Consider using a pressure relief valve on liquid lines that could be blocked in, usually by closing valves. Liquids are nearly incompressible, and thermal expansion can create unsafe pressures. The source of heat that expands the liquid can include: steam on the shell side of a blocked-in heat exchanger, heat tracing, solar radiation, etc.

Bravo and Beatty showed that, for most piping cases, a ³/₄" × 1" relief valve with a 0.06 in.² orifice is large enough to protect up to 10,000 feet of 4-inch pipe [4]. The most common practice is to set the relief valve at the design pressure of the pipe.

The formula most often used to compute the relieving rate is [4]:

$$GPM = \frac{\alpha\,H}{500\,SG\,C} \tag{20-1}$$

Where:

GPM = relieving capacity in gal/m
α = cubical expansion coefficient per deg of temperature rise, $1/^{\circ}F$
H = heat transfer rate or flux, Btu/h
SG = specific gravity of liquid, dimensionless
C = specific heat of trapped liquid, Btu/lb-$^{\circ}F$

The expansion coefficient is calculated with:

$$\alpha = \left(\frac{v_2 - v_1}{v_{avg}}\right)\left(\frac{1}{t_2 - t_1}\right) \qquad (20\text{-}2)$$

Where:

v = specific volume of liquid at two temperatures and the average
t = temperatures corresponding to the specific volume data

Although compressibility of the liquid, and even the pipe, can be incorporated into the evaluation (see [4]), this refinement is unnecessary for a typical engineering analysis.

Check to be sure that vaporization will not occur by determining if the temperature of the heat source is below the boiling point of the trapped fluid at the relieving pressure. If this is not the case, assume that the pressure relief will be all vapor and calculate the rate using:

$$W_{cold} = \frac{Q_{hot}}{L_{cold}} \qquad (20\text{-}3)$$

Where:

W_{cold} = relieving rate, lb/h
Q_{hot} = heat input from the hot source, Btu/h
L_{cold} = heat of vaporization of the trapped fluid at the relieving pressure, Btu/lb

Flash Point of Binary Mixtures

The flash point is one of the most important fire-safety physical properties. It is defined as the temperature at which sufficient quantity of vapor exists, in equilibrium with air at 760 mm Hg pressure, to form a combustible mixture. The flash point is determined experimentally using either a closed container ("closed-cup test") or an open container ("open-cup test"). The closed-cup value is usually a few degrees cooler than the open-cup result, and is preferred when making critical safety-related decisions.

The Lower Explosive Limit (LEL) is the concentration (mole % or volume %) of the flammable material in air at its flash point.

The flash point of pure materials is readily found in Material Safety Data Sheets (MSDS) and other tabulations in books and on the Internet. However, the flash points of mixtures may be difficult to find. This section provides an easy way to estimate the flash points of mixtures if the flash points of the individual components are known, and gives a useful example for aqueous alcohol solutions.

Flash point estimation follows a procedure that is very similar to the vapor-liquid equilibrium (VLE) calculations discussed in Chapter 3. Recall that Raoult's Law predicts VLE for "ideal" mixtures, and that activity coefficients are used for non-ideal solutions. Polar compounds, such as water and alcohol, are non-ideal and require the use of activity coefficients for satisfactory results. For the equations given in this section, set the activity coefficient to unity for ideal solutions.

Hristova, et al. published an article in 2010 from which the material in this section is derived [10].

The general equation for estimating the flash point of a binary mixture is:

$$\frac{x_1\gamma_1 P_1}{P_{1,fp}} + \frac{x_2\gamma_2 P_2}{P_{2,fp}} = 1 \qquad (20\text{-}4)$$

Where:

x = mole fraction of components 1 and 2 (which will sum to 1)
γ = activity coefficient of components 1 and 2
P = vapor pressure of components 1 and 2
P_{fp} = vapor pressure at the flash point of pure components 1 and 2

For the case where one of the components is not flammable, such as water, the second term is eliminated and the first term relates only to the flammable compound such as an alcohol. Then, using a goal seeking algorithm in Excel or MathLab, along with the Antoine Equation to relate vapor pressure to temperature, the temperature can be found that solves the equation.

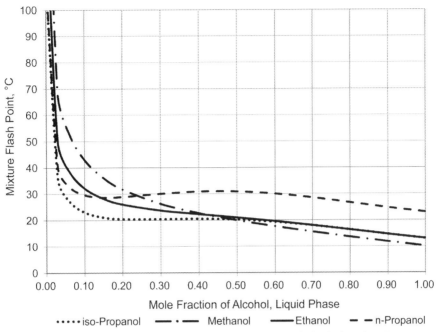

Figure 20-2. Estimated flash point of mixtures of alcohol and water.

The companion Excel workbook for this chapter solves the problem for the four alcohols given in the referenced article. Results are shown in Figure 20-2.

Deflagration Prevention or Control

Deflagrations are rapidly burning fires in which the combustion zone propagates at a velocity that is slower than the speed of sound. When the deflagration occurs inside an enclosure, an explosion results if the enclosure ruptures due to development of internal pressure from the deflagration. NFPA® 69 provides minimum systems requirements, intended to prevent explosions in enclosures, and this should be consulted whenever one of the methods described in this section is contemplated [11]. Table 20-5 and Table 20-6 list the various methods.

Table 20-5
Methods based on prevention of combustion

Method	Considerations	Limitations
Oxidant concentration reduction (see Purging, page 331) Combustible concentration reduction	Oxygen reduction is an asphyxiation hazard Design for startup, normal operation, shutdown, temporary operations, and emergency shutdown process conditions Achieved through dilution with clean air (e.g., in a room by introducing fresh air) or by reducing the amount of flammable species (e.g., catalytic destruction) Except under certain circumstances, the concentration must be maintained below 25% of the Lower Explosive Limit (LEL)	Best suited to totally enclosed spaces, especially inside equipment, where people are not present and purge gas is easily controlled Generally not an appropriate method for the inside of process equipment

Table 20-6
Methods based on the prevention or limitation of damage

Method	Considerations	Limitations
Predeflagration detection and control of ignition sources	Applicable to systems handling combustible particulate solids Optical sensing equipment senses radiation from hot or glowing particles then actuates an extinguishing medium such as water, steam, carbon dioxide, or another method. Gas sensing systems detect the formation of gaseous thermal decomposition products and actuate a means of control or extinguishment such as automated shutdown or release of an extinguishing medium. Generally installed in conjunction with deflagration suppression or venting measures	Not suitable for systems with flammable gases or mixtures of flammable gases with particulates
Deflagration suppression	Within equipment, a deflagration is detected due to the onset of a rapid pressure rise or flame presence; a suppressant is quickly dispersed into the equipment which prevents the full impact of the deflagration from developing Suppressants include chemicals, halon, and water An alternative is to contain the deflagration within the equipment and prevent it from traveling through pipes or ducts to other equipment by using fast acting valves	Requires equipment that can withstand the pressure rise prior to effective suppressant application Can typically react only once; if flammable conditions recur after the suppressant has been discharged there is no further protection
Isolation methods	Passive isolation interrupts flames, deflagration pressures, pressure piling, or flame-jet ignition between enclosures that are connected by pipe or duct. Several types of devices are used: Flame front diverters Passive float valves Material chokes (rotary valves) Static dry flame arresters Hydraulic (liquid seal) flame arresters Liquid product flame arresters	Each type of device is limited to specific applications. The equipment or enclosure where the deflagration originates is unprotected; however, in some cases the device protects equipment — for example, a flame arrester installed on the vent from a storage tank can prevent an ignition of vent gases outside the vessel from propagating into the vessel
Deflagration pressure containment	An entire vessel and its appurtenances are protected from damage by specifying a design pressure that is higher than the maximum pressures resulting from an internal deflagration	Requires additional controls if two or more vessels are connected by large diameter pipes or ducts Not applicable if detonation can occur

Prevent Backflow

Englund cites preventing backflow as a very important aspect of plant safety that lacks readily available information. Where there is flow of liquid or gases in pipes, it is usually important to prevent material from flowing in a direction opposite to that intended. This "backflow" can happen for many reasons, such as when the pressure in a line that is normally flowing into another pipe or system becomes less than in the line or system to which it is connected, and the valves are open or leaking. Four aspects of backflow prevention should be considered [7]:

1. *Detection* usually involves using suitable control and sensing devices to detect conditions that indicate backflow could happen, is about to happen, or is happening. Detection and subsequent prevention of backflow usually involves process control.

2. *Operation.* Proper operation of equipment can reduce the possibility of backflow by starting pumps and opening valves in the right order, stopping pumps and closing valves in the right order, reducing pressure downstream, etc.

3. *Action.* The action to be taken if backflow conditions are detected during operation usually involves automatically closing valves, shutting down pumps, or relieving pressure.

4. *Design.* Good equipment design can reduce the possibility of backflow. Backflow prevention is a feature that process engineers often incorporate into the design of plants and may not involve the detection of conditions that could lead to

backflow. Examples of backflow prevention: the use of check valves, double or triple valves in series, some kinds of pumps, double block valves with a vent between, and suitable selection of design pressures.

Equipment Purging

Purging with inert gas (e.g., nitrogen) is commonly used to reduce the oxygen concentration to below that needed to support combustion. Most hydrocarbons cannot burn if the oxygen concentration is less than about 11% by volume. To provide a safety factor, the concentration should never exceed 8% in the vapor space of a tank or pipe that contains a hydrocarbon liquid.

At least four methods are used to purge equipment and piping: dilution, displacement, pressure cycle, and evacuation-replacement. Each method is briefly described here, along with the method for estimating the amount of purge gas required.

Dilution Purging

For dilution purging, an inert gas is swept through the vapor space of a tank at a rate that is sufficient to provide good mixing with the resident air. The entrance velocity of the gas must be sufficient for it to reach to the bottom of the vapor space in the tank, and the exit gas should be located far from the entry. This method may not be suitable for systems with significant deadlegs or pockets, since the dilution gas may not successfully displace the air in those cavities; this could lead to local hot spots.

The minimum amount of inert gas, assuming it is oxygen-free, is calculated from:

$$V_n = \ln\left(\frac{C_i}{C_f}\right) \tag{20-5}$$

Where:

V_n = number of vessel-volumes of inert gas required (theoretical)
C_i = initial concentration of oxygen in the vapor space (e.g., 21% for pure air)
C_f = final concentration of oxygen in the vapor space

Example:
A 15,000-gallon storage tank is inerted with nitrogen to 3% oxygen concentration prior to pumping alcohol into the tank. What is the theoretical minimum quantity of nitrogen required? The tank pressure is maintained at 0.25 psig and the temperature is 90 °F.
Solution:

$$V_n = \ln\left(\frac{21}{3}\right) = 1.95$$

Actual cubic feet of nitrogen required = 1.95 (15,000 gal) (1 ft^3/7.48 gal) = 3,900 ft^3
Standard cubic feet of nitrogen required = (3,900 acf) (14.7 + 0.25)/14.7 (460 + 32)/(460 + 90) = 3,550 scf

Displacement Purging

For displacement purging, inert gas is slowly admitted to the tank or pipe and air is displaced and removed from another port. The method assumes little or no mixing between the purge gas and the displaced air so it is only applicable to tall and narrow vessels, such as a distillation column, and pipelines. For pipes the method may include the use of a "pig" that precedes the nitrogen and pushes the air out like a plunger or piston. The minimum amount of inert gas required is equal to the volume of the equipment. However, a safety factor of 20% should be added.

Pressure-Cycle Purging

By alternately pressurizing then venting the vessel using inert gas, the oxygen concentration is reduced by the factor equal to the ratio of the vented and pressurized absolute pressures. It assumes good mixing in the vapor phase, and requires that the system be built to withstand the pressure.

Assuming the vented pressure is 1 atmosphere (14.7 psia), the concentration in the vapor space is calculated with:

$$C_f = C_i \left(\frac{1}{P}\right)^n \tag{20-6}$$

Where (see other nomenclature above):

P = pressure of inert gas cycle, atm
n = number of cycles (integer)

The volume of inert gas required is:

$$V_n = n\,(P - 1) \tag{20-7}$$

Evacuation-Replacement Purging

This is similar to the pressure-cycle method, but uses a vacuum source to first evacuate the vessel and then break the vacuum using the inert gas. The equipment must be designed to withstand the vacuum. This method is very effective for equipment that contains many deadlegs and pockets.

The formulae are inverted from those used for pressure-cycle purging. Again assuming that the vented pressure is 1 atmosphere and the evacuation pressure is given in atmospheres (e.g., 0.4 atmospheres absolute pressure = 304 mm Hg):

$$C_f = C_i \left(\frac{P}{1}\right)^n \tag{20-8}$$

$$V_n = n\,(1 - P) \tag{20-9}$$

Electrical Area Classification

In the US, electrical area classification is defined with a Class/Division structure in accordance with the National Electric Code (NEC®). Outside the US, the zone system, defined by IEC, is generally used. NEC also allows the use of the zone system as defined in Article 505 of the Code [12]. Although there are strong parallels between the IEC and US zone systems, they are not identical. The specific requirements for a particular location must be determined by the engineer, including the applicable edition of the Code. See Tables 20-7 and 20-8.

In addition to the area classification, the hazardous materials are grouped in accordance with Table 20-9.

Table 20-7
Zone and class/division electrical area classifications for flammable gases or vapors

Condition	Zone System	Class/Division System
Ignitable concentrations of flammable gases or vapors are present continuously or for long periods of time	Class I, Zone 0	Class I, Division 1
Ignitable concentrations of flammable gases or vapors are likely under normal operating conditions, may frequently exist because of repair or maintenance operations, or may frequently exist due to leakage. Equipment breakdown could cause release of ignitable concentrations of flammable gases or vapors while simultaneously causing failure of electrical equipment such that the equipment could become an ignition source. Is adjacent to a Class 1, Zone 0 location from which ignitable concentrations of gases could be communicated unless adequate safeguards are in effect	Class I, Zone 1	
Ignitable concentrations of gases or vapors are not likely to occur in normal operations, but if they do occur will exist for a short period. Flammable gases or vapors are fully contained in containers or equipment from which they can only escape in the event of accidental rupture or as a result of abnormal operation. Ignitable concentrations are prevented by mechanical ventilation (i.e., dilution with fresh air to keep the concentration below 25% of the LEL), but which could become ignitable in the event of failure of the mechanical system. Is adjacent to a Class 1, Zone 1 location from which ignitable concentrations of gases could be communicated unless adequate safeguards are in effect	Class I, Zone 2	Class I, Division 2

Adapted from [12].

Table 20-8
Zone and class/division electrical area classifications for combustible dusts or flyings

Condition	Zone System	Class/Division System
Ignitable concentrations of combustible dusts or flyings are present continuously or for long periods of time	Zone 20	Class II, Division 1
Ignitable concentrations of combustible dusts or flyings are likely under normal operating conditions, may frequently exist because of repair or maintenance operations, or may frequently exist due to release. Equipment breakdown could cause release of ignitable concentrations of combustible dusts or flyings while simultaneously causing failure of electrical equipment such that the equipment could become an ignition source. Is adjacent to a Class 1, Zone 0 location from which ignitable concentrations of dusts could be communicated unless adequate safeguards are in effect	Zone 21	
Ignitable concentrations of combustible dusts or flyings are not likely to occur in normal operations, but if they do occur will only persist for a short period. Combustible dusts or flyings are fully confined in containers or equipment from which they can only escape in the event of accidental rupture or as a result of abnormal operation. Ignitable concentrations are prevented by mechanical ventilation (i.e., dilution with fresh air to keep the concentration below 25% of the LEL), but which could become ignitable in the event of failure of the mechanical system. Is adjacent to a Class 1, Zone 1 location from which ignitable concentrations of dusts could be communicated unless adequate safeguards are in effect	Zone 22	Class I, Division 2

Adapted from [12].

Table 20-9
Material groups for gases and dusts

Material	Zone System	Class/Division System
Acetylene	IIC	A
Hydrogen, or any vapor that may burn or explode with a maximum experimental safe gap (MESG) value less than or equal to 0.45 mm or a minimum igniting current ratio (MIC ratio) less than or equal to 0.40		B
Ethylene, or any vapor that may burn or explode with a maximum experimental safe gap (MESG) value from 0.45 mm to 0.75 mm or a minimum igniting current ratio (MIC ratio) from 0.40 to 0.80	IIB	C
Propane, and most other flammable gases that may burn or explode with a maximum experimental safe gap (MESG) value greater than 0.75 mm or a minimum igniting current ratio (MIC ratio) greater than 0.80	IIA	D
Atmospheres containing combustible metal dusts, including aluminum, magnesium, their alloys, and others that present similar hazards	Not Applicable	E
Atmospheres containing combustible carbonaceous dusts that have more than 8% total entrapped volatiles. Examples: coal, carbon black, charcoal, coke dusts		F
Other combustible dusts, such as flour, grain, wood, plastic, and chemicals		G

Adapted from [12].

Process engineers should fully understand and appreciate the electrical area classifications for their projects. Classifications are normally depicted on drawings that show rooms and equipment in plan and elevation. It is often possible to deliberately locate electrical equipment – typically, instrumentation and motors – so that it is in a less hazardous location. This practice saves money for the initial purchase of the equipment, as well as its installation and maintenance.

The most hazardous locations are those where flammable vapors or combustible dusts are routinely present at a concentration near the flammable range. Examples include: the inside of equipment containing flammable liquids, near openings to tanks that are routinely opened (such as manways), at low points where vapors might accumulate (e.g., sumps, trench drains), and around transfers from portable containers (e.g., drums). Due to environmental and health concerns, there is a trend toward eliminating open processing thereby reducing the occurrence of Division I locations.

Causes for Loss of Containment

This checklist (Table 20-10) is a guide to failures that could result in process materials inadvertently discharging to the environment. It is adapted from Appendix A, [6], and is not presumed to be an exhaustive list.

Table 20-10
Loss of containment checklist [6]

Category	Causes
"Open-end" route to atmosphere	• Planned process relief or venting • Inappropriate operation of equipment in service, such as a spurious relief valve operation or rupture disk failure • Operator error, such as leaving a drain or vent valve open, overfilling a tank, or opening a unit that is under pressure

Imperfections in equipment	• Imperfections arising prior to commissioning and not detected before start-up due to poor inspection or testing procedures
	  • Equipment inadequately designed for proposed duty
	  • Defects arising during manufacture of equipment
	  • Equipment damaged or deteriorated during transit or storage
	  • Defects arising during facility construction and equipment installation
	• Imperfections due to equipment deterioration in service and not detected before the effect becomes significant
	  • Normal wear and tear on pump or agitator seals, valve packing, flange gaskets, etc.
	  • Corrosion, including stress corrosion cracking
	  • Erosion or thinning
	  • Metal fatigue or vibration effects
	  • Previous periods of gross maloperation
	  • Hydrogen embrittlement
	• Imperfections arising from routine maintenance or minor modifications not carried out correctly

External factors	• Impact damage, such as by cranes, vehicles, and machinery • Damage by confined explosions due to accumulation and ignition of flammable mixtures arising from small process leaks • Settlement of structural supports • Damage to portable containers (including tank trucks, rail cars, and Totes) during transport of materials on- or off-site • Fire exposure • Blast effects from a nearby explosion (including blast overpressure, projectiles, and structural damage) • Natural events such as windstorms, earthquakes, floods, lightning, hurricanes, etc.

Deviations in plant operating conditions beyond the design limits	• Overpressuring of equipment
	  • Due to a connected pressure source
	  • Due to rising process temperature
	  • Due to an internal explosion
	  • Due to physically or mechanically induced forces (e.g., thermal expansion)
	• Underpressuring of equipment (for equipment not capable of withstanding vacuum)
	  • By direct connection to an ejector set or equipment normally running under vacuum
	  • Due to movement or transfer of liquids
	  • Due to cooling of gases or vapors
	  • Due to solubility effects (dissolution of gases in liquids)
	• High metal temperature (causing loss of strength)
	  • Fire under equipment
	  • Flame impingement causing local overheating
	  • Overheating by electric heaters
	  • Inadequate flow of fluid via heated equipment
	  • Higher flow rate or higher temperature of the hotter stream or lower flow rate or higher temperature of the colder stream
	• Low metal temperature (causing cold embrittlement and overstressing)
	  • Overcooling by refrigeration units
	  • Incomplete vaporization and/or inadequate heating of refrigerated material before transfer into equipment of inadequate temperature rating
	  • Loss of system pressure on units handling low boiling point liquids

Table 20-10
Loss of containment checklist [6]—cont'd

Category	Causes
	• Wrong process materials or abnormal impurities
	• Variations in stream compositions outside design limits
	• Abnormal impurities introduced with raw materials or wrong raw materials
	• Byproducts of abnormal chemical reactions
	• Oxygen, chlorides, or other impurities remaining in equipment at start-up due to inadequate evacuation or decontamination
	• Impurities entering process from atmosphere, service connections, tube leaks, etc., during operation

Dust Explosion Hazards

Dust explosions can result in very significant property damage and serious injuries to personnel. Combustible dusts are defined as:

"A combustible particulate solid that presents a fire or explosion hazard when suspended in air or the process-specific oxidizing medium over a range of concentrations, regardless of particle size or shape." [13]

This differs from older definitions that specified that the particle size is less than 420 micrometers; the current definition relies on testing, and recognizes that dusts may have an odd shape that would prevent combustible particles from passing through a screening sieve.

In addition to NFPA 654, OSHA is planning a new standard for combustible dusts and FM Global has a comprehensive *Data Sheet* on the topic [9]. Each of these references should be consulted when designing a system that processes, or could produce as a byproduct, a combustible dust. FM Global Data Sheet 7-76 is particularly useful, with specific recommendations for a variety of equipment such as dust collectors, silos, and spray dryers. FM Global also maintains a proprietary computer program, *DustCalc*, that predicts explosion pressures based on specifics from a given situation (www.fmglobal.com). *DustCalc* reports results from the FM Global algorithms, and also from methods presented in NFPA 68 and the German VDI calculation methods (VDI 3673).

FM Global recommends that owners:

Treat all equipment that handles combustible dusts, as well as any rooms or buildings where combustible dusts can be present and might be put into suspension, as having a dust explosion hazard. [9]

Potential sources of dust include:

- "As received" fine powder
- Course material that contains fines
- Fines generated as a result of coarse material attrition during handling and/or processing
- Fines generated during machining operations on finished parts

An explosive dust cloud is orders of magnitude more concentrated than the amount that would be troublesome from a hygienic viewpoint. For example, a typical Personnel Exposure Limit for dust is less than $0.01 \, g/m^3$, whereas a typical minimum explosive concentration (MEC) is over $100 \, g/m^3$. This is why catastrophic dust explosions often seem to be associated with silos and other large unoccupied spaces. However, building explosions do occur when a smaller primary explosion, inside a piece of process equipment such as a dust collector for example, creates a shock wave that causes powder to become suspended in the building resulting in a secondary explosion.

Design recommendations include [9]:

- Control fugitive dust releases using enclosures, collection systems, or equipment design.
- Locate dust producing operations in areas separated from other occupancies.
- Minimize chances for dust accumulation in buildings.
- If fugitive dusts will exist, design the structure to safely vent potential explosions using damage-limiting construction techniques.

- Locate the highest hazard equipment – dust collectors – outside, away from important buildings and utilities.
- Construct equipment that processes or transfers combustible particles to contain or safely vent a potential explosion.

- Where explosion containment or venting of equipment is not possible, eliminate the oxygen in the system with inerting, or install an explosion-suppression system.

References

[1] American Petroleum Institute. *Sizing, Selection, and Installation of Pressure-Relieving Devices in Refineries, Part 1 – Sizing and Selection, API Recommended Practice 520.* 7th ed, January, 2000.
[2] American Petroleum Institute. *Venting Atmospheric and Low-Pressure Storage Tanks, API Standard 2000.* 6th ed. November, 2009.
[3] Arendt J, Lorenzo D, Lusby A. *Evaluating Process Safety in the Chemical Industry: A Manager's Guide to Quantitative Risk Assessment.* Washington, DC: Chemical Manufacturers Association; June, 1989.
[4] Bravo F, Beatty B. Decide Whether to Use Thermal Relief Valves. *Chemical Engineering Progress,* December 1993:35.
[5] Center for Chemical Process Safety. *Essential Practices for Managing Chemical Reactivity Hazards.* New York: American Institute of Chemical Engineers; 2003.
[6] Center for Chemical Process Safety. *Guidelines for Chemical Process Quantitative Risk Analysis.* New York: American Institute of Chemical Engineers; 1989.
[7] Englund S. Inherently Safer Plants: Practical Applications. Denver, Colorado: paper presented at the AIChE Summer National Meeting; August 16, 1994.
[8] Eskridge C. Assessing Pressure Relief Needs. *Chemical Processing,* August 15, 2001.
[9] FM Global Property Loss Prevention Data Sheets 7-76. Prevention and Mitigation of Combustible Dust Explosion and Fire. FM Global; March 2009.
[10] Hristova M, Damgaliev D, Popova D. Estimation of Water-Alcohol Mixture Flash Point. *Journal of the University of Chemical Technology and Metallurgy,* Sofia, Bulgaria 2010;45(1):19–24.
[11] NFPA 69. *Standard on Explosion Prevention Systems.* Quincy, Massachusetts: National Fire Protection Association (NFPA); 2011.
[12] NFPA 70. *National Electric Code.* Quincy, Massachusetts: National Fire Protection Association (NFPA); 2011.
[13] NFPA 654. *Standard for the Prevention of Fire and Dust Explosions from the Manufacturing, Processing, and Handling of Combustible Particulate Solids.* Quincy, Massachusetts: National Fire Protection Association (NFPA); 2012 (draft).
[14] Occupational Safety & Health Administration (OSHA). Flammable and Combustible Liquids. 29 CFR Part 1910.106. September 2005.
[15] Occupational Safety & Health Administration (OSHA). Process Safety Management of Highly Hazardous Chemicals. 29 CFR Part 1910.119. May 1992.
[16] Ulrich G, Vasudevan P. Predesign With Safety in Mind. *Chemical Engineering Progress,* July, 2006:27–37.

21

Controls

Introduction

This chapter provides a glimpse into the world of instrumentation and control. Concepts in this area that every process engineer needs to know include process control objectives, terminology, and basic control constructs. The field is changing rapidly with technological advances, but underlying principles persist unshaken.

Here are some of the important objectives for a process control system [2]:

1. **Safety.** Instruments and controls protect people, the environment, and equipment. Sometimes there are conflicting approaches, where equipment is sacrificed in favor of personnel safety for example. Conflicts are best resolved by using risk assessment tools such as FMEA or HazOp. Regulatory expectations often provide the minimum acceptable level of instrumentation and control. Safety Instrumented Systems (SIS) assure that potentially catastrophic failures are mitigated to an appropriate degree.

2. **Profit.** Process automation benefits the bottom line by maintaining consistent operation, leading to: meeting final product specifications, minimizing waste production, minimizing environmental impact, minimizing energy use, and maximizing overall production rate.

3. **Reliability.** Equipment and systems are more reliable today than in the past, in part due to advances in instrument and control technologies. Well controlled systems operate with low variability.

4. **Data collection.** Distributed control systems routinely gather, store, and analyze copious amounts of plant data. This leads to better process understanding and increasingly effective control strategies and algorithms.

Basic Nomenclature

Control loops strive to keep a measured process variable at a certain set point by manipulating another process variable using a final control element. The control loop is required because any number of unmeasured disturbances affect the process.

In Figure 21-1, the operator enters a set point for the process stream temperature into the Temperature Indicating Controller (TIC). The thermocouple (TE) is connected to the Temperature Indicating Transmitter (TIT) that sends the measured process stream temperature to the TIC. The difference between the measured value and the set point is called the controller error. The magnitude of the error is used by the controller to calculate an output value called the controller output. This is usually an analog value in the range of 4 to 20 mA. In the example, a pneumatic controller output is shown; in the US, pneumatic signals usually range from 3 to 15 psig. The Temperature Control Valve (TCV) opens in proportion to the controller output, thus manipulating the flow of steam into the exchanger. Unmeasured disturbances include change to the temperature, flow, or composition of the process stream where it enters the heat exchanger. If the heat exchanger is outdoors, an unmeasured disturbance might result from rainfall or wind.

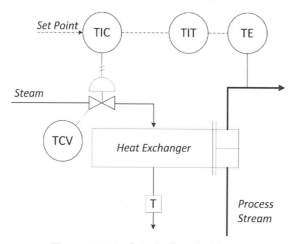

Figure 21-1. A typical control loop.

Definitions of Control Modes

In the most basic arrangement, controllers send an analog output signal that is calculated from a measured value and a set point. The difference between the measurement and the set point is called the error. There are three types of control mode, defined below [8]. Together, these are known as PID control.

Proportional Control. This is a mode of control that causes the output of a controller to change in a linear fashion to the error signal.

Integral (Reset) Control. This is a control algorithm that attempts to eliminate the offset (caused by proportional control) between the measurement and the set point of the controlled process variable. This control mode "remembers" how long the measurement has been off the set point. The controller output is proportional to the time integral of the error signal.

Derivative Control. This is a mode of control that anticipates when a process variable will reach its desired control point by sensing its rate of change. This allows a controlled change to take place before the process variable overshoots the desired control point. The controller output is proportional to the rate of change of the input.

Generally, process engineers specify the need for a control loop on the P&ID. They also indicate the presence of more advanced control loops such as cascade control (where the output from one controller is used as the set point for a second controller) and ratio control (where a second controller output is calculated as a ratio of the first controller's output). Instrumentation and control engineers determine the control mode and detailed settings for the controllers.

Cascade Control

Consider using cascade control instead of basic single-loop control when one of the following conditions exist [11]:

- The process reacts slowly in response to a change in the valve position (when a valve is the final control element).
- The process variable drifts around the set point.
- Process changes appear as disturbances on the measured value.
- One or more variables directly related to the set point can be disturbed.
- With a change in set point, a quick parallel tracking between the recorded and desired value is required with minimal overshoot.

Figure 21-2 is an example of cascade control. The tank level set point is entered into the Level Indicating Controller (LIC) which is the "master" controller. With a steady pressure in the pipe, the master controller could manipulate the flow control valve directly. However, if the header pressure fluctuates, due to other use points opening and closing for example, then cascade control is used as illustrated. By measuring and controlling the flow rate into the tank, changes in the header pressure are quickly

compensated. On the other hand, if LIC controlled the flow valve, the response to changes in the header pressure would be slower.

Consider these principles when designing a cascade control system [11]:

1. Be sure the secondary loop receives the maximum disturbances. In the example, if the header pressure is

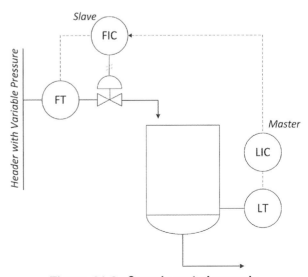

Figure 21-2. Cascade control example.

steady but the flow rate out of the bottom of the tank fluctuates, then the cascade loop shown in the example would be inappropriate, since the maximum disturbances would affect the primary loop.

2. The secondary loop must include the worst expected disturbances since those are the ones to be eliminated.

3. The secondary loop should have a quick reaction to each change, whether or not the change is desired. The slave must have a faster reaction time than the master.

4. The secondary loop must have a direct influence on the primary control set point. In the example, changing the flow rate into the tank directly influences the level in the tank.

5. There must be a direct, proportional relationship between the primary and secondary loops. This means that if the primary variable has a linear scale (level, in the example), then the secondary variable must also have a linear scale. In the example, if the flow element is an orifice, then it has a square root scale and its value would have to be linearized with a square root extractor.

Specifying Instruments to Measure Values

Measured values include: pressure, differential pressure, temperature, flow rate, level, weight, composition, and many others. Decisions include: what to measure, where to measure it, and critical characteristics of the measuring instrument.

Consider these characteristics when specifying a measuring instrument:

- **Functions.** Measuring instruments can indicate locally, transmit the value, indicate and transmit, or be combined with a control function. Multi-function instruments may also include calculation capability that condition the raw value; for example, on-line density measurement with integral temperature correction. Other functional characteristics that could narrow the selection of a particular instrument are: power source (e.g., 24 VDC), analog or digital output signal (4–20 mA, 3–15 psig, capillary, FOUNDATION™ fieldbus, Modbus, etc.), intelligence (e.g., HART protocol), wired or wireless signaling, electrical hazard classification, and compatibility with corrosive or hygienic process fluids.
- **Accuracy** is the difference between the sensor's reading and the actual process value. Accuracy is usually stated as the maximum error that is expected

under certain circumstances. For instance, the accuracy may be reported as +/− 2% of the full scale reading. In this case, a pressure gage with a range of 0 psig to 100 psig +/− 2% may be expected to read within 2 psig of the actual value; a reading of 20 psig would indicate that the actual process value is between 18 psig and 22 psig, a potential error of 10%.
- **Repeatability** is the sensor's ability to report the same value with multiple readings when the process value is constant. For example, imagine a pressure gage that can be isolated from the process with a valve. Record pressure, close valve, relieve gage pressure, re-open valve. Initial gage reading repeated?
- **Precision, or resolution,** is the number of significant digits available in the signal. For example, a load cell with 3,000 kg capacity and 0.02% accuracy (+/−0.6 kg) may offer a resolution of 1 kg. Therefore, due to randing an actual load of 100 kg may be reported as 99, 100, or 101 kg.
- **Range** refers to the limits of the measured value, such as 0 to 100 psig.
- **Response time.** The measured value should update quickly in comparison with the operation of the final control element.

Control Ranges

For many systems, it's important to distinguish between the design range and the process requirement range. A simple example is controlling the air temperature in a room. If the process requires that the room temperature

be within the range of 15 °C to 25 °C, the engineer may decide to specify a control system that is capable of maintaining a set point from 18 °C to 22 °C. This gives a design cushion so that slight excursions would not cause a process problem. Alarm set points would be established to provide a warning that conditions are approaching the process limits, and an action point should the temperature reach the limits. Figure 21-3 illustrates the concept.

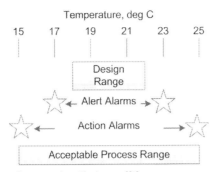

Figure 21-3. Use alarms for warning that conditions are approaching acceptable limits.

Drawing Conventions

Process engineers are generally responsible for creating Piping and Instrumentation Diagrams (P&IDs). Instrument engineers, electrical engineers, and CAD designers often assist.

P&IDs must be as complete and accurate as practicable. The project team relies on them when the detailed construction documents are prepared, the plant is built, and when it is commissioned. When in operation, plant engineers turn to the P&IDs to troubleshoot problems and to evaluate potential changes.

The example in Figure 21-1 is suitable for a very preliminary P&ID or a Process Flow Diagram (PFD). A finished P&ID contains much more information. Standard drawing symbols and nomenclature are usually defined on a Lead Sheet. It's Good Engineering Practice to establish the symbols for a plant or entire company, and use them consistently.

Figure 21-4 illustrates a range of symbols and nomenclature that are typically found on a finished P&ID. Notice how much more crowded this diagram is compared to the same basic control loop shown in Figure 21-1. It's important to anticipate this effect, and plan the P&IDs accordingly. There is often a tradeoff between including sufficient equipment on a drawing to tell a story or limiting the amount of equipment so that the instrumentation and other information can be clearly depicted.

The example includes features described below:

- Each instrument is called out with a "bubble". The diamond-in-square instrument symbol indicates that it is within a Programmable Logic Controller (PLC).

The instruments are numbered to make it easy to reference them on lists, specifications, datasheets, and other drawings. Instruments associated with each other are assigned the same number; if there are multiple instruments in a loop that have the same letter designation, then the numbers are appended with letters (e.g., FCV-211A and FCV-211B).

- The valve's non-energized, or failure, position is indicated. "FC" means "fail closed".

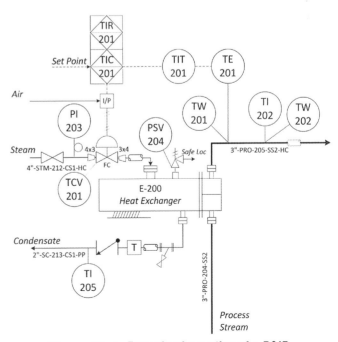

Figure 21-4. Example of a portion of a P&ID.

- The lines connecting instrument bubbles indicate the type of signal: solid for direct connection to the process, dotted for electrical, and hashed for pneumatic.
- Pipelines are numbered using a system that is specific to the company, plant, or project. In this example, the line numbering system is: size-fluid-number-pipe specification-insulation. The fluid field identifies what is inside the pipe and is chosen from a list on the Lead Sheet (STM = steam, SC = condensate, PRO = process). The pipe specification field references the detailed pipe specification that is used for the line. Insulation type indicates whether the insulation is intended for heat conservation (HC), cold conservation (CC), or personnel protection (PP). Insulation thickness might be included in the line number, or it can be found in the insulation specifications based on the insulation type and line size.
- Reducers shown at the control valve indicate that the valve body is 3 inches, installed in the 4-inch pipeline. Based on the symbol used, these are concentric reducers.
- Insulation is shown on the heat exchanger. An insulation symbol is also shown on the insulated pipes although this is redundant with the line number indication.
- The pressure gage is shown with a steam siphon, also called a pigtail, used to trap steam and isolate it from the gage. The siphon makes a liquid seal in the line to the gage and prevents the high temperature of the steam from reaching the gage internals.

Artistic Expression

P&IDs are an important communication tool that demand attention to detail. The minimum acceptable standard should be that they are 100% accurate to the extent that information is depicted. Every piece of process equipment, every pipeline above a threshold line size, and every tagged instrument should be shown. Components such as blind flanges, hoses, spray balls, lubricators, siphons, funnels, and sight glasses should be depicted. Notes and legends should explain any ambiguous or special situation. The goal should be that others can glean all needed data from the drawings without need for the process engineer to explain.

Beyond the basic expectations, there are several practices that enhance legibility, thus reducing misinterpretation and error:

- Arrange equipment on the drawing in a logical fashion. This often means locating it in a manner that is consistent with intended layout and elevation. Draw the equipment to imply relative size, but not to scale. Consider the main process pipelines when laying out the equipment to ensure there is sufficient space for valves and instruments, and to minimize line crosses.
- Use at least three line weights to draw the pipelines. Main process lines should be heaviest, minor process lines and major utilities medium weight, and minor utilities light weight. When drawing the lines, try to minimize those of the same weight crossing over each other. Where lines do cross, break the lighter weight line, or consistently break the line running horizontally (or vertically) when the two lines have the same weight.
- Ensure that instrument bubbles and lettering are sized appropriately. If the drawing is intended to be printed at large size (e.g., A1, or 32" × 42"), consider the result when it is reduced to A3 (11" × 17"). See Figure 21-5.

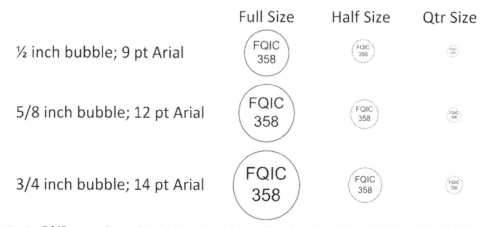

Figure 21-5. P&IDs are often printed at a reduced size. Consider how the reduction will affect readability.

Safety Instrumented Systems

An essential aspect of process control is to ensure that people, the environment, and equipment are adequately protected from harm. The control systems which are implemented to keep a process in optimal tune are not always adequate to prevent disastrous consequences when there are excursions beyond the ranges expected by the control system. Nor will the process control system function appropriately when failures occur such as loss of compressed air or power.

Undertake comprehensive qualitative and/or quantitative risk analyses to assess the potential hazards from a process, the pathways that could lead to the hazards, and the risk control design and control features built into the process. The risk assessments usually lead to improved understanding of the process, and often point to the need to add or modify instrumentation and control systems to further mitigate risks.

The protection levels identified in the risk assessment are defined in terms of specific safety control loops (with sensors, logic solvers, and control elements). These loops are called "Safety Instrumented Systems (SIS)". See Table 21-1.

International, US, and European Standards define appropriate risk mitigation behaviors, including the classification of safety control loops (Table 21-2). Devices (i.e., sensors, controllers, and final control elements) are certified against the classifications to ensure that their availability and reliability are consistent with the needs of the classification. However, the *combination* of the devices into a functional safety *system* must follow the requirements in the Standards in order to achieve the overall risk abatement that is desired.

IEC 61508 [6] gives a framework that leads to four Safety Integrity Levels (SIL). The target SIL for each SIS is determined from the risk assessment, and the specific components within the SIS are designed accordingly. The key concept for SIL is *availability*; these factors are used to design a functional safety system to achieve the desired SIL [3]:

- Device integrity
- Diagnostics

Table 21-1
Safety integrity levels are based on the availability of the safety function

SIL	Low Demand Mode Average Probability of Failure on Demand	High Demand Mode Probability of Dangerous Failure per Hour	Comment [1]
SIL1	0.01 to 0.1	0.000001 to 0.00001	95% of all Safety Instrumented Systems
SIL2	0.001 to 0.01	0.0000001 to 0.000001	5% of all Safety Instrumented Systems
SIL3	0.0001 to 0.001	0.00000001 to 0.0000001	Not likely in refineries, but possible in off-shore platforms and nuclear
SIL4	0.00001 to 0.0001	0.000000001 to 0.00000001	Highest risk, only seen in the nuclear industry

Table 21-2
Major standards pertaining to safety instrumented systems

Standard	Scope
IEC 61508 [6]	Defines appropriate means for achieving functional safety in systems it covers. Users determine the need for functional safety with a hazards analysis, which is not covered by the standard. The scope includes the entire safety life cycle.
ISO 13849 [7]	Provides safety requirements and guidance on the principles for the design and integration of safety-related parts of control systems, including design of software. It specifies characteristics that include the safety level required for carrying out safety functions.
ANSI/ISA 84.01 [4]	International standard addresses the application of safety instrumented systems for the process industries. Safety instrumented systems include sensors, logic solvers, and final elements. Developed as a process sector implementation of IEC 61508.
OSHA 1910.119 [10]	Applicable to processes containing threshold quantities of flammable, toxic, or reactive highly hazardous compounds, Process Safety Management is intended to prevent or minimize the consequences of a catastrophic release. It requires that process hazards analyses be conducted.

- Systemic and common cause failures
- Testing
- Operation
- Maintenance

The availability of the system depends on:

- Failure rates and failure modes of the components
- Redundancy
- Voting scheme(s) adopted
- Testing frequency

Enclosures: NEMA and IP

Instruments and electrical components are usually mounted inside an enclosure that is specified in accordance with its ability to protect the device from the environment. Two commonly used enclosure standards are IEC and NEMA. The following outlines the meaning of the standards' nomenclature.

IEC 60529, the "Ingress Protection" Code (also known as the "International Protection Rating"), uses a two-digit numerical system, followed by an optional letter, to describe an enclosure: solids, liquids, and impact [5]. The first two digits, if not used, are replaced with "X." See Tables 21-3, 21-4 and 21-5. For example:

Only the first digit is meaningful: IP 3X
Only the second digit is meaningful: IP X8
Both characters are meaningful: IP 65

The following tables define the IP digits

Table 21-3
First digit, protection from foreign bodies (solids)

Digit	Protection Against Human or Tools Contact	Protection Against Foreign Bodies
0	No special protection	No special protection
1	Back of hand; fist. No protection against deliberate contact with body part	Large bodies, >50 mm diameter (2 inches)
2	Finger	Medium-sized bodies, >12.5 mm (1/2 inch)
3	Tools and wires with a thickness greater than 2.5 mm (0.1 inch)	Small foreign bodies, >2.5 mm (0.1 inch)
4	Tools, screws, and wires with a thickness greater than 1.0 mm	Granular bodies, >1.0 mm
5	Complete protection against contact. Sufficient dust protection to ensure equipment functions satisfactorily	Dust protected; dust deposits permitted but their volume must not affect the function of the unit
6	Complete protection against contact and no ingress of dust	Dust tight

Table 21-4
Second digit, protection from water and liquids

Digit	Protection Against Water	Test Details
0	No special protection	NA
1	Water dripping or falling vertically	10-minute duration; water equivalent to 1 mm rainfall per minute
2	Water dripping vertically with enclosure tilted up to 15° from its normal position	10-minute duration; water equivalent to 3 mm rainfall per minute
3	Water spraying from any direction up to 60° from the vertical	5-minute duration; 0.7 l/m with pressure 80 to 100 kN/m^2
4	Water splashing from any direction	5-minute duration; 10 l/m with pressure 80 to 100 kN/m^2
5	Low pressure water jets (6.3-mm nozzle) from all directions	3-minute duration; 12.5 l/m with pressure 30 kN/m^2 from a distance of 3 m
6	Powerful water jets (12.5-mm nozzle) from all directions; limited ingress permitted	3-minute duration; 100 l/m with pressure 100 kN/m^2 from a distance of 3 m
7	Temporary immersion, 15 cm to 1 m (6 inches to 3 feet)	30-minute duration; immersion in tank at a depth of 1 m
8	Permanent immersion, under conditions specified by the manufacturer	Continuous immersion in water at a depth specified by the manufacturer

Table 21-5
Letters optionally appended to classify only the level of protection against access by persons, or to provide additional information about the protection

Letter	Access by Persons	Additional Information
A	Back of hand	NA
B	Finger	NA
C	Tool	NA
D	Wire	NA
H	NA	High voltage device
M	NA	Device moving during water test
S	NA	Device standing still during water test
W	NA	Weather conditions

Table 21-6
NEMA enclosure classifications

NEMA Designation	Intended Use and Description
1	**General purpose indoor** use to provide a degree of protection to personnel against incidental contact with the enclosed equipment and to provide a degree of protection against falling dirt.
2	**Drip-proof indoor** use to provide a degree of protection to personnel against incidental contact with the enclosed equipment, to provide a degree of protection against falling dirt, and to provide a degree of protection against dripping and light splashing of non-corrosive liquids.
3	**Dust tight, rain tight, and ice/sleet resistant for either indoor or outdoor** use to provide a degree of protection to personnel against incidental contact with the enclosed equipment; to provide a degree of protection against falling dirt, rain, sleet, snow, and windblown dust; and that will be undamaged by the external formation of ice on the enclosure.
3R	**Rain proof and ice/sleet proof for either indoor or outdoor** use to provide a degree of protection to personnel against incidental contact with the enclosed equipment; to provide a degree of protection against falling dirt, rain, sleet, and snow; and that will be undamaged by the external formation of ice on the enclosure.
3S	**Dust tight, rain tight, and ice/sleet proof for either indoor or outdoor** use to provide a degree of protection to personnel against incidental contact with the enclosed equipment; to provide a degree of protection against falling dirt, rain, sleet, snow, and windblown dust; and in which the external mechanisms remain operable when ice laden.
4	**Water and dust tight for either indoor or outdoor** use to provide a degree of protection to personnel against incidental contact with the enclosed equipment; to provide a degree of protection against falling dirt, rain, sleet, snow, windblown dust, splashing water, and hose-directed water; and that will be undamaged by the external formation of ice on the enclosure.
4X	**Water and dust tight, and corrosion resistant, for either indoor or outdoor** use to provide a degree of protection to personnel against incidental contact with the enclosed equipment; to provide a degree of protection against falling dirt, rain, sleet, snow, windblown dust, splashing water, hose-directed water, and corrosion; and that will be undamaged by the external formation of ice on the enclosure.
5	Superseded by NEMA 12.
6	**Submersible, water tight, dust tight, and ice/sleet resistant for either indoor or outdoor** use to provide a degree of protection to personnel against incidental contact with the enclosed equipment; to provide a degree of protection against falling dirt, rain, sleet, snow, hose-directed water and the entry of water during occasional temporary submersion at a limited depth; and that will be undamaged by the external formation of ice on the enclosure.
7	**Underwriters Lab (UL) Class 1, Groups C and D (Explosion Proof) for indoor** use in hazardous (Classified) locations classified as Class I, Division 1, Groups A, B, C, or D as defined in NFPA 70. Capable of withstanding the pressures generated by an internal explosion of specified gases, and contain such an explosion sufficiently that an explosive air-gas mixture surrounding the enclosure will not be ignited.
8	Same as NEMA 7, but for indoor or outdoor use
9	**Underwriters Lab (UL) Class II, Groups E, F, and G (dusts) for indoor** use in hazardous (Classified) locations classified as Class II, Division 1, Groups E, F, or G as defined in NFPA 70.
10	Enclosures constructed to meet the requirements of the US Mine Safety and Health Administration, 30 CFR, Part 18.
12	**Industrial use, dust and drip tight, indoor** use to provide a degree of protection to personnel against incidental contact with the enclosed equipment; to provide a degree of protection against falling dirt; against circulating dust, lint, fibers, and flyings; against dripping and light splashing of non-corrosive liquids; and against light splashing and consequent seepage of oil and non-corrosive coolants.
13	**Oil and dust tight indoor** use to provide a degree of protection to personnel against incidental contact with the enclosed equipment; to provide a degree of protection against falling dirt; against circulating dust, lint, fibers, and flyings; and against the spraying, splashing, and seepage of water, oil, and non-corrosive coolants.

Table 21-7
Approximate cross reference comparison of IP and NEMA rating

NEMA	IP23	IP30	IP32	IP55	IP64	IP65	IP66	IP67
1	X							
2		X						
3					X			
4							X	
4X							X	
6								X
12				X		X		
13						X		

A special case is designated IP69K by German standard DIN 40050-9. IP69K is rated for high pressure, high temperature wash down conditions. Enclosures must be dust tight (IP6X) and able to withstand high-pressure and steam cleaning.

In the US, the National Electrical Manufacturers Association (NEMA) defines enclosures with numerical terms as generally listed in Table 21-6.

NEMA published an excellent overview of the various enclosure types that includes comparison with the International Protection Ratings [9]. An approximate cross-reference is given in Table 21-7.

References

[1] Bergstrom J. Safety Instrumented Systems (SIS) and Safety Life Cycle. Presentation retrieved from www.processengr.com; September, 2009.

[2] Cooper D. Motivation and Terminology of Automatic Process Control. Downloaded from, www.controlguru.com; July, 2011.

[3] Gillespie S. Safety Instrumented Systems. widely published on the Internet, undated.

[4] Instrument Society of America (ISA). Functional Safety: Safety Instrumented Systems for the Process Industry Sector. ANSI/ISA 84.01, Parts 1, 2, and 3; 2004.

[5] International Electrotechnical Commission. IEC 60529: Degrees of protection provided by enclosures (IP Code). IEC 60529 ed2.1 Consol. With am1, Geneva. Feb 27, 2001.

[6] International Electrotechnical Commission, IEC 61508: Functional safety of electrical/electronic/ programmable electronic safety-related systems. IEC 61508 ed 2.0 (Parts 1 to 7), Geneva. April 30, 2010.

[7] International Organization for Standardization (ISO). Safety of Machinery – Safety-Related Parts of Control Systems. ISO-13849-1:2006 Part 1: General principles for design and ISO-13849-2:2003 Part 2: Validation.

[8] Gas Processors Suppliers Association (GPSA). *Engineering Data Book. SI Version*, vol. 1. 12th ed. 2004.

[9] National Electrical Manufacturers Association (NEMA). NEMA Enclosure Types, http://www.nema.org/prod/be/enclosures/upload/NEMA_Enclosure_Types.pdf; November, 2005.

[10] Occupational Safety & Health Administration (OSHA). Process Safety Management of Highly Hazardous Chemicals. 29 CFR Part 1910.119, May, 1992.

[11] Verhaegen S. When to Use Cascade Control. *InTech*, October, 1991:38–40.

22

Troubleshooting

Introduction

Troubleshooting is a form of problem solving where the objective is to return a system (or deliver it) to its normal operating state. The "problem" is often expressed in terms of symptoms such as deviations in instrument readings (e.g., low flow) or chronic equipment failures (e.g., seal leakage). Effective troubleshooting resolves the symptoms by determining and correcting the *root cause*, which often is disconnected from the equipment or devices where the symptoms are observed. The troubleshooter is a detective, who gathers facts, analyzes data, forms hypotheses, and systematically eliminates suspects until the underlying issue is uncovered.

Trouble comes in many forms. It ranges from sudden failure to chronic decay. It may be localized to a specific device or spread across an entire unit. It can be life threatening or be a minor irritation. The financial implications can be huge (plant shuts down) or inconsequential. The approach taken, and resources applied, to troubleshooting should account for these factors.

This chapter discusses general approaches to troubleshooting, and provides tips and checklists for troubleshooting a variety of equipment types. Since no single approach or checklist can solve every problem, the information presented here is intended to help engineers reach reasonable conclusions without purporting to be complete or applicable for all situations.

General Troubleshooting Approach

Troubleshooting is a form of problem solving, and the techniques for identifying and solving problems apply. Two critical aspects of troubleshooting can interfere with a textbook solution:

1. Safety. Chemical plants and equipment are inherently dangerous, and the fact that troubleshooting is required increases the likelihood of an environmental, health, or equipment-damaging occurrence. Above all, troubleshooters must ensure that safe conditions are maintained.
2. Time. Intense pressure may be exerted to solve the problem quickly, particularly if the problem has shut down or slowed production. This can result in jumping to conclusions and incorrect resolution of the situation. While it may not be possible to deny the extreme urgency of the breakdown, advance preparation and careful methodology can lead to a successful outcome more quickly than a purely reactive approach.

Troubleshooting is a team effort. The team must possess sufficient technical knowledge to identify and solve the problems, and access to relevant information and data are required. Often, an engineer is given accountability for solving the problem. His or her leadership behaviors will determine how the team functions, and this may be instrumental in its ultimate success. Interpersonal skills and personal styles play vital roles. Successful troubleshooting depends on much more than technical savvy.

Professor Donald Woods is a frequent lecturer on the topic of process troubleshooting, at AIChE conferences and at his university. The tips that follow in this section are reported from his webinar [18].

Woods teaches a six-stage method for process troubleshooting. The six stages are not linear; there is considerable bouncing back-and-forth between the stages, and the same overall strategy is typically used multiple times during a troubleshooting exercise. Essentially, the troubleshooting team defines the problem, plans a solution, carries out the plan, then evaluates the results. The steps are repeated because the approach begins with simple and obvious targets, then uses the findings to dig deeper on subsequent passes.

The six stages are depicted in Figure 22-1. Define the task by adopting a positive, open-minded attitude. Use all available information to identify symptoms, deviations from normal conditions, and desired outcomes. Discuss to confirm that everyone has a shared understanding of the problem, and ensure there are no competing viewpoints of the problem definition.

Plan a solution by first brainstorming a list of hypotheses and checking them against the observed facts. On the first pass through the stages, simple solutions should be identified and tested. This gives the team an opportunity to;

a) possibly solve the problem immediately,
b) learn new facts from the simple exercise that can be applied in the next pass, and
c) determine the *root cause* of the trouble.

Woods showed that teams which jump to a single solution and spend most of their time executing that solution tend to be unsuccessful. On the other hand, teams that iterate through the six stages and end up spending proportionately more time in the definition phase are more successful.

Solve the problem on the next pass through the procedure. Using all of the available information, the team should develop a comprehensive set of working hypotheses and systematically test them. Woods suggests keeping three to six hypotheses in play until a solution is reached.

After the problem is resolved, the team should use the strategy one last time to ensure that the issue will not recur. This includes another look at the instrument readings and follow-up interviews with operators to confirm that the symptoms are gone. Plan and execute any corrective action that has been identified, such as adding monitoring devices, writing new procedures, or updating spare parts and maintenance databases.

The troubleshooters should aim for these targets:

- Use the six-stage approach outlined above
- Ensure that safety is the number one concern
- Keep the big picture and overall system in perspective; don't get blinded by symptoms manifested at a specific piece of equipment
- Bring technical know-how to the team with knowledge of chemical reactions, stream properties, and causes that could result in the observed symptoms
- Manage bias, such as coloring facts with opinion and jumping to conclusions with insufficient data
- If the lead hypothesis indicates a single, but rarely occurring, cause, then consider that multiple more plausible causes may be combining to result in the problem
- Collect data, test hypotheses, and explore alternatives
- Manage personal style. If possible, combine people that have different styles to keep the team centered
- Use probability analysis
- Interact with team members and stakeholders in a professional and respectful manner

A very useful tool for evaluating brainstormed hypotheses is shown in Figure 22-2. This table is partially filled out with example data from the webinar. Each hypothesis is listed in the first column; there should be many identified. Columns are provided for all of the symptoms collected from the plant, such as deviations from normal instrument readings or mechanical failures. The symptoms are put into a lettered list (a, b, c, etc.). For each hypothesis, the team decides whether each of the observed symptoms supports, disproves, or says nothing about the possible cause. Then, the list is sorted, or prioritized, to bring the hypotheses with the preponderance of supporting evidence to the top. Diagnostic tests are identified and listed (A, B, C, etc.) and the tests that pertain to each hypothesis checked off. A roadmap is now in hand for performing tests that will, hopefully, narrow the list of working hypotheses to a small number.

Problems can usually be classified as process issues (e.g., fluid properties changed, fouling occurred) or equipment failures. It may be obvious from the observed facts which category fits, but equipment failures could be the result of an underlying process issue, in which case a simple repair will not solve the problem.

Woods reported that equipment failures are nearly equally divided between heat exchangers, rotating

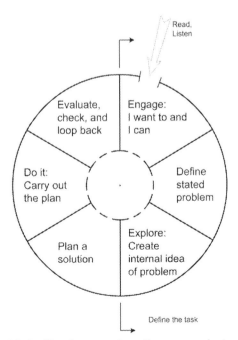

Figure 22-1. Six-stage systematic approach to troubleshooting. The first three activities comprise defining the task at hand [18].

Working Hypotheses	Initial Evidence					Diagnostic Actions			
	a	b	c	d	e	A	B	C	D
1. Entrainment from the vaporizer	S	S	S	S		X	X		
2. Liquid drain line from drum too small or plugged	S	S	S	S			X	X	X
3. Liquid reverse flows from vaporizer	S	S	S	S			X	X	X
4. Manual glove valve is stuck part way open	D	N	N	S					
S = Supports; D = Disproves; N = Neutral									

Figure 22-2. Worksheet for evaluating hypotheses and planning diagnostic actions [18].

equipment, and vessels, followed closely by towers and piping. For plants that experience trouble after they have already been operating normally, process issues are usually associated with plant fluids (e.g., properties such as viscosity or composition) when the fluids are processed at ambient temperatures. But for high temperature operations, the issues are more likely to be caused by materials failures, such as corrosion or seal leaks.

Before any field work is done, be sure to engage the co-operation of the plant operators and supervisors. The plant is their turf, and they must be informed of any activities planned, and consulted for their own observations and opinions.

Ensure that safety precautions are taken, including communicating intentions to everyone working in the area, removing process fluids from systems, following lock-out/tag-out procedures, and wearing appropriate protective equipment.

Checklists are useful for evaluating the performance of specific types of equipment. Examples are given in the remaining sections of this chapter.

Diagnostic Tools

Tools used to troubleshoot problems may be built into the plant, added to it after the problem occurs, or applied on a transient basis as part of the engineer's arsenal. In an ideal world, permanent instrumentation is installed around every unit operation, to measure and record all of the process parameters used to design the equipment. Typically, flow rates, temperatures, and pressures would be monitored.

Here are some practical suggestions:

- Provide as much on-line diagnostic instrumentation as practicable when a plant is originally designed and constructed. Use good engineering judgment to identify the location and type of instrumentation. Consider the project budget and criticality of each piece of equipment or unit operation to prioritize the type and level of instrumentation. Justify the much higher cost for capturing a variable in the data historian compared with having only local indication. Determine which process variables will be needed to troubleshoot a unit (e.g., pressure, flow, and temperature into and out of a heat exchanger for both streams) and document where those values could be obtained in the plant.
- Label equipment, instruments, components, pipes, and wires for easy identification in the field.
- Utilize self-diagnostic options for instruments (e.g., HART protocol).
- If permanent instrumentation cannot be justified, consider installing locations where gages could be installed at a future date, even on a temporary basis.
- Use non-invasive instruments if necessary to collect plant data. These include infrared temperature sensors and clamp-on ultrasonic flow monitors.
- Consider whether process variables for points that are not captured in a data historian should instead be manually recorded on operator log sheets or in batch records.
- Compile equipment history files, which should be part of every facility anyway, that include the original design criteria, specifications, certified "as built"

vendor drawings, and maintenance history. Data collected during commissioning tests can be stored here. Files should be controlled to ensure their accuracy, completeness, and integrity.

- Use process simulation programs to predict process conditions based on particular setpoints or assumptions. Compare actual plant data with simulated results to help identify the source of observed problems.

Instruments

Process measurements are key to troubleshooting. Generally, instrument readings should be trusted once some corroborating evidence has been found. Table 22-1 provides tips for troubleshooting symptoms observed with instrument readings.

Table 22-1
Common measurement problems [5]

Variable	Symptom	Problem Source	Solution
Pressure	Zero shift, air leaks in signal lines	Excessive vibration from positive displacement equipment	Use independent transmitter mounting with flexible process connection lines. Use liquid filled gage
	Variable energy consumption under temperature control	Change in atmospheric pressure	Use absolute pressure transmitter
	Unpredictable transmitter output	Wet instrument air	Mount local dryer. Use regulator with sump, slope air line away from transmitter
	Permanent zero shift	Overpressure	Install pressure snubber for spikes
Level (differential pressure)	Transmitter does not agree with level	Liquid gravity change	Gravity compensate measurement, or recalibrate
	Zero shifts, high level indicated	Water in process absorbed by glycol seal liquid	Use transmitter with integral remote seals
	Zero shift, low level indicated	Condensable gas above liquid	Heat trace vapor leg. Mount transmitter above connections and slope vapor line away from transmitter
	Noisy measurements, high level indicated	Liquid boils at ambient temperature	Insulate liquid leg
Flow	Low mass flow indicated	Liquid droplets in gas	Install demister upstream; heat gas upstream of sensor
	Mass flow error	Static pressure change in gas	Add pressure recording pen
	Transmitter zero shift	Free water in fluid	Mount transmitter above taps
	Measurement is high	Pulsation in flow	Add process pulsation damper
	Measurement error	Non-standard pipe runs	Estimate limits of error
Temperature	Measurement shift	Ambient temperature change	Increase immersion length; insulate surface
	Measurement not representative of process	Fast changing process temperature	Use quick response or low thermal time constant device
	Indicator reading varies second to second	Electrical power wires near thermocouple extension wires	Use shielded, twisted pair thermocouple extension wire and/or install in conduit

Control Valves (Process Control)

When process control is unstable, or desired flow rates are unachievable, control valve malfunction may be identified as the possible cause. Gather and assess operational data before concluding that a valve is the culprit (Table 22-2). Before disassembling a suspected control valve, recalculate the required valve and actuator sizes using the actual plant data to confirm that the valve is appropriately sized for the service.

If a control valve is suspected of malfunctioning, also consider alternatives including: the actuator, control transmitter, and process controller software or hardware (and tuning of the control loop).

Table 22-2
Operational data used to troubleshoot process control problems [1]

Data Type	How to Use to Troubleshoot Process Control Problem
Pressure data	Compare the specified pressures upstream and downstream of the suspect valve with plant readings. Data should encompass the range of conditions that the valve experiences, not just steady-state. If pressure sensors are not located near the valve, remote sensors may be used (at vessels upstream and downstream of the valve, for example), and the pressure drop across the valve estimated by calculating the pressure loss through connected piping.
Fluid data	Determine the actual temperature, composition, vapor pressure, viscosity, and other critical constants of the fluid being handled by the suspect valve and compare with the design assumptions and/or historical data. Problems can occur when feedstocks change.
	Changes in composition can lower the fluid vapor pressure, leading to cavitation.
	Information about fluid characteristics will help diagnose cavitation, sticking, galling, leaky gaskets, and packing problems. Impurities in the fluid can cause sticking in cage-guided valves. Corrosives can cause failure of the gasket, packing, or metal parts.
Flow rates and corresponding pressure drops	Calculate the valve flow coefficient (C_v) from the fluid data, and measured flow rates and pressure drop. An undersized valve will experience either excessive pressure drop or will fail to deliver the expected flow. An oversized valve (more often the case) will fail to deliver stable control.
Valve service history	Review maintenance records. Interview operating and maintenance personnel from all shifts. Try to determine when the problem was first observed, how the malfunction changes with fluctuating operating conditions, and whether the problem occurs during startup or at maximum capacity operation.
Actuator information	Investigate the signal to the actuator from the control system, pressure of the air supply to the actuator, and size of the air supply line. Inspect potential obstructions to the air supply such as regulators, filters, volume boosters, lock-up systems, and solenoid valves.
Valve noise	Loud "white noise" indicates excessive pipe velocity or excessive pressure drop across the valve.
	Discrete frequency noise, such as whistling or screeching, indicates aerodynamic resonances. Spectrum analysis can be useful for diagnosing extremely troublesome problems, especially where mechanical vibration is also encountered.
	"Snap, crackle, and pop" sound, like bacon frying in a pan, indicates cavitation in a small control valve.
	Cavitation in larger valves sounds like gravel flowing through the valve.

Shell-and-Tube Heat Exchangers

Gulley discusses thermal problems with heat exchangers [6]. Some of his recommendations are summarized in this section. Collect plant information for the problem exchanger, including temperatures, flow rates, and pressure drops. Obtain the original specification data sheets, if possible, and fabrication drawings for the heat exchanger. Problems and possible causes are listed in Table 22-3.

Table 22-3
Shell-and-tube heat exchanger troubleshooting

Process Issue	Troubleshooting Tips
Pressure drop lower than expected	Indicates fluid bypassing. If on the tube side, examine the pass plates and tubesheet gasket. Look for corrosion, gasket problems, and manufacturing defects.
	If on the shell side, improper bundle sealing is indicated. Inspect the bundle seal strips. For two-pass shells, fluid may be bypassing the long baffle if it isn't welded in. Long baffles with leaf seals do not give a perfect seal, and are subject to damage.

Table 22-3
Shell-and-tube heat exchanger troubleshooting—cont'd

Process Issue	Troubleshooting Tips
Pressure drop higher than expected	Causes include: • Improper venting — check this first, especially for condensers • High fouling • Debris from start up • Freezing of the process stream • Slug flow for two-phase systems • Fabrication problems
Fouling	Indicated if there is a gradual decline in thermal performance. Check the exchanger's operating history to determine if there are deviations from design conditions or periods of operation when the flow is lower than design. If fouling is expected on the cooling water side, look for evidence that water flows were reduced during cold weather conditions.
Debris	Especially for new exchangers, check for a strainer at the inlet to the exchanger. If there is no strainer, debris such as rocks, trash, tools, gloves, pencils, etc. may be lodged in the exchanger.
Excess surface area	Most exchangers are designed for the fouled condition. When operating clean they may transfer too much heat, resulting in problems due to high temperature or freezing. Cure excess surface problems by plugging tubes; tapered metal plugs are most common.

Plate-and-Frame Heat Exchangers

Sloan recommends the tips and procedures given here for troubleshooting plate-and-frame heat exchangers (PHEs) [16]. Some of these also pertain to shell-and-tube exchangers.

Use information from temperature and pressure sensors to evaluate the thermal and hydraulic performance of the exchanger. A gradual decline in thermal performance (often with a gradually increasing pressure drop) indicates a build-up of fouling deposits. Consider using chemicals to clean a partially fouled exchanger where the increase in pressure drop is no more than 40% to 50% of the original design pressure drop. Alkaline cleaners are used for organic materials such as biological or hydrocarbon-based deposits. Acids can be used to remove calcium carbonate scale that is typical in cooling tower applications.

A significant, or sudden, increase in pressure drop may indicate a particulate obstruction. Particulates might be removed by backflushing the exchanger. If this problem occurs, consider installing a filter or strainer upstream of the exchanger.

If cross-contamination between fluids, or leaks to the environment, is suspected, simple tests can be conducted. Before conducting tests that require opening the exchanger, be sure that personnel are qualified and trained.

Test for a cross-contaminating leak:

1. Remove the piping from the bottom port connection to enable a visual inspection inside the connection.

2. If there is residual liquid in the bottom port connection, vacuum it out.

3. Using water, pressurize the other side of the exchanger to about 100 psig and hold it for a hydrostatic test. If the pressure gage shows any decrease in pressure, use a flashlight to visually inspect the open port, looking for a rising level of liquid between the channel plates and possibly lapping over from one channel to another. If the plates have experienced a chloride stress crack it might take as long as 30 minutes for leakage to gather in the lower port area.

4. If an internal leak is verified, count the number of plates from the inside of the stationary cover to identify the defective plate. If the leak is farther back (making it too difficult to count plates) in the plate pack, measure to the leak with a tape measure then transfer the measurement to the outside of the unit to approximately identify the leaking plate(s).

Test for an external leak by pressurizing the entire exchanger (both sides) with water, preferably to the pressure listed on the nameplate, for 15 to 20 minutes. If the unit shows some signs of slow external leaks, placing cardboard underneath the unit during the test can make it easier to approximate the location of the leak.

These procedures allow maintenance technicians to focus their inspection time on repairing or replacing defective plates instead of visually inspecting many plates that show no signs of leakage.

Fractionation Towers

Due to the extremely high value of fractionation problems and the complexity of the systems, a huge body of literature exists to aid in troubleshooting them. Kister's tome covers over 1,200 case histories of problems, diagnoses, solutions, and key issues for distillation systems [12]; engineers responsible for troubleshooting towers should have his book.

Simple checks should be done first. These include column liquid level, temperature profile, pressure profile, and stream flow rates. In addition, a pressure drop survey using test gages should be made in the field and if not by the troubleshooter then under his or her direct supervision.

Table 22-4 presents a checklist for troubleshooting distillation columns. It can be used as a starting point for a specific unit. The order will depend on which process variables are most important. Again, the simple checks, which can be made quickly, should be done first, even if they do not appear at the time to be that important.

Table 22-4
Fractionation column troubleshooting checklist [4]

1. LEVELS
   a. Bottom
   b. OVH Accumulator
   c. Other

2. PRESSURE
   a. Top
   b. Bottom
   c. Delta-P Rectifier
   d. Delta-P Stripper
   e. Other

3. TEMPERATURE
   a. Top
   b. Bottom
   c. Feed
   d. Trays
   e. Pumparounds

4. COMPOSITION
   a. Feed
   b. OVH
   c. Bottoms
   d. Side Stream

5. FLOWS
   a. Feed
   b. OVH
   c. Bottoms
   d. Side Stream Draws

6. INSTRUMENT STABILITY
   a. Last Calibration
   b. Instrument Settings
   c. Process Stability

7. FEED LOCATION(S) – VALVE POSITIONS

8. PURGES
   a. Inerts
   b. Contaminants

9. GAMMA SCANS
   a. Flooding
   b. Weeping or entraining
   c. Plugging
   d. Mechanical Damage

10. HISTORICAL DATA
    a. Previous Worked at Rate
    b. Possible Upsets
    c. Modifications / Turnarounds

11. SAMPLING / ANALYTICAL
    a. Sample Location
    b. Sample Technique
    c. Analytical Technique (method, calibration)

12. MATERIAL BALANCE
    a. Overall Mass
    b. Component

13. ENERGY BALANCE

14. REFLUX RATIO
    a. Design
    b. Operating

15. ACTUAL / THEORETICAL STAGES

16. INTERNAL DESIGN LIMITATIONS
    a. Percent Flood
    b. Internal Feeds
    c. Internal Bottlenecks

17. OTHER CHECKS
    a. Cooling Water Temp / Air Temp
    b. Leaking Bypass on Condenser / Reboiler
    c. Steam Quality
    d. Leaking Exchanger(s)
    e. Bypass Line Open / Cracked
    f. Control Valve Positions
    g. Pump Purges / Seals
    h. Internal Chemical Reaction
    i. System Leaks

Fractionation Case Histories

This section compiles some case histories from the literature. These troubleshooting examples provide several useful general principles.

Pinch

Problem [13]: Fractionator with 140 trays was a bottleneck. A revamp was proposed that would not have worked.

Cause: Computer simulation indicated design was okay, but didn't consider pinch point.

Solution: Constructed McCabe-Thiele diagram. Pinch found.

Moral: Check computer designs with a graphical method. McCabe-Thiele recommended. Can use computer simulation to help construct McCabe-Thiele diagram.

Key Components

Problem [13]: Stripper computer simulation for a base case did not match test data. Proper simulation was needed for revamp design checking.

Cause: VLE data inaccuracies explained part of the problem. The computer simulation didn't allow understanding of the column operation. Different components were the effective key components in different parts of the column.

Solution: Hengstebeck [9] diagrams were produced based upon using different key component pairs. These showed where pinch points were occurring and the resulting understanding allowed a proper computer simulation to be selected.

Moral: Check computer designs with a graphical method. In this case the Hengstebeck method was good for the breaking out of key components from the multi-component mix. The Hengstebeck method uses a McCabe-Thiele type diagram.

Profiles

Problem [13]: A structured packing vacuum tower had too much heavy key in a vapor side-draw between the feed and the column bottom. The side-draw heavy key concentration was several times the design value.

Cause: The packing height was set by an optimistic HETP. It was delivering 6–7 stages versus 8 design.

Solution: Even though packing was only short by 1–2 stages, this was enough to cause the problem during off-design conditions. Plotting the column heavy key profiles showed a very steep and unforgiving slope. The heavy key was designed to drop from 55% to 1% in only 5 stages. Loss of 1–2 stages, out of 8 total required, was devastating.

Moral: Examine column profiles before investing money.

Troubleshooting Technique

Problem [8]: A 100-tray vacuum distillation column was run in blocked operation mode. After a run on a previous product the column would not run properly for a new product.

Cause: The operators thought the cause was plugged trays, but a careful engineer looked deeper and found the water supply valve to the condenser was only one-quarter open. The previous product run didn't require any more condenser cooling than this.

Solution: Opened the valve.

Moral: Conduct a proper troubleshooting technique as described in the article:

1. The current and past operating data were compared and the timing of the operating problems was defined.

2. The probable causes were compared to the data available and the physical conditions of equipment were checked. Probable causes were identified.

3. Insufficient data were available to confirm the exact source of the problem. So, a program to collect the additional data was conducted.

4. The new data directed the investigation to the correct part of the system, and the trouble was quickly identified.

Cutting Corners

Problem [17]: A deisobutanizer would not produce the required isobutane removal from the bottoms. The bottoms product ran 17% instead of 5%.

Cause: Several errors were made in the original design which were described as "cutting-corners."

Solution: After thorough analysis, including heat and material balances and hydraulic calculations, the initial design flaws were corrected including:

1. Installing the standard antijump baffles for certain inboard downcomer trays to keep droplets (produced by tray "blowing") from entering the downcomer.
2. "Picket fence" weirs were installed on other inboard downcomers also for shielding any blowing.
3. Added a feed distributor (omitted originally) to allow the proper lower feed point to be used for one feed stream.
4. Added a feed preheat exchanger so that feed was not subcooled. This unloaded the critical bottom tower section.

Moral: Cutting corners for a design saves pennies and costs big dollars over the years.

Trapped Water

Problem [10]: A de-ethanizer overhead chiller experienced tubeside plugging due to freezing.

Cause: A recent process change had inadvertently trapped water in an endless loop. This allowed the water to build up, whereas before the change the water had a way out.

Solution: Added a small package TEG dehydrator to the stream ahead of the chiller.

Moral: The designer must look beyond any modification itself to see how it interacts with the existing system.

Cocurrent

Problem [10]: A gas plant absorber/de-ethanizer train was not achieving design separation. The absorber and the de-ethanizer seemed to be operating at poor efficiencies.

Cause: A cross exchanger designed to cool absorber lean oil while heating de-ethanizer rich oil feed was not doing its job. It was found to be fabricated for cocurrent flow instead of countercurrent.

Solution: Repiped one side of the exchanger to convert it to countercurrent.

Moral: Check design details carefully.

Foaming

Problem [10]: An amine absorber was carrying over due to foaming.

Cause: The amine had strong foaming tendencies and antifoam had not been added. When antifoam was added batchwise, the foaming became worse because too high an antifoam concentration actually causes foaming.

Solution: A dilution method was employed to achieve the correct antifoam concentration. An injection pump was installed for slow, rather than batch, antifoam addition.

Moral: Add antifoam slowly with an injection pump.

Stacking Packing

Problem [10]: A packed, direct contact, water spray tower cooled acetylene furnace effluent. The bottom one foot or so of the bed would plug with polymer material. This is the hottest part of the bed.

Cause: Polymer deposits became stuck in the random packing interstices. When the bed was hand stacked in a staggered arrangement, rather than random packed, the vapor and liquid channeled and the gas was not cooled.

Solution: Only the bottom one-foot was hand stacked and the rest of the bed was randomly packed using a wet-packing technique.

Moral: Sometimes conventional methods have to be modified to fit the conditions.

Tray Supports

Problem [10]: Absorber bubble caps were replaced with sieve trays. A severe flow upset dislodged the new trays.

Cause: The sieve trays had weaker connections at the support ring than the old bubble cap trays. No welding was allowed (heavy-metal acetylides present).

Solution: A unique support connection was designed and installed as shown in the article.

Moral: Check modifications for abnormal conditions to the extent possible.

Boilup Control

Problem [10]: An ethane/ethylene splitter exhibited poor control of overhead and bottoms concentrations.

Cause: Because the top product is very pure, the top tray temperature is insensitive to concentration. Therefore, the temperature difference between the top tray and tray 50 was the design control signal to adjust reflux. This was very sluggish, because a change in reflux is slowly reflected down the column in changed liquid overflows from tray to tray.

Solution: Because vapor rate changes are reflected up and down the column much faster than liquid rate changes, the temperature difference controller was disconnected and the tower was controlled instead by boilup. A temperature 10 trays from the bottom set reboiler heating medium and the reflux was put on flow control.

Moral: In large superfractionators, the fast response of boilup manipulation is advantageous for composition control.

Condenser Velocity

Problem [10]: A knockback condenser mounted on a C_3 splitter reflux drum exhibited liquid carryover (as evidenced by the vent line icing up). This indicated product loss from liquid carrying over rather than dripping back into the reflux drum. Also, the vent line metallurgy would not withstand the cold temperatures produced.

Cause: The velocity was too high in the vent condenser, thus causing the vapor to entrain liquid.

Solution: A valve limiter was installed on the vent control valve to limit opening.

Moral: Excessive vapor flows in knockback condensers lead to entrainment. (Reference 3 shows how to predict the maximum allowable velocity.)

Blind Blind

Problem [10]: A fired reboiler outlet temperature would not rise above 350°F no matter how hard the heater was fired.

Cause: A blind had been left in the heater outlet circuit. The blind had no handle, so it was difficult to find this error. Troubleshooting was difficult because certain instruments had not been commissioned and the available ones indicated everything was okay except for the 350°F maximum.

Solution: Removed the blind.

Morals: Blinds should have long handles and tags. Instrumentation must be operational before the system is commissioned. In a troubleshooting investigation, the obvious interpretation of an observation may not be the correct one.

Temperature Control of Both Ends

Problem [10]: A lean oil still had unstable control and erratic operation.

Cause: The instrumentation was attempting to control the temperature at both ends of the still which prevented steady state operation.

Solution: Disconnected the top temperature controller and used the more important bottom controller alone.

Moral: Don't attempt to control the temperature of both ends of a column.

Nozzle Bottleneck

Problem [4]: Three chemical plant recovery train towers were limited to half of design rates by bottoms pump cavitation and high tower pressure drop.

Cause: A quick review (less than 30 minutes) of the vessel bottom nozzle indicated that the 8-inch nozzle was not adequate for design liquid rates.

Solution: Replaced the 8 in. nozzle with a 15 in. nozzle.

Moral: Check the simple things first.

Hatless Riser

Problem [7]: A 2-ft diameter packed scrubber, removing acetic acid from process offgas, had excessive acetic acid emissions causing unacceptable losses and odors.

Cause: A pan distributor had a centered vapor riser with no hat. The liquid feed entered directly above the riser and poured right through the riser, bypassing the distributor.

Solution: Since the distributor annular space was capable of handling the vapor flow, the riser was simply blanked off. Adding a hat or relocating the feed pipe were possible alternative solutions.

Moral: Poor distribution is often the culprit in packed column problems.

Tilted Distributor

Problem [7]: A packed column was designed to strip methanol and water from ethylene glycol. When all the methanol and water were stripped out of the glycol, excessive glycol carryover occurred.

Cause: As was suspected, the reflux distributor was tilted allowing all the liquid to flow down one side of the column. The distributor had not been securely attached to its support ring because the installation drawings didn't specify a method of attachment. A clever troubleshooter confirmed the hypothesis prior to shutdown by measuring vessel wall temperatures with a contact pyrometer stuck through the insulation. One side of the column was, of course, colder than the other for several feet below the reflux distributor.

Solution: The distributor was securely and evenly clamped to its support ring.

Moral: Be creative to test hypotheses.

Hammer

Problem [7]: There was a severe water-hammer type pounding at a column feedpoint.

Cause: The feed was sub-cooled and at a rate of 30% of design. The hypothesis for the problem proved true, namely that the oversized feed sparger allowed all the liquid to run out of the first several upstream orifices. Consequently vapor would enter the remaining orifices and condense in the cold sparger. This caused hydraulic hammering.

Solution: The sparger (feedpipe) was turned so that the holes were on top rather than the bottom. This ensured that the sparger remained full of liquid even at low feed rates. A deflector was installed above the orifices to keep feed from impinging on the tray above. This solution was better than the alternate of plugging holes, because the hole area would then have been undersized at the higher design feed rates.

Moral: Collapsing vapor can produce severe hammering.

Aeration

Problem [14]: Reflux flow from a vertical accumulator was erratic. Reflux flowed by gravity to the distillation column on flow control with no accumulator level controller.

Cause: Liquid flow to the accumulator poured in through an open-ended pipe and had a 6-ft fall to the liquid surface. This aerated the falling liquid with fine vapor bubbles. The accumulator liquid level had to be held at only 10% as the operators attempted to dampen flow fluctuations. Therefore the liquid didn't have time to deaerate, and bubbles passed out with the reflux to the control valve and some became trapped at the control valve inlet.

Solution: The accumulator inlet line was rerouted to near the vessel bottom, turned up internally at the center line and provided with a flat hat. This eliminated the 6-ft "waterfall" and the aeration.

Moral: The solution was found by equipment capacity and hydraulic analysis, thus combining basic calculations with field operations and tests.

Siphoning

Problem [14]: A horizontal liquid/liquid separator experienced siphoning in a drawoff loop seal in the heavy phase bottom drawoff line. This produced erratic flow and level so the siphoning needed to be stopped. A 50-ft drop from the top of the loop seal to grade produced a strong suction.

Cause: The original design provided a 1-inch vapor balancing line from the top of the loop seal to the vessel as a siphon breaker, but it proved to be too small. Hydraulic calculations indicated that the loop seal, even with a larger balancing line, was marginal at best.

Solution: The loop seal was replaced by a more positive siphon breaking system, consisting of a 1- by 4-ft vertical drum and a 3-inch balancing line. The heavy phase exited the new drum from a side sump with an inlet chordal weir. This prevented any siphoning in the new heavy phase draw line. Also a weir was installed just upstream of the light phase outlet nozzle. Good operations resulted.

Moral: A piping system that appears capable of siphoning needs to be carefully checked.

Hot-Vapor Rules Bypassed

Problem [14]: A hot-vapor bypass pressure control system for a new debutanizer failed to work.

Cause: The piping hookup for the hot-vapor bypass system didn't follow proven principles:
1. Bypass vapor must enter the vapor space of the reflux drum.
2. The bypass line should be free of pockets where liquid can accumulate.
3. Any horizontal pipe runs should drain into the reflux drum.
4. Liquid from the condenser(s) must enter the reflux drum well below the liquid surface.

Solution: The piping system was changed to follow the proven rules.

Moral: Don't rediscover the wheel.

Trapped

Problem [11]: This is a general problem statement rather than a specific case. An intermediate component can get trapped and accumulate in a column. Water in a hydrocarbon tower can also cause a variety of problems.

Cause: The intermediate component can get trapped when it can't get out the top (too cold) or the bottom (too hot). Water can cause problems such as hydrate formation, corrosion, and low water flowrates in the draw boot making boot level control difficult, fouling, trapping, etc.

Solution: Five classes of cures to tower accumulation problems.
1. Reducing the column temperature difference.
2. Removing the accumulated component from the tower.
3. Removing the accumulated component from the feed.
4. Modifying tower and internals.
5. Living with the problem.

Moral: Go with the expert on this one.

Centrifugal Pumps

Pumps are singled out when desired flow rates are not achieved. While the pump *may* be the cause of the trouble, the root cause might be far removed from the pump. To troubleshoot this situation, use plant data to recreate the system curve and compare it with the design conditions used when the pump was sized and specified. Consider questions such as:

- Does the pump operate smoothly and quietly? Noise or vibration could signal a problem with the pump, or be the result of cavitation.
- Did the problem occur suddenly, or did it develop gradually? Sudden problems indicate an operator error (such as closing a valve), or a mechanical failure including something being lodged in the pipeline.
- Was maintenance performed recently? Parts may have been replaced with incorrect spares (e.g., a different size impeller).

- What are the characteristics of the process stream? Changes in specific gravity, viscosity, or vapor pressure might affect the performance of the pump.
- Has each restriction in the pipeline been inspected? Sticking check valves, partially closed block valves, and fouled heat exchangers will reduce the flow from the pump.
- How is the pump controlled, and is the control loop functioning normally? Check the drive if it's a variable speed pump. Confirm that tuning parameters are correctly configured.
- Are the terminal pressures at expected values? The elevation of the suction and discharge points, and environmental pressure at those points (e.g., head pressure in the feed tank), directly affect the flow rate of a centrifugal pump.

Compressors

Tables 22-5 and 22-6 give checklists for troubleshooting compressor problems.

Table 22-5
Probable causes of reciprocating compressor trouble [5]

Trouble	Possible Cause(s)
Compressor will not start	1. Power supply failure 2. Switchgear or starting panel 3. Low oil pressure shut down switch 4. Control panel

(Continued)

Table 22-5
Probable causes of reciprocating compressor trouble [5]—cont'd

Trouble	Possible Cause(s)
Motor will not synchronize	1. Low voltage 2. Excessive starting torque 3. Incorrect power factor 4. Excitation voltage failure
Low oil pressure	1. Oil pump failure 2. Oil foaming from counterweights striking oil surface 3. Cold oil 4. Dirty oil filter 5. Interior frame oil leaks 6. Excessive leakage at bearing shim tabs and/or bearings 7. Improper low oil pressure switch setting 8. Low gear oil pump bypass/relief valve setting 9. Defective pressure gage 10. Plugged oil sump strainer 11. Defective oil relief valve
Noise in cylinder	1. Loose piston 2. Piston hitting outer head or frame end of cylinder 3. Loose crosshead lock nut 4. Broken or leaking valve(s) 5. Worn or broken piston rings or expanders 6. Valve improperly seated / damaged seat gasket 7. Free air unloader plunger chattering
Excessive packing leakage	1. Worn packing rings 2. Improper lube oil and / or insufficient lube rate (blue rings) 3. Dirt in packing 4. Excessive rate of pressure increase 5. Packing rings assembled incorrectly 6. Improper ring side or end gap clearance 7. Plugged packing vent system 8. Scored piston rod 9. Excessive piston rod run-out
Packing overheating	1. Lubrication failure 2. Improper lube oil and / or insufficient lube rate 3. Insufficient cooling
Excessive carbon on valves	1. Excessive lube oil 2. Improper lube oil (too light, high carbon residue) 3. Oil carryover from inlet system or previous stage 4. Broken or leaking valves causing high temperature 5. Excessive temperature due to high pressure ratio across cylinders
Relief valve popping	1. Faulty relief valve 2. Leaking suction valves or rings on next higher stage 3. Obstruction (foreign material, rags), blind or valve closed in discharge line
High discharge temperature	1. Excessive ratio on cylinder due to leaking inlet valves or rings on next higher stage 2. Fouled intercooler / piping 3. Leaking discharge valves or piston rings 4. High inlet temperature 5. Fouled water jackets on cylinder 6. Improper lube oil and / or lube rate
Frame knocks	1. Loose crosshead pin, pin caps, or crosshead shoes 2. Loose / worn main, crankpin, or crosshead bearings 3. Low oil pressure 4. Cold oil 5. Incorrect oil 6. Knock is actually from cylinder end
Crankshaft oil seal leaks	1. Faulty seal installation 2. Clogged drain hole
Piston rod oil scraper leaks	1. Worn scraper rings 2. Scrapers incorrectly assembled 3. Worn / scored rod 4. Improper fit of rings to rod / side clearance

Table 22-6
Probable causes of centrifugal compressor trouble [5]

Trouble	Possible Cause(s)
Low discharge pressure	1. Compressor not up to speed 2. Excessive compressor inlet temperature 3. Low inlet pressure 4. Leak in discharge piping 5. Excessive system demand from compressor
Compressor surge	1. Inadequate flow through the compressor 2. Change in system resistance due to obstruction in the discharge piping or improper valve position 3. Deposit buildup on rotor or diffusers restricting gas flow
Low lube oil pressure	1. Faulty lube oil pressure gage or switch 2. Low level in oil reservoir 3. Oil pump suction plugged 4. Leak in oil pump suction piping 5. Clogged oil strainers or filters 6. Failure of both main and auxiliary oil pumps 7. Operation at a low speed without the auxiliary oil pump running (if main oil pump is shaft-driven) 8. Relief valve improperly set or stuck open 9. Leaks in the oil system 10. Incorrect pressure control valve setting or operation 11. Bearing lube oil orifices missing or plugged
Shaft misalignment	1. Piping strain 2. Warped bedplate, compressor, or driver 3. Warped foundation 4. Loose or broken foundation bolts 5. Defective grouting
High bearing oil temperature (lube oil temperature leaving bearings should never be permitted to exceed 180°F)	1. Inadequate or restricted flow of lube oil to bearings 2. Poor conditions of lube oil or dirt or gummy deposits in bearings 3. Inadequate cooling water flow lube oil cooler 4. Fouled lube oil cooler 5. Wiped bearing 6. High oil viscosity 7. Excessive vibration 8. Water in lube oil 9. Rough journal surface
Excessive vibration (Vibration may be transmitted from the couple machine. To localize vibration, disconnect coupling and operate driver alone. This should help to indicate whether driver or driven machine is causing vibration.)	1. Improperly assembled parts 2. Loose or broken bolting 3. Piping strain 4. Shaft misalignment 5. Worn or damaged coupling 6. Dry coupling (if continuously lubricated type is used) 7. Warped shaft caused by uneven heating or cooling 8. Damaged rotor or bent shaft 9. Unbalanced rotor or warped shaft due to severe rubbing 10. Uneven buildup of depsits on rotor wheels, causing unbalance 11. Excessive bearing clearance 12. Loose wheel(s) (rare case) 13. Operating at or near critical speed 14. Operating in surge region 15. Liquid "slugs" striking wheels 16. Excessive vibration of adjacent machinery (sympathetic vibration)
Water in lube oil	1. Condensation in oil reservoir 2. Leak in lube oil cooler tubes or tubesheet

Steam Jet Vacuum Systems

Birgenheir, et. al. offer troubleshooting tips for steam jet vacuum systems [2]. First, to locate the source of the problem, perform these steps:

1. Determine if any changes were made to the process served by the steam jet system.
2. Determine if the pressure and temperature of the steam or condensing water differ from system specifications.
3. Determine if recent process changes have altered the feed rate of the vapor stream evacuated from the process vessel.
4. Determine if the problem developed gradually or suddenly. As a general rule, a gradual loss of vacuum indicates deterioration of the vacuum system, while a sudden loss is due to a change in utilities, increase in backpressure, or a system leak.
5. Review the maintenance history and determine if any recent modifications were made.
6. Review records of previous problems.

If it is determined that the correct utility flows, pressures, and temperatures are in use, and there is not excessive backpressure after the final ejector stage, then pinpoint internal problems by systematically testing each stage. Start by blanking off the ejector system and operating it against the blocked source. Measure the shut-off pressure at the inlet to the first stage; compare with values in Table 22-7. If the shut-off pressure is near the expected range, shift focus to the process conditions upstream of the ejector system.

Additional tests on the ejector system include:

1. Hydrostatic test to check for air leakage. Use water only if the system is designed to contain and support the extra pressure and weight of the water. Otherwise, use low pressure air, roughly 5 psig, for the test.

Table 22-7
Approximate shut-off pressure for diagnosing steam jet ejector performance [2]

Stages	Approximate Shut-Off Pressure
One	50 mm Hg absolute
Two	4 to 10 mm Hg
Three	0.8 to 1.55 mm Hg
Four	0.1 to 0.2 mm Hg
Five	0.01 to 0.02 mm Hg
Six	0.001 to 0.003 mm Hg

2. For systems operating under vacuum, shaving cream can be sprayed onto joints and other potential leakage points; if there is a leak the shaving cream will be sucked into the system, which is easily observed.
3. Check internals for damage or wear. Disassemble the ejector and check for deposits, scaling, and wear in the nozzle and diffuser.
4. For multistage ejectors, starting with the final stage, check the threads of the nozzle. Look for white or tan streaks which indicate a steam leak through the threaded connection.
5. Remove deposits from the suction chamber and make sure it is not cracked, rusted, or corroded. Shine a small light through the diffuse to make sure it is completely free from scale and is not pitted, grooved, or cut.
6. After cleaning, measure the throat diameters of the nozzles and diffusers as accurately as possible. Compare with the original dimensions to determine wear. Replace any part with a 7% increase in diameter.

Refrigeration

Table 22-8 gives a troubleshooting checklist for refrigeration.

Table 22-8
Refrigeration system checklist [5]

Indication	Checklist
High compressor discharge pressure	Check the accumulator temperature. If the accumulator temperature is high, check: 1. Condenser operation for fouling 2. High air or water temperature 3. Low fan speed or pitch 4. Low water circulation If condensing temperature is normal, check for: 1. Noncondensables in refrigerant 2. Restriction in system which is creating pressure drop
High process temperature	Check refrigerant temperature from chiller. If refrigerant temperature is high and approach temperature on chiller is normal, check: 1. Chiller pressure 2. Refrigerant composition for heavy ends contamination 3. Refrigerant circulation or kettle level (possible inadequate flow resulting in superheating of refrigerant) 4. Process overload of refrigerant system If refrigerant temperature is normal, and approach to process temperature is high, check: 1. Fouling on refrigerant side (lube oil or moisture) 2. Fouling on process side (wax or hydrates) 3. Process overload of chiller capacity
Inadequate compressor capacity	Check: 1. Process overload of refrigerant system 2. Premature opening of hot gas bypass 3. Compressor valve failure 4. Compressor suction pressure restriction 5. Low compressor speed
Inadequate refrigerant flow to economizer or chiller	Check: 1. Low accumulator level 2. Expansion valve capacity 3. Chiller or economizer level control malfunction 4. Restriction in refrigerant flow (hydrates or ice)

References

[1] Barnes R, Doak R. Troubleshooting Control Valves: Learn the Tricks, Avoid the Pitfalls. *Chemical Engineering*, May 1, 1990.

[2] Birgenheier D, Butzbach T, Bolt D, Bhatnagar R, Ojala R, Aglitz J. Designing Steam Jet Vacuum Systems. *Chemical Engineering*, July 1, 1993.

[3] Diehl J, Koppany C. *Chemical Engineering Progress Symposium Series*, 1969;92(65):77.

[4] France J, Sulzer Chemtech USA, Inc. Troubleshooting Distillation Columns, presented to the Rio Grande Chapter of AIChE *April 20, 1999*.

[5] Gas Processors Suppliers Association (GPSA). *Engineering Data Book, SI Version*. 12th ed. vol. 1; 2004.

[6] Gulley D. Troubleshooting Shell-and-Tube Heat Exchangers. *Hydrocarbon Processing*, September 1996:91–8.

[7] Harrison M, France J. Troubleshooting Distillation Columns. *Four-article series, Chemical Engineering*, March, April, May, June 1989.

[8] Hasbrouck J, Kinesh J, Smith V. Successfully Troubleshoot Distillation Towers. *Chemical Engineering Progress*, March 1993.

[9] Hengstebeck R. An Improved Shortcut for Calculating Difficult Multicomponent Distillations. *Chemical Engineering*, January 13, 1969:115.

[10] Hower T, Kister H. Solve Process Column Problems. *Hydrocarbon Processing*, Part 1 – May 1991, Part 2 – June 1991.

[11] Kister H. Component Trapping in Distillation Towers: Causes, Symptoms and Cures. *Chemical Engineering Progress*, August 2004.

[12] Kister H. *Distillation Troubleshooting*. Wiley; 2006.

[13] Kister H. Troubleshoot Distillation Column Simulations. *Chemical Engineering Progress*, June, 1995.

[14] Kister H, Litchfield J. Distillation: Diagnosing Instabilities in the Column Overhead. *Chemical Engineering*, September 2004.

[15] McCabe W, Thiele E. Graphical Design of Fractionating Columns. *Ind Eng Chem*, 1925;17: 605.

[16] Sloan M. A Practical Approach to PHE Maintenance. *Process Cooling*, January/February 2009: 19–23.

[17] Sloley A, Golden S. Analysis Key to Correcting Debutanizer Design Flaws. *Oil and Gas Journal*, February 8, 1993.

[18] Woods D. Process Troubleshooting. Webinar presented on October 6, 2010 at www.aiche.org

23

Startup

Introduction

Startup begins when a project is "mechanically complete," and ends when the plant is turned over to the manufacturing department. It can apply to a single piece of equipment, to a unit operation, or to a complete facility. Startup is needed after maintenance, replacement-in-kind, renovation, and new installation.

Subsets or synonyms to startup include "set to work," "commission" and "qualify." In every case, startup intends to achieve stable operation of the equipment, unit, or operation and to find and correct deficiencies. Those deficiencies have various root causes that usually fall into the broad categories of: poor design, failure to procure and/or install in accordance with design specifications, improper operation, changed requirements, or human error.

This chapter puts startup into the context of the larger project (design – construct – startup), and presents approaches that help achieve a successful outcome. When done methodically, startup is most dependent on the foundation of the process science. But even the most well understood unit operations may be sabotaged if startup is haphazard or incompetently staffed.

Myers shows that startup surprises (measured by excessive time and/or cost to complete) are strongly influenced by poor or incomplete process understanding (measured by the number of units in the plant that use new or innovative technology) [8]. Similarly, more startup problems occur if the heat and material balances are incomplete. And overlapping Detailed Design with Construction leads to more issues since construction errors are more likely. Figure 23-1 illustrates this hierarchy.

The chapter messages are:

- Think big. Startup is an integral process that is affected by everything that comes before it.
- Plan early. Work out the management and staffing structure, create a realistic schedule, allocate sufficient resources, and start the startup long before the project is mechanically complete.

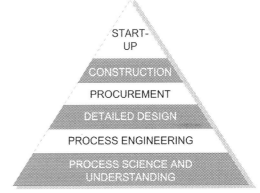

Figure 23-1. Each phase in the project relies on all of the portions that came before.

Startup Management

Informed and professional project management is essential for startup. The project manager is accountable for:

- Establishing the measurable goals and acceptance criteria for the construction and startup activities.
- Defining stage gates, such as "mechanical completion" in unambiguous terms.
- Ensuring participation from all stakeholders, including engineers, vendors, contractors, operators, instrument and maintenance technicians, R&D scientists, and quality assurance groups.
- Assembling startup teams.
- Establishing plans and schedules.
- Communicating plans and progress to team members, stakeholders, and management.
- Identifying deviations from plans as the work is underway.
- Determining corrective action that brings the work back to plan, or adjusts plans, when deviations are identified.

The project manager is also responsible for cost and budget. These are often thought of as the highest priority for assessing the PM's performance, but it's a mistake to create an environment where the cost of one project element (startup in this case) is allowed to dictate behaviors that may have far-reaching consequences.

Cutting corners in commissioning and startup can result in unintended expense, including accidents that cause serious injury to personnel. Startup must be well planned, adequately staffed, and appropriately documented. It's critical that sufficient resources be allocated for the work, and that contingency for unplanned issues is provided.

Change Control During Startup

The startup teams will identify possible changes, from the initial system walkdowns through performance testing with process materials. The changes usually fall into one of these categories:

- Deficiencies in the installation, defined as deviations from the contract documents. These include obvious problems, such as components installed in a sequence that differs from the P&IDs or installation drawings, to subtle issues such as valves installed with the marked direction flow opposite to the process stream.
- Design errors that are embedded in the contract documents that must be corrected for the system to perform as intended. These include issues like component sizes, materials of construction, and missing check valves.
- Communication issues, where the stakeholder's vision of the result differs from that which was designed and installed. For example, a manual shutoff valve located in overhead piping where a ladder is needed for access may have inadvertently received the stakeholder's approval during design because he misinterpreted the drawing.
- Changed owner requirements, such as the need to process a material at a higher temperature, or the desire for an in-line redundant pump. These changes can result from changing business needs, organizational changes, or to exploit technical developments. Regulatory actions or implementation of new national or global standards can result in changed owner requirements.
- Scope changes, closely related to changed owner requirements, distinguished by the notion that additional features are desired for the project that, for instance, may have previously been earmarked for a separate initiative. Project managers should pay close attention to managing scope, since allowing it to change mid-course usually means added cost, greater risk, and increased duration.

The changes can be further divided into those that must be implemented before a system can be safely operated, those that must be implemented prior to normal operation but don't affect startup activities, and those that are considered optional and subject to business rather than technical approval.

All types of changes should be documented with a formal Change Control procedure. A good procedure will ensure that possible changes are appropriately evaluated in the context of operation, safety, and business need. Documents requiring updating to reflect the change will be identified, and unintended consequences from the change will be minimized. If the system was previously assessed for potential risks and hazards, impacts to the risk assessment should be determined.

The procedure should include clear guidelines for categorizing changes, and it should identify who may authorize implementation. Authorization might depend on the categorization of the change, or whether it could affect a completed risk assessment.

Ideally, every proposed change should be pre-approved with the Change Control procedure. This is not always possible, so the procedure should provide for "emergency" changes that must be introduced immediately, usually for safety reasons but sometimes due to cost or schedule.

Startup Cost

It is very difficult to accurately predict the cost of startup, especially if the estimate is made before a shovel even hits the ground. There are too many unknown factors. As a rule of thumb, the cost ranges from <5% to 20% of the total direct costs (i.e., the cost for engineering, equipment, installation, and construction management). Established processes tend toward the bottom of the range, and highly novel processes to the top [7]. Small facilities tend to have a higher startup cost when expressed as a percentage of direct costs. Large facilities benefit from economies of scale and repetition to decrease the cost on a percentage basis.

Another factor that influences the *reported* cost for startup is how the *scope* of the effort is defined. In Myers's survey, some facilities reported zero time and cost for startup [8]. Startup for those plants might have been buried into the contractor's scope, so the owner simply walked in and started operating a plant that had undergone startup that was paid for as part of the direct cost. At the back end, the point where startup transitions to operations may be early (the plant is safe and functional) or late (the plant achieves its nameplate throughput). The startup scope should be clearly defined at the beginning of the project to reduce surprises and improve the accuracy of cost and schedule estimates.

Startup costs fall into three categories [8]:

1. installation of new equipment items to remedy deficiencies uncovered during startup,
2. labor and expenses for startup activities, and
3. normal operating costs for plant operators and raw materials.

The third category blurs the actual cost of startup, especially if those costs are incurred whether or not startup is in progress. The important thing is to understand how the company's accounting policies and procedures capture and report the data, and to structure cost estimates and comparisons on a consistent basis. The startup cost

percentages given above (i.e., <5% to 20%) represent only the first two categories.

Solids processing plants are particularly susceptible to physical or mechanical difficulties leading to lengthy startup time and excessive cost. Such plants are also more prone to incorrect equipment sizing due to the difficulty in preparing accurate heat and material balances [6].

If reliable data are available, startup costs can be estimated using the approach developed by Derrick and Sutor (as reported by Myers) [2,3,8]. Estimate each of these sub-activities (then sum):

- New fixed capital and maintenance costs in excess of normal costs during the startup period
- Salaries for startup personnel
- Engineering assistance during startup, usually from outside contractors and equipment vendors, but also from internal engineering groups if the company's accounting policies assign costs to the project
- Training of manufacturing supervisors and operators
- Yield loss and disposal of non-conforming product. This assumes any conforming product is salable, which may not be valid for regulated industries (i.e., pharmaceuticals) or products with limited shelf life that are produced during startup in advance of commercial launch (e.g., processed food)
- Utilities consumed in excess of normal utility costs during the startup period
- Any other operating costs in excess of normal costs during the period, such as supplies, analytical costs, and rentals

Systems Approach

Systems are assemblies of equipment or unit operations that work together to provide a defined function. For example, a boiler system takes treated water and condensate and produces steam. A distillation system splits a feedstock into two or more product streams. A pharmaceutical water distribution system collects purified water in a tank, keeps it hot, and circulates it through a piping system. System building blocks, when put together, make a plant.

When properly defined, every piece of equipment and every component in the plant belongs to a system. A massive P&ID that depicts the entire process on one drawing could be cut up, like a jigsaw puzzle, into pieces that represent each of the systems in the facility. Every

element on the drawing belongs to only one system. Nothing is left out.

Figure 23-2 shows the boundaries of a simple "Buffer Hold Tank System". Although designed to be piped and instrumented after installing the tank, bottom-mounted mixer, vent filter, and discharge transfer panel in the plant, this entire system could alternatively be fabricated as a module in a shop then shipped to site for connection to the adjacent systems. For this example, the adjacent systems are plant-wide utilities that span many P&IDs (e.g., Clean Steam, Clean Air, Waste Collection), plus two process systems (Buffer Prep, Temperature Control Module) that have dedicated P&IDs similar to this one.

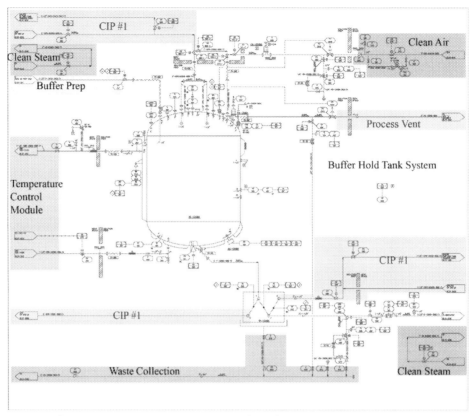

Figure 23-2. System boundaries identified on a P&ID. Shaded areas belang to adjacent systems.

System definitions offer significant advantages from the perspective of plant startup:

- Prioritization. Determine the optimum sequence for starting the systems. In the example, the Buffer Tank Hold System could be started before Buffer Prep, but it might be easier and more efficient to have Buffer Prep first so a source of water is available to fill the Buffer Hold tank. The Waste Collection system is clearly needed before starting Buffer Hold since water used during startup must be disposed of.
- Organization. Keep the information pertaining to each system together. Associate foundation documents (e.g., user requirements, process data, heat and material balances, calculations, specifications, design drawings), as-built information (e.g., shop drawings, cut sheets, material certifications, performance curves), and test data (e.g., installation and operational protocols and checklists, testing results, reports). This provides a straightforward way to assess progress and identify gaps during the startup program.
- Staffing. Use the system definitions to determine skill sets needed for startup. This includes departmental representation within the company (e.g., R&D, engineering, maintenance, manufacturing, safety, quality) and external resources such as process consultants and vendor representatives.
- Efficiency. The systems approach promotes a "do it once" mindset. Although there are interactions between systems, and many systems may have to be started simultaneously, this approach compartmentalizes the process and helps management structure the work for the most resource-effective result.

The benefits of using a systems approach are compounded when the project is initiated on a systems basis. Process design, specifications, and construction can all follow the same systems approach. This would, for example, ensure that a foundation system such as steam is completely constructed and ready for early turn-over and startup. Contrast with a construction trades approach, where design packages are organized by the type of work (e.g., piping, insulation) with no thought to systems turnover. In the trades approach, a missing critical component could prevent the turnover of a system, but the oversight would be difficult to discern early in the project.

Construction Quality

Beck reported that a rigorous construction quality program greatly reduces the cost and time for startup [1]. The program engages all subcontractors to identify and rectify deviations from the contract documents (i.e., drawings and specifications) on an ongoing basis. The result is that fewer construction errors are discovered during startup.

Challenges to the project's quality include [1]:

- Field execution: adherence to drawings and specifications
- Design and construction misalignment: accurately conveying the design intent to contractors
- Construction and commissioning misalignment: commissioning of process systems is poorly understood by construction team
- Material integrity: counterfeiting is increasing, leading to downstream consequences in safety, reliability, and quality
- Alternative suppliers: the economy forces projects to use new suppliers, and for suppliers to use new sources for materials and components
- Global projects: differences in language and culture challenge the ability to accurately communicate expectations and work processes

The recommended construction quality program emphasizes *self-inspection* for contractors and subcontractors. The goal is to prevent deficiencies (stemming from any of the factors listed above) to minimize defective work, and strive for a punch list with zero critical items. A critical item is defined as one that would prevent commissioning from proceeding until it is corrected. Ultimately, good construction quality results in trouble-free commissioning and startup.

The construction quality program consists of the following key elements:

- Pragmatic and not overly burdensome to the contractors
- Utilize contractors that are qualified and capable of meeting the quality challenge
- Require that contractors have a written quality program in place
- Plan and manage the quality program to meet expected standards and specifications. This means that the program is objective and measurable, not dependent on personal opinions of what constitutes acceptable work.
- Model after the contractor safety program which, typically, shares many of the same methods and goals

The program is managed on a day-to-day basis by assigning people to continuously monitor work as it is performed. The monitors identify and track issues using a spreadsheet or purpose-designed software (see Latista, [5], for example). Ideally, the contractors employ the monitors who are there to help the tradesmen do good work. In addition to performing field observations, the program includes testing materials for conformance with specifications, reviewing training records for the workers, and reviewing documentation (e.g., welding logs) as they are produced. This is not a punitive system, but one designed to focus on compliance issues that may be missed in the pressure to erect the project.

As with a safety program, everyone is engaged in the quality initiative. Regular (e.g., weekly) issues resolution meetings ensure that the issues that the inspectors identify are prioritized, assigned, and resolved on a timely basis. Metrics are collected and communicated.

Beck says that, typically, 35% of the issues identified during construction would impact commissioning if not corrected prior to mechanical completion. On one project, with about 1.5 million construction labor hours, the team identified 15,000 quality issues. Since they were identified early, only 2% of them remained unresolved when commissioning was undertaken, and many of those were intentionally left open (such as completing insulation of certain pipes) because it was advantageous to the project.

Another benefit is the effect on startup costs. As stated previously (page 367), equipment installed during startup typically is a significant portion of the total startup cost. Deficiencies discovered during startup are often paid for by the owner, because the project is already accepted and contractors have left the site. However, if the same deficiencies are found during a routine construction monitoring program, they are often remedied by the contractor at no cost to the owner [1].

Startup Checklists

Checklists are essential tools when planning and executing startup tasks. There are many sources for lists, but it's essential to specifically tailor checklists to the project's unique requirements. Use lists from equipment manuals, books and articles, the internet, and previous projects as starting points. Then review carefully with the startup team to account for everyone's know-how and know-what.

Some of the benefits of using a checklist:

- Standardization. Ensure outputs from multiple people are uniform
- Prevent cognitive overload. Document routine tasks to free minds for critical thinking
- Continuous improvement. Update checklists with new insights or attributes
- Collective learning. Ensure all team members are engaged

At the lowest level ("Level 3"), use checklists to confirm the installation and operation of specific components and equipment items. Each entry on this type of checklist must be precise, with little or no ambiguity. The start-up engineer affirms that the objective tests are completed, and reports any deviations from expected results. The engineer also collects evidence that the tests are satisfactory, such as output from the process control system or witnessed observations of performance. Deviations are escalated to a resolution procedure that ensures corrective action is performed.

On an intermediate level, use checklists to guide the creation of the Level 3 tests. For example, each entry in Table 23-1 could result in many pages of Level 3 checklists depending on the number of systems and pipelines in the plant.

The Level 1 checklists are comprised of the systems list with cross-references to the test categories used for each. For example, a matrix table lists each of the systems in the first column. Each category of tests, such as equipment setting, piping, and instruments heads a column, and check marks indicate the systems for which the tests are applicable. Additional information pertinent to the systems approach is listed, including the name of the team leader, priority, and schedule.

Other general checklists are used to ensure that systems are ready for startup. See Table 23-2 and Table 23-3 for examples.

Table 23-1
Typical checklist listing the specific procedures, and responsibilities, required for piping prior to starting up associated systems [9]

	Pre-Commissioning Tests for Piping Systems	Contractor	Owner
a.	Hydrostatically or pneumatically test all piping as required by the drawings or specifications	X	
b.	Notify Owner of test schedule	X	
c.	Witness field pressure tests when notified by Contractor		X
d.	Provide at owner's expense any special media for test purposes and facilities for their disposal	X	
e.	Check pipehangers, supports, guides and pipe specialities for the removal of all shipping and erection stops, and for the correctness of cold settings for the design service. In addition, provide Owner with instructions for hot settings	X	
f.	Check pipehangers, supports, guides and pipe specialities for hot settings and make minor adjustments as necessary	X	
g.	Install carseals on valves where necessary		X
h.	Check and record position of all carsealed valves		X
i.	Check steam tracing for tightness and continuity (use steam when available) and check proper tagging of supply and returns	X	
j.	Install all strainers (temporary and permanent), spades, and spectacle blinds as called for on drawings	X	
k.	Waterflush and/or air-blow small bore piping	X	
l.	Carry out chemical cleaning as required per job specification	X	
m.	Operate steam and water systems for flushing, cleaning and tracing		X
n.	Blowdown/clean steam lines with steam e.g. turbine inlets		X
o.	Witness flushing and cleaning		X
p.	Clean strainers after flushing and remove if warranted	X	
q.	Maintain records of blinds installed		X
r.	Obtain Owner's approval and permits and make tie ins as required by applicable specification	X	
s.	Isolate all systems to allow tie ins to be made		X
t.	Remove blinds as required during startup	X	

Table 23-2

Checklist for utilities and service systems can be expanded to suit any plant [4]

Step	Description	Constraints	Safety Precautions	Prerequisite Steps	Remarks
1. Activate electrical systems	Handled by electrical maintenance.		Clear with electrical group before using any equipment.	Removal of lockouts, completion of pre-operational maintenance work.	
2. Activate fire, process, and potable water systems	Start water systems, venting and testing hydrants.	Control valves operated manually until air is available.		Ensure piping is ready.	Vent and test deluge systems, checking for leaks.
3. Fill cooling-water system and circulate		Same as in Step 2.		Same as in Step 2.	Vent exchanger high points; check for leaks; charge chemicals.
4. Activate instrument and plant air systems	After checkout, pressurize systems; with Instrument Dept., activate air to instruments.		Check control valves to see that valves' action does not affect process equipment when air is activated.	Same as in Step 2.	Check for leaks; drain water from air headers and piping; check driers and filters, check dewpoint.
5. Activate steam and condensate-return systems	Introduce steam to high-pressure header; with it in service, pressurize low-pressure systems.	Boiler feed water must be available, and instruments activated.	Vent inerts from steam systems; slowly pressurize to avoid hammer; do not over-pressure reduced-pressure systems.	Check all headers, traps, etc. before heating steam lines.	Check for leaks.
6. Activate nitrogen system	Purge and pressurize N_2 headers.	Vessel entry procedures must be enforced before introduction of N_2.	Purge to less than 1% O_2 by successive pressuring and venting; noted unless N_2 inlets are blanked.	Ensure piping is complete.	Check dewpoint.
7. Waste disposal	Water lines to battery limits must be functional.	Outside-battery-limit lines and facilities must be ready.			

Table 23-3
Procedural checklist ensures safe, complete, pressure testing and purging of process equipment [4]

Step	Description	Constraints	Safety Precautions	Prerequisite Steps	Remarks
1. Unit checkout	Check that required mechanical work has been completed, tags and blinds pulled, and temporary piping disconnected.	Plant supervision must certify completion of work.	Cancel all entry and work permits.	Utility system has been commissioned.	Check blind list and inspect lines; close bleed, drain and sample valves.
2. Flange taping	Apply masking tape to all pipe and vessel flanges.			Same as in Step 1.	Seal space between flanges and punch 1.5mm hole in tape for monitoring leaks.
3. Nitrogen pressuring	With N_2, pressurize all process systems to designated test pressures.	Test at pressures near to normal; avoid lifting relief valves.	Manual valves requiring it must be carsealed after being set.	Same as in Step 1.	Prepare lists for equipment groupings and test pressures; isolate leaking sections, if possible.
4. Locate leaks	Apply soap solution to tape bleed holes and threaded connections; depressure when tests are done.	Block in N_2 and check for loss of pressure.	Check all connections.	Same as in Step 1.	Tag flanges to identify leaks; tighten or repair leaks; each time a system is pressurized, drain condensate at low-point bleeds.
5. Puring (O_2 removal)	Alternately depressure and repressure system to test (or header) pressure, until O_2 by meter is less than 3.0%.		Regularly check O_2 meter to ensure its accuracy.	Passing leak check at test (or N_2 header) pressure.	

Operating Modes

Once a plant or system is considered to be ready for use, the actual introduction of chemicals and energy may commence. There are different "modes" that define distinct operating regimes, each of which requires its own plan. Although this chapter primarily discusses initial commissioning, the other modes should also be planned for in advance of startup since those modes may be entered, interrupting startup, for various reasons.

- Normal startup
- Normal operation
- Normal shutdown
- Emergency shutdown
- Prolonged shutdown

Normal Startup

Plants are started from various initial conditions ranging from the very first startup to restarting after a hot standby period. Startup instructions and checklists should be prepared that define the checks and procedures to be followed, depending on the initial condition.

A pre-startup hazards analysis should be conducted, especially for new systems or after a prolonged or emergency shutdown. The entire startup team should participate, reviewing requirements, settings, potential hazards, team make-up, training status, and emergency response procedures. Potential hazards include: exposure to chemicals, leaks and spills, overpressure, overflow, mechanical failures, electrical failures, instrument failures, and other process excursions. During startup, there is a much higher than average risk of getting unwanted materials into the plant such as air or water. Particular attention should be given to these hazards.

Pre-startup checks include: walkdowns to confirm that all components are installed (e.g., filter elements, hoses, jumpers, etc.), valves are positioned correctly, blinds and temporary strainers are removed, and systems are energized.

The availability of utilities is confirmed, including drainage and collection systems. Safety equipment, including personnel protective equipment (PPE) and fire/spill response apparatus must be checked for presence and operation.

Normal Operation

The plant is designed for normal operation, and most of the documentation is geared toward steady-state conditions. Operating instructions, control systems, and emergency response criteria are all focused on normal operation. Operators are generally trained to optimize the plant's throughput.

The other operation modes mentioned in this section interface with normal operation, and those interfaces should be clearly understood, documented, and communicated to the operations staff. Some considerations:

- Ensure that the plant operating objectives and constraints are clearly defined. Some of these may be refined during startup, as the plant systems are exercised to their limits.
- Ensure that operators understand the operation of alarm and shutdown systems, and the response expected of them should alert or action levels be reached. Also provide information and training that helps operators keep the process from reaching those alarm levels, by identifying potential sets of conditions that could lead to the operating excursions.
- Plan for holding points, or static running conditions, where the plant can be moved in the event of a problem rather than resorting to a shutdown. For example, there may be safe operating zones that are far from the optimized control point, but that provide very stable conditions.
- Safe operation should always take precedence over all other objectives.

Normal Shutdown

Since normal shutdown is, by definition, a planned activity, the shutdown procedure should be clear, prescriptive, and well understood. The specific steps often mirror those taken during normal startup and include consideration of the effect on utilities and other connected process units.

Normal shutdown includes steps to render the systems safe, such as removal of hazardous process materials and inert (asphyxiating) gases. The systems might be cleaned as part of the shutdown; cleaning is often a process unto itself requiring its own set of startup, operation, and shutdown procedures.

Emergency Shutdown

Emergency shutdowns may be initiated automatically when certain limiting conditions exist (e.g., high temperature in a reactor), or manually when hazardous conditions are detected (e.g., a spill). The shutdown might be localized to one system or area, or be enacted plant-wide. The goal for an emergency shutdown is to react to conditions that are leading to a catastrophic event, before the catastrophe occurs. Because these situations usually arise unexpectedly, operators must be trained to react with clearly defined actions. Typical emergency shutdown actions include shutting off energy inputs (heating systems, electrical power) and depressurizing equipment.

Emergency procedures and training should cover:

- Leak detection and characterization, isolation of leaks, and handling of leaks and spills
- Action against fire
- Emergency equipment including protective clothing, breathing apparatus, tool kits, and firefighting equipment
- Communications, and "raising the alarm"
- Special features of particular chemicals
- Actions in response to "alerts" and "alarms" issued by the automated control system

Prolonged Shutdown

If a shutdown is expected to be prolonged, precautions are taken to protect the systems and equipment from deterioration. This may include cleaning residues, drying the equipment to avoid corrosion from standing water or high moisture content, lubricating equipment, and even providing further protection such as plastic wraps. If the systems are exposed to below-freezing weather, freeze protection is provided by draining water or filling with non-freezing solutions. Bypass lines may be used to allow utilities to continue to flow while isolating process equipment for disassembly and maintenance.

In addition, some equipment may be partially disassembled. Electrical systems are turned off and locked out. Disposable components, such as filter elements, are removed.

References

[1] Beck B. Construction Quality Management: Case Studies in Implementation. Orlando, FL: presented at the International Society for Pharmaceutical Engineering (ISPE) Annual Meeting; November 2010.

[2] Derrick G. Estimation of Industrial Chemical Plant Startup Costs. *1974 Transactions of the American Association of Cost Engineers*, 1974:169–75.

[3] Derrick G, Sutor W. Estimation of Industrial Chemical Plant Startup Costs. *1975 Transactions of the American Association of Cost Engineers*, 1975:308–15.

[4] Gans M, Kiorpes S, Fitzgerald F. Plant Startup – Step-by-Step. *Chemical Engineering*, October 3, 1983:80–92.

[5] Latista Field software, www.latista.com.

[6] Merrow E. A Quantitative Assessment of R&D Requirements for Solids Processing Technology. Rand Corporation for the Private Sector Sponsors Program, R-3216-DOE/PSSP July 1986.

[7] Mukherjee S. Preparations for Initial Startup of a Process Unit. *Chemical Engineering*, January, 2005:36.

[8] Myers C, Shangraw R. Understanding Process Plant Schedule Slippage and Startup Costs. Rand Corporation for the Private Sector Sponsors Program, R-3215-PSSP/RC June 1986.

[9] Red-Bag Engineering. Checklist for Preparation for Start-Up. Downloaded from www.red-bag.com; 2011.

24

Energy Conservation

Introduction

Energy conservation and management is an important aspect of every chemical engineer's function. The engineer's role in plant design, operation, and maintenance is crucial to optimizing energy consumption. Material and energy balances, prepared by chemical engineers, underpin a sound energy usage assessment. Payoffs come from an integrated analysis that identifies opportunities to reduce consumption or recover energy for reuse elsewhere in the process.

There are many resources available, since energy usage is now a government priority in most developed countries. These include technical papers, computer software, and individualized expert advice. Financial incentives may also be available in the form of tax credits, grants, and rebates. However, positive results require aggressive commitment from the executives, engineers, and operators responsible for the plant.

Steam is the most significant component of a typical chemical plant or petroleum refinery energy profile, generally accounting for approximately 50% of the total energy [9]. There is wide variation in the chemical industry segment, ranging from 30% to 70% of total energy use. Therefore, a plant's steam system may offer the most fruitful opportunity for optimizing energy consumption.

Every plant is unique, but these generalizations apply to most:

- Identify and empower an ongoing cross-functional energy conservation team. Provide resources (time and budget).
- Assemble data including: material and energy balances, with stream temperatures, for all chemical processes; transient conditions such as start-up and hot standby; operating and maintenance manuals for major equipment; costs for purchased utilities such as electricity, gas, water, and wastewater treatment; staffing levels and costs; expected remaining useful life of equipment; etc.
- Determine the average and opportunity costs for generating utilities such as steam and compressed air
- Conduct energy audits. Find energy waste such as compressed air leaks, malfunctioning steam traps, missing or inappropriate insulation, and inefficient operating practices – "low hanging fruit."
- Use process integration methodologies (i.e., Pinch Analysis) to identify opportunities for energy optimization
- Define and justify specific projects, obtain funding, and implement

Calculating Steam Cost

The US Department of Energy (DOE) published a useful Best Practices Brief that explains how to calculate the cost of steam [5]. According to the Brief, process heating accounts for an average of more than 60% of thermal energy use, predominantly in the form of steam. It also accounts for a significant portion of controllable operating costs and is an area of opportunity for reducing operating costs and improving profits.

Evaluate steam cost on an overall average basis, and also on a marginal basis (the cost for adding or subtracting increments of generation). In most plants there are multiple steam sources and multiple fuels. There are also multiple steam pressure levels with multiple paths by which the steam pressure is reduced. DOE shows that low pressure steam that is produced by

reducing the pressure with a pressure reducing valve (PRV) costs about the same as the high pressure steam from which it is derived. However, low pressure steam produced with a turbine with byproduct electricity production (i.e., co-generation) is much cheaper than the high pressure steam due to the energy recovery feature of the turbine. Consider the use of a turbine when the low pressure steam flow is over 50,000 lb/h.

Although it is convenient to report steam cost as the average cost at a particular generation rate, this is not particularly useful for managing the steam system to minimize costs. DOE highly recommends that steam systems be simulated with a computerized model. This is particularly important when there are multiple boilers and multiple steam pressures, each with highly variable consumption rates. In an example model, DOE shows that the cost of low pressure steam from a single facility can vary by a factor of about 3½ to 1 depending on the rate of consumption.

Steam cost calculation includes these elements:

- Fuel (type and cost)
- Raw water supply
- Boiler feed water treatment
- Feedwater pumping power
- Combustion air fan power
- Sewer charges for boiler blowdown
- Ash disposal
- Environmental emissions control
- Maintenance materials and labor

Fuel cost is the dominant component, accounting for as much as 90% of the total [5]. Calculate it with:

$$C_F = a_F \frac{(H_S - h_W)}{1000 \, \eta_B} \qquad (24\text{-}1)$$

Where (in US Units):

C_F = Cost of fuel, \$/Klb steam generated
a_F = Cost boiler fuel, \$/MMBtu
H_S = enthalpy of steam, Btu/lb
h_W = enthalpy of boiler feedwater, Btu/lb
η_B = overall boiler efficiency, fractional

DOE states that, in principle, the individual cost components should be rigorously calculated for the site-specific conditions. However, it is usually sufficient to use this approximation for the average steam generation cost (C_G):

$$C_G = C_F \, (1 + 0.30) \qquad (24\text{-}2)$$

The number 0.30 represents a typical value for the sum of cost components other than fuel. It could be more in smaller facilities, or in those that use coal or biomass fuel.

Steam System Rules of Thumb

CEERE lists these general rules of thumb [1]:

- The average cost of electricity is \$0.05/kWhr (\$15/MMBtu)
- The average cost of natural gas is \$0.35/CCF
- The average cost of #2 fuel oil is \$4/MMBtu
- There are 2,000 hours per year per shift (based on the assumption that one shift is 8 hours per day, 5 days per week, 50 weeks per year)
- A typical boiler or furnace has a combustion efficiency of 80%
- 90% of the heat loss from a hot, uninsulated surface can be economically eliminated by installing insulation.

- Cost of high pressure (125 psig) steam leaks are in the order of \$150 to \$500/leak/shift/year
- Cost of low pressure (15 psig) steam leaks are in the order of \$30 to \$110/leak/shift/year
- Cost of heat lost through hot, uninsulated pipes: (associated per 100 feet of uninsulated pipe)
   25psig: \$375/100 ft/shift/year
   50psig: \$430/100 ft/shift/year
   75psig: \$480/100 ft/shift/year
   100 psig: \$515/100 ft/shift/year
- Switching from electric heat to natural gas or #2 fuel oil can reduce heating costs by 78%

Equivalent Cost of Fuels

Use Figure 24-1 to quickly estimate the equivalent cost of fuels. Fuels are ranked according to their gross heating value. This was adapted from [7].

The US Government reports the cost for a variety of fuels on a monthly basis. See [12].

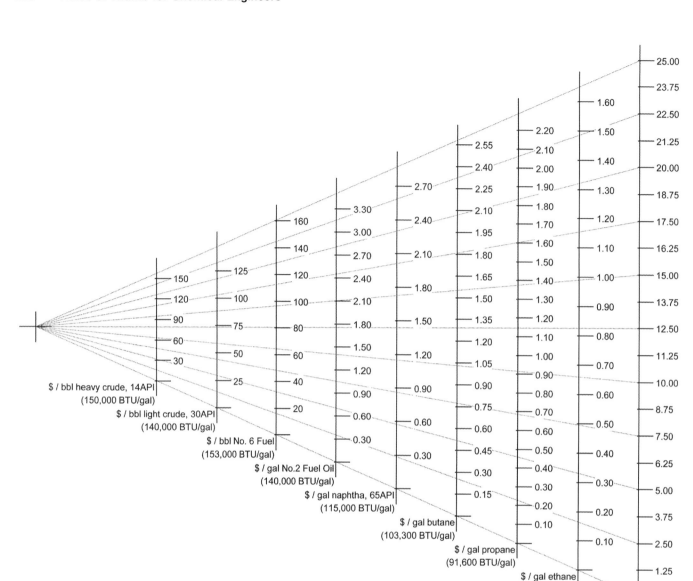

Figure 24-1. Equivalent values for selected hydrocarbons compares the higher heating value with no regard to consumption efficiency. For example, if the cost of natural gas is $5.00 per MMBtu then equivalent value for No. 2 Fuel Oil is about $0.35 per gallon: gas is a better value if No. 2 costs more than $0.35, but No. 2 is a better value if it costs less than $0.35.

Steam Traps

Steam traps inevitably fail. They fail open, blowing steam directly to the condensate system. They fail closed, potentially resulting in condensate backing into steam pipes and process equipment and causing mechanical damage or reduced heat transfer. They also fail to perform optimally if fouled, poorly sized, the wrong design, or installed incorrectly.

Experts agree that at any particular moment a high percentage of traps in a typical facility are in failure mode. Estimates range to 30% for facilities that have not been maintained for several years to less than 5% when a rigorous inspection and maintenance program is in place. EPRI estimates that [3]:

Table 24-1
Typical steam loss values with condensate in discharge [3]

Trap Type	Pressure Range	Steam Loss
Steam Drip Traps	600 – 1000 psi	100 – 150 lb/h
Trace Line Traps	50 – 150 psi	15 – 45 lb/h
Liquid Heater Traps	50 – 150 psi	130 – 330 lb/h
Air Heater Traps	40 – 150 psi	110 – 330 lb/h
Other Traps (overall value)	All pressures	90 – 100 lb/h

At any given time, 80% – 90% of the installed trap population might be determined to be operating normally or as expected.

A typical defective trap in a steam distribution system wastes 50 million Btu/yr at a cost of $100 to $1000 [1], but calculations using factors or equations in this section give higher results.

To estimate the steam loss through a trap that has failed open, many sources recommend using a form of the Napier equation. This method calculates steam flow through an orifice (the orifice size given by the manufacturer for certain trap types) at a given pressure differential across the trap. The problem with this approach is that the degree of trap failure may be significantly less than 100% so the computation is likely to be overly pessimistic.

EPRI has a much simpler method to estimate steam loss. Assuming the trap is failed open 75% of the time, there is condensate present in the discharge, and the failed traps are reasonably sized and selected. Use Table 24-1 to assign a steam loss value to each trap that is ascertained to be failed.

To use the Napier equation instead of Table 24-1, obtain the orifice diameter of the steam trap from the manufacturer. Solve the appropriate equation (below), and then multiply by 0.67 to account for the fact that actual flow will be less than theoretical flow [10].

The Napier equation is valid for steam flowing through an orifice when the discharge pressure is *less than* 58% of the steam pressure [2]. For steam discharging to atmosphere, the equation is good for steam pressures above about 75 kPag (10 psig):

$$W = 0.413\,P\,d^2 \tag{24-3}$$

W = steam flow, kg/h
P = steam pressure, kPa absolute
d = orifice diameter, cm

$$W = 40.4P\,d^2 \tag{24-4}$$

W = steam flow, lb/h
P = steam pressure, psia
d = orifice diameter, inch

Assuming the failed trap behaves like a thin plate orifice, use this equation if the discharge pressure is *greater than* 58% of the steam pressure [2]:

$$W = 0.576\,d^2\sqrt{(P-\Delta P)\,\Delta P} \tag{24-5}$$

W = steam flow, kg/h
d = orifice diameter, cm
P = steam pressure, kPa absolute
ΔP = difference in pressure between inlet and discharge, kPa

$$W = 56.3\,d^2\sqrt{(P-\Delta P)\,\Delta P} \tag{24-6}$$

W = steam flow, lb/h
d = orifice diameter, inch
P = steam pressure, psia
ΔP = difference in pressure between inlet and discharge, psi

Energy Efficient Design for Fractionators

Minimization of fuel use by fractionators is a common design goal. GPSA discusses several approaches as listed in Table 24-2.

Table 24-2
Methods for conserving energy in fractionators [4]

Conservation Method	Design Considerations
Optimize temperature of coolant in overhead condenser	Use air or cooling water (tower) if practicable. For columns requiring refrigerated coolant, temperature is very important for energy usage. Lower temperature coolant increases both the capital and operating costs.
Recover energy from feed and product streams	Preheat the feed stream with the bottoms product in a heat exchanger. This will generally reduce the reboiler duty, but also increase the overhead condenser duty. The net effect depends on many system parameters, but feed/product exchange is usually an attractive energy conservation measure.
Add energy with a side heater	Due to the temperature gradient in the column, side heaters can use lower temperature heating media. This offers the opportunity to use any stream that needs cooling that is of sufficient temperature to be useful. The bottom product is often used to side-heat the column. Additional column height is needed for each side heater, approximately 1.8 m to 2.4 m (6 ft to 8 ft), to accommodate the liquid draw-off tray and vapor-liquid disengagement at the return. A good rule of thumb is to limit the side heat duty to less than 50% of the total heat requirement.
Remove energy with a side cooler or condenser	Remove heat with a side cooler located above the feed tray. Due to the temperature gradient in the column, side coolers can use higher temperature cooling media than the overhead condenser. This can be particularly attractive if the overhead condenser uses refrigeration and the side cooler can use tower water.
Use a heat pump to transfer energy from the overhead condenser to the reboiler	A heat transfer fluid is compressed to a temperature that is higher than that needed by the reboiler. After it leaves the reboiler the fluid is flashed and used to condense the overheads. The main operating cost is the compressor. This method may not be feasible for a single column. However if there are a number of columns operating at different temperatures a heat pump might be effectively used to link the reboiler and condenser of two different columns.

Pinch Analysis

Pinch analysis is a systematic method that helps engineers optimize the energy usage of process plants. It consists of a group of techniques that:

- Analyze heat flows
- Establish energy targets
- Identify inefficiencies
- Define process improvements

With little more than a process flow diagram and heat and material balance, engineers can use pinch analysis to determine overall energy usage targets and to pinpoint design goals for individual utilities. Originally intended for use when designing new process plants, the method is also useful for planning retrofits to existing facilities.

Pinch analysis looks at process units, or entire plants, and identifies opportunities for resource recovery. It is then up to the process engineer to determine how those opportunities could be realized through piping and equipment configuration changes as well as changes to operating conditions such as temperature, flow, and pressure.

The fundamental tool in pinch analysis is the representation of a resource as a composite curve (CC). In its most elementary form, the CC represents a single process

stream that must be heated or cooled. For example, assume a plug flow reactor receives 100 kg/min of hexane at 50 °C. The hexane is stored in a tank that is normally 20 °C. With a heat capacity of 197.7 J/mol-K, and temperature increase of 30 °C, the enthalpy-flow of the hexane stream increases by

$$\Delta H_1 = \left(100 \ \frac{kg}{min}\right)\left(197.7 \ \frac{J}{mol - °C}\right)$$
$$\times \left(\frac{30 \ °C \ mol}{86.18 \ g}\right)\left(\frac{1000 \ g}{kg}\right)\left(\frac{0.0167 \ kW}{1000 \ J/min}\right)$$
$$= 115 \ kW$$

The enthalpy increase can be divided among any number of temperature increments. For increments of 10 °C, the increase is 38.3 kW from 20 ° to 30 °, 38.3 kW from 30 ° to 40 °, and 38.3 kW from 40 ° to 50 °.

Similarly, the temperature-enthalpies for all process streams that are heated (the "cold streams") are calculated and allocated to temperature a consistent set of temperature increments. Then, a composite curve is plotted. The same procedure is used for the streams that are cooled (the "hot streams"), yielding another composite curve.

Specialized software is used to perform a pinch analysis. The example given here is intended to illustrate the basic concepts.

Example

Tower pump-around, product, and overhead streams, listed in Table 24-3, are cooled. The only plant streams listed are those that are candidates for heat integration. The total enthalpy change for each stream is calculated as shown above, and the results are given in the last column. After establishing temperature ranges, the enthalpy for each stream is allocated as shown in Table 24-4. A similar procedure is used to tabulate the cold stream enthalpy changes over the same ranges (data not shown).

The data are plotted in Figure 24-2 as composite curves for the hot and cold streams. A specified minimum temperature approach for heat recovery exchangers is defined, typically in the 30 °C to 40 °C range (50 °F to 70 °F) for process stream to process stream heat recovery [11]. The example uses an approach of about 35 °C (65 °F). The hot and cold composite curves are positioned so they are separated by about 35 °C at their closest approach, called the pinch point or pinch area.

In the region above the pinch area, some heat integration is possible, but there is a net heat deficit and an external heat source is required. Any heat integration in this region will reduce the demand on the external source; no external cooling should be used.

In the region below the pinch area, some heat integration is possible, but there is a net heat surplus and an external heat sink (or cooling source) is required. No external heating should be used.

Table 24-3
Example data for hot streams

Stream	Flow, lb/min	T Inlet	T Outlet	c_p, Btu/lb-°F	ΔH, MBtu/h
Atmos Bottom Pump Around	3,600	626.9	518.0	0.71	16.7
Atmos Mid Pump Around	8,200	550.1	403.1	0.73	52.5
Atmos Top Pump Around	8,200	327.6	185.0	0.71	49.7
Atmos Tower Overheads	9,500	281.3	107.6	0.84	83.6
Atmospheric Gas Oil Product	1,500	524.6	363.6	0.71	10.3
Diesel Side Product	3,000	475.2	150.0	0.92	53.7
Kerosene Side Product	2,500	376.3	90.0	0.46	19.8
Vacuum Residual	2,600	693.0	383.9	0.52	25.3
Vacuum Gas Oil Pump Around	15,600	481.8	310.0	0.67	107.6
Vacuum Gas Oil Product	4,000	341.5	200.0	0.69	23.6

Adapted from [11].

Table 24-4
Enthalpy change for hot streams, MBtu/h, for tabulated temperature ranges (entire analysis covers temperature range from 90 °F to 700 °F for the example)

Stream	Temperature Range, °F				
	245–280	280–315	315–350	350–385	385–420
Atmos Bottom Pump Around	–	–	–	–	–
Atmos Mid Pump Around	–	–	–	–	6.0
Atmos Top Pump Around	12.2	12.2	4.4	–	–
Atmos Tower Overheads	16.8	0.6	–	–	–
Atmospheric Gas Oil Product	–	–	–	1.4	2.2
Diesel Side Product	5.8	5.8	5.8	5.8	5.8
Kerosene Side Product	2.4	2.4	2.4	1.8	–
Vacuum Residual	–	–	–	0.1	2.9
Vacuum Gas Oil Pump Around	–	3.1	21.9	21.9	21.9
Vacuum Gas Oil Product	5.8	5.8	4.4	–	–
Total Enthalpy Change, MBtu/h	43.1	30.0	38.9	31.0	38.8

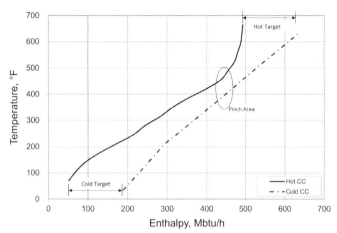

Figure 24-2. Composite curves for the hot and cold streams. Move the composite curves horizontally until the minimum distance between them is the minimum temperature difference for heat recovery (approximately 65 °F in this example). The potential energy savings for the primary heat source (e.g., boiler) is the "hot target"; the potential savings for the cold source (e.g., cooling tower) is the "cold target."

Energy targets, shown in the example plot, are determined in practice using a method called the "problem table algorithm" (PTA). Comparison of the energy targets with the total heat and cooling loads without any heat integration provides an estimate for the amount of energy savings possible. In the example, the hot target is roughly 130 MBtu/h. If the total heat load without integration is 160 MBtu/h, the potential saving is 30 MBtu/h. It is then up to the engineer to match up the various streams and decide if it is feasible to provide heat exchangers to realize the potential. Use factors such as the location and relative flow rates of the hot and cold streams to decide which streams to integrate.

Pinch analysis was developed in the 1990s for designing heat exchanger networks (HEN). This is why the composite curves relate temperature and enthalpy. Kumana [6] describes several ways to use the same methodology to solve these plant problems:

- Power conservation and recovery, through expansion of high pressure gases through turbines. Kumana offers three strategies for power conservation: reduce the flow, reduce the pressure drop, and reduce the inlet temperature of a gas to a compressor. He reports that when process

modifications for power conservation are followed up with thermal pinch analysis, total energy savings exceeding 30% can be achieved.

- Total site energy analysis. Unlike HEN which typically addresses a single process unit, combining all process units together in an analysis provides an opportunity to understand integration possibilities between processes through utilities, appropriate steam levels and loads, cooling water duty, refrigeration levels and duties, optimum cogeneration strategy, fuel use, etc. Kumana claims that a large site can achieve 10 to 20% efficiency improvements.
- Water consumption and distributed effluent treatment. The axes on the composite curves are water purity and flow. Optimization involves reuse, recycling, and treatment, possibly to multiple levels of purity.
- Integrated process debottlenecking. Pinch analysis provides an integrated design approach, combining hydraulic analysis, thermodynamic analysis, column targeting, and pressure drop vs. heat recovery network design.
- Batch process optimization. Kumana says that traditional scheduling techniques (such as Critical Path Method) combined with pinch concepts leads to a systematic approach for optimizing batch process schedules. The analysis includes energy pinch curves, cascade analysis, and batch utility curves and can also incorporate labor utilization and environmental considerations.
- Oil refinery hydrogen and sulfur optimization. Using tools such as hydrogen surplus diagrams, short-cut simulations, and LP optimization, refiners use pinch analysis to find solutions that lead to reduced capital outlay, lower operating costs, lower emissions, improved product quality, and increased yields. Kumana reports typical savings of 5 to 10% of fresh hydrogen consumption.
- Complex distillation systems. Using pinch analysis in conjunction with conventional optimization techniques leads to 30% energy savings and 15 to 20% cost reductions for new designs.
- Combined reaction/separation systems. Methodologies are being developed which set performance targets for yield and catalyst selectivity. Rather than trying to reduce energy consumption, the objective is to increase process yield for a given energy input.

Insulation

The Whole Building Design Guide (WBDG) includes an extremely practical resource for the application of insulation to piping, vessels, and buildings [8]. It includes web-based calculators that compute the minimum thickness of insulation for condensation control, equipment, piping, personnel protection, and other conditions. See Table 24-5 for a summary of design considerations presented in the Design Guide.

Other design considerations discussed by WBDG are:

- Abuse resistance
- Corrosion under the insulation
- Indoor air quality
- Maintainability
- Regulatory compliance
- Service and location
- Service life

To calculate the time, in hours, required for a liquid to cool in a pipe with no flow, use:

$$\theta = \rho \, c_p \, \pi \left(\frac{D_1}{2}\right)^2 L \, R_T \ln\left[\frac{(t_i - t_a)}{(t_f - t_a)}\right] \quad (24\text{-}7)$$

Where:

θ = time to reach final temperature, hours
ρ = density, lb/ft^2
c_p = specific heat, Btu/lb-°F

Table 24-5
Design objective for pipe and equipment insulation [8]

Design Goal	Summary of Design Considerations
Condensation control: minimize condensation and the potential for mold growth by keeping surface temperature above the dew point of surrounding air	• Often the overriding design objective for systems operated at below-ambient temperatures • Two separate issues: 1) avoiding surface condensation on outer surface of insulation, and 2) minimizing water vapor infiltration • Calculating surface temperature is relatively simple, but determining appropriate design conditions is confusing because satisfying the "worst case" condition is often impossible • For outdoor conditions, use a design relative humidity of 90% and install vapor-retarding jacket or mastic on the outside of the insulation
Energy conservation to minimize the use of scarce natural resources, maximize return on investment, and/or minimize emissions associated with energy usage	• National and local energy Codes may dictate the minimum allowable thickness of insulation • Properly designed and installed insulation systems can reduce heat loss (or gain) by 90% to 98%. Insulation systems can quickly pay for themselves in reduced energy costs • Economic thickness is defined as the thickness that minimizes the total life cycle cost (initial capital cost plus net present value of energy loss and maintenance over the expected lifetime of the insulation) • Insulation contributes to sustainability primarily by reducing energy usage and the associated greenhouse gas emissions • Insulation can also result in smaller size equipment, which also contributes to sustainability
Fire safety: protect critical elements from fire and slow the spread of fire	• Specific insulation designs are tested and assigned hourly ratings based on performance; ratings are assigned to complete insulated systems, not specific insulation materials or products • Piping insulation: Codes generally limit maximum flame spread index to 25 and maximum smoke developed index to 450
Freeze protection: minimize energy required for heat tracing and/or extend the time to freezing in the event of system failure	• Insulation retards heat flow; it does not stop it completely • See equations 24-7 and 24-8 • When unusual conditions make it impractical to maintain protection with insulation or flow, heat tracing (steam or electric) is required; the heat tracing supplies the heat lost through the insulation
Personnel protection: control surface temperature to prevent contact burns (hot or cold)	• Standard industry practice is to specify a maximum surface temperature of 140 °F for surfaces that may be contacted by personnel

(Continued)

Table 24-5
Design objective for pipe and equipment insulation [8]—cont'd

Design Goal	Summary of Design Considerations
	• This temperature produces no more than a first degree burn if contact is limited to 5 seconds
	• When calculating surface temperature, use engineering judgment to determine ambient design conditions (temperature and wind)
	• Choice of jacketing strongly influences a surface's relative safety. For example, at 175 °F stainless steel blisters skin more severely than a non-metallic jacket at equal contact time
Process control: minimize temperature change in processes where close control is needed	• Insulation minimizes temperature change of a fluid when it is piped from one location to another
	• Temperature change calculators are available at the WBDG website
Noise control: reduce noise in mechanical systems	• Noise radiating from pipes can be reduced by adding an absorptive layer of insulation and jacketing material
	• "Pipe insertion loss" is a measurement of the reduction in sound pressure level from a pipe as a result of application of insulation; the larger the value, the larger the amount of noise reduction
	• HVAC ducts conduct noise from mechanical equipment (e.g., air handling units). Lined metal ducts and fibrous glass rigid ducts are used to reduce transmission noise.
	• Breakout noise is from vibration of the duct wall that is caused by air pressure fluctuations

D_1 = inside diameter of pipe, ft
L = length of pipe, ft
R_T = combined thermal resistance of pipe wall, insulation, and exterior air film, h-°F/Btu
t_i = initial temperature of fluid, °F
t_f = final temperature of fluid, °F
t_a = ambient air temperature, °F

The thermal resistance of the pipe wall and the exterior film resistance are normally neglected for a conservative calculation. The combined thermal resistance is then calculated with:

$$R_T = 12 \frac{\ln\left(\dfrac{D_3}{D_2}\right)}{2\pi k L} \qquad (24\text{-}8)$$

Where:

D_3 = outside diameter of the insulation, ft
D_2 = inside diameter of the insulation, ft (for tight insulation = outside diameter of pipe)
k = thermal conductivity of insulation material, Btu-in./h-ft²-°F

Nomenclature

a_F	=	Cost boiler fuel, \$/MMBtu
C_F	=	Cost of fuel, \$/Klb steam generated
C_G	=	Total cost of generation, \$/Klb steam generated
c_p	=	specific heat, Btu/lb-°F
D_1	=	inside diameter of pipe, ft
D_2	=	inside diameter of the insulation, ft (for tight insulation = outside diameter of pipe)
D_3	=	outside diameter of the insulation, ft
H_S	=	enthalpy of steam, Btu/lb
h_W	=	enthalpy of boiler feedwater, Btu/lb
k	=	thermal conductivity of insulation material, Btu-in/h-ft²-°F
L	=	length of pipe, ft
R_T	=	combined thermal resistance of pipe wall, insulation, and exterior air film, h-°F/Btu
t_a	=	ambient air temperature, °F
t_i	=	initial temperature of fluid, °F
t_f	=	final temperature of fluid, °F
η_B	=	overall boiler efficiency, fractional
ρ	=	density, lb/ft²
θ	=	time to reach final temperature, hours

References

[1] Center for Energy Efficiency and Renewable Energy (CEERE). *Steam Delivery System Upgrade and Repair*. Industrial Assessment Center, http://www.ceere.org/iac/assessment%20tool/ARC2213.html; 2006.

[2] Babcock and Wilcox. Steam: its Generation and Use, www.guttenberg.org.

[3] Electric Power Research Institute (ERPI). Application and Maintenance of Steam Traps. TR-105853s; December 1996.

[4] Gas Processors Suppliers Association (GPSA). *Engineering Data Book, SI Version*. 12th ed. vol 2; 2004.

[5] Kumana & Associates. How To Calculate the True Cost of Steam. U.S. Department of Energy, Industrial Technologies Program; September 2003. DOE/GO-102003–1736.

[6] Kumana J, Al-Qahtani A. Optimization of Process Topology Using Pinch Analysis. *Saudi Aramco Journal of Technology*, Spring 2005.

[7] Leffler WL. Fuel Value Nomogram Updated. *Oil and Gas Journal*, March 31, 1980:150.

[8] National Mechanical Insulation Committee (NMIC). *Mechanical Insulation Design Guide*. National Institute of Building Science, www.wbdg.org; 2011. Accessed December.

[9] Office of Energy Efficiency and Renewable Energy. U.S. Department of Energy, Steam System Opportunity Assessment for the Pulp and Paper, Chemical Manufacturing, and Petroleum Refining Industries; October, 2002. DOE/GO-102002–1639.

[10] Personal communication with Tom Lago. Sr. Application Engineer. Armstrong International Inc., Three Rivers, Michigan, November 2010.

[11] Rossiter A. Improve Energy Efficiency via Heat Integration. *Chemical Engineering Progress*, December, 2010:33–42.

[12] U.S. Energy Information Administration. Independent Statistics and Analysis, www.eia.gov.

25
Process Modeling

Introduction

The purpose of this chapter is to differentiate the various types of process modeling tools. Ranging from equipment sizing and design programs to full-plant simulation environments, tools are available to solve nearly every problem. Some programs require only a small amount of input data, while others need extensive configuration to produce meaningful results. Stand-alone programs solve specific problems; linked modular programs model process manufacturing facilities from raw materials through to finished product.

The cost to procure the software, learn to use it, and then implement it on actual design problems is high. The investment should pay off with a better design solution, optimized for both the initial capital cost and ongoing operating expense. However, the risk that users will become locked into a single solution – due to the monetary, time, and emotional investment – indicates a need to tread carefully when selecting software for the first time.

Table 25-2 (page 395) lists various categories of software that may solve problems that are considered to be "simulation." These are just a small sample of a large number of available programs. The annual Winter Simulation Conference (WSC, http://wintersim.org) is a tremendous resource for exploring the simulation universe. Abstracts, and many full texts, of the papers presented at WSC since 1996 can be found at this website. (See, for example, http://informs-sim.org/wsc05papers/abstracts05/CS.htm#halls26876p for an abstract of Hall's 2005 paper, "Predicting the Effect of Process Time Variability on Annual Production in a Cell Culture Manufacturing Facility".)

To choose software for a particular project, it's usually best to consider how the output from the program will be used, then determine how much information is, or could be, available to prepare input to the program. Select the software by analyzing the data transformation that is required and comparing potential packages against that process.

While the steps just iterated are seemingly straightforward, there are a few pitfalls to be wary of:

- Most full-featured simulation software is expensive. Licensing and maintenance fees are often levied on a per-user basis. Users invest a great amount of time and effort in learning the program. Consultants might be engaged to do the initial setup, with updates and cases run by the owner. These costs in time and money magnify the pain if the software ultimately fails to fulfill expectations. However, if the costs are understood and the software helps achieve the project goals then the results should easily return the investment.
- Consultants are often familiar with only one family of programs (e.g., ChemCad or Aspen, but not both). If outside assistance is anticipated for the project, the choice of consultants may therefore be limited by the choice of simulation software. Conversely, if a consultant is engaged prior to choosing software, the consultant's bias may dictate the direction. Be careful if the consultant is "pushing" the only solution they are currently supporting; this may be the simulator that truly is a good fit for the problem, but it could be just an adequate choice with a far superior tool being available somewhere else.
- After making an investment in a package, there will be strong pressure to utilize the same software for all projects, even if it isn't the right fit. This pressure comes from the desire to maximize return on the financial investment, and also from emotional attachment to the program by the users. This is an argument for choosing one of the large solutions that has many modules rather than a single purpose program that can't solve other types of problems.

Concepts and Terminology

Static modeling provides data such as material transfers (quantity and time), volumes during and after each material transfer, heat transfer requirements, etc. When a plant representation is *static* it means that the results are the *same* and *predictable* from one batch to the next and one moment to the next (i.e., steady-state). Static models are applied to batch or continuous processes, but they are generally more appropriate for continuous operations.

Dynamic modeling adds time-dependent features to the simulation. For example, the rate of heat transfer is related to the temperature difference between two bodies. As a vessel is cooled, the rate of heat transfer decreases if the coolant temperature remains constant. With dynamic modeling, the cooling demand is accurately calculated over the course of the tank cooling and reported in a utility histogram. Dynamic modeling is used for transient conditions in a continuous process, such as during startup and shut down.

Discrete event modeling is used when defined units of material are processed. It is often associated with filling and packaging operations, where the defined units are bottles and cartons. For example, the model may simulate multiple lines operating in parallel, serviced by a fleet of fork trucks, and feeding into a few carton palletizer machines. In this case, discrete event modeling helps optimize the number of fork trucks and number of palletizer machines to support the fixed number of packaging lines.

Stochastic modeling introduces randomness, with selected inputs varied according to rules (usually statistical distributions). Each time the simulation is run, the results may differ due to changing inputs. Thus, stochastic models are usually run many times (the collection of runs is called an "experiment"), with the results collected and analyzed statistically. For example, if a fermentation is said to have a 95% success rate, a stochastic simulation might use a random number generator to decide, each time the fermentation is encountered, if it succeeds. Stochastic models are sometimes called "Monte Carlo" simulations.

Successful simulations are often built using these steps:

1. Formulate the problem. Set objectives, plan the effort, and conceptualize a simulation model.
2. Collect data. Determine the key assumptions and inputs; collect real-world data if possible, qualify the data.
3. Analyze the data and model the inputs to the simulation. Determine how the input data will interact with the simulation and perform sensitivity analyses. For data that have a relatively big impact on the simulation, model the inputs using an appropriate statistical correlation.
4. Build the simulation. Start simple and then build to the necessary complexity.
5. Verify the simulation model. Ensure the computer code generates results that are consistent with the problem objectives, and validate the model against the real world.
6. Run design cases (called "experiments"). Decide which sets of input data to use, what output to collect, and how to report the results; document both the computer code and the input/results.

When a working plant is available for verification of the model, choose experiments that provide the verification. For example, the TANKJKT model described in the next section calculates the inside and outside heat transfer coefficients for a jacketed vessel. If the program predicts that the inside coefficient is much lower than the outside coefficient then the heat transfer should be relatively insensitive to changes in the flow rate of the jacket fluid. The program returns that result: tripling the jacket coolant flow rate has virtually no effect on the rate of cooling if the inside coefficient is 20% that of the outside coefficient. If a similar tank is available for testing, try tripling the jacket flow rate and observing the change in heat transfer; if there is no change then confidence in the model is improved.

Inputs are also referred to as "boundary conditions" or "assumptions" by end users or model builders.

The old adage, "Garbage In = Garbage Out", is true for simulation. Step 3, modeling the inputs, is extremely important. This is not necessarily as simple as filling out an input data sheet, because the best input values may differ from "actual" values for the simulation to accurately model the process. When the simulation model is verified (Step 5), it may be necessary to "fudge" some input values to get the correct results. This isn't necessarily a failure of the model per se, but could be a way to account for effects, such as heat loss through insulation, that aren't otherwise built into the calculations.

Equipment Design

Process equipment is usually designed on the basis of specific parameters, as discussed throughout this book. For equipment that must satisfy multiple process cases, the equipment is sized on the basis of one set of criteria, chosen because this is thought to be the most common or critical case. Then, the equipment is "rated" using other assumptions, often selected because they represent worst-case scenarios, to confirm that the equipment will operate satisfactorily under those conditions.

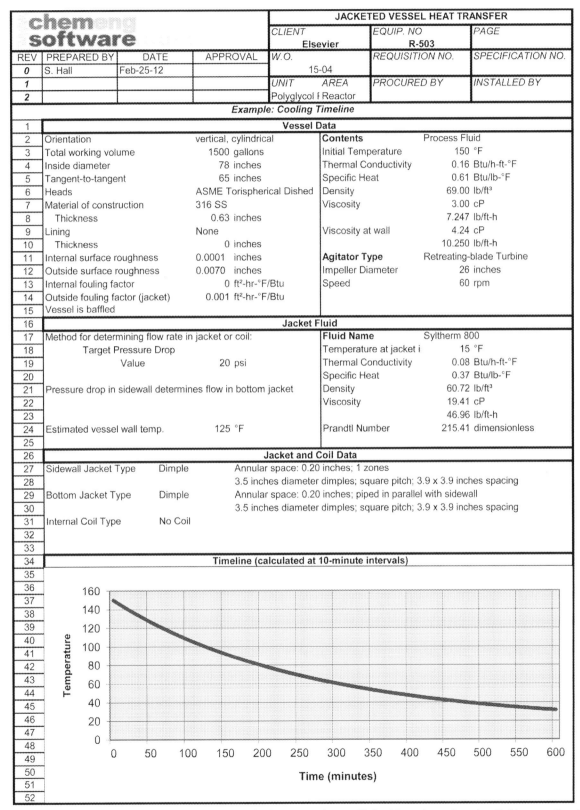

Figure 25-1. Reactor cooling simulation recalculates the heat transfer coefficient based on changing fluid properties in a step-wise simulation (source: www.chemengsoftware.com).

Equipment design modules are included in the major simulation software suites like ChemCAD and Aspen. Data from the general process simulation can be transferred to the equipment modules to reduce the amount of duplicate data entry. For two reasons, recalculation of the equipment details is generally performed only as needed, not on each recalculation of the overall module: 1) the computational overhead is high, and 2) the equipment details are not needed to solve the overall material and energy balances. For example, fractionation calculations for the overall process simulation determine the separation given compositions and flow rates, but the detailed design of trays and downcomers is performed only after the overall design is established.

Equipment modeling is indicated where dynamic conditions are expected to influence the equipment's performance. Consider the example of a jacketed, stirred tank where the heating or cooling time must be predicted. The physical properties of the tank content change with temperature, especially viscosity which is an important factor in the calculation of the inside heat transfer coefficient. The author's spreadsheet solution simulates heating or cooling by recalculating the coefficient at user-defined time intervals, applying the coefficient to the heat transfer over the next time increment, and continuing the step-wise calculation until a predetermined total simulated time has elapsed. Figure 25-1 gives the result from a typical simulation, clearly showing how the rate of heat transfer slows as the reactor content cools.

More examples where equipment design simulation is useful include:

- Transients occur, such as water hammer when a valve is closed quickly
- Tight control is needed, where the interaction between a control valve and pump curve determine a flow rate
- Complex mass and/or energy transfers occur, such as in mixing or blending operations

Mass and Energy Transfer

Computational Fluid Dynamics (CFD) software is used to create and visualize highly detailed mass and energy transfer models. Characterized by the colorful three-dimensional output plots, CFD is used to design reactors, agitators, compressors, pumps, air handling systems, heat exchangers, and many other operations. The programs work by dividing physical space into small, 3D cells, usually cubes. Rules define how mass or energy transfers through the cell; the user defines the size of the cells and the boundary conditions (for example, the speed of an impeller). Momentum moves through each cell according to the rules, and is output to the next cell in all directions.

Complex geometries can be modeled in CFD, but each cell operates as uniform modules within the whole. A typical unit operation, such as a reactor, is modeled with about one million cells; this number will increase as computing power increases. There are fewer than 100 variables per cell; again, the practical limit is computing power. Variables consist of things like velocity vectors for multiple components in three dimensions, kinetic energy, and temperature.

The leading CFD software is Ansys (formerly Fluent). There is an ample library of case histories and tutorials in the literature.

Flowsheet Simulation – Continuous Process

The major simulation packages – CC Steady State (formerly known as ChemCAD), Aspen Plus®, Aspen HYSIS®, Pro/II – are adept at calculating material and energy balances for unit operations and complete plants. They incorporate extensive thermodynamic and physical property data, and can predict properties when published data are lacking. These programs are powerful, easy to use, and well accepted.

Hill gives these tips for planning and building a simulation [5]. See Table 25-1.

Shethna, Dziuk, and Nandigam presented an approach for simulating a complex facility – a refinery in their case – by first modeling the unit operations as simple blocks [9]. Using simple assumptions for the performance of each block, the overall flowsheet converges quickly and provides a platform for a more in-depth study. Any of the

Table 25-1
Tips for planning and building a simulation

Project Phase	Tips and Best Practices
Planning a simulation	• Establish clear and specific goals for the simulation.
	• Sketch out the process on paper to ensure there are no obvious flaws and that the goals are achievable. Do some simple calculations and indicate rough flows and temperatures on the sketch to help visualize the problem.
	• Scrutinize the input data and confirm it is the best available information.
	• Identify major chemicals in the process. Include atmospheric air (often modeled as nitrogen) if process streams may be aerated. Trace chemicals can often be ignored or grouped together as a pseudo chemical.
	• Select the thermodynamic models that are appropriate for the systems under study.
	• Obtain plant data if an existing plant exists, or lab data if the process is being developed. Avoid allowing the program to estimate properties if possible by entering data for proprietary materials.
	• Make preliminary calculations in the simulator to get a feel for expected results and to aid in checking the validity of assumptions.
Building a simulation	• Start small, stay focused. Building a model slowly makes it easier to isolate problem areas.
	• Avoid connecting recycle streams until the simulation has been refined. Instead, use separate feed and product streams.
	• Set the engineering units and be careful that all data are input in consistent units.
	• Make notes that document the steps taken to develop the simulation.
	• Specify outcomes first. Begin setting specifications with the highest quality data. For example, specify the output pressure for a pump rather than entering a performance curve; the curve can be entered later after the simulation is refined.
	• For columns, start with ideal stage models before moving to mass transfer. Ideal-stage calculations are faster, and the column is easier to troubleshoot if convergence problems appear.
	• Run, revise, recycle. Run the simulation and review the results. Continue to revise and refine it until the results are close to expectations. Then, connect recycle loops one by one and save the simulation after each successful recycle connection.
Making the most of the simulation	• Question everything. Compare outcomes to reality; focus on heat duties, recycle flowrates, and convergence. If the simulator does not converge, the problem must be resolved before results can be considered in any way valid. Loosening a tolerance to force convergence can lead to seriously flawed results.
	• Perform sensitivity analyses by varying inputs. If the results seem to be very sensitive to inputs then process control may be difficult. Consider the effect of seasonal variations, such as the temperature of cooling water. Also see the effect when varying throughput.
	• Store the simulation in a logical place (on a network drive, for example) along with supporting documents such as explanatory notes. Document any learnings from the simulation.

simple blocks can be replaced with a robust model of that unit to provide more information and better accuracy. They call the block model "low fidelity" and the robust simulations "high fidelity."

They cite these challenges to simulating a refinery as reasons for taking a low fidelity approach:

• Good definition of the crude oil assay is difficult to obtain, and the assay changes frequently.
• Each process unit presents challenges to accurately build a high-fidelity model; a high degree of expertise is required.
• Large high fidelity models face computational hurdles: speed, convergence, and robustness.
• The "component slate" representing the physical properties of the oil fractions changes through the refinery. This is a technical challenge that makes a "superset" component slate impractical.
• Some unit operations are not included in the simulation software.
• Maintenance of a refinery-wide high fidelity model requires overall expertise in all process units.

In the low fidelity block approach, each unit operation is solved using simplified or short-cut models. Simplified versions are used whenever the overall model is not affected; for many unit operations, the refinery's throughput and product mix are accurately simulated even though those specific operations are simplified. For example, rather than using rigorous tray-to-tray distillation models, equations like those presented in Chapter 3 are utilized. The simplified models are "trained" to simulate rigorous behavior, by comparing the outputs against rigorous analogs and adjusting the parameters to

better fit the specific range of conditions found in the plant.

Flowsheet Simulation – Dynamic Process

Dynamic simulators are flowsheet simulators that add the time dimension to steady-state material and energy balances. They are especially useful for testing and checking process control strategies. For batch processes, in particular those with highly exothermic reactions, dynamic simulators are valuable because they enable analysis of thermal behaviors, enabling engineers to study alternative designs for heat transfer equipment and chemical reactors [6].

IDEAS is a dynamic simulation package that is built on the ExtendSim platform. Users quickly build flowsheet models, then add dynamic elements such as control loops, with templates that describe the behavior of unit operations. Templates are provided with the software, and users can also make their own. Although it is possible to build templates that use thermodynamic properties for distillation or reaction operations, this is not IDEAS' strength; a program like Aspen Plus® or CC-Steady State would be an appropriate choice for such simulations. Instead, use IDEAS to test control loop strategies as plant variables like tank levels, flowrates, and utility consumptions are changed. Employ the program to help train operators; they can see the effect of their simulated actions such as oscillations around a set point.

Aspen HYSIS® is the tool of choice for refinery simulations, for both continuous and dynamic modeling. Bhattacharya presented a good example of a dynamic simulation [2]. The hydro-processing reactor normally operates above 1,000 kPa (150 psig). When upset conditions occur such as a power outage, fire event, runaway reaction, or major leak, the reactor unit must be depressurized in a controlled manner. This is done through a control valve; the purpose of the simulation is to determine the appropriate size of the valve.

The hydro-processing unit is depicted in Figure 25-2. During depressurization, the control valve opens, reactor effluent cooling is reduced, the reactor outlet temperature changes (up or down, depending on the process), and gases to the separator are hotter but at lower pressure than during normal operation. These factors complicate the calculation for the venting rate needed to safely depressurize the unit. In addition to valve sizing, the simulation

also provides peak temperatures in the integrated reactor and stripper circuit over an extended time period (used for checking design conditions), and the stripper feed enthalpy and operational conditions to calculate the stripper relief load.

To obtain these answers, the simulation requires careful characterization of the inputs and assumptions. For this example, Bhattacharya reported that a simple reactor model was used, but rigorous calculation of the reactor void volume and hold-up in the heat exchangers and piping were needed because these volumes determine the time required to depressurize the system. The heat exchangers were fully designed, with the heat transfer coefficients recalculated as flow rates changed, to provide a realistic estimate of enthalpy change in the system. Other complications were also included in the model, using the software's built-in functions or using the modeler's algebraic equations. Many different scenarios were simulated; the results showed that a power outage was the worst case scenario and the valve was sized to accommodate this case with turndown to safely depressurize the other scenarios.

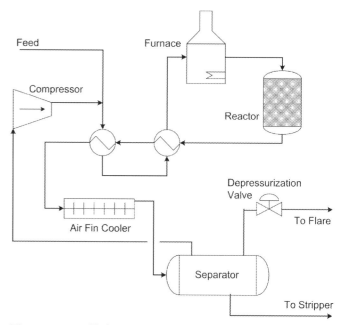

Figure 25-2. Hydro-processing unit for dynamic simulation example (after [2]).

Discrete Process Simulation

Discrete event simulation models individual elements that occur at given time intervals, accounting for available resources, constraints, and interaction rules. For chemical processes, these programs are most often used for the filling and packaging operations, where easily identifiable elements such as bottles, boxes, pallets, and fork trucks come together. Aspects such as mechanical breakdowns, shift changes, operator fatigue, in-process inspections, and shipping schedules can be modeled using statistical distributions or defined schedules. Discrete event simulation is used throughout the supply chain, covering raw materials through shipment of finished product.

Cope argues that discrete event simulation can provide insights for a chemical plant, answering questions such as [3]:

- What are the effects on chemical production by adding or removing suppliers, transportation vendors, or bulk storage tanks (for raw materials, intermediates, or finished product)?
- What are the minimum requirements to allow for a continuous production outflow?

- How is a process affected by a debottlenecking effort?
- Will a process require additional resources or inventory when seasonal or other variability is taken into account?

Cope also gives these hurdles to using simulation:

- Difficult for non-specialists to grasp. Understanding the methodology behind the simulation is often a prerequisite to stakeholder buy-in.
- Significant effort for data collection and analysis. Large amounts of data are usually required to obtain statistically relevant distributions.
- Modeling efforts often require the knowledge of a statistics or simulation expert.
- Huge investments in time, resources, and money for initial implementation. Once developed, models must undergo validation and verification to ensure usefulness and accuracy.

General Purpose Simulation Environments

Shelden and Dunn describe the use of relatively simple dynamic simulation environments for modeling a variety of engineering problems with relatively complex and nonlinear models [8]. The programs are quick and convenient to use for studying system behavior. With this software, the user models the system of interest with a set of differential equations, and the software solves the equation set. A big advantage is that the engineer is building the model based on first principles rather than relying on the equations built into a flowsheet-based simulation package. This is also a disadvantage since it's necessary to understand and apply the mathematics of the model.

A tool on nearly every desktop, Microsoft Excel, can be used to create sophisticated simulation models. Hall described how to use Excel with VBA to simulate water consumption in a pharmaceutical facility [4]. Probability distributions are used to vary the water consumption rate, time of day, and duration for each of the water users. The level in a water storage tank is calculated over the course

of a simulated day (24 hours). Each time the model is run, a new consumption profile is computed based on the probabilities. The model's primary output is a chart showing the level in the tank (Figure 25-3). The spreadsheet model is available for download from the web site accompanying this chapter. Its algorithm is shown in Figure 25-4. Many add-in modules are available for Excel that simplify the creation of simulations (Table 25-2). References [1] and [7] are excellent resources.

The following VBA listing is a Function subroutine that returns a normally distributed random number. In Hall's algorithm, the user inputs a nominal duration (NT) and standard deviation (SD) for each of the water users. When the model is run, the duration (T) for each instance of water use is calculated with:

$$T = Gauss^*SD + NT$$

If T computes to a value less than zero, it is reset to zero meaning the usage wouldn't occur that time.

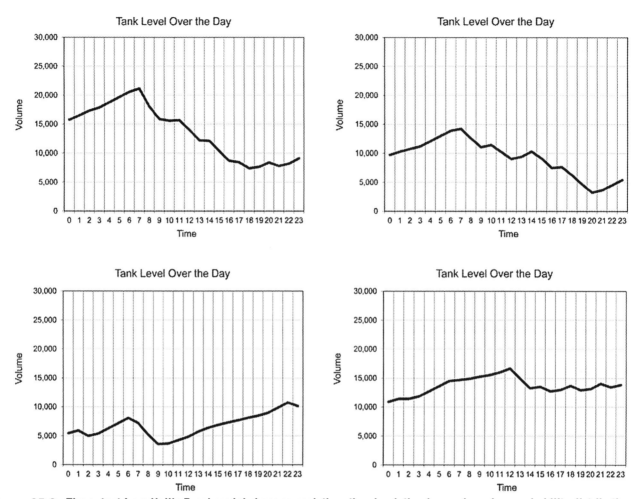

Figure 25-3. The output from Hall's Excel model changes each time the simulation is run, based on probability distributions for water consumption that are established by the user.

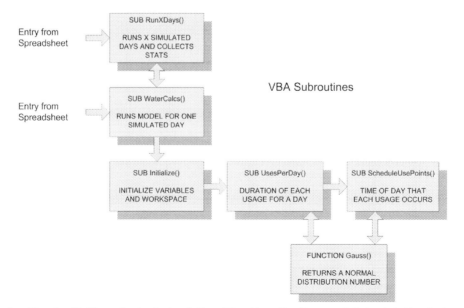

Figure 25-4. The algorithm for Hall's water tank simulation. The UsesPerDay and ScheduleUsePoints subroutines place their results directly on the Excel spreadsheet. Formulae in the spreadsheet use the results to feed the tank level chart.

```
Function Gauss()
    ' Function returns a normal distribution
random number
    ' Source: http://www.anthony-vba.kefra.com/
    Dim fac As Double, r As Double, V1 As
Double, V2 As Double
    10 V1 = 2 * Rnd - 1
    V2 = 2 * Rnd - 1
    r = V1 ^ 2 + V2 ^ 2
    If (r >= 1) Then GoTo 10
    fac = Sqr(-2 * Log(r) / r)
    Gauss = V2 * fac
End Function
```

Software List

Table 25-2
Simulation software vendors

Program Name	Category	Vendor	Web Site
SuperPro Designer	Batch flowsheet	Intelligen, Inc.	www.intelligen.com
K-TOPS	Batch flowsheet with dynamic optimization	Foster Wheeler Biokinetics	www.fwbiok.com
Arena	Discrete event, dynamic	Rockwell Automation	www.arenasimulation.com
Witness	Discrete event, dynamic	Lanner	www.lanner.com
CC-DYNAMICS	Dynamic flowsheet	Chemstations	www.chemstations.com
IDEAS	Dynamic flowsheet	Andritz Automation	www.andritzautomation.com
Aspen Exchanger Design and Rating	Equipment design	AspenTech	www.aspentech.com
CC-THERM	Equipment design	Chemstations	www.chemstations.com
TANKJKT	Equipment design	ChemEngSoftware	www.chemengsoftware.com
Crystal Ball	Excel add-in	Oracle	www.oracle.com
Ersatz	Excel add-in	EpiGear International	www.epigear.com
ModelRisk	Excel add-in	Vose Software	www.vosesoftware.com
ANSYS CFD	Fluid dynamics (CFD)	Ansys	www.ansys.com
ExtendSim	General purpose – continuous, discrete event, and dynamic	Imagine That Inc.	www.extendsim.com
acslX	General purpose – dynamic	AEgis Technologies	www.acslX.com
Berkeley Madonna	General purpose – dynamic	Macey & Oster	www.berkeleymadonna.com
gPROMS	General purpose – dynamic	Process Systems Enterprise	www.psenterprise.com
MATLAB®-Simulink®	General purpose – dynamic	MathWorks	www.mathworks.com
STELLA®	General purpose – dynamic	isee systems, Inc.	www.hps-inc.com
aspenONE® Engineering with Aspen Plus®	Steady-state and dynamic flowsheet	AspenTech	www.aspentech.com
Design II	Steady-state and dynamic flowsheet	WinSim	www.winsim.com
Aspen HYSIS®	Steady-state and dynamic flowsheet – oil & gas	AspenTech	www.aspentech.com
CC-Steady State	Steady-state flowsheet	Chemstations	www.chemstations.com
Pro/II	Steady-state flowsheet	Invensys Operations Management	http://iom.invensys.com

References

[1] Albright SC. *VBA for Modelers: Developing Decision Support Systems Using Microsoft® Excel.* 2nd ed. South-Western College Pub; 2006.

[2] Bhattacharya D, Valappil J, Chaplin D, Tekumalla R. Dynamic Simulation: An Efficient Approach to Hydro-Processing Unit Design, presented at the Spring Meeting of AIChE, April 27, 2009.

[3] Cope D. Consider Discrete Event Simulation. *Chemical Processing*, October, 2010:27.

[4] Hall S. Process Simulation Using VBA Code with Variable Input Data. *Pharmaceutical Engineering*, September/October, 2010.

[5] Hill D. Process Simulation from the Ground Up. *Chemical Engineering Progress*, April, 2009:50–3.

[6] Feliu J, Grau I, Alós M, Hernandez J. Match Your Process Constraints Using Dynamic Simulation. *Chemical Engineering Progress*, December, 2003:42–8.

[7] Powell SB, Baker KR. *Management Science: The Art of Modeling with Spreadsheets. Excel 2007 Update.* 2nd ed. Wiley; 2008.

[8] Shelden R, Dunn I. Dynamic Simulation – Modeling Processes, the Environment, the World. *Chemical Engineering Progress*, December, 2001:44–8.

[9] Shethna H, Dziuk S, Nandigam R. Use of Low and High Fidelity Models in Refinery-wide Simulation, presented at the Spring Meeting of AIChE April 27, 2009.

26
Conversion Factors and Constants

397

Significant Figures and Rounding

Follow the rules shown in Tables 26-1 and 26-2 to round the results of calculations to the appropriate number of significant figures (also called significant digits).

Table 26-1
Significant figure rules

Rule	Example	Number of Significant Figures in Example
All non-zero digits are significant	51.3	3
Zeros between non-zero digits are significant	9010.05	6
Zeros to the left of the first non-zero digit are not significant	0.0041	2
Zeros to the right of a decimal point are significant if they are at the end of a number	391.0 0.00870	4 3
Trailing zeros when there is no explicit decimal point may or may not be significant; assume zeros are not significant unless otherwise specified. Scientific notation should be used to eliminate uncertainty	1500 1.500×10^3	2, 3, or 4 4
When adding or subtracting, the result should be reported with the same number of significant figures as the number with the least number of decimal places	$6.842 - 0.19841$ $= 6.644$	4
When multiplying or dividing, the product or quotient should contain no more than the least number of significant figures in the calculation	0.2865×9.2 $= 2.6$	2
Exact numbers have no uncertainty. Therefore, exact numbers never limit the number of significant figures in a calculation	1 inch exactly equals 2.54 cm 12.34 in. $\times$ 2.54 cm/in. = 31.29 cm	4

Table 26-2
Rounding rules

Rule	Example	Number of Significant Figures in Example
Consider the digit to the right of the last significant figure	Round 59.34 to 3 figures = 59.3	3
Do not use cumulative rounding	Round 3.648 to 2 figures = 3.6 NOT rounded to 3.65 then to 3.7	2
If the digit to the right of the last significant figure is 0 to 4, drop it	7.44 rounds to 7.4	2
If the digit to the right of the last significant figure is 5 to 9, drop it and increase the last significant digit by 1	9.78 round to 9.8 1.296 rounds to 1.30	2 3
Do not round values until the final result is calculated	$(87.254 \times 0.118) \div 4.00 = 2.573993$ rounds to 2.57	3

Gas Constant

The gas constant, R, is expressed in various units below. A comprehensive list can be found at [1].

$$8.3145 \ \frac{J}{K \ gmol}$$

$$1.9859 \ \frac{cal}{K \ gmol}$$

$$8.3145 \ \frac{m^3 \ Pa}{K \ gmol}$$

$$0.000083145 \ \frac{m^3 \ bar}{K \ gmol}$$

$$62.364 \ \frac{l \ mmHg}{K \ gmol}$$

$$10.732 \ \frac{ft^3 \ psi}{R \ lbmol}$$

$$1545 \ \frac{ft_{lbf} \ psi}{R \ lbmol}$$

$$0.7302 \ \frac{ft^3 \ atm}{R \ lbmol}$$

$$1.986 \ \frac{Btu}{R \ lbmol}$$

Standard Gas Volume

scf = standard cubic feet measured at 60°F and atmospheric pressure

379.5 scf/lb-mole (perfect gas)

Nm^3 = normal cubic meters measured at 0°C and atmospheric pressure

0.02214 Nm^3/g-mole (perfect gas)

1 Nm^3 = 37.326 scf

Exception: some organizations, such as the US Environmental Protection Agency, use a reference temperature of 20°C (68°F) for both scf and Nm^3.

Atmospheric Pressure

The standard atmosphere (atm) is defined as 101,325 Pa. A similar unit, the bar, is 100,000 Pa. Since these differ by only about 1% they are often used interchangeably. Conversion factors are shown in Table 26-3.

Design conditions at any altitude are defined with an equation set adopted by the US Government [2]. See Table 26-4. The temperature and density values are intended for aircraft design, so may not be appropriate for use in ground-level plant design projects.

Table 26-3
Conversion factors for atmospheric pressure

	1 atm	1 bar
Pa	101,325	100,000
kPa	101.325	100.000
bar	1.01325	1
mbar	1013.25	1000
torr	760	750.06
mm-Hg	760	750.06
in.-Hg	29.92	29.53
cm-H_2O	1033.2	1019.7
in.-H_2O	406.78	401.46
lb/in.^{2} (psi)	14.696	14.504
lb/ft^2 (psf)	2116.2	2088.5

Table 26-4
Standard atmosphere table

Altitude m	Pressure kPa	Temp °C	Density kg/m^3	Altitude feet	Pressure psi	Temp °F	Density lb/ft^3
0	101.32	15.0	1.225	0	14.70	59.0	0.07648
304.8	97.72	13.0	1.190	1,000	14.17	55.4	0.07427
609.6	94.21	11.0	1.155	2,000	13.66	51.8	0.07210
914.4	90.81	9.1	1.121	3,000	13.17	48.3	0.06999
1,219.2	87.51	7.1	1.088	4,000	12.69	44.7	0.06792
1,524.0	84.31	5.1	1.056	5,000	12.23	41.1	0.06590
1,828.8	81.20	3.1	1.024	6,000	11.78	37.6	0.06393
2,133.6	78.19	1.1	0.993	7,000	11.34	34.0	0.06200
2,438.4	75.27	(0.8)	0.963	8,000	10.92	30.5	0.06012
2,743.2	72.44	(2.8)	0.934	9,000	10.51	26.9	0.05828
3,048.0	69.69	(4.8)	0.905	10,000	10.11	23.3	0.05649

Irrigation Conversion

1 acre-foot = 325,851 gallons

1 acre-foot = 43,560 cubic feet

0.001 acre-foot = 325.9 gallons

1 acre-inch = 27,154 gallons

1 acre-inch = 3,360 cubic feet

450 gallons/minute = 1 acre-inch in 1 hour

References

[1] Katmar Software, Values of the Universal Gas Constant, R, at www.katmarsoftware.com/gconvals.htm

[2] NOAA, NASA, USAF. U. S. Standard Atmosphere 1976. U.S. Government Printing Office NOAA-S/T 76–1562; October, 1976.

27
Properties

Introduction

This chapter provides a variety of physical properties, correlations, and conversion factors that may be difficult to find elsewhere. They range from approximations that are useful for quick conceptual calculations, to the precise equations-of-state for water and steam.

Properties for pure compounds are readily available in handbooks and on the web. It is more difficult to find correlation coefficients for temperature dependent properties such as viscosity and vapor pressure. If accurate values at three temperatures are known for these properties, it is easy to write equations that fit the three points that are quite sufficient for general process design; the procedures are given in this chapter.

Properties for natural substances (e.g., foods and oils) may vary significantly depending on the source of the material and its purity. Major properties can be found in specialty handbooks and are sometimes available from the equipment manufacturers who specialize in processing the materials.

VBA functions are included in the accompanying Excel workbook for:

- Computing the relative humidity of air, given atmospheric pressure, dry-bulb temperature, and wet-bulb temperature.
- Solving the International Association for the Properties of Water and Steam (IAPWS) correlations for density, enthalpy, entropy, heat capacity, viscosity, thermal conductivity, boiling point, and vapor pressure of saturated water and steam.
- Estimating a gas compressibility factor using the Redlich-Kwong Equation of State.
- Estimating the equilibrium solubility of inorganic gases in petroleum liquids.

When undertaking a new project, it can save time to compile physical properties for the materials that will be used, including all of the process streams that will appear on material and flow balances and utilities. A data sheet is provided to help organize the data.

Approximate Physical Properties

Gravity of a boiling hydrocarbon liquid mixture at all pressures:

4 lb/gal or 30 lb/ft^3

Therefore a useful relationship is:

1 mmHg $\approx$ 0.5 in H_2O $\approx$ 1 in boiling hydrocarbon mixture

The higher heating value (HHV) is defined as the amount of heat released from a fuel at 25°C once it is

Table 27-1
Liquid volume contribution of dissolved gases in hydrocarbon mixtures

Component	MW	Liters/kgmole	Gal/lbmole
O_2	32	28.1	3.37
CO	28	34.9	4.19
CO_2	44	53.1	6.38
SO_2	64	45.8	5.50
H_2S	34	43.1	5.17
N_2	28	34.6	4.16
H_2	2	28.1	3.38

Table 27-2
Liquid volume contribution of light hydrocarbons in hydrocarbon mixtures

Component	MW	Liters/kgmole	Gal/lbmole
C_1	16	53.3	6.4
$C_2^=$	30	84.3	10.12
C_2	28	80.3	9.64
$C_3^=$	44	86.8	10.42
C_3	42	80.5	9.67
iC_4	58	103.1	12.38
nC_4	58	99.4	11.93
iC_5	72	115.3	13.85
nC_5	72	114.2	13.71
iC_6	86	129.1	15.50
C_6	86	129.7	15.57
iC_7	100	143.2	17.2
C_7	100	145.4	17.46
C_8	114	161.5	19.39

combusted and the products of combustion have returned to 25°C. This takes into account the heat of vaporization of water (as a combustion product) which assumes that the water returns to the liquid state.

Table 27-3
Higher heating values of solid fuels

Solid Fuel	HHV, MJ/kg	HHV, Btu/lb
Low volatile bituminous coal and high volatile anthracite coal	25.5 – 32.5	11,000 – 14,000
Subbituminous coal	20.9 – 25.5	9,000 – 11,000
Lignite	16.3 – 18.6	7,000 – 8,000
Coke and coke breeze	27.9 – 30.2	12,000 – 13,000
Wood	18.6 – 20.9	8,000 – 9,000
Bagasse (dry)	16.3 – 20.9	7,000 – 9,000

Table 27-4
Higher heating values of gaseous fuels

Gaseous Fuel	HHV, MJ/scm	HHV, Btu/scf
Coke oven gas	18.6 – 26.1	500 – 700
Blast furnace gas	3.4 – 4.5	90 – 120
Natural gas	33.5 – 48.4	900 – 1,300
Refinery off gas	41.0 – 74.5	1,100 – 2,000
Hydrogen	141.9	3,808

Table 27-5
Higher heating values of liquid fuels

Liquid Fuel	HHV, MJ/m³	HHV, Btu/gal
Heavy crude (14°API)	41,800	150,000
Light crude (30°API)	39,000	140,000
Fuel Oils No. 1	38,300	137,400
No. 2	38,900	139,600
No. 3	39,500	141,800
No. 4	40,400	145,100
No. 5	41,500	148,800
No. 6	42,500	152,400
15°API	49,800	178,710
Kerosine (45°API)	44,300	159,000
Gasoline (60°API)	41,800	150,000
Naphtha (65°API)	32,100	115,000
Butane	28,800	103,300
Propane	25,500	91,600
Ethane	18,400	66,000

Assumed gravities used commercially in determining rail tariffs:

Propane or butane	4.7 lb/gal
Natural gasoline	6.6 lb/gal

Volume of a "perfect gas":

76 ft³/lb mol at 76°F and 76 psia

This volume is directly proportional to the absolute temperature = °F plus 460. It is inversely proportional to the absolute pressure.

Molal heat of vaporization at the normal boiling point (1 atm):

8,000 cal/gmol or 14,400 Btu/lbmol

The latent heat of water is about five times that for an organic liquid.

Heat Capacities, Btu/lb°F

Water	1.0
Steam/Hydrocarbon Vapors	0.5
Air	0.25
Solids	0.25

For structural strength:

Weight of concrete = weight of equipment supported

For desert landscaping (home or office) – contractors use:

yards (yds³) rock = ft² covered/100
For water analysis: 1 mg/liter = 1 ppm

Source for this section: [4]

Vapor Pressure Equation

The Antoine Equation is often used to model vapor pressure.

$$Log_{10}(P) = A - \frac{B}{C+t} \qquad (27\text{-}1)$$

Where:

P = Vapor Pressure, mm Hg
t = temperature, °C

A, B, C are coefficients.

The accuracy of the expression is related to the behavior of the liquid and the vapor pressure range being modeled. Rather than regress a large number of data points, coefficients can be calculated directly from three known values of vapor pressure. Over a range of 1 mm Hg to 760 mm Hg, the accuracy is usually within 3% for all types of liquids – even mixtures. With a wider range, to 30 atm, the accuracy is often within 5%.

Here's how to compute the constants, with the three data points termed P_1, t_1, P_2, t_2, P_3, and t_3. An intermediate calculation simplifies the expression to obtain the coefficient C.

$$X = \frac{Log_{10}\left(\frac{P_2}{P_3}\right)(t_1 - t_2)}{Log_{10}\left(\frac{P_1}{P_2}\right)(t_2 - t_3)}$$

$$C = \frac{t_1 - (t_3\,X)}{X - 1}$$

$$B = \frac{Log_{10}\left(\frac{P_2}{P_3}\right)}{1/(t_3 + C) - 1/(t_2 + C)}$$

$$A = Log_{10}(P_2)\,\frac{B}{t_2 + C}$$

Example

Viscosity values for toluene are listed in Table 27-6 [16]. Using the data points for vapor pressures 1 mm Hg, 400 mm Hg, and 30,400 mm Hg,

$X = 0.365975$
$C = 226.25$
$B = 1410.9$
$A = 7.0705$

Vapor pressures predicted by the equation are compared with the published values at each value. The maximum difference of 2.74% at 5 mm Hg could indicate a slight error in the published values rather than a failure of the correlation. When the exercise is repeated using the data point for vapor pressure 5 mm Hg instead of 1 mm Hg, the maximum error is reduced to 1.21%. Notice that the predicted values exactly match the source data at the three vapor pressure points used to calculate the parameters.

Table 27-6
Vapor pressure of toluene

Published Data		Predicted From Equation	
Vapor Pressure, mm Hg	Temperature, °C	Vapor Pressure, mm Hg	Difference
1.0	−26.7	1.0	0.00%
5.0	−4.4	5.1	2.74%
10.0	6.4	10.1	1.38%
20.0	18.4	20.1	0.55%
40.0	31.8	40.1	0.19%
60.0	40.3	59.9	0.20%
100.0	51.9	99.5	0.46%
200.0	69.5	199.5	0.26%
400.0	89.5	400.0	0.00%
760.0	110.6	762.0	0.26%
1,520.0	136.5	1,517.1	0.19%
3,800.0	178.0	3,804.7	0.12%
7,600.0	215.8	7,564.3	0.47%
15,200.0	262.5	15,266.5	0.44%
22,800.0	292.8	22,503.9	1.30%
30,400.0	319.0	30,400.0	0.00%

Liquid Viscosity

Dynamic (or Absolute) Viscosity

Water at 20.2°C (68.4°F) has an absolute viscosity of 1 cP. Table 27-7 lists conversion factors.

Kinematic Viscosity

Kinematic viscosity is the ratio of dynamic viscosity to density. Table 27-8 lists conversion factors.

$$v = \frac{\mu}{\rho} \qquad (27\text{-}2)$$

Where:

v = kinematic viscosity, mm^2/s (= cSt)
μ = dynamic viscosity, mPa-s (= .001 g/mm-s, = cP)
ρ = density, g/cm^3

Example
Ethylene Glycol at 20°C is approximately 12 cP and 1.11 g/cm^3, therefore:

v = (.012 g/mm) / [1.11 g/cm^3 * (1 cm^3/ 1000 mm^3)]
   = 10.8 cSt

Table 27-7
Dynamic viscosity conversion factors

Multiply by

Units From	cP	Poise	mPa-s	Pa-s	kg/m-s	N-s/m^2	lb_f-s/in.^{2}	lb_f-s/ft^2	lb_m/ft-s	lb_m/ft-h	slug/ft-s	dyne-s/cm^2	g/cm-s
cP	1	0.01	1	0.001	0.001	0.001	0.0000001450	0.00002089	0.0006721	2.419	0.00002089	0.01	0.01
Poise	100	1	100	0.1	0.1	0.1	0.00001450	0.002089	0.06721	241.9	0.002089	1	1
mPa-s	1	0.01	1	0.001	0.001	0.001	0.0000001450	0.00002089	0.0006721	2.419	0.00002089	0.01	0.01
Pa-s	1,000	10	1,000	1	1	1	0.0001450	0.02089	0.6721	2,419	0.02089	10	10
kg/m-s	1,000	10	1,000	1	1	1	0.0001450	0.02089	0.6721	2,419	0.02089	10	10
N-s/m^2	1,000	10	1,000	1	1	1	0.0001450	0.02089	0.6721	2,419	0.02089	10	10
lb_f-s/in.^{2}	6,897,000	68,970	6,897,000	6,897	6,897	6,897	1	144.1	4,635	16,680,000	144.1	68,970	68,970
lb_f-s/ft^2	47,870	478.7	47,870	47.87	47.87	47.87	0.006941	1	32.17	115,800	1	478.7	478.7
lb_m/ft-s	1,488	14.88	1,488	1.488	1.488	1.488	0.0002157	0.03108	1	3,599	0.03108	14.88	14.88
lb_m/ft-h	0.4134	0.004134	0.4134	0.0004134	0.0004134	0.0004134	0.00000005994	0.000008636	0.0002778	1	0.000008636	0.004134	0.004134
slug/ft-s	47,870	478.7	47,870	47.87	47.87	47.87	0.006941	1	32.17	115,800	1	478.7	478.7
dyne-s/cm^2	100	1	100	0.1	0.1	0.1	0.00001450	0.002089	0.06721	241.9	0.002089	1	1
g/cm-s	100	1	100	0.1	0.1	0.1	0.00001450	0.002089	0.06721	241.9	0.002089	1	1
Units To	cP	Poise	mPa-s	Pa-s	kg/m-s	N-s/m^2	lb_f-s/in.^{2}	lb_f-s/ft^2	lb_m/ft-s	lb_m/ft-h	slug/ft-s	dyne-s/cm^2	g/cm-s

Table 27-8
Kinematic viscosity conversion factors

Multiply by

Units From	mm^2/s	m^2/s	m^2/h	cSt	St (Stokes)	in.^{2}/s	ft^2/s	ft^2/h
mm^2/s	1	0.000001	0.003600	1	0.01	0.001550	0.00001076	0.03875
m^2/s	1,000,000	1	3,600	1,000,000	10,000	1,550	10.76	38,750
m^2/h	277.8	0.0002778	1	277.8	2.778	0.4306	0.002990	10.76
cSt	1	0.000001	0.0036	1	0.01	0.001550	0.00001076	0.03875
St (Stokes)	100	0.0001	0.36	100	1	0.1550	0.001076	3.875
in.^{2}/s	645.2	0.0006452	2.323	645.2	6.452	1	0.006944	25
ft^2/s	92,900	0.09290	334.5	92,900	929.0	144	1	3,600
ft^2/h	25.81	0.00002581	0.09290	25.81	0.2581	0.04000	0.0002778	1
Units To	mm^2/s	m^2/s	m^2/h	cSt	St (Stokes)	in.^{2}/s	ft^2/s	ft^2/h

Viscosity of Mixtures

For nonpolar mixtures, the following equation gives mixture viscosity usually within ±5% to 10%: [16]. Do not use with polar compounds, including water.

$$\ln(\mu_{mix}) = \sum W_i \ln(\mu_i) \tag{27-3}$$

Where:

μ_{mix} = mixture viscosity
μ_i = viscosity of pure component i
W_i = mass fraction of pure component i

Emulsions

The viscosity of an emulsion of nonmiscible liquids can be estimated with [21]:

$$\mu = \frac{\mu_C}{\delta_C} \left(1 + \frac{1.5 \, \mu_D \, \delta_D}{\mu_D + \mu_C}\right) \tag{27-4}$$

Where:

μ = viscosity, any units
δ = volume fraction of liquid phase

Subscripts:
D = dispersed phase
C = continuous phase

The relationship is valid for the following ranges of variables:

Viscosity of continuous phase = 0.5 to 50 cP
Viscosity of dispersed phase = 0.5 to 100 cP
Volume fraction of continuous phase = 0.1 to 1.0
Volume fraction of dispersed phase = 0.0 to 0.9

Saybolt Universal Seconds

An obsolete method for measuring kinematic viscosity in the petroleum industry is by use of the Saybolt Viscometer. The method, described in ASTM D88-07 [1], measures the time for 60 ml of a petroleum product to flow through an orifice at a controlled temperature. The result is expressed in terms of Saybolt Universal Seconds (SUS) and Saybolt Furol Seconds (SFS). ASTM D2161-05e1 [2] establishes the equations relating SUS and SFS to SI kinematic viscosity units (mm²/s).

Relative Humidity

Here is an equation for relative humidity (RH) [14] and a short Visual Basic for Applications (VBA) function subroutine that enhances its value. The average difference between the equation and actual values runs only 0.33 percentage points and the highest absolute difference seen in comparisons was 2 percentage points.

$$RH = \frac{100 \, (P_w - P_m)}{P_d} \tag{27-5}$$

Where:

$$P_d = P_c \, 10^{K_d \, (1 - T_c/T_d)}$$

$$P_w = P_c \, 10^{K_w \, (1 - T_c/T_w)}$$

$$P_m = 0.000367 \left(1 + \frac{T_w - 491.67}{1571}\right) P_b \, (T_d - T_w)$$

K_d and K_w are functions of dry and wet bulb temperatures, T_d and T_w.

```
Function RH(Td, Tw, Pb)
 ' calculates relative humidity
 ' Source: Pallady, P.H., "Compute Relative
Humidity Quickly"
 ' Chemical Engineering, November 1989, p. 255.
 '
 ' Td = Dry-bulb temperature, deg C - see
comment below
 ' Tw = Wet-bulb temperature, deg C - see
comment below
 ' Pb = Atmospheric pressure, mm Hg
 '
 ' Convert temperatures to absolute, deg R
 ' If inputs are in deg F instead of deg C,
then omit the next two lines
   Td = Td * 1.8 + 32 'converts to deg F
   Tw = Tw * 1.8 + 32 'converts to deg F
   Td = Td + 459.67 'converts to deg R
   Tw = Tw + 459.67 'converts to deg R
 ' Constants
```

```
PC = 166818 'critical Pressure, mm Hg
TC = 1165.67 'critical Temperature, deg R
' Correlation coefficients
C0 = 4.39553
C1 = -0.0034691
C2 = 0.0000030721
C3 = -0.00000000088331
K0 = 0.000367
K1 = 1571
K2 = 491.67
' Calculate Kd and Kw
Kd = C0 + C1 * Td + C2 * Td ^ 2 + C3 * Td ^ 3
Kw = C0 + C1 * Tw + C2 * Tw ^ 2 + C3 * Tw ^ 3
' Calculate Pd, Pw, and Pm
Pd = PC * 10 ^ (Kd * (1 - TC / Td))
Pw = PC * 10 ^ (Kw * (1 - TC / Tw))
Pm = K0 * (1 + (Tw - K2) / K1) * Pb * (Td - Tw)
' Calculate Relative Humidity
RH = 100 * ((Pw - Pm) / Pd)
' Check to see if RH is in the range of
0 to 100
' If not, set to the limits
If RH < 0 Then RH = 0
If RH > 100 Then RH = 100
End Function
```

Nomenclature

K = Temperature-dependent parameter, dimensionless; K_d = value at T_d; Kw = value at T_w

P = Pressure of water vapor at any temperature T(R), mm Hg

P_b = Barometric pressure, mm Hg

P_c = Critical pressure of water, 166,818 mm Hg

P_d = Saturation pressure of water at the dry-bulb temperature, mm Hg

P_m = Partial pressure of water vapor due to solution depression ($T_d - T_w$), mm Hg

P_W = Saturation pressure of water at the wet-bulb temperature, mm Hg

RH = Relative humidity, %

T_c = cCritical temperature of water = 1,165.67R

T_d = Dry-bulb temperature, R

T_w = Dry-bulb temperature, R

Surface Tension

For a liquid with known critical temperature and surface tension at one temperature, use this equation to estimate its surface tension at other temperatures [15]. It is accurate within a few percent for non-polar compounds in regions well removed from the critical point. For polar liquids, the exponent 1.2 may be as small as 1.0.

$$\sigma_2 = \sigma_1 \left(\frac{T_C - T_2}{T_C - T_1}\right)^{1.2} \tag{27-6}$$

Thermodynamic Properties of Water and Steam

The International Association for the Properties of Water and Steam (IAPWS) published a set of correlations that give properties for water and steam throughout the range from the melting-pressure curve to 1,273 K at pressures to 1,000 MPa [11]. Agreement with experimental data ranges from +/−0.1% to about 2% depending on the property and the temperature and pressure at which it is evaluated. IAPWS gives charts that map the uncertainties.

Due to the large number of coefficients, the correlations require a computer for efficient evaluation. Spang wrote a set of Visual Basic for Applications function subroutines that solve the IAPWS equation set as a function of temperature and pressure [20].

Spang's VBA functions are located on the *Rules of Thumb for Chemical Engineers Excel Spreadsheets*. The function calls are listed in Table 27-9.

Table 27-9
VBA functions for water and steam properties

Property	VBA Function
Density, single-phase, kg/m^3	=densW (T, P)
Specific internal energy, single-phase, kJ/kg	=energyW (T, P)
Specific enthalpy, single-phase, kJ/kg	=enthalpyW (T, P)
Specific entropy, single-phase, kJ/kg	=entropyW (T, P)
Specific isobaric heat capacity, single-phase, kJ/(kg-K)	=cpW (T, P)
Specific isochoric heat capacity, single-phase, kJ/(kg-K)	=cvW (T, P)
Dynamic viscosity, single-phase, Pa-s	=viscW (T, P)
Thermal conductivity, single-phase, W/(m-K)	=thconW (T, P)
Boiling point as function of pressure, K	=tSatW (P)
Vapor pressure, bar absolute	=pSatW (T)
Density in saturation state, kg/m^3	=densSatLiqTW (T)
Boiling water as function of temperature	=densSatLiqPW (P)
Boiling water as function of pressure	=densSatVapTW (T)
Saturated steam as function of temperature	=densSatVapPW (P)
Saturated steam as function of pressure	
Specific internal energy in saturation state, kJ/kg	=energySatLiqTW (T)
Boiling water as function of temperature	=energySatLiqPW (P)
Boiling water as function of pressure	=energySatVapTW (T)
Saturated steam as function of temperature	=energySatVapPW (P)
Saturated steam as function of pressure	
Specific enthalpy in saturation state, kJ/kg	=enthalpySatLiqTW (T)
Boiling water as function of temperature	=enthalpySatLiqPW (P)
Boiling water as function of pressure	=enthalpySatVapTW (T)
Saturated steam as function of temperature	=enthalpySatVapPW (P)
Saturated steam as function of pressure	
Specific entropy in saturation state, kJ/kg	=entropySatLiqTW (T)
Boiling water as function of temperature	=entropySatLiqPW (P)
Boiling water as function of pressure	=entropySatVapTW (T)
Saturated steam as function of temperature	=entropySatVapPW (P)
Saturated steam as function of pressure	
Specific isobaric heat capacity in saturation state, kJ/(kg-K)	=cpSatLiqTW (T)
Boiling water as function of temperature	=cpSatLiqPW (P)
Boiling water as function of pressure	=cpSatVapTW (T)
Saturated steam as function of temperature	=cpSatVapPW (P)
Saturated steam as function of pressure	
Specific isochoric heat capacity in saturation state, kJ/(kg-K)	=cvSatLiqTW (T)
Boiling water as function of temperature	=cvSatLiqPW (P)
Boiling water as function of pressure	=cvSatVapTW (T)
Saturated steam as function of temperature	=cvSatVapPW (P)
Saturated steam as function of pressure	
Dynamic viscosity in saturation state, Pa-s	=viscSatLiqTW (T)
Boiling water as function of temperature	=viscSatLiqPW (P)
Boiling water as function of pressure	=viscSatVapTW (T)
Saturated steam as function of temperature	=viscSatVapPW (P)
Saturated steam as function of pressure	
Thermal conductivity in saturation state, W/(m-K)	=thconSatLiqTW (T)
Boiling water as function of temperature	=thconSatLiqPW (P)
Boiling water as function of pressure	=thconSatVapTW (T)
Saturated steam as function of temperature	=thconSatVapPW (P)
Saturated steam as function of pressure	

Arguments are Temperature (K) and Pressure (bar, absolute). "Single-phase" may be liquid water or gaseous steam, but not a mixture. An answer of "−1" is returned if the input data is outside the allowable or applicable range.

Gas Diffusion Coefficients

Due to the scarcity of reliable experimental data, investigators have developed various methods for estimating binary gas diffusion coefficients. This one, by Singh and Singh [19] requires input data that is easy to find. Comparisons with known data and other correlations show it to have an average error of 17.3%. The authors say that experimental data is difficult to obtain and it is often 5% to 10% erroneous.

$$D_{AB} = 2790 \; \frac{T^{1.622} \left(\frac{1}{M_A} + \frac{1}{M_B} \right)^{0.5}}{P \left(V_A^{1/3} + V_B^{1/3} \right)^2} \qquad (27\text{-}7)$$

Where:

D_{AB} = diffusion coefficient, cm^2/s
T = temperature, K
P = pressure, atm
M_A, M_B = molecular weight
V_A, V_B = molecular volume at the normal boiling point (from boiling point density), cc/gmol

If the diffusion coefficient at one temperature is known, the coefficient at another temperature is found with:

$$\frac{D_1}{D_0} = \left(\frac{T_1}{T_0} \right)^{1.622} \qquad (27\text{-}8)$$

Bulk Modulus

Values for bulk modulus, a measure of a fluid's compressibility, are listed in Table 27-10.

Table 27-10
Bulk modulus

Fluid	Temp. °C	SI Units (10^9 Pa)	US Units (10^5 lb$_f$/in.^{2})	Source
Acetic Acid	20	1.10	1.60	5
Acetone	20	0.78	1.13	5
Aniline	20	2.21	3.21	5
Benzene	20	1.07	1.55	5
Benzene, chloro-	20	1.34	1.94	5
n-Butyl alcohol	0	1.23	1.78	5
Carbon disulfide	20	1.08	1.57	5
Carbon tetrachloride	20	0.97	1.41	5
Chloroform	20	1.01	1.46	5
Cyclohexane	25	0.90	1.31	5
Dodecane	60	0.88	1.28	5
Ethyl alcohol	20	0.89	1.29	5
Ethyl bromide	20	0.77	1.12	5
Ethylene chloride	20	1.25	1.81	5
Ethyl ether	20	0.54	0.78	5
Gasoline	Not specified	1.3	1.9	7
Glycol	25	2.69	3.90	5
n-Heptane	25	0.70	1.02	5
1-Heptanol	0	1.42	2.06	5
n-Hexane	25	0.62	0.90	5
1-Hexanol	0	1.34	1.94	5
Mercury	20	25.00	36.26	5
Methyl alcohol	20	0.83	1.20	5

(Continued)

Table 27-10
Bulk modulus—cont'd

Fluid	Temp. °C	SI Units (10^9 Pa)	US Units (10^5 lb$_f$/in.^{2})	Source
Methylene bromide	16.9	1.55	2.25	5
Methyl iodide	16.9	1.03	1.49	5
n-Octane	25	0.83	1.20	5
1-Octanol	0	1.47	2.13	5
Paraffin Oil	Not specified	1.66	2.41	7
n-Pentadecane	60	0.98	1.42	5
Petrol	Not specified	1.07 to 1.49	1.55 to 2.16	7
Phenol	60	1.65	2.39	5
n-Propyl alcohol	0	1.19	1.73	5
SAE 30 Oil	Not specified	1.5	2.2	7
Toluene	20	1.12	1.62	5
Water	20	2.18	3.16	5
Seawater	Not specified	2.34	3.39	7
m-Xylene	20	1.18	1.71	5

Water and Hydrocarbons

Water in Natural Gas

Figure 27-1 shows the water content of lean, sweet natural gas. It can also be used for gases that have as much as 10% CO_2 and/or H_2S if the pressure is below 500 psia. Above 500 psia, acid gases must be accounted for by rigorous three-phase flash calculations [12] or approximation methods [10].

Hydrocarbons in Water

An easy-to-use nomograph has been developed for the solubility of liquid hydrocarbons in water at ambient conditions (25°C). The accuracy of the nomograph has been checked against available solubility data. Performance of the nomograph has been compared with the predictions given by two available analytical correlations. The nomograph is much simpler to use and far more accurate than either of the analytical methods.

Two main sources of data were used to develop the nomograph: McAuliffe [13] and Price [17]. The hydrocarbons were divided into 14 homologous series as listed in Table 27-11. Solubilities at 25°C were then regressed with the carbon numbers of the hydrocarbons in order to obtain the best fit for each homologous series. A second order polynomial equation fits the data very well:

$$-\log S = a + b\, C^2 \qquad (27\text{-}9)$$

Where:

S = solubility, mole fraction
a, b = regression constants
C = carbon number of the hydrocarbon

The constants a and b for the different homologous series are given in Table 27-11. The nomograph shown in Figure 27-2 was then built with the help of equations for the 14 homologous series. The X and Y coordinates for the different homologous series are listed in Table 27-11. The carbon number range of the nomograph is 4 to 10 while the mole fraction range is 10^{-9} to 10^{-4}. Its accuracy beyond these ranges is not known.

As an example of the use of the nomograph, the line is shown which would be drawn to determine the solubility of n-hexane in water at 25°C. The coordinates given in Table 27-11 for normal paraffins have been used: X = 15.0, Y = 20.0. The predicted solubility is 2×10^{-6} mole fraction: The experimental value is 1.98×10^{-6} as given by McAuliffe [13] and Price [17].

Methanol Injection

Methanol is frequently used to inhibit hydrate formation in natural gas, so we have included

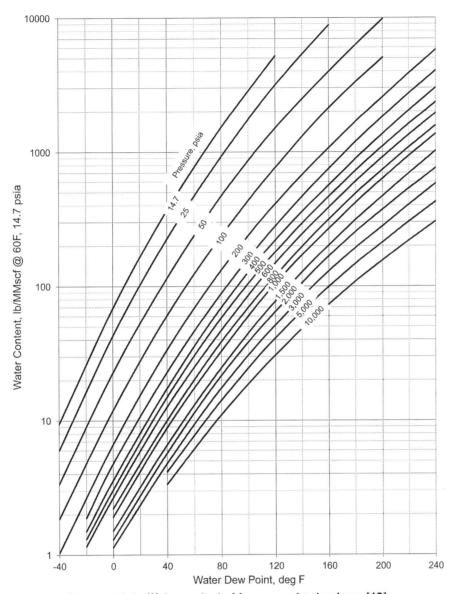

Figure 27-1. Water content of lean, sweet natural gas [12].

information on the effects of methanol on liquid phase equilibria. Shariat, Moshfeghian, and Erbar [6,18] have used a relatively new equation of state, and extensive calculations to produce interesting results on the effect of methanol. Their starting assumptions are the gas composition in Table 27-12, the pipeline pressure/temperature profile in Table 27-13 and methanol concentrations sufficient to produce a 24°F hydrate-formation-temperature depression. Resulting phase concentrations are shown in Tables 27-14, 27-15, and 27-16. Methanol effects on CO_2 and hydrocarbon

solubility in liquid water are shown in Figure 27-3 and Figure 27-4. Sources for this section are Refs [6], [10], and [18].

Here are some conclusions from the study:

1. Water or water plus methanol has a negligible effect on predicted liquid hydrocarbon knockout.
2. Methanol causes substantial enhancement of the solubility of CO_2 in the water phase (Figure 27-3). Watch for increased corrosion.

Table 27-11
Regression constants and nomograph coordinates for the solubilities of liquid hydrocarbons in water at 25°C

Homologous Series	No. of Compounds	Regression Constants		Nomograph Coordinates	
		a	b	X	Y
Normal paraffins	7	3.81	0.0511	15.0	20.0
Isoparaffins − monosubstituted	8	3.91	0.0476	18.0	21.5
Isoparaffins − disubstituted	7	4.02	0.0409	23.2	25.1
Isoparaffins − trisubstituted	3	5.23	0.0186	47.1	20.8
Naphthenes (nonsubstituted)	4	3.45	0.0378	26.0	40.0
Naphthenes − monosubstituted	4	3.48	0.0441	20.7	33.1
Naphthenes − di- and trisubstituted	4	4.75	0.0228	41.8	27.2
Alkenes − normal and monosubstituted	10	3.25	0.0479	18.0	34.0
Dienes	5	2.79	0.0460	19.1	43.9
Cycloalkenes	4	2.69	0.0460	19.1	46.1
Acetylenes	6	2.32	0.0469	18.9	52.5
Diynes	2	1.51	0.0391	25.0	76.8
Aromatics − single ring, mono-and disubstituted by n-paraffins	6	1.91	0.0418	23.5	66.1
Aromatics − single ring, mono-substituted by isoparaffins	2	2.99	0.0264	37.5	62.6

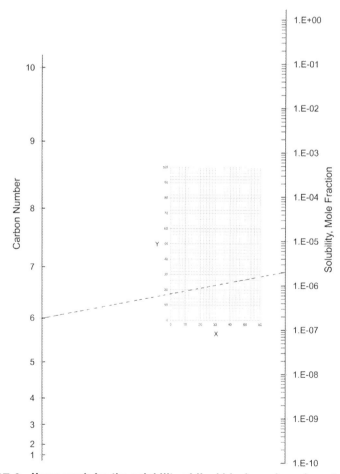

Figure 27-2. Nomograph for the solubility of liquid hydrocarbons in water at 25°C.

Table 27-12
Gas composition

Component	Mole %
N_2	2.85
C_1	79.33
CO_2	5.46
C_2	4.19
C_3	2.00
iC_4	0.44
nC_4	0.78
iC_5	0.40
nC_5	0.34
Lt. arom.	0.34
C_6^*	3.87
Total	100.00

Table 27-13
Pipeline pressure temperature profile*

P, psia	T, °F
2,000**	150**
1,800	100
1,600	80
1,400	70
1,200	60
1,000***	60***

* primary separator operates at 150°F and 2,000 psi.
** Inlet from separator.
*** Discharge from pipeline.

Table 27-14
Results based on dry hydrocarbon flash

Pressure / Temperature psia / °F	Hydrocarbons bbl/MMscf	Water, lb/MMscf		Methanol	
		Liquid	Vapor	Water rich liq., wt%	Vapor phase, lb/MMscf
2,000 / 150	0	—	131.0	—	86.1
1,800 / 100	3.36	95.0	36.0	—	—
1,600 / 80	4.57	107.0	24.0	—	—
1,400 / 70	5.19	114.5	16.5	—	—
1,200 / 60	5.75	118.8	12.2	—	—
1,000 / 60	5.59	117.8	13.2	24.75	47.03

Table 27-15
Results based on wet hydrocarbon flash

Pressure / Temperature psia / °F	Hydrocarbons bbl/MMscf	Water, lb/MMscf		Methanol	
		Liquid	Vapor	Water rich liq., wt%	Vapor phase, lb/MMscf
2,000 / 150	0	0	137.8	—	86.1
1,800 / 100	3.36	94.5	43.3	—	—
1,600 / 80	4.56	111.4	26.4	—	—
1,400 / 70	5.18	116.9	20.9	—	—
1,200 / 60	5.72	121.3	16.5	—	—
1,000 / 60	5.58	119.7	18.1	24.75	47.03

Table 27-16
Results for wet methanol-hydrocarbon flashes

Pressure / Temperature psia / °F	Hydrocarbons bbl/MMscf	Water, lb/MMscf		Methanol	
		Liquid	Vapor	Water rich liq., wt%	Vapor phase, lb/MMscf
2,000 / 150	0	0	137.8	–	85.15
1,800 / 100	3.36	98.8	39.0	16.22	64.43
1,600 / 80	4.69	114.9	22.9	20.40	52.07
1,400 / 70	5.33	120.1	17.7	22.68	45.24
1,200 / 60	5.90	124.1	13.7	24.78	38.20
1,000 / 60	5.74	122.8	15.0	24.77	38.56

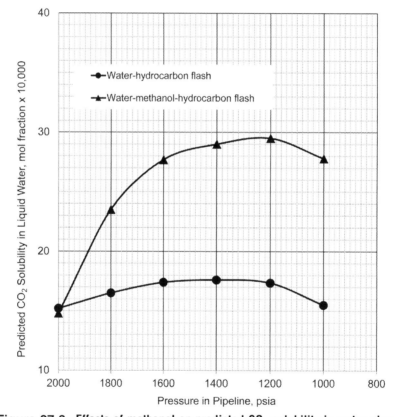

Figure 27-3. Effects of methanol on predicted CO_2 solubility in water phase.

3. Methanol concentration has essentially no effect on the predicted water content of the liquid hydrocarbon phase. The water content (not shown in the tables) was about 0.02 mol%.

4. For this problem, methanol had no practical effect on the solubility of hydrocarbons in the water phase (Figure 27-4).

5. Methanol causes a 10 to 15% reduction in the predicted water content of the pipeline gas.

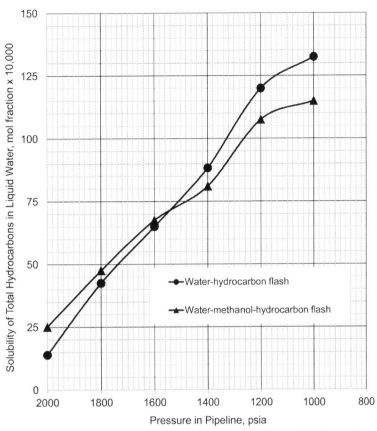

Figure 27-4. Effects of methanol on predicted hydrocarbon solubility in liquid water.

Compressibility Factor

For gas phase calculations, such as pressure drop (Chapter 1) and compressor performance (Chapter 6), the compressibility factor can be adequately estimated using the Redlich-Kwong Equation of State.

$$P = \frac{R\,T}{V_m - b} - \frac{a}{V_m\,(V_m + b)\,T^{0.5}} \qquad (27\text{-}10)$$

Where:

$$a = \frac{R^2\,T_c^{2.5}}{2.3393\,P_c}$$

$$b = \frac{R\,T_c}{11.5420\,P_c}$$

$$Z = \frac{V_m}{V_m - b} - \frac{a}{R\,T\,(V_m + b)\,T^{0.5}} \qquad (27\text{-}11)$$

Knowing the pressure, temperature, critical pressure, and critical temperature of the gas, solve Equation 27-5 for V_m, the molecular volume. Use Goal Seek in Excel, or the Excel VBA subroutine listed below. If pressure is in atm units, and temperature in K, the value of R is 0.08206.

```
Function ZSI(P, T, Pc, Tc)
' calculates compressibility factor, Z,
using SI units
' P = gas pressure, kPa
' T = gas temperature, C
' Pc = gas pseudo-critical pressure, atm
' Tc = gas pseudo-critical temperature, C
' a = Redlich-Kwong parameter with Pc in
atm and Tc in K
' b = Redlich-Kwong parameter with Pc in
atm and Tc in K
' convert P to units atm
P = P * 0.009869233
' convert T to units K
T = T + 273.15
```

```
Tc = Tc + 273.15
R = 0.08206 'gas constant
' calculate Redlich-Kwong parameters
a = (R ^ 2 * Tc ^ 2.5) / (2.3393 * Pc)
b = R * Tc / (11.542 * Pc)
' calculate the molecular volume, Vm
VmLow = 0.000001
VmHigh = 100
Vm = (VmLow + VmHigh) / 2
j = 0
Do
j = j + 1
RHS = R * T / (Vm - b) - a / (Vm * (Vm +
b) * T ^ 0.5)
If RHS < P Then
```

```
VmHigh = Vm
Vm = (VmLow + VmHigh) / 2
Else
VmLow = Vm
Vm = (VmLow + VmHigh) / 2
End If
Loop While Abs((RHS - P) / P) > 0.001 And
j < 1000
ZSI = Vm / (Vm - b) - a / (R * T * (Vm +
b) * T ^ 0.5)
If j = 1000 Then ZSI = j
End Function
```

Inorganic Gases in Petroleum

ASTM provides a standard method for estimating the equilibrium solubility of common inorganic gases in petroleum liquids [3]. The method applies to petroleum liquids having densities from 0.63 to 0.90 grams per milliliter at 15.5°C (approximately 93° to 25° API at 60°F). The results are given as parts per million by weight for the specified partial pressure of the gas and for specified temperatures in the range from −45°C to 150°C (−50°F to 300°F).

The following short Visual Basic for Applications (VBA) function subroutine calculates the solubility. After entering the code into an Excel Module, call it from an Excel worksheet with the formula:

= GasSol(OC, MW, T, D, P, VP)

Where:

OC = Ostwald Coefficient (see Table 27-17)
MW = molecular weight
T = temperature, °C

D = density, kg/m^3
P = pressure above the liquid surface, atm absolute
VP = vapor pressure of the liquid, atm absolute

Table 27-17
Ostwald coefficient V_o for 0°C

Air	0.095
Ammonia	2.80
Argon	0.23
Carbon Dioxide	1.00
Carbon Monoxide	0.10
Hydrogen	0.039
Krypton	1.30
Nitrogen	0.075
Oxygen	0.15
Xenon	2.00

```
Function GasSol(OC, MW, T, D, P, VP)
'Estimates gas solubility in petroleum
liquid
'Limited to liquid density between 630 and
900 kg/m3
'
'OC = Ostwald Coefficient at 0C
'MW = gas molecular weight
'T = temperature, degC (default is 0)
'D = density of liquid, kg/m3 (default is 850)
'P = total pressure, atm abs (default is 1)
'VP = vapor pressure of liquid, atm abs
(default is 0)
T = T + 273.15
If D = 0 Then D = 850
If P = 0 Then P = 1
' convert density to g/ml
D = D / 1000
' correct for temperature
V = 0.3 * Exp(0.639 * ((700 - T) / T) *
Log(3.333 * OC))
' correct for liquid density
If D <> 850 Then V = V * 7.7 * (0.98 - D)
' correct for pressure
B = 273 * P * V / T
G = (B * MW / 0.0224) * (D * (1 - 0.000595 *
((T - 288.6) / D ^ 1.21))) ^ (-1)
```

```
' correct for vapor pressure
If VP > 0.1 * P Then G = ((P - VP) / P) * G
GasSol = G
End Function
```

Foam Density

Zanker gives this equation for estimating the density of foam on distillation and absorption column trays [23]. He says it agrees well with published data.

$$d_F = d_L \left[1 - 0.46 \log \left(\frac{f}{0.0073} \right) \right] \qquad (27\text{-}12)$$

Where:

$$f = u \left[\frac{d_V}{(d_L - d_V)} \right]^{0.5}$$

d_F = foam density, lb/ft^3
d_L = liquid density, lb/ft^3
d_V = vapor density, lb/ft^3
u = superficial vapor velocity, ft/s

Equivalent Diameter

Zanker's article, reprinted here, gives a way to estimate the elusive equivalent diameter of solids [22]. For further reading, refer to: http://ciks.cbt.nist.gov/~garbocz/monograph/3-D_Particles/dimensions.htm

Estimating Equivalent Diameter of Solids

Particle diameter is a primary variable important to many chemical engineering calculations, including settling, slurry flow, fluidized beds, packed reactors, and packed distillation towers. Unfortunately, this dimension is usually difficult or impossible to measure, because the particles are small or irregular. Consequently, chemical engineers have become familiar with the notion of "equivalent diameter" of a particle, which is the diameter of a sphere that has a volume equal to that of the particle.

The equivalent diameter can be calculated from the dimensions of regular particles, such as cubes, pyramids, cylinders, cones, etc. For irregular particles, the equivalent diameter is most conveniently determined from such data as the fractional free volume and the specific surface or area of the particle in a bed of the particles, as determined experimentally from measurements that determine particle volume as a displaced fluid.

These two nomographs provide a convenient means of estimating the equivalent diameter of almost any type of particle: Figure 27-5 for regular particles from their dimensions, and Figure 27-6 for irregular particles from fractional free volume, specific surface, and shape.

Also, in cases where the dimensions of a regular particle vary throughout a bed of such particles or are not known, but where the fractional free volume and specific surface can be measured or calculated, the shape factor can be calculated and the equivalent diameter of the regular particle determined from Table 27-18.

In Figure 27-5, the equivalent diameter is related to regular shapes through equivalent volumes by the following formulae:

$$D_e = \sqrt[3]{6\,V/\pi} = \frac{6\,(1 - \varepsilon)}{\phi\,s} \qquad (27\text{-}13)$$

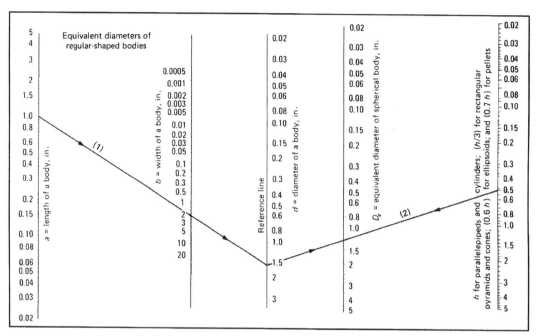

Figure 27-5. Equivalent diameters of regular-shaped bodies.

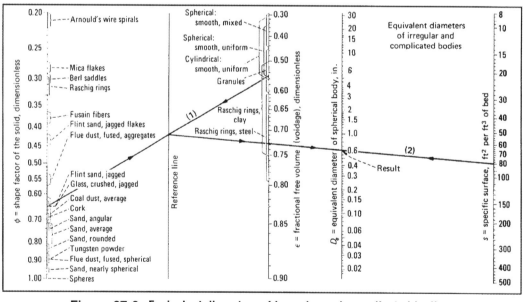

Figure 27-6. Equivalent diameters of irregular and complicated bodies.

Where the terms are in consistent units of length:

V = volume of shape
   (See
   a = length of base edge
   b = width of base measured perpendicular to a

d = diameter
h = height measured perpendicular to plane ab
   Table 27-18)
ε = fractional free volume, dimensionless

					PHYSICAL PROPERTIES				

CLIENT : client name   PAGE : 1
REV | PREPARED BY | DATE | APPROVAL
0 | your name | date |
W.O. : wo number
1 | | | | UNIT : unit   AREA : area
2 | | |

New Project

		CAS #	7732185	67561	64175	8001227	8001307
1							
2				Methyl alcohol			
3		Compound	Water		Ethyl alcohol	Soybean oil	Corn oil
4	Property	Units					
5	Formula		H2O	CH3OH	C2H5OH		
6	Formula Weight		18.02	32.04	46.07		
7	Specific Gravity	Water @ 4°C =1	0.9982 @ 20°C	0.791 @ 20°C	0.789 @ 20°C	0.9 @ 20°C	0.92 @ 20°C
8	Weight per Volume	kg/cu.m.	998.17	790.98	788.98	899.97	919.97
9	Boiling Point	°C	100.00	64.67	78.33		
10	Freezing Point	°C	0.00	-97.67	-114.06	-3.89	-10.00
11	Heat of Vaporization	kJ/kg	2,259	1,100	842		
12	Heat of Fusion	kJ/kg	333	99	108		
13	Specific Heat of Liquid	joules/kg-°C	4.19	2.52	2.40		
14	Viscosity	Pa-Sec @ ~20°C	0.00100	0.00058	0.00115	0.06930	0.07500
15	Thermal Conductivity	W/m-°C	0.399	0.144	0.116		
16	Flash Point	°C		12	13	171	254
17	Autoignition Point	°C		385	363	445	393
18	Surface Temp. Classification			T2	T2	T2	T2
19	Lower Flammable Limit	%		6.7	4.3		
20	Upper Flammable Limit	%		36.0	19.0		
21	Lower Oxygen Conc. Limit	%		23.3	24.4		
22	Heat of Combustion	kJ/kg		22,353	29,773	42,798	
23	Solubility in Water		miscible	miscible	miscible		insoluble
24	Solubility in Alcohol		miscible	miscible	miscible		
25	Surface Tension	mN/m	75.6	21.9	21.8		
26	Critical Temperature	°C	374	239	243		
27	Critical Pressure	kPa	22,113.8	8.093.9	6,381.9		
28	Normal Physical State (1 atm, 20°C)		liquid	liquid	liquid	liquid	liquid
29							
30	Vapor pressure data found for Water, Methyl alcohol, Ethyl alcohol						

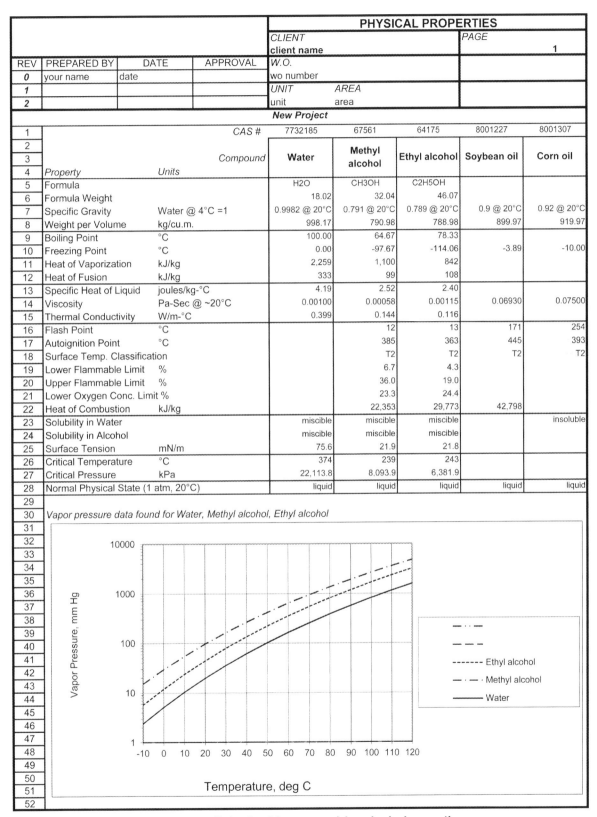

Figure 27-7. Data sheet for summarizing physical properties.

Table 27-18
Formulae to determine volume of various shapes

Shape	Formula
Parallel piped	$V = a\,b\,c$
Cube	$V = a^3$
Rectangular pyramid	$V = (a\,b)\,h/3$
Cylinder	$V = h\,(d^2\,\pi/4)$
Cone	$V = (d^2\,\pi/4)\,(h/3)$
Pellet	$V \cong 0.7\,h\,(d^2\,\pi/4)$
Ellipsoid	$V \cong 0.6\,h\,(d^2\,\pi/4)$

ϕ = shape factor = area of sphere divided by the surface area of a particle of equal volume

s = specific surface = particle surface per unit volume of bed

a = length of base edge

b = width of base measured perpendicular to a

d = diameter

h = height measured perpendicular to plane ab

Examples

What is the equivalent diameter of a rectangular pyramid with base sides 1 and 2 in., and a height of 1.5 in.? On Figure 27-5, align 1.0 (or 2.0) on the a scale with 2.0 (or 1.0) on the b scale and note the intersection of this line on the reference line. Align this intersection with $h/3 = 0.5$ and read $D_e = 1.24$ in.

What is the equivalent diameter of a pellet with a diameter of 0.5 in. and a length of 0.5 in.? On Figure 27-5, align 0.5 on the d scale with 0.35 (0.7 h), and read $D_e = 0.51$.

What is the equivalent diameter of crushed glass (f = 0.65) with a fractional free volume of 0.55 and a specific surface of 80? On Figure 27-6, align f = 0.65 with e = 0.55 and note the intersection on the reference line. Align this with s = 80, and read $D_e = 0.62$.

Properties Data Sheet

Assemble the applicable physical properties in the project's document management system, whether it's paper-based or digital. A data sheet makes a convenient tool for capturing basic data, ensuring that team members use the same information and helping to identify missing values. See Figure 27-7 for an example.

Nomenclature

a = length of base edge

b = width of base measured perpendicular to a

C = carbon number of the hydrocarbon

D_{AB} = diffusion coefficient, cm^2/s

d = diameter

d_F = foam density, lb/ft^3

d_L = liquid density, lb/ft^3

d_V = vapor density, lb/ft^3

h = height measured perpendicular to plane ab

P = Vapor Pressure, mm Hg

P = Pressure, atm

S = solubility, mole fraction

s = specific surface = particle surface per unit volume of bed

t = temperature, $^\circ C$

u = superficial vapor velocity, ft/s

V = volume of shape

W_i = mass fraction of pure component i

Z = compressibility factor

δ = volume fraction of liquid phase

υ = kinematic viscosity, mm^2/s (= cSt)

μ = dynamic viscosity, mPa-s (= .001g/mm-s, = cP)

μ_i = viscosity of pure component i

μ_{mix} = mixture viscosity

ε = fractional free volume, dimensionless

φ = shape factor = area of sphere divided by the surface area of a particle of equal volume

ρ = density, g/cm^3

σ = surface tension

a, b = regression constants

δ = volume fraction of liquid phase

References

[1] ASTM Standard D88-07. Standard Test Method for Saybolt Viscosity. West Conshohocken, PA: ASTM International, www.astm.org; 2007. DOI: 10.1520/D0088-07.

[2] ASTM Standard D2161-05e1. Standard Practice for Conversion of Kinematic Viscosity to Saybolt Universal Viscosity or Saybolt Furol Viscosity. West Conshohocken, PA: ASTM International; 2005. DOI: 10.1520/D2161-05.

[3] ASTM Standard D2779-92(2007). Standard Test Method for Estimation of Solubility of Gases in Petroleum Liquids. West Conshohocken, PA: ASTM International, www.astm.org; 2007. DOI: 10.1520/D2779-92R07.

[4] Branan C. *The Process Engineer's Pocket Handbook*, vol. 3, Gulf Publishing Co.

[5] Chemical Rubber Co. *Handbook of Chemistry and Physics*, 50th ed; 1969.

[6] Erbar J, Maddox R. Water-Hydrocarbon System Behavior. *Oil and Gas Journal*, March 16, 1981:75.

[7] The Engineering Toolbox. Bulk Modulus Elasticity, www.engineeringtoolbox.com; 2010. Downloaded September.

[8] Fredenslund A, Gmehling J, Rasmussen P. *Vapor-Liquid Equilibria Using UNIFAC: A Group Contribution Method*. Elsevier Scientific Pub. Co.; 1977.

[9] Fredenslund A, Jones R, Prausnitz J. Group-Contribution Estimation of Activity Coefficients in Non-Ideal Liquid Mixtures. *AIChE Journal* 1975;21:1086–99.

[10] Gas Processors Suppliers Association (GPSA). *Engineering Data Book*. SI Version, 12th ed. vol. 2; 2004.

[11] The International Association for the Properties of Water and Steam (IAPWS). Revised Release on the IAPWS Formulation 1995 for the Thermodynamic Properties of Ordinary Water Substance for General and Scientific Use. The Netherlands: Doorwerth, www.iapws.org; 2009.

[12] Kabadi V, Danner R. Nomograph Solves for Solubilities of Hydrocarbons in Water. *Hydrocarbon Processing*, May, 1979:245.

[13] McAuliffe CJ. *J Phys Chem*, 1966;70:1267–75.

[14] Pallady PH. Compute Relative Humidity Quickly. *Chemical Engineering*, November 1989:255.

[15] Perry RH, Chilton CH, editors. Perry's Chemical Engineers' Handbook. 5th ed. McGraw-Hill; 1973.

[16] Perry RH, Green DW, editors. *Perry's Chemical Engineers' Handbook*. 6th ed. McGraw-Hill; 1984.

[17] Price L. *American Association of Petroleum Geologists Bulletin*, 1976;60:213.

[18] Shariat A, Moshfeghian M, Erbar J. Predicting Water Knockout in Gas Lines. *Oil and Gas Journal*, November 19, 1979:126.

[19] Singh PC, Singh S. Development of a New Correlation for Binary Gas Dispersion Coefficients. *Int Comm Heat Mass Transfer*, 1983;10:123–40. doi:10.1016/0735-1933(83)90039-8.

[20] Spang B. Excel Add-In for Properties of Water and Steam in SI Units, www.cheresources.com/staff.shtml.

[21] Zanker A. Find Emulsion Viscosity Through Use of Nomograph,. *Hydrocarbon Processing*, January, 1969:178.

[22] Zanker A. Estimating Equivalent Diameters of Solids. *Chemical Engineering*, July 5, 1976:101.

[23] Zanker A. Quick Calculation for Foam Densities. *Chemical Engineering*, February 17, 1975.

Appendix A
Resources

Useful organizations, publications, and conferences are listed here. Find complete information at the given internet addresses.

Table A-1 lists a few of the major associations along with their leading publications. Each of these groups has annual (or more frequent) conferences. They provide numerous benefits to members; every chemical engineer should consider joining. For example, AIChE and IChemE provide access to the Knovel digital library containing dozens of books for chemical engineers.

Table A-2 is organized like the chapters in this book, with trade and manufacturer organizations dedicated to specific technologies. Most of these groups publish a magazine or journal and provide technical resources for their members, and in some cases to all interested parties. They generally host an annual conference; many conduct seminars.

Table A-1
General organizations

Organization	Description	Publications	Availability
American Institute of Chemical Engineers (AIChE) New York, NY *www.aiche.org*	The world's leading organization for chemical engineering professionals, with over 40,000 members from over 90 countries.	*Chemical Engineering Progress*	Free to members; download articles since January 2001. Anyone can search back issues
Institution of Chemical Engineers (IChemE) Rugby, UK *www.icheme.org*	Global professional engineering institution with over 30,000 members in 113 countries. The only organization to award Chartered Chemical Engineer status.	*AIChE Journal* *tce today*	By subscription Members only

Table A-1
General organizations—cont'd

Organization	Description	Publications	Availability
European Federation of Chemical Engineering (EFCE) www.efce.info	An association of professional societies in Europe and India that are concerned with chemical engineering. 40 societies in 30 countries linking 162,000 chemical engineers	*Chemical Engineering Research and Design*	Available at ScienceDirect
Construction Industry Institute® (CII) Austin, TX www.construction-institute.org	Consortium of more than 100 leading owner, engineering-contractor, and supplier firms joined to enhance the business effectiveness and sustainability of the capital facility life cycle	*CII Best Practices*	Free to members; available for purchase by anyone

Table A-3 identifies organizations that are associated with specific segments of the CPI. These groups have various aims ranging from lobbying to dissemination of technical information. They all have annual conferences, publish magazines, and provide resources for their members.

Table A-4 lists some of the leading magazines intended for technical audiences working in the chemical process industries. These are not associated with specific non-profit associations.

Table A-5 lists major trade shows and exhibitions that cater to chemical engineers.

Table A-6 shows some of the major user group conferences where users of specific tools, often a proprietary brand, gather to share case histories, application notes, add-on products, etc.

Table A-2
Organizations specializing in specific technologies

Organization	Description	Publications	Availability
Compressors			
Compressed Air Challenge (CAC) www.compressedairchallenge.org	CAC is a trade organization that provides resources that educate industry about optimizing their compressed air systems	*Best Practices for Compressed Air Systems*	Purchase on line from CAC
Pumps			
Hydraulic Institute Boston, MA www.pumps.org www.pumplearning.org	Association of pump industry manufacturers in North America; provides product standards and forum for information exchange.	Pump Standards *Hydraulic Institute Engineering Data Book* Many other books and seminars	Purchase in hardcopy, CD-ROM, Web-based format Available for purchase
Heat Exchangers			
Heat Transfer Research, Inc. (HTRI) College Station, TX www.htri.net	The global leader in process heat transfer and heat exchanger technology. Industrial research and development consortium serves over 1000 corporate member sites.	*The Exchanger* *X*changer computer software suite	Distributed to members. Anyone can browse and download back issues Available for purchase
Tubular Exchanger Manufacturers Association Tarrytown, NY www.tema.org	Trade association of leading manufacturers of shell-and-tube heat exchangers	*Standards of the Tubular Exchanger Manufacturers Association* Computer software for heat exchanger design	Purchase from TEMA Demonstration and licensed versions

Table A-2
Organizations specializing in specific technologies—cont'd

Organization	Description	Publications	Availability
Fractionators			
Fractionation Research, Inc. Stillwater, OK *www.fri.org*	Non-profit research consortium supported by membership	Articles and presentations in public domain	Listed at fri.org
		Fractionation Design Handbooks (5 vols)	For members only
		Computerized rating programs and databases	For members only
Vessels			
Steel Tank Institute / Steel Plate Fabricators Association (STI/ SPFA) Lake Zurich, IL *www.steeltank.com*	Trade association that has as its members fabricators of steel construction products (and their suppliers)	Specifications, spreadsheets, and technical documents	For members
		Standards (see website)	Available for purchase
Welding Research Council (WRC) / Pressure Vessel Research Council (PVRC) Cleveland, OH *www.forengineers.org*	WRC publications — records of important progress in welding and pressure vessel research — are applications oriented and provide specific recommendations based on validated technology. Whether on fitness for service, weld crack prevention or toughness, over 475 WRC Bulletins convey the knowledge gained from diverse sources at the cost of millions of dollars.	Bulletins	Available for purchase
Boilers			
American Boiler Manufacturers Association (ABMA) Vienna, VA *www.abma.com*	Trade association of commercial, institutional, industrial, and electricity-generating boiler system manufacturing companies, dedicated to advancement and growth of the industry	*Today's Boiler*	Download the current issue of this semi-annual magazine from the ABMA web site
American Society of Mechanical Engineers (ASME)			
Cooling Towers			
Cooling Technology Institute (CTI) Houston, TX *www.cti.org*	Technical association dedicated to improvement in technology, design, performance, and maintenance of evaporative heat transfer systems.	*CTI Journal*	Free subscription. Semi-annual. Back issues free to download from CTI web site
Refrigeration			
American Society of Heating, Refrigerating, and Air-Conditioning Engineers (ASHRAE) Atlanta, GA *www.ashrae.org*	International organization advancing heating, ventilation, air conditioning, and refrigeration to serve humanity and promote a sustainable world	*ASHRAE Journal* *ASHRAE Handbook — Refrigeration* — Fundamentals — HVAC Systems and Equipment — HVAC Applications	For members only For members only; each volume is updated every four years
Air-Conditioning, Heating, and Refrigeration Institute (AHRI) Arlington, VA *www.ahrinet.org*	Trade association representing more than 300 heating, ventilation, air-conditioning, and commercial refrigeration manufacturers within the global HVACR industry	*AHRI Standards* *HVACR Industry Guidelines*	Free downloads of more than 70 standards Free downloads of more than 10 guidance documents

Table A-2
Organizations specializing in specific technologies—cont'd

Organization	Description	Publications	Availability
Blending and Agitation			
North American Mixing Forum (NAMF) www.1.mixing.net	NAMF is a multidisciplinary, international group promoting scholarship, research, and education in the field of mixing	Book: *Handbook of Industrial Mixing: Science and Practice*	Purchase at Amazon or Wiley
Process Evaluation			
Association for the Advancement of Cost Engineering (AACE International) Morgantown, WV www.aacei.org	The members of AACE enable organizations around the world to achieve their investment expectations by managing and controlling projects, programs, and portfolios	*Cost Engineering Journal* *AACE International Recommended Practices*	Free to members. Monthly journal. Available for subscription. Free to anyone from website
Project Management Institute (PMI) Newtown Square, PA www.pmi.org	The world's leading not-for-profit membership association for the project management profession with more than 500,000 members	*PMBOK® Guide* and Standards *PM Network* *Project Management Journal*	Purchase from PMI Monthly magazine for members only Peer reviewed quarterly journal for members
Metallurgy			
NACE International Houston, TX www.nace.org	The largest organization in the world committed to the study of corrosion	*Materials Performance Corrosion* Technical research journal Conference Papers	Monthly. Free with membership Paid subscription; single articles available for purchase Purchase CD-ROM sets by year
The Minerals, Metals & Materials Society (TMS) Warrendale, PA www.tms.org	International in membership and activities, TMS encompasses the entire range of materials and engineering, from minerals processing and primary metals production to basic research and the advanced applications of materials	*JOM: The Member Journal of TMS*	A benefit of membership. Anyone can browse the table of contents of past issues. A few articles are available as open access
Safety			
National Fire Protection Association (NFPA) www.nfpa.org			
Controls			
International Society of Automation (ISA) Research Triangle Park, NC www.isa.org	Global, nonprofit organization that sets the standard for automation by helping professionals solve difficult technical problems. 30,000 members	*InTech*	Bimonthly member benefit. Others may apply for free subscription. Anyone can read back issues online.

Table A-3
Organizations specific to CPI sectors

Organization	Description	Publications	Availability
Pharmaceuticals and Bioprocessing			
International Society for Pharmaceutical Engineering (ISPE) Tampa, FL *www.ispe.org*	The world's largest not-for-profit association dedicated to educating and advancing pharmaceutical manufacturing professionals and their industry.	*Pharmaceutical Engineering Baseline®, GAMPS®, and Good Practice Guides*	Free to members Purchase from ISPE
Petroleum — Oil and Gas			
Society of Petroleum Engineers (SPE) Richardson, TX *www.spe.org*	Collects, disseminates, and exchanges technical knowledge concerning the exploration, development, and production of oil and gas resources	*Journal of Petroleum Technology*	Provided with membership; non-member subscriptions available (monthly)
American Petroleum Institute (API) Washington, DC *www.api.org*	The only US trade associate that represents all aspects of America's oil and natural gas industry. 400 corporate members. Leads the development of petroleum and petrochemical equipment and operating standards	Specifications Recommended Practices Standards	Available for purchase. Certain US government-cited standards and safety publications are freely available for reading online (no printing or downloading)
Gas Processors Association Tulsa, OK *www.gasprocessors.com*	Association of individuals and companies with interest in the midstream industry: gathering, compression, treating, processing, marketing, and storage of natural gas	Proceedings from annual conference	Browse and purchase papers presented since 1995. Members receive a discount.
Gas Processors Suppliers Association (GPSA) Tulsa, OK *http://gpsa.gpaglobal.org*	Membership is open to any company or individual that provides services or supplies for the natural gas, gas processing, or related industries.	*GPSA Data Book*, 2 volumes and CD in inch-pound or SI units	Available for purchase; members receive a discount
Specialty Organic Chemicals			
Society of Chemical Manufacturers & Affiliates (SOCMA) Washington, DC *www.socma.com*	The only US based trade association dedicated to the batch, custom, and specialty chemical industry. Advocates US laws and regulations.	ChemStewards® EH&S management program	For member companies
Power			
Association of Energy Engineers (AEE) *www.aeecenter.org*			
Food and Beverage			
Food Processing Suppliers Association (FPMSA) McLean VA *www.fpsa.org*	Members supply a wide range of products and services to the food and beverage industries, covering each link in the supply chain	Process Expo is the premier worldwide trade show for the food and beverage industry, with education sessions	Annual (fall); Chicago

Table A-3
Organizations specific to CPI sectors—cont'd

Organization	Description	Publications	Availability
Pulp and Paper			
Technical Association of the Pulp and Paper Industry (TAPPI) Norcross, GA www.tappi.org	The leading association for the worldwide pulp, paper, packaging, and converting industries	*Paper 360* *TAPPI Journal*	Anyone can read current issue. Members download back issues free; fee for others Member benefit; others may purchase single articles
Polymers and Resins			
Society of Plastics Engineers (SPE) Newtown, CT www.4spe.org	The objective of the Society is to promote the scientific and engineering knowledge relating to plastics.	*Plastics Engineering*	Membership required. Anyone can search the technical library; content is an SPE benefit
Fine Chemicals			
World Batch Forum (WFF) Chandler, AZ www.wbf.org	WBF is dedicated to supporting the process automation and operations needs of the technical and management professions in process manufacturing	Whitepapers and reports, primarily about ISA S-88 automation and XML	Membership required
Environmental Protection			
Air and Waste Management Association (AWMA) Pittsburgh, PA www.awma.org	Nonprofit professional organization that enhances knowledge and expertise for more than 8000 environmental professionals in 65 countries	*Journal of the Air & Waste Management Association*	Membership required; archives may be searched by anyone – articles may be purchased

Table A-4
For-profit magazines

Organization	Description	Publications	Availability
Access Intelligence New York, NY www.che.com	For over 125 years, Chemical Engineering has provided a timely mix of technical news reporting and practical, expert information on all aspects of the chemical engineering practice.	Chemical Engineering	Subscribers download articles from archive; anyone can search
Chemical Processing.com www.chemicalprocessing.com	Basic mission is to keep the key decision makers in the chemical community abreast of the latest developments in new products, new applications, new technologies and new solutions	Chemical Processing	Registration required
Gulf Publishing Company www.hydrocarbonprocessing.com	Through its monthly magazine, website and e-newsletters, Hydrocarbon Processing covers technological advances, processes and optimization developments from throughout the global Hydrocarbon Processing Industry (HPI).	Hydrocarbon Processing	Subscription service; current issue can be viewed online for free

Table A-5
Industry exhibitions

Organization	Description	Exhibition	Schedule
International Exposition Co. Westport, CT *www.chemshow.com*	Showcasing the latest process products and technology; with educational programs.	Chem Show	Biannual , odd years (usually in the fall) in NYC
Reed Expositions Norwalk, CT *www.interphex.com*	Tradeshow for pharmaceutical equipment and suppliers; includes an educational track	INTERPHEX	Annual (usually NYC in the spring) Annual (Puerto Rico in fall or winter)
DECHEMA Frankfurt am Main, Germany *www.achema.de*	The world forum of the process industry and the trend-setting technology summit for chemical engineering, environmental protection, and biotechnology	ACHEMA	Triannual (2012, 2015, …) in Frankfurt am Main, Germany
UBM International Media Princeton, NJ *www.informex.com*	The premier annual tradeshow for bringing together motivated buyers and sellers of chemicals	INFORMEX USA®	Annual (winter)
Access Intelligence New York, NY *www.che.com*	Features the CPI's most comprehensive conference program, a cutting-edge solutions exhibition, technology demonstrations and multiple co-located events	ChemInnovations Conference and Expo	Annual (TX in the fall)

Table A-6
User group conferences

Organization	Description	Conference	Schedule
Winter Simulation Conference Foundation *http://wintersim.org*	WSC is the premier international forum for disseminating recent advances in the field of system simulation. Past papers free for download	Winter Simulation Conference	Annual (December, various locations in the US and Germany)

Appendix B
Tank Cleaning

Dale Seiberling wrote this interesting narrative in an email. It has been edited for clarity.

October 16, 2010

I am delighted to have the opportunity to share the information about the "rules of thumb" on which my CIP engineering design and practice has been based for nearly 60 years. Most of it was not exactly, to use your term, "scientific", but it was based on logic and it worked. The first Russian space travel happened on April 12, 1961 and the first NASA trip to the moon in December 1968. My initial studies of vertical tank CIP occurred in the 1962 to 1963 era and I did not have today's technology available at that time.

But let's back up 10 years, and the beginning work in what became a lifetime profession, and still is.

I had completed my six quarter MSc program at Ohio State University in December 1951 and was hired as an Instructor to teach Dairy Technology 401 (beginning course for Dairy Tech students and elective for others on campus). I was also teaching a course in Dairy Refrigeration, Pumps and Boilers and a course in Dairy Plant Layout and Equipment, and my responsibilities included the further development of those courses. But CIP had come into my life in March of 1950 when I wrote a term paper on this developing technology for a senior seminar course. My Department Chairman, however, thought CIP

was a lot of nonsense and I was not encouraged to work in this field on the campus. I had found another way, working occasional nights and weekends as a consultant for Frank Wish, owner of Hopewell Dairy in Bellefontaine, OK. This was a small, family-owned and operated dairy, like tens of thousands of others across the USA, and receiving, processing, packaging and delivering 200 gallons of milk daily (seven days per week) to the customers doorstep provided employment for 40 workers. In 1952 to 1953 I guided the conversion of the dairy from batch pasteurization to HTST, with a two speed homogenizer as the timing pump to process milk (600 gph) and ice cream mix (300 gph) and converted the takedown piping system to be cleaned in-place with the HTST, via the first automated CIP system in the world, at least to be found in the literature.

These 2000 gallon tanks and a 1500 gallon farm bulk milk pickup tanker were the tanks in which my spray development study began. The tanker was cleaned outside the dairy, on a concrete pad pitched for unloading and solution return purposes, and the tanker was also fitted with a permanently installed spray, making it unnecessary to open the manway for cleaning. A second, manually operated, CIP system was connected to the three sprays via "spool pieces". The CIP pump drew recirculated solutions directly from the raw milk tank being cleaned, but a return pump was used for the tanker. Heating was via direct steam injection into the recycled water, under control of a thermostatically controlled valve. This was "mickey mouse technology" compared to today's norm, but "state-of-the-art" in a dairy in 1955 to 1957. Tanks and tankers in other dairies were still being hand cleaned, and many would be for the next 30 years. Breweries, wineries, and food processors used hoses for product transfer, not piping, and these were mostly just rinsed, though some were "solutioned", meaning a mild alkali was pumped through the hose occasionally. Pharmaceutical processing was done in the hospital laboratory at that time and the word "biotech" had not yet been coined – and would wait for perhaps another 25 to 30 years.

My spray-based CIP work began as an attempt to develop fixed spray devices that could be permanently installed in conventional tanks in use in the dairy industry in 1954 to 1955. In the mid to late 50s, dairy tanks ranged in size from 500 to 5000 gallons, mostly cylindrical horizontal, though some were of rectangular construction. Diameter was limited to 7 foot 6 inches by regulatory agencies, this being the height that could be brush washed by an average man with an 18-inch gong brush, by entering the tank with a 3 gallon pail of mild alkali solution, a brush, and a hose. Mr. Wish and I began the spray development work in 1954. The first patent on what became the KlenzSpray was issued in 1956, and Frank Wish and I licensed Klenzade Products to market that spray in the fall of 1956. It was introduced to the dairy industry at the DFISA show and convention in Atlantic City in the fall of 1956 and Safeway Stores (a food supermarket chain) bought the first 20 sprays for a new milk plant then under construction in Los Angeles. The early sprays were drilled for 80 gpm as that was about the maximum flow that could be drawn from a tank with a 2-inch outlet without creating a substantial puddle of quiescent nature within the vessel. Mr. Wish and I spent a lot of time in the tanks in bathing suits and swim goggles and we quickly learned that there was no value in spraying the entire tank, for this would provide physical cleaning of only 100 small spots at 0.8 gpm per hole. The pressure was set at 25 psi and we learned that a variation from 20 psi (64 gpm) to 30 psi (110 gpm) would not affect reach or coverage significantly. We found that pressures above 30 psi caused the streams to atomize and carry a lesser distance, from small holes in thin wall balls. We quickly decided that all of the water should be sprayed on the upper third of the tank area, the remaining coverage

being obtained via run-down of a larger quantity of water than would occur if it was uniformly distributed over the entire inner surface. We used the bacteriological swab method for the evaluation of cleanliness. We also employed UV light observation for organic films. A lot of the early work was accomplished by spraying heavy, viscous, cultured buttermilk on the surfaces and drying with a heat lamp before doing spray evaluation tests. This is where we began to understand the power of chemical, rather than physical cleaning.

A 5000 gallon tank would contain approximately 400 ft^2 of surface, and our early recommendations were to spray 0.2 gpm to 0.3 gpm (for larger to smaller tanks) per ft^2 of total surface, applying this flow to approximately the upper third of the surface area. A spray flow rate of 80 gpm was applied in all vessels and all new vessels were specified to have at least 2-inch outlets to handle this flow. When we moved into transport tankers approximating 40 feet in length, we found that two 40 gpm balls, close to the bottom of the tanker would each cover one end and 20 to 22 feet of cylinder to the manway, at the same 80 gpm and 25 psi. Cleaning the same tanker with a drop in spray was solved by the SB-8 which combined a ball for the center two-thirds of the length with seven taper drilled and reamed holes in each of two nozzles on the ball, operated at 55 to 60 psi for reach at 105 to120 gpm. The heavy streams would drive through the ricochet from the ball coverage, impact below the top of the head and bounce back to the upper fourth of the cylinder.

From 1960 onwards I did not have to worry about proving coverage. Every Dairy Quality Control Supervisor and the regulatory agencies responsible for the plant were more than willing to accept the required cleaning verification work. I got involved only when failures occurred, always due to human or equipment failures.

My spray CIP experience involved 100% automated CIP from 1960 onwards. I trusted engineering design and automatic control of time, temperature and concentration. I had no confidence in people and manual control of anything. The permanently welded-in double balls and drop-in SB-8 did materialize in about 1961. Then, they became the standard of the industry, and still were when I retired (somewhat) in 1997.

But your question is based on vertical tanks and my first experience was with the dairy silo-type tank. I have only one technical article on Silo-Type Tanks, written by the *Maunufactured Milk Products* Journal editor, with input from many, in August 1962. I believe, however, that my work established the "rule of thumb".

Dean Foods in Rockford, IL was a significant customer of Klenzade's at that time and asked for my help in gaining approval for a 25,000 gallon tank under the jurisdiction of the Chicago Board of Health, a very tough agency. I was involved in developing a spray distribution device (not a spray ball), and the ongoing work was in two other projects I then had underway, each new plants with all storage in silo type tanks. My associate and I considered a ball to be unsuitable in the top of a 15 to 30 foot tall tank, which had to be accessed from the outside, a terrible situation in Minnesota in the winter. Crepaco had turned to a welded-in disc distributor. I was developing the Klenz-Spray SD-6 to function in the same manner, but to be installed through a 3" nozzle. CIP flush, wash and rinse solutions were delivered through a $1/8$" slot around the bottom of a 2" tube and a drilled hole in the bottom disc sprayed the opening of a 3" or 4" sidearm vent pipe. We fabricated 4 or 5 sprays for test purposes, and spent many hours in a 10,000 gallon silo 10 foot ID, in a bathing suit, watching the water distributed on the dish head flow to and down the sidewall. This could have been done in a business suit after the first three seconds to develop full distribution pressure and flow, if we had rubber protection on our shoes. As we varied the flow rate on many trials we learned that at a flow equal to 1.7 gpm/foot of circumference, surface tension would cause the sheet of falling water to pull together near the bottom, leaving dry areas on the sidewall. A flow rate of 2.0 gpm/foot of circumference would always maintain a solid film. We set this as the required minimum for all types of dish or disc distributors and it has worked in small tanks, medium size tanks and even tanks up to 26 feet in diameter in breweries. The typical large dairy silo is 11 feet 6 inches ID, making the circumference 36.1 feet. Our spray was designed to deliver 80 gpm through the slot to the sidewall and sprays 10 gpm to the inlet of the Vent/Overflow line. It works as well in short tanks as in tanks 65 feet to 70 feet high, validating the comments I had made in early years about how easy it would be to spray CIP transport tankers if we could stand them on end.

Our tests with our spray device gave us the confidence to participate in the Dean test for the Chicago Health department, assuming Crepaco would accept our flow rate recommendation, which they did. The evaluation was by bacteriological swabbing in accordance with Standard Methods. Once per month for six months, Dean employees passed all the pieces of a 25 foot scaffold through the manhole door and assembled it to a point near the 33 foot top. Then, George Sikorsky, a senior inspector

for Chicago, and I climbed to the top and swabbed areas of the high sidewall, knuckle radius of the dish head, the dome itself, and the CIP distributor device. Before the scaffold was installed we had both swabbed easily accessible low level and bottom areas. All bacteriological work was by the Dean Foods lab personnel, acceptable to the City of Chicago. My purpose in being there was to make certain that Sikorsky adhered to Standard Methods, in which I had been well trained by Dr. L.H. Burgwald and Dr. W.J. Harper at Ohio State University Department of Dairy Technology.

I don't have the hard copy to share, but I can tell you that seven of ten swabs pulled from this tank over six months revealed no growth. That only says that no organisms were recovered, not that they weren't there. The other three of ten averaged counts of 7 to 10 organisms per swab against an allowable 100 organisms for every swab. CIP was at 80 gpm in an 11.5 foot diameter tank, using 2500 ppm caustic with 30 ppm chlorine for the wash at 135°F for 10 minutes, followed by a cold acid rinse perhaps 90 to 120 seconds, using Phosphoric acid at pH 6.5, primarily to neutralize the caustic which would not rinse off easily. The Chicago Health Department fully accepted the silo tank based on this work, and in 1963 I guided the design, installation and start-up of a new 250,000/day dairy for the Borden Company in Woodstock, Il. This dairy used twenty 11.5 foot silos ranging from 10,000 to mostly 25,000 gallons, and ten 2000 gallon vertical processors for buttermilk, yogurt and sour cream, cleaned to the same criteria but with ball sprays.

Since 1962, when this work was done, my evaluation of silo tanks has been via high powered UV light and, if considered necessary, low level swabbing of easily reached areas.

I will observe that in 10 years, CIP automation in America had enabled the design and operation of a new facility that processed as much milk in the first six minutes each morning as Hopewell Dairy (where I started) handled in the entire day (2000 gallons). And the 3-inch lines in the dairy from the receiving room to and from the raw silos and on to the three 60,000 lb/h HTST systems held half of that 2000 gallons in the pipes.

When I began working in Pharm/Bio in the late 70's initially, and then with greater activity in the late 80's and early 90's, I based my approach to cleaning the pharmaceutical tank on the dairy silo work, but recognized the need to add flow (streams) for nozzles and agitator and manway collars. My most recent (15 years at least) "rule-of-thumb" in the pharm/bio industry has been 2.5 to 3.0 gpm per foot of circumference. Design work under my direction starts at 2.0 gpm/foot of circumference for the vessel sidewall and heads, and then adds two streams per nozzle plus the same circumference based value for the manway and agitator collar, based on diameter.

Sincerely,
Dale A. Seiberling

Index

Printed in the United States
By Bookmasters